Gmelin Handbook of Inorganic and Organometallic Chemistry

8th Edition

Gmelin Handbook of Inorganic and Organometallic Chemistry

8th Edition

Gmelin Handbuch der Anorganischen Chemie

Achte, völlig neu bearbeitete Auflage

PREPARED AND ISSUED BY
Gmelin-Institut für Anorganische Chemie
der Max-Planck-Gesellschaft
zur Förderung der Wissenschaften
Director: Ekkehard Fluck

FOUNDED BY
Leopold Gmelin

8TH EDITION
8th Edition begun under the auspices of the Deutsche Chemische Gesellschaft by R. J. Meyer

CONTINUED BY
E. H. E. Pietsch and A. Kotowski, and by Margot Becke-Goehring

Springer-Verlag
Berlin · Heidelberg · New York · London · Paris · Tokyo · Hong Kong · Barcelona · Budapest · Milan 1996

GMELIN HANDBOOK

Dr. J. von Jouanne

Dr. L. Berg, Dr. H. Bergmann, Dr. J. Faust, Dr. H. Katscher, Dr. A. Kubny, Dr. P. Merlet, Prof. Dr. W. Petz, Dr. H. Schäfer

Dr. R. Albrecht, D. Barthel, Dr. N. Baumann, Dr. K. Behrends, Dr. W. Behrendt, D. Benzaid, Dr. R. Bohrer, K. D. Bonn, Dr. U. Busch, A.-K. Castro, Dipl.-Ing. V. A. Chavizon, A. Dittmar, Dipl.-Geol. R. Ditz, R. Dowideit, Dr. H.-J. Fachmann, B. Fischer, Dr. D. Fischer, Dr. K. Greiner, Dipl.-Bibl. W. Grieser, Dr. R. Haubold, Dipl.-Min. H. Hein, H.-P. Hente, Dr. G. Hönes, Dr. W. Huisl, Dr. M. Irmler, B. Jaeger, Dr. R. Jotter, Dipl.-Chem. P. Kämpf, Dr. B. Kalbskopf, Dipl.-Phys. H. Keller-Rudek, Dipl.-Chem. C. Koeppel, R. Kolb, Dr. M. Kotowski, E. Krawczyk, Dr. W. Kurtz, Dr. B. Ledüc, H. Mathis, M. Meßer, C. Metz, K. Meyer, Dipl.-Chem. B. Mohsin, Dr. U. Neu-Becker, K. Nöring, Dipl.-Min. U. Nohl, Dr. U. Ohms-Bredemann, Dr. H. Pscheidl, Dipl.-Phys. H.-J. Richter-Ditten, Dr. B. Sarbas, Dr. R. Schemm, Dr. D. Schiöberg, V. Schlicht, A. Schwärzel, Dr. B. Schwager, Dr. F. Stein, Dr. C. Strametz, Dr. D. Tille, Dipl.-Phys. J. Wagner, R. Wagner, M. Walter, Dr. E. Warkentin, Dr. C. Weber, Dr. A. Wietelmann, Dr. M. Winter, Dr. B. Wöbke, K. Wolff

GMELIN ONLINE

Dr. G. Olbrich

Dr. P. Kuhn

Dr. R. Baier, Dr. B. Becker, Dipl.-Phys. R. Bost, Dr. A. Brandl, Dr. R. Braun, R. Hanz, Dipl.-Phys. C. Heinrich-Sterzel, Dr. M. Körfer, Dipl.-Chem. H. Köttelwesch, Dr. V. Kruppa, Dr. M. Kunz, Dipl.-Chem. R. Maass, Dr. A. Nebel, Dipl.-Chem. R. Nohl, Dr. B. Rempfer

Organometallic Compounds of Transition Metals

The following listing indicates in which volumes these compounds are discussed or are referred to:

Ag	Silber B5 (1975)
Au	Organogold Compounds (1980)
Co	Kobalt-Organische Verbindungen 1, 2 (1973), Kobalt Erg.-Bd. A (1961), B 1 (1963), B 2 (1964)
Cr	Chrom-Organische Verbindungen (1971)
Cu	Organocopper Compounds 1 (1985), 2 (1983), 3 (1986), 4 (1987), Index (1987)
Fe	Eisen-Organische Verbindungen A 1 (1974), A 2 (1977), A 3 (1978), A 4 (1980), A 5 (1981), A 6 (1977), A 7 (1980), Organoiron Compounds A 8 (1986), A 9 (1989), A 10 (1991), A 11 (1995), Eisen-Organische Verbindungen B 1 (partly in English; 1976), Organoiron Compounds B 2 (1978), Eisen-Organische Verbindungen B 3 (partly in English; 1979), B 4, B 5 (1978), Organoiron Compounds B 6, B 7 (1981), B 8, B 9 (1985), B 10 (1986), B 11 (1983), B 12 (1984), B 13 (1988), B 14, B 15 (1989), B 16a, B 16b, B 17 (1990), B 18 (1991), B 19 (1992), Eisen-Organische Verbindungen C 1, C 2 (1979), Organoiron Compounds C 3 (1980), C 4, C 5 (1981), C 6a (1991), C 6b (1992), C 7 (1985), and Eisen B (1929–1932)
Hf	Organohafnium Compounds (1973)
Mo	Organomolybdenum Compounds 5 (1992), 6 (1990), 7 (1991), 8 (1992), 9 (1993), 10 (1995), 11 (1996), 12 (1994), 13 (1996)
Nb	Niob B 4 (1973)
Ni	Nickel-Organische Verbindungen 1 (1975), 2 (1974), Register (1975), Nickel B3 (1966), and C1 (1968), C2 (1969), Organonickel Compounds Suppl. Vol. 1 (1993), 2 (1994), 3 (1996)
Np, Pu	Transurane C (partly in English; 1972)
Os	Organoosmium Compounds A 1 (1992), A 2 (1993), B 3 (1994), B 4a (1995), B 5 (1994), B 6 (1993), B 8 (1995), B 9 (1995)
Pt	Platin C (1939) and D (1957)
Re	Organorhenium Compounds 1, 2 (1989), 3 (1992), 4 (1996), 5 (1994), 7 (1996) **present volume**
Ru	Ruthenium Erg.-Bd. (1970)
Sc, Y, La to Lu	Rare Earth Elements D 6 (1983)
Ta	Tantal B 2 (1971)
Ti	Titan-Organische Verbindungen 1 (1977), 2 (1980), Organotitanium Compounds 3 (1984), 4 and Register (1984), 5 (1990)
U	Uranium Suppl. Vol. E 2 (1980)
V	Vanadium-Organische Verbindungen (1971); Vanadium B (1967)
Zr	Organozirconium Compounds (1973)

Gmelin Handbook of Inorganic and Organometallic Chemistry

8th Edition

Re

Organorhenium Compounds

Part 7

With 131 illustrations

AUTHOR Reinhard Albrecht

FORMULA INDEX Bernd Kalbskopf, Paul Kämpf

EDITOR Adolf Slawisch

CHIEF EDITOR Adolf Slawisch

Springer-Verlag
Berlin · Heidelberg · New York · London · Paris · Tokyo · Hong Kong · Barcelona · Budapest · Milan 1996

LITERATURE CLOSING DATE: END OF 1995
IN MANY CASES MORE RECENT DATA HAVE BEEN CONSIDERED

Library of Congress Catalog Card Number: Agr 25-1383

ISBN 3-540-93740-4 Springer-Verlag, Berlin · Heidelberg · New York · London · Paris · Tokyo · Hong Kong · Barcelona · Budapest · Milan
ISBN 0-387-93740-4 Springer-Verlag, New York · Heidelberg · Berlin · London · Paris · Tokyo · Hong Kong · Barcelona · Budapest · Milan

Typesetting, printing, and bookbinding: Wiesbadener Graphische Betriebe GmbH, Wiesbaden

Preface

The present volume belongs to a series of handbooks dealing with organorhenium compounds. It covers the literature up to the end of 1995, but many recent data published in 1996 have also been considered. Patents, conference reports, and dissertations generally were not reviewed. An empirical formula index, a ligand formula index, and a transition metal cross reference table provide access to all compounds covered. The following comments may be helpful to rapidly find the compound(s) on which you wish to get information.

In the Gmelin series "Organometallic Compounds", the term "organometallic" is reserved for all compounds containing at least one carbon-to-metal bond. For a list of all volumes published in this series, see p. VI.

The series on organorhenium compounds started with the description of mononuclear compounds in Volumes 1 to 4 and was continued with the description of binuclear compounds in Volumes 5 to 7. **Volume 5** covers binuclear $^1L_nRe_2$ compounds (^1L = alkyl, aryl) and $(CO)_nRe_2$-type compounds where n = 1 to 10 (except $(CO)_{10}Re_2$) which may contain additional ^nD, X, E, and ^nD-Y ligands. **Volume 6** (to be published) describes $(CO)_{10}Re_2$ and binuclear compounds with isocyanide, carbene, and carbyne ligands. **Volume 7** (this volume) describes the remaining binuclear organorhenium compounds with ^nL ligands ($n > 1$).

Rhenium is the last element (No. 70) in the Gmelin system. When applying the "system of the last position", compounds containing rhenium atoms and any heterometal atoms are considered **organorhenium compounds**, if there is at least one C atom of an organic ligand bonded to these Re atoms (cyano and cyanato groups as well as thiocyano and thiocyanato groups are considered inorganic). The term "**binuclear**" refers to the number of rhenium atoms within the molecule to which organic ligands are bonded via C atoms. Re atoms within the molecule not bonded to C are treated as heterometals. Therefore, some Re_n complexes ($n > 2$), that were hitherto considered to be tri- or even tetranuclear such as $(C_5(CH_3)_5)_2$-$Re_2(OReO_3)_2(O)(\mu\text{-}O)_2$, are compiled in the present volume (see p. 192). Otherwise, a compound like $(CO)_5Re\cdots F\cdots ReF_5$ is considered mononuclear. Ionic compounds with rhenium-containing cations and anions are classified based on the ion containing the larger number of C-bonded Re atoms.

In the present volume the compounds are arranged by the type and number of **^{n}L ligands** (n = 2 to 12) where ^nL denotes an organic ligand bonded by n C atoms to rhenium. Terminal ^nL ligands are bonded to one Re atom only, and bridging ^nL ligands are bonded to both Re atoms. Be aware that organic ligands bridging between an Re and a heterometal atom (e.g. CO, C≡CR) make the whole heterometal-containing unit become an L ligand, e.g., the $(C_5(CH_3)_5)W(O)(C{\equiv}CC_6H_5)H$ unit in $[\mu\text{-}(C_5(CH_3)_5)W(O)(C{\equiv}CC_6H_5)H]Re_2(CO)_6(\mu\text{-}H)$ (see p. 99) is a ^2L ligand. The compounds are further classified by the type and number of Re-bonded ^nD, X, and E ligands. **^{n}D ligands** are donor ligands (n = number of electrons donated to Re atoms, n = 2, 4, 6, 8; i.e. $P(CH_3)_3$ is a ^2D ligand). **X ligands** are negatively charged ligands such as H, Cl, or OR. Terminal X ligands donate one electron, whereas bridging ones (μ-X) donate one (H) or three (Cl, OR) electrons. Bridging **E ligands** are atoms or larger units linking Re atoms by two covalent bonds (S, SnR_2); terminal ones are bonded to Re by double bonds (=S, =NR). **^{n}L-^{m}D** ligands are polydentate ligands (terminal or bridging) bonded simultaneously via n C atoms and donor groups D.

Many of the data, particularly in tables, are given in an abbreviated form without the dimension; for explanations, see p. X. Additional remarks, if necessary, are made under the headings of the tables.

Frankfurt am Main
November 1996

Reinhard Albrecht
Adolf Slawisch

Remarks on Abbreviations and Dimensions

Many compounds in this volume are presented in tables which contain numerous abbreviations and values (with dimensions omitted). This necessitates the following clarifications:

Abbreviations used with **temperature data** (usually given in °C, only below −195 °C in K) are m.p. for melting point, dec.p. for decomposition point, and b.p. for boiling point. **Densities** D are given in g/cm^3; D_{calc} and D_{meas} stand for calculated and experimental densities.

Solvents are given by their common name (e.g. acetone; "ether" means diethyl ether) or formula (C_6H_6 = benzene, C_6H_{12} = cyclohexane, C_5H_5N = pyridine) or by one of the following abbreviations: diglyme (diethyleneglycoldimethylether), DMF (N,N-dimethylformamide), DMSO (dimethyl sulfoxide), THF (tetrahydrofuran).

Abbreviations used with **workup methods** are TLC (thin-layer chromatography), GC (gas chromatography), PLC (preparative layer chromatography), and HPLC (high-pressure liquid chromatography).

NMR stands for **nuclear magnetic resonance**. Noise decoupling is indicated by braces { }. Chemical shifts are given in terms of δ values in ppm with the positive sign for downfield shifts. Reference substances are $(CH_3)_4Si$ for ^1H and ^{13}C, H_3PO_4 for ^{31}P, $CFCl_3$ for ^{19}F, if not otherwise stated. For other nuclei, reference substances are given in the tables. Multiplicities are abbreviated s, d, t, q, qui, sext, sept (singlet to septet), m (multiplet); br means broad. Also used are terms such as dt, meaning "doublet of triplets", and v.t., meaning "virtual triplet". Assignments referring to labeled structures are given in a form such as H-α, H-3,4, or C_{ipso}. Coupling constants ^nJ (n is the number of bonds between the coupled nuclei) are given after the multiplicity and assignment as J(P,H) or J(H-1,3) in Hz; also used is J_{gem} which means the geminal coupling constant.

ESR means **electron spin resonance**. Hyperfine interactions are characterized by a (nucleus), e.g. a (Re), a_{iso} (Cu).

Optical spectra are labeled IR (infrared), Raman, and UV (electronic spectrum including the visible region). IR and Raman bands (in cm^{-1}) are followed by their assignments, where the symbols ν and δ stand for stretching and deformation vibrations, and the indexes sym and asym mean symmetrical and asymmetrical. Intensity data usually are omitted; if given, they are indicated qualitatively (vs, s, m, w, vw). Stretching force and interaction constants are denoted k_1, k_2, and k_i, respectively. UV absorption maxima, λ_{max}, are given in nm, occasionally followed by the band shape (br: broad, sh: shoulder), extinction coefficient ($L \cdot cm^{-1} \cdot mol^{-1}$) ε or log ε, and assignment in parentheses. MLCT defines the metal-to-ligand charge transfer transition.

CV denotes **cyclic voltammetry**. Values labeled $E_{1/2}$ (red./ox.), $E_{p,a}$, or $E_{p,c}$ are given in V. Reference electrodes are Ag/AgCl, $(C_5H_5)_2Fe^{0/I}$, or SCE (saturated calomel electrode).

Solvents or the physical state of the sample and the temperature (none is given if room temperature applies) are given in parentheses immediately after the spectral symbol, for example: Raman (solid), ^1H NMR ($CDCl_3$, 50 °C), or CV (CH_3CN/0.1 M $[N(C_4H_9\text{-}n)_4]PF_6$).

Mass-spectral data (MS) are given as the fragment ion composition; m/e values are not listed. $[M]^+$ represents the molecular ion. Methods of ionization are EI (electron impact), FAB (fast atom bombardment), FD (field desorption), or CI (chemical ionization). "MS" solely usually stands for the EI mass spectrum.

Molecular weights are given in g/mol. The abbreviation calc. is used for calculated.

Figures illustrating the results of single-crystal structure determinations give only selected data. Bond lengths are always given in Å, bond angles are given in °.

Table of Contents

Organorhenium Compounds

Binuclear Compounds 3

2.2 Compounds with Ligands Bonded to Rhenium by Two C Atoms (^{2}L Compounds)

2.2.1 Compounds with σ,σ-Bonded ^{2}L Ligands

There are many compounds with σ,σ-bonded ^{2}L ligands, but in only a few of them the ^{2}L ligand is attached to only one Re atom ("terminal"). In all other compounds the ^{2}L ligand is connected to both Re atoms ("bridging"), thereby forming a large variety of structures.

2.2.1.1 Compounds with Terminal ^{2}L Ligands

The structures of the compounds are shown in Formulas I to III.

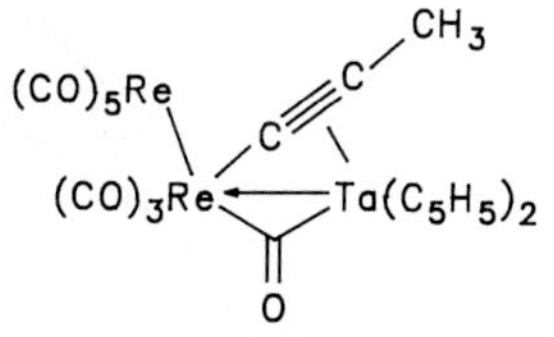

I

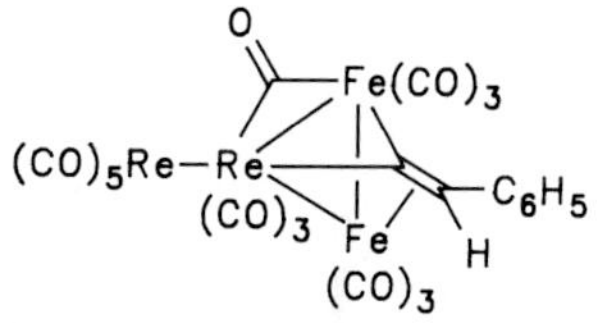

II

[$(C_5H_5)_2$Ta(C≡CCH_3)CO]$Re_2(CO)_8$ (see Formula I). Treatment of $(CO)_{10}Re_2$ with the Schrock carbene $(C_5H_5)_2Ta(=CH_2)CH_3$ (THF, 5 °C, ca. 1 h) followed by evaporation of the mixture gave a crude compound which could be recrystallized from THF/pentane at −40 °C as a deep orange-red solid. Yield: 62%. By-products were $(C_5H_5)_2Ta(O)CH_3$ and CH_4 [3, 4]. Combination of $(^{13}CO)_{10}Re_2$ with $(C_5H_5)_2Ta(=CH_2)CH_3$ [3, 4] or combination of $(CO)_{10}Re_2$ with $(C_5H_5)_2Ta(=^{13}CH_2)^{13}CH_3$ [4] resulted in products with the isotopic pattern Re-^{13}C≡^{12}C-$^{12}CH_3$ or Re-^{12}C≡^{13}C-$^{13}CH_3$, respectively [3, 4]. Thus, the Re-bonded C atom of the methylacetylide ligand originates from a CO group initially having been attached to $(CO)_{10}Re_2$ [3, 4]. Furthermore, monitoring of the progress of the reaction by low-temperature ^{1}H NMR spectroscopy established the initial formation of the zwitterionic intermediate $(CO)_9Re_2^-C(=CH_2)OTa^+$-$(C_5H_5)_2CH_3$, which did not decompose until ca. 0 °C [4].

^{1}H NMR spectrum (THF-d_8): δ = 3.01 (s, CH_3), 5.27 (s, C_5H_5) ppm. ^{13}C NMR spectrum (THF-d_8): δ = 18.71 (m, CH_3; J(C,H) = 128.8, J(C,C) = 5.0 Hz), 102.35 (m, C_5H_5; ^{1}J(C,H) = 178, ^{2}J(C,H) = 7.0, ^{3}J(C,H) = 6.0 Hz), 127.74 (m, **C**CH_3; J(C,H) = 8.0, ^{1}J(C,C) = 101.4 Hz), 139.69 (m, TaCRe; ^{3}J(C,H) = 6.0, ^{2}J(C,C) = 5.0 Hz); 197.40, 197.65, 200.57 (CO); 266.53 (μ-CO) ppm [3, 4]. IR spectrum: 1713 ($\nu(CO_\mu)$) [3, 4]; 1847, 1921, 1980, 1994, 2007, 2030, 2070, 2097 (ν(CO)) [4] cm^{-1}. FAB mass spectrum: $[M - n\ CO]^+$ (n = 0; 2 to 5; 7, 8), $[M - 5\ CO - CCH_3]^+$ [4].

The molecular geometry along with several bond lengths and angles is depicted in **Fig. 1** (lattice parameters are not given) [3, 4].

References on p. 3

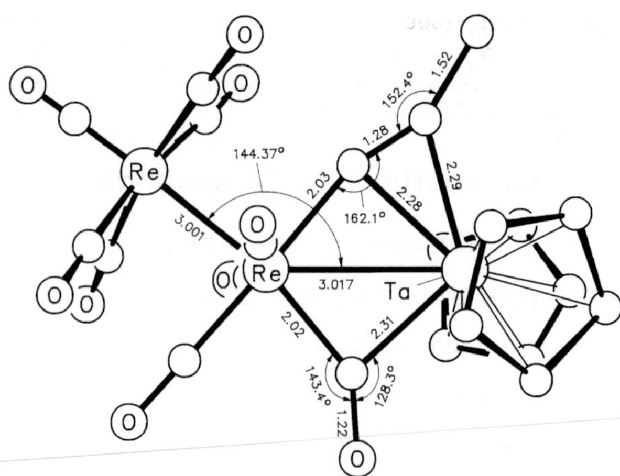

Fig. 1. Molecular structure of $[(C_5H_5)_2Ta(C{\equiv}CCH_3)CO]Re_2(CO)_8$ [3, 4].

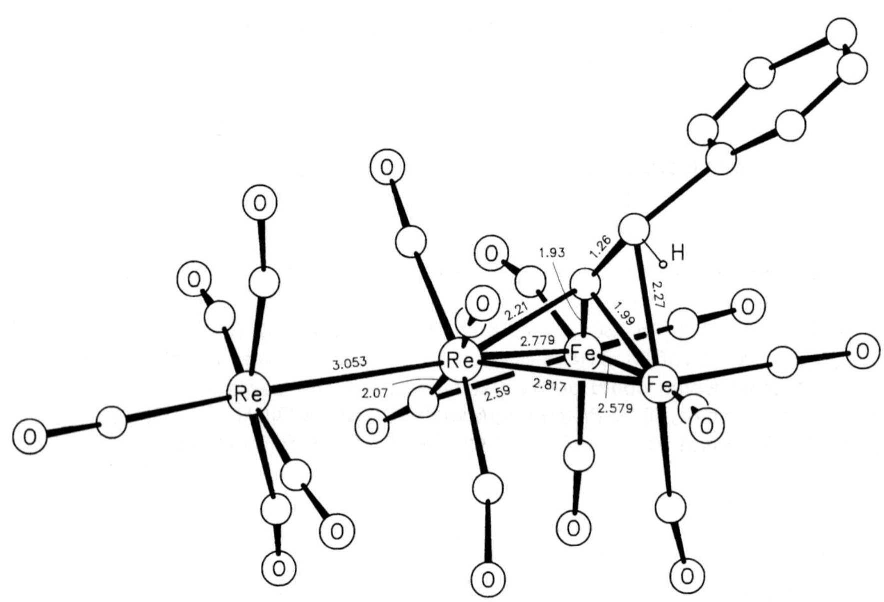

Fig. 2. Molecular structure of $[(CO)_6Fe_2(C{=}CHC_6H_5)CO]Re_2(CO)_8$ [2].

$[(CO)_6Fe_2(C{=}CHC_6H_5)CO]Re_2(CO)_8 \cdot 0.5\ C_6H_6$ (see Formula II) was obtained by treating $(\mu\text{-}\eta^{2:1}\text{-}C_6H_5C{\equiv}C)Re_2(CO)_8(\mu\text{-}H)$ with 0.5 equivalent $(CO)_{12}Fe_3$ (benzene, reflux, 10 h). Concentration of the mixture and cooling to −5°C gave cherry-brown prisms. Yield: 43%.

IR spectrum (KBr): 650, 695, 750; 1995, 2015, 2030, 2065, 2120 (ν(CO)) cm^{-1}.

The semisolvate crystallizes in the monoclinic space group $P2_1/c-C_{2h}^5$ (No. 14) with a = 12.03(4), b = 14.205(6), c = 17.513(8) Å, β = 94.39(3)°; Z = 4 formula units per unit cell. The cluster molecule is illustrated in **Fig. 2** [2].

References on p. 3

$[(C_6H_4(CH_2)_2)_2ReO]_2Mg(OC_4H_8)_4$ (see Formula III) was obtained by treating the salt $[ReO_2(P(CH_3)_3)_4][B(C_6H_5)_4]$ with a 4-fold excess of 1,2-$C_6H_4(CH_2MgCl)_2$ in THF suspension at −78 °C. After room temperature was reached, the solvent was removed, and the residue extracted with toluene. Concentration and cooling of the extract followed by recrystallization from THF/toluene (1 : 1) yielded 22% of orange-red, air-sensitive needles; m.p. > 300 °C.

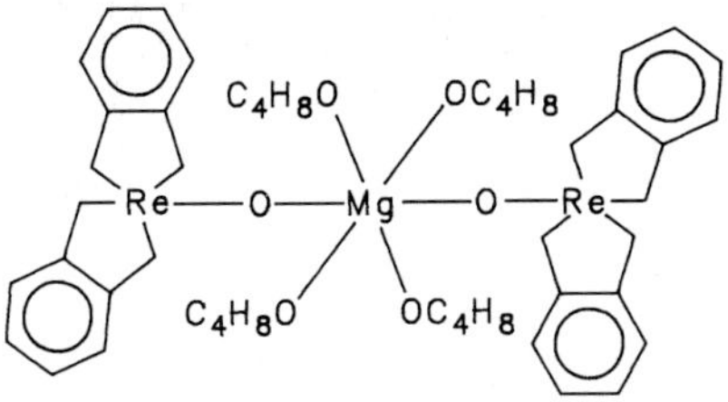

III

^1H NMR spectrum: δ = 1.40 (THF), 2.10 (s, CH_2), 3.56 (THF); 6.92, 7.03 (C_6H_4) ppm. ^{13}C {^1H} NMR spectrum: δ = 25.22 (THF), 36.15 (CH_2), 70.81 (THF); 125.45, 128.79, 137.28 (C_6) ppm. IR spectrum (Nujol): 936 (ν(ReO)); others: 320, 370, 450, 730, 740, 880, 915, 925, 1010, 1023, 1032, 1140, 1160, 1210, 1260, 1290 cm^{-1}.

The compound is readily soluble in ethers and mixtures of THF with aromatic hydrocarbons, but insoluble in hexane. Crystals lose THF solvent molecules upon heating under vacuum at 130 °C and thereby disintegrate to a red powder [1].

References:

[1] Stavropoulous, P.; Edwards, P. G.; Wilkinson, G.; Motevalli, M.; Malik, K. M. A.; Hursthouse, M. B. (J. Chem. Soc. Dalton Trans. **1985** 2167/75).

[2] Pasynskii, A. A.; Eremenko, I. L.; Nefedov, S. E.; Yanovskii, A. I.; Struchkov, Yu. T.; Shaposhnikova, A. D.; Stadnichenko, R. A. (Zh. Neorg. Khim. **38** [1993] 455/65; Russ. J. Inorg. Chem. [Engl. Transl.] **38** [1993] 423/32).

[3] Proulx, G.; Bergman, R. G. (Science [Washington, D.C.] **259** [1993] 661/3).

[4] Proulx, G.; Bergman, R. G. (J. Am. Chem. Soc. **118** [1996] 1981/96).

2.2.1.2 Compounds with Bridging ^2L Ligands

2.2.1.2.1 Compounds of the Type $[(\mu\text{-}COC(R){=}CH)Re_2(^1L)((C_6H_5)_2PCH_2P(C_6H_5)_2)_2X_3]^y$

General. Structure. The structure of the compounds dealt with in this section is shown in Formula I, where ^1L represents CO or RNC (R = 2,6-$(CH_3)_2C_6H_3$), and X stands for a halogen (Cl, Br). The compounds are cationic (y = 1+) or neutral. All cationic compounds were obtained as PF_6 salts; a few cations were also isolated with a BF_4 or $B(C_6H_5)_4$ counteranion.

The bonding within the metalated metallafuran can be described by the canonical forms IIa, IIb. The bond lengths determined by X-ray crystallography on several derivatives are consistent with the 3-metallafuran ring, whereby the overall structure to a large extent is represented by formulation IIa; however, the shortness of the C_μ-O bond length implies that the formulation IIb also makes a significant contribution [1, 2].

References on p. 9

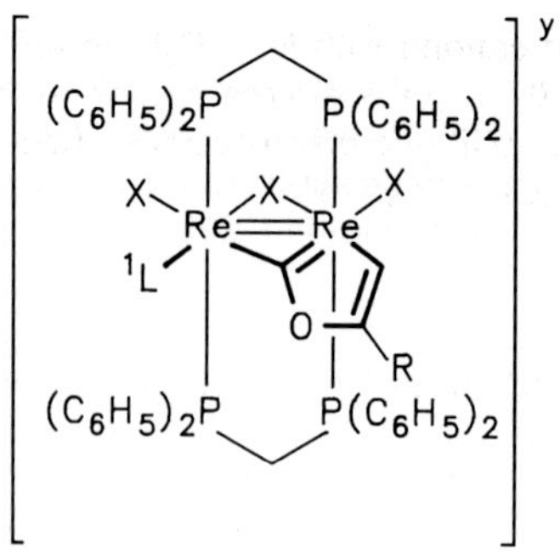

I

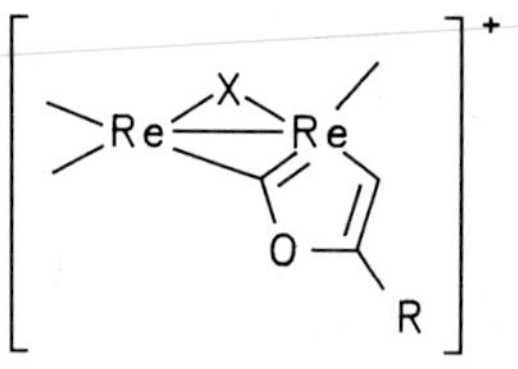

IIa 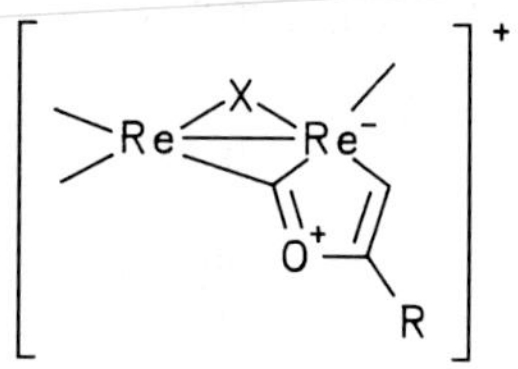 IIb

Preparation. The compounds were prepared as follows:

Method I: Treatment of $(CO)_2Re_2((C_6H_5)_2PCH_2P(C_6H_5)_2)_2X_4$ (X = Cl, Br) or 2,6-$(CH_3)_2$-$C_6H_3NCRe_2(CO)((C_6H_5)_2PCH_2P(C_6H_5)_2)_2Cl_4$ with the terminal alkyne HC≡CR (CH_2Cl_2, room temperature) in the presence of $TlPF_6$. After filtration and evaporation, the residue was recrystallized from CH_2Cl_2/ether [1, 2].

Method II: Reduction of $[(\mu\text{-}COC(R)=CH)Re_2(^1L)((C_6H_5)_2PCH_2P(C_6H_5)_2)_2X_3]PF_6$ with $(C_5H_5)_2Co$ (acetone, room temperature). The product separated from the solution [2].

The reaction described in Method I did not proceed with internal alkynes [1, 2]. Reactions starting from $(CO)_2Re_2((C_6H_5)_2PCH_2P(C_6H_5)_2)_2X_4$ were complete within ca. 5 h, whereas reactions starting from 2,6-$(CH_3)_2C_6H_3NCRe_2(CO)((C_6H_5)_2PCH_2P(C_6H_5)_2)_2Cl_4$ required longer reaction times (10 h) [2].

Spectroscopy. ^1H and $^{31}P\{^1H\}$ NMR spectra were recorded on almost all derivatives. In the ^1H NMR spectra, the CH protons of the rhenafuran ring show up as singlets. The $^{31}P\{^1H\}$ NMR spectra show well-resolved AA′BB′ coupling patterns in accord with the presence of pairs of inequivalent P atoms. The IR spectra show one absorption band for the terminal CO or isocyanide ligand due to ν(CO) or ν(CN), respectively. The analogous bands in the spectra of the neutral compounds are significantly shifted to lower wave numbers [2].

Electrochemistry. Cyclic voltammograms were recorded on all cationic compounds in a CH_2Cl_2/0.1 M $[N(C_4H_9\text{-}n)_4]PF_6$ solution at a Pt bead electrode. All data are referenced to the Ag/AgCl electrode. All are very similar and reveal one reversible one-electron oxidation and two one-electron reduction steps. These processes are shifted to more negative potentials in compounds where ^1L = 2,6-$(CH_3)_2C_6H_3NC$ [2].

References on p. 9

Thermolysis. Thermolysis of cations with R = C_3H_7-n and C_4H_9-n for ^1L = CO and 2,6-$(CH_3)_2C_6H_3NC$ (Nos. 2, 4, 9, 10) in polar solvents such as acetone, CH_3CN, or methanol caused ring-opening reactions giving carbyne compounds of the type $[RCH_2C{\equiv}Re_2(^1L)(\mu\text{-}CO)((C_6H_5)_2PCH_2P(C_6H_5)_2)_2Cl_3]PF_6$ [3, 4].

Table 1
Compounds of the Type $[(\mu\text{-}COC(R){=}CH)Re_2(^1L)((C_6H_5)_2PCH_2P(C_6H_5)_2)_2X_3]^y$.
An asterisk indicates further information at the end of the table.
For explanations, abbreviations, and units see p. X.

No.	R	X	method of preparation (yield) properties and remarks
the type $[(\mu\text{-}COC(R){=}CH)Re_2(CO)((C_6H_5)_2PCH_2P(C_6H_5)_2)_2X_3]PF_6$			
*1	H	Cl	I (with gaseous HC≡CH; reaction time: 2 h; yield: 72%) [1, 2] ^{1}H NMR (CD_2Cl_2): 3.38, 4.27 (m's, 2 H); 7.05 to 7.85 (m, 42 H) [1] ^{13}C {^{1}H} NMR (CD_2Cl_2): 188.6 (CO), 331.5 (C_μ) [2] ^{31}P {^{1}H} NMR (CD_2Cl_2): −15.4, −12.0 [1, 2] IR (Nujol): 2032 (ν(CO)) [1, 2] conductivity (acetone) Λ_M = ca. 110 $cm^2 \cdot \Omega^{-1} \cdot mol^{-1}$ CV: $E_{1/2}$ (red.) = −0.90, 0.10, $E_{1/2}$ (ox.) = 1.56 [2]
2	C_3H_7-n	Cl	I (62%) ^{1}H NMR (CD_2Cl_2): 1.01 (t); 1.12, 3.36 (m's); 4.02 (br), 4.23 (m), 7.05 to 7.80 (m) ^{31}P {^{1}H} NMR (CD_2Cl_2): −16.4, −12.9 IR (Nujol): 2032 (ν(CO)) CV: $E_{1/2}$ (red.) = −0.94, 0.05, $E_{1/2}$ (ox.) = 1.40 reduction with $(C_5H_5)_2Co$ gave No. 7 [2]
3	C_3H_7-n	Br	I (64%) IR (Nujol): 2034 (ν(CO)) CV: $E_{1/2}$ (red.) = −0.91, 0.08, $E_{1/2}$ (ox.) = 1.42 [2]
4	C_4H_9-n	Cl	I (63%) ^{1}H NMR (CD_2Cl_2): 0.97 (t); 1.02, 1.40, 3.43 (m's); 4.10 (br), 4.38 (m), 7.00 to 7.80 (m) ^{31}P {^{1}H} NMR (CD_2Cl_2): −16.4, −12.8 IR (Nujol): 2035 (ν(CO)) CV: $E_{1/2}$ (red.) = −0.95, 0.04, $E_{1/2}$ (ox.) = 1.40 [2]
5	C_6H_5	Cl	I (66%) ^{1}H NMR (CD_2Cl_2): 3.39, 4.24 (m's); 6.95 to 7.80 (m), 8.35 (s) ^{31}P {^{1}H} NMR (CD_2Cl_2): −15.3, −13.1 IR (Nujol): 2027 (ν(CO)) CV: $E_{1/2}$ (red.) = −0.89, 0.11, $E_{1/2}$ (ox.) = 1.32 [2]
6	$C_6H_4CH_3$-4	Cl	I (67%) ^{1}H NMR (CD_2Cl_2): 2.82 (s); 3.30, 4.22 (m's); 6.22, 6.87 (d's); 6.80 to 6.90 (m); 8.32 (s) ^{31}P {^{1}H} NMR (CD_2Cl_2): −16.1, −13.2

References on p. 9

Table 1 (continued)

No.	R	X	method of preparation (yield) properties and remarks
6 (continued)			IR (Nujol): 2032 (ν(CO)) CV: $E_{1/2}$ (red.) = −0.89, 0.10, $E_{1/2}$ (ox.) = 1.26 [2]
the type (μ-COC(R)=CH)Re_2(CO)((C_6H_5)$_2$PCH_2P(C_6H_5)$_2$)$_2$$X_3$			
7	C_3H_7-n	Cl	II (84%) yellow-brown solid IR (Nujol): 1938 (ν(CO)) [2]
the type [(μ-COC(R)=CH)Re_2(CNC_6H_3(CH_3)$_2$-2,6)((C_6H_5)$_2$PCH_2P(C_6H_5)$_2$)$_2$$X_3$]$PF_6$			
*8	H	Cl	I (76%); see also "Further information" ^{1}H NMR ($CDCl_3$): 2.0 (br, 6 H); 3.54, 4.11 (m's, 2 H); 6.80 to 7.85 (m, 45 H); broad signal at 2 ppm collapses upon lowering the temperature (T_{coal} ca. −22°C) and splits into two sharp singlets at 0.83 and 2.91 at −80°C ^{31}P {^{1}H} NMR (CD_2Cl_2): −13.6, −12.2 IR (Nujol): 2162 (ν(CN)) CV: $E_{1/2}$ (red.) = −1.13, −0.17, $E_{1/2}$ (ox.) = 1.28 [2]
*9	C_3H_7-n	Cl	I (88%) ^{1}H NMR (CD_2Cl_2): 1.04 (t, 3 H), 1.26 (m, 2 H), 2.0 (br, 6 H); 3.41, 3.71, 4.11 (m's, 2 H); 6.80 to 7.85 (m, 43 H) ^{31}P {^{1}H} NMR (CD_2Cl_2): −14.4, −12.9 IR (Nujol): 2159 (ν(CN)) CV: $E_{1/2}$ (red.) = −1.19, −0.22, $E_{1/2}$ (ox.) = 1.13 [2]
*10	C_4H_9-n	Cl	I (89%) ^{1}H NMR (CD_2Cl_2): 1.01 (t); 1.18, 1.42 (m's, 2 H); 2.0 (v br, 6 H); 3.42, 3.76, 4.11 (m's, 2 H); 6.80 to 7.85 (m, 43 H) ^{31}P {^{1}H} NMR (CD_2Cl_2): −14.4, −12.7 IR (Nujol): 2162 (ν(CN)) CV: $E_{1/2}$ (red.) = −1.18, −0.22, $E_{1/2}$ (ox.) = 1.14 [2]
11	C_6H_5	Cl	
	PF_6 salt		I (73%) ^{1}H NMR (CD_2Cl_2): 2.2 (br, 6 H); 3.45, 4.09 (m's, 2 H), 6.80 to 7.85 (m, 48 H), 8.21 (s, 1 H) ^{31}P {^{1}H} NMR (CD_2Cl_2): −12.8 (br s) IR (Nujol): 2147 (ν(CN)) CV: $E_{1/2}$ (red.) = −1.07, −0.13, $E_{1/2}$ (ox.) = 1.12; additional $E_{p,a}$ wave between 1.58 and 1.69 reduction with (C_5H_5)$_2$Co gave No. 13 [2]
	BF_4 salt		according to Method I, except that the reaction filtrate was treated with $HBF_4 \cdot O(C_2H_5)_2$. Subsequent concentration and addition of ether separated the product; yield: 59%

References on p. 9

Table 1 (continued)

No.	R	X	method of preparation (yield) properties and remarks
			^1H NMR (CD_2Cl_2): 2.13 (br, 6 H); 3.46, 4.11 (m's, 2 H); 6.80 to 7.90 (m, 48 H); 8.21 (s, 1 H) $^{31}P\{^1H\}$ NMR (CD_2Cl_2): −12.8 (br s) IR (Nujol): 1058 (ν(BF)), 2151 (ν(CN)) CV: $E_{1/2}$ (red.) = −1.10, −0.14, $E_{1/2}$ (ox.) = 1.11; additional $E_{p,a}$ wave between 1.58 and 1.69 [2]
	$B(C_6H_5)_4$ salt		from the PF_6 salt and $Na[B(C_6H_5)_4]$ (acetone, 20 h); concentration and addition of ether yielded 93% product brown-green solid from CH_2Cl_2/pentane ^1H NMR (CD_2Cl_2): 2.13 (br, 6 H); 3.42, 4.08 (m's, 2 H); 6.80 to 7.90 (m, 48 H); 8.25 (s, 1 H) $^{31}P\{^1H\}$ NMR (CD_2Cl_2): −12.9 (br s) IR (Nujol): 2152 (ν(CN)) CV: $E_{1/2}$ (red.) = −1.10, −0.15, $E_{p,a}$ = 0.91 (oxidation of the anion), $E_{1/2}$ (ox.) = 1.10; additional $E_{p,a}$ wave between 1.58 and 1.69 [2]
*12	$C_6H_4CH_3$-4	Cl	
	PF_6 salt		I (87%) ^1H NMR (CD_2Cl_2): 2.15 (br, 6 H), 2.71 (s); 3.43, 4.09 (m's, 2 H); 6.80 to 7.90 (m, 47 H); 8.16 (s, 1 H) $^{31}P\{^1H\}$ NMR (CD_2Cl_2): −13.3 (symmetrical m) IR (Nujol): 2162 (ν(CN)) CV: $E_{1/2}$ (red.) = −1.17, −0.14, $E_{1/2}$ (ox.) = 1.06; additional $E_{p,a}$ wave between 1.58 and 1.69 reduction with $(C_5H_5)_2Co$ gave No. 14 [2]
	BF_4 salt		according to Method I, except that the reaction filtrate was treated with $HBF_4 \cdot O(C_2H_5)_2$. Subsequent concentration and addition of ether precipitated the product ^1H NMR ($CDCl_3$): 2.1 (br, 6 H), 2.68 (s, 3 H); 3.49, 4.17 (m's, 2 H); 6.80 to 7.90 (m, 47 H); 8.14 (s, 1 H) $^{31}P\{^1H\}$ NMR ($CDCl_3$): −12.6 (symmetrical m) IR (Nujol): 1056 (ν(BF)), 2159 (ν(CN)) CV: $E_{1/2}$ (red.) = −1.11, −0.15, $E_{1/2}$ (ox.) = 1.06; $E_{p,a}$ between 1.58 and 1.69 [2]

the type (μ-COC(R)=CH)$Re_2(CNC_6H_3(CH_3)_2$-2,6)$((C_6H_5)_2PCH_2P(C_6H_5)_2)_2X_3$

No.	R	X	method of preparation (yield) properties and remarks
13	C_6H_5	Cl	II (81%) yellow-brown solid IR (Nujol): 2101 (ν(CN)) [2]
14	$C_6H_4CH_3$-4	Cl	II (78%) IR (Nujol): 2094 (ν(CN)) [2]

References on p. 9

* Further information:

$[(\mu\text{-COCH=CH})Re_2(CO)((C_6H_5)_2PCH_2P(C_6H_5)_2)_2Cl_3]PF_6$ (Table **1**, No. **1**) crystallizes in the monoclinic space group Cc$-C_s^4$ (No. 9) with a = 19.396(2), b = 15.495(2), c = 20.001(7) Å, β = 99.66(2)°; Z = 4 formula units per unit cell, D_{calc} = 1.654 g/cm³ [2].

$[(\mu\text{-COCH=CH})Re_2(CNC_6H_3(CH_3)_2\text{-}2,6)((C_6H_5)_2PCH_2P(C_6H_5)_2)_2Cl_3]PF_6$ (Table **1**, No. **8**). When a reaction time of only 5 h was employed, the intermediate $[\pi\text{-HC≡CHRe}_2(CNC_6H_3\text{-}(CH_3)_2\text{-}2,6)(CO)((C_6H_5)_2PCH_2P(C_6H_5)_2)_2Cl_3]PF_6$ (see p. 113) could be separated from the mixture. This salt almost quantitatively (83%) converted into No. 8, when it was stirred in CH_2Cl_2 for 2 d [2].

$[(\mu\text{-COC}(C_3H_7\text{-n})\text{=CH})Re_2(CNC_6H_3(CH_3)_2\text{-}2,6)((C_6H_5)_2PCH_2P(C_6H_5)_2)_2Cl_3]PF_6$ (Table **1**, No. **9**) was also crystallographically studied. Due to an extensive disorder of the anion the structure could not successfully be refined [1, 2]. The Re-Re distance is ca. 2.56 Å [2].

$[(\mu\text{-COC}(C_4H_9\text{-n})\text{=CH})Re_2(CNC_6H_3(CH_3)_2\text{-}2,6)((C_6H_5)_2PCH_2P(C_6H_5)_2)_2Cl_3]PF_6$ (Table **1**, No. **10**). Treatment with 60% aqueous HPF_6 or 85% $HBF_4 \cdot O(C_2H_5)_2$ in acetone at reflux yielded $[n\text{-}C_4H_9CH_2C≡Re_2(CNC_6H_3(CH_3)_2\text{-}2,6)(\mu\text{-CO})((C_6H_5)_2PCH_2P(C_6H_5)_2)_2Cl_3]X$ with X = PF_6 or BF_4. Thermolysis in CH_3OD or CD_3OD incorporated deuterium giving $[n\text{-}C_4H_9(D)\text{-}(H)CC≡Re_2(CNC_6H_3(CH_3)_2\text{-}2,6)(\mu\text{-CO})((C_6H_5)_2PCH_2P(C_6H_5)_2)_2Cl_3]PF_6$ [4].

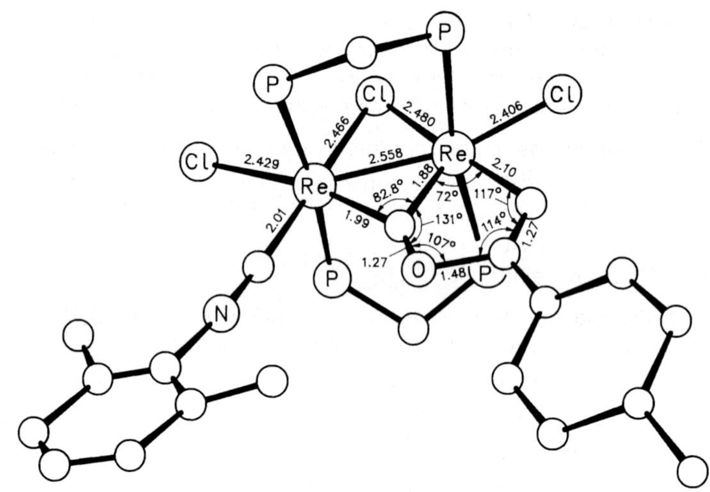

Fig. 3. Molecular structure of
$[(\mu\text{-COC}(C_6H_4CH_3\text{-}4)\text{=CH})Re_2(CNC_6H_3(CH_3)_2\text{-}2,6)((C_6H_5)_2PCH_2P(C_6H_5)_2)_2Cl_3]^+$
(phenyl groups at P not shown) [1, 2].

$[(\mu\text{-COC}(C_6H_4CH_3\text{-}4)\text{=CH})Re_2(CNC_6H_3(CH_3)_2\text{-}2,6)((C_6H_5)_2PCH_2P(C_6H_5)_2)_2Cl_3]BF_4 \cdot 0.5\ H_2O$ (Table **1**, No. **12**) crystallizes in the monoclinic space group C2/c$-C_{2h}^6$ (No. 15) with a = 37.402(8), b = 15.302(4), c = 28.578(8) Å, β = 119.06(3)°; Z = 8 formula units per unit cell, D_{calc} = 1.504 g/cm³. The structure of the cation is illustrated in **Fig. 3** [1, 2].

References on p. 9

References:

[1] Shih, K.-Y.; Fanwick, P. E.; Walton, R. A. (J. Am. Chem. Soc. **115** [1993] 9319/20).
[2] Shih, K.-Y.; Fanwick, P. E.; Walton, R. A. (Organometallics **13** [1994] 1235/42).
[3] Shih, K.-Y.; Fanwick, P. E.; Walton, R. A. (J. Chem. Soc. Chem. Commun. **1994** 861/2).
[4] Kort, D. A.; Shih, K.-Y.; Wu, W.; Fanwick, P. E.; Walton, R. A. (Organometallics **14** [1995] 448/55).

2.2.1.2.2 Compounds of the Type $(CO)_5Re(\mu\text{-}\eta^{1:1}\text{-}^2L)Re(CO)_5$ and Derivatives Originating from Formal CO Substitution at Rhenium (NO, ^2D, ^4D, ^6D, CNR)

General. This section presents compounds where the Re centers of two $(CO)Re(^6D)NO$, $(CO)_3Re(^4D)$, $(CO)_4Re(^2D)$, $(CO)_5Re$, or $RNCRe(CO)_4$ fragments are connected by a σ,σ-bonding ^2L ligand (a large majority of the compounds is of the type $(CO)_5Re-(\mu\text{-}^2L)-Re(CO)_5$). The ^2L ligand belongs to widely different classes such as alkyls, alkenes, alkynes, or acyls; heterometal-containing moieties are also included.

Preparation. The compounds were prepared by the following methods:

Method I: Treatment of cis-$Ru(N_2C_{10}H_6R_2)_2Cl_2$ ($N_2C_{10}H_6R_2$ = 4,4′-disubstituted 2,2′-bipyridine with R = H or CO_2) with $(CO)_3Re(N_2C_{10}H_8)CN$ in alkaline acetone (reflux, 3 h) followed by evaporation and recrystallization [31].

Method II: Treatment of the bisacyl compounds $(CO)_5Re-C(O)(CH_2)_nC(O)-Re(CO)_5$ (n = 6, 7) with 2 or 4 equivalents CH_3Li in THF [7].

Method III: Treatment of $(CO)_5Re-(CH_2)_4-Re(CO)_5$ with PR_3 in THF (2.5 to 6 d under reflux). The mixture was evaporated, and the residue was extracted into CH_2Cl_2. Re-evaporation of the extract and trituration of the residue with hexane gave a white solid which was recrystallized from CH_2Cl_2/hexane [24].

Method IV: Treatment of the respective bis(trifluorosulfonato)-substituted hydrocarbon $CF_3SO_3-A-O_3SCF_3$ with 2 equivalents $Na[(CO)_5Re]$ in THF at −20°C followed by warming the solution to room temperature [23, 24, 32, 33, 36, 40].

Method V: Treatment of $[(\pi\text{-}CH_2{=}CHR)Re(CO)_5]^+$ (R = H, CH_3) with $Na[(CO)_5Re]$ in CH_3CN at low temperatures [10, 11].

Method VI: Decarbonylation of the compounds $(CO)_5Re-C(O)-A-C(O)-Re(CO)_5$ (A = $(CH_2)_n$ [24, 42] or C_6H_4 [2, 27]) in refluxing toluene.

Method VII: Treatment of $[(\pi\text{-}CH_2{=}CH_2)Re(CO)_5]^+$ with the dianionic metal carbonyls $Na_2[(CO)_8Fe_2]$ [20], $Li_2[(CO)_6Fe_2S_2]$, $Na_2[(CO)_6Fe_2Te_2]$ [44], $Na_2[(CO)_4Os]$ [18], or $Na_2[(CO)_{12}Os_3]$ [25] in THF at low temperatures.

Method VIII: Treatment of fumaric acid dichloride or mesaconic acid dichloride with $Na[(CO)_5Re]$ in THF at −78°C, followed by warming to room temperature and evaporation, extraction into CH_2Cl_2, and addition of hexane [41].

Method IX: Carbonylation of the rhenacycles $(CO)_4ReO{=}(CH_3O)CC[Re(CO)_5]{=}CH$-cyclo [30] and $(CO)_4ReO{=}C(OC_2H_5)C[{=}C(CO_2C_2H_5)Re(CO)_5]$-cyclo [38] (see Table 3, Nos. 2 and 4; pp. 40/1) under pressure (80 atm, room temperature) caused ring-opening. Workup by chromatography with CH_2Cl_2/hexane.

Method X: Treatment of the respective bisacylchloride $Cl(O)C-A-C(O)Cl$ with 2 equivalents $Na[(CO)_5Re]$ in THF. The solvent was removed, and the product could usually be extracted into CH_2Cl_2 and crystallized [2, 7, 13, 24, 27, 42].

References on pp. 38/9

Method XI: Treatment of $[t-C_4H_9NCRe(CO)_5]BF_4$ with the diamine $H_2N-A-NH_2$ in ether. The mixture was filtered, the filtrate evaporated, and the residue triturated with hexane at −78 °C leaving a colorless solid [17].

In Table 2 the compounds are arranged by the number of Re-bonded CO groups. Compounds with additional Re-bonded isocyanide ligands are listed at the end of the table.

Table 2
Compounds of the Type $(CO)_5Re(\mu-\eta^{1:1}-^2L)Re(CO)_5$ and Derivatives Originating from Formal CO Substitution at Rhenium (NO, ^2D, ^4D, ^6D, Isocyanides).
An asterisk indicates further information at the end of the table.
For explanations, abbreviations, and units see p. X.

No.	compound	method of preparation (yield) properties and remarks
	compound with 2 CO groups	
*1	$[O(-CH_2Re(CO)(NO)N_3C_6H_{15})_2]I_2$ 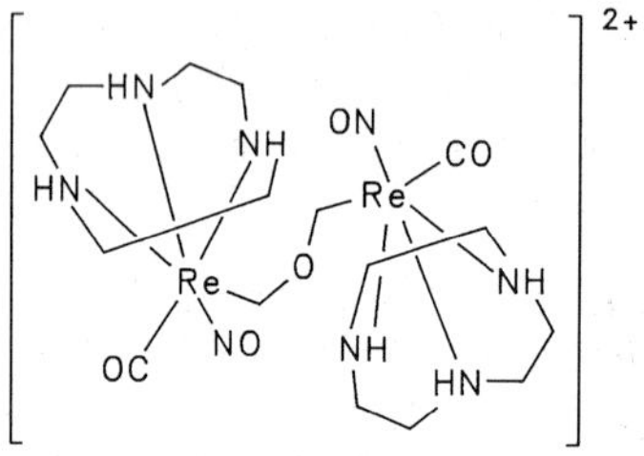	for preparation see "Further information" orange-red solid ^1H NMR (DMSO-d_6): 2.80 to 3.50 (m, NCH_2); 3.99, 4.30, 4.62, 4.98 (d's, OCH_2; J = 10.11); 6.08, 6.38, 6.62, 6.97 (s's, 4 NH); 7.48 (s, 2 NH); δ(NH) peaks disappear when D_2O is added $^{13}C\{^1H\}$ NMR (DMSO-d_6): 46.7 to 53.8 (NCH_2); 71.7, 72.1 (OCH_2); 220.11, 220.92 (CO) IR (KBr): 1680 (ν(NO)), 1920 (ν(CO)), 2750 to 2990 (br, CH), 3060 (ν(NH)) UV (H_2O): λ_{max} (ε) = 482 (89) [16]
	compounds with 6 CO groups	
*2	$[(CO)_3Re((C_6H_5)_2P(CH_2)_3P(C_6H_5)_2)C(O)OX]_2 \cdot C_6X_6$ (X = H or D) 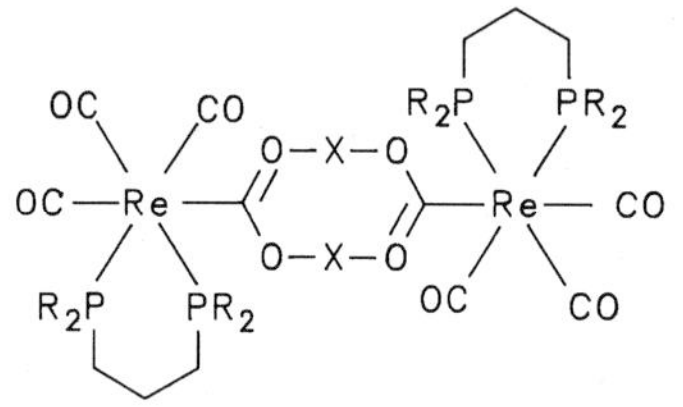 (X = H, D; R = C_6H_5)	separated from the solution when dissolving $(CO)_3Re((C_6H_5)_2P(CH_2)_3$-$P(C_6H_5)_2)CO_2X$ in C_6X_6 (X = H, D); yield: 35 or 41% (X = H or D) colorless prisms; m.p. 147 to 148 or 153 to 154 °C in the cases X = H, D; other data for X = D: ^1H NMR (DMSO-d_6): 2.40 (m, CH_2), 7.58 (m, C_6H_5); further peak for X = H: 9.87 (s, O···H···O) $^{13}C\{^1H\}$ NMR (DMSO-d_6): 19.6 (s, CH_2), 23.4 (t, CH_2; J = 15); 127.8 to 138.0 (C_6H_5); 128.6 (C_6D_6); 192.8, 194.3, 201.7 (t's, CO; J = 18, 8, 11, respectively) IR (Nujol): 1578; (C_6H_6): 1914, 1939, 2022 (ν(CO)) [34]

Table 2 (continued)

No.	compound	method of preparation (yield) properties and remarks
*3	$(C_{44}H_{28}N_4)Sn[C{\equiv}Re(CO)_3]_2$ $C_{44}H_{28}N_4$ = meso-tetraphenylporphyrin	by heating $(CO)_{10}Re_2$ and $C_{44}H_{28}N_4SnCl_2$ (mole ratio 1:1, 160 to 165 °C (not above), 10 d) dark red solid IR (KBr): 1920, 1995, 2000 (ν(CO)) UV (CH_2Cl_2): λ_{max} = 402, 420 (Soret), 512, 552, 591, 628 [8]
4	$[((CO)_3Re^{I}(N_2C_{10}H_8)C{\equiv}N)_2Ru^{II}(N_2C_{10}H_8)][CF_3SO_3]_2$ $N_2C_{10}H_8$ = 2,2′-bipyridine	I (poorly reproducible) IR: 1890, 1925, 2030 (ν(CO)) [31]
*5	$((CO)_3Re^{I}(N_2C_{10}H_8)C{\equiv}N)_2Ru^{II}(N_2C_{10}H_6(CO_2H)_2)N_2C_{10}H_6(CO_2)_2 \cdot 10\ H_2O$ $N_2C_{10}H_8$ = 2,2′-bipyridine (N N = $C_{10}H_6(CO_2H)_2$ and $C_{10}H_6(CO_2)_2$)	I (reprecipitation from H_2O/dilute $HClO_4$; yield: 60%); compound also exists in the dicationic or dianionic form (see "Further information") IR (KBr, CH_2Cl_2?): 1890, 1920, 2030 (ν(CO)) CV (DMF/0.1 M $[N(C_4H_9\text{-}n)_4]ClO_4$): $E_{1/2}$ (red.) = −1.70 ($Ru^4D^{0/-I}$), −1.30 ($Re^4D^{0/-I}$, two-electron wave); $E_{p,a}$ = 0.75 ($Ru^{II/III}$), >1.40 ($Re^{I/II}$) (vs. SCE) [31]
*6	$(CO)_3Re(N_2C_{10}H_6(C_4H_9\text{-}t)_2)-C{\equiv}CC{\equiv}C-Re(CO)_3N_2C_{10}H_6(C_4H_9\text{-}t)_2 \cdot 2\ O{=}C(CH_3)_2$ $N_2C_{10}H_6(C_4H_9\text{-}t)_2$ = 4,4′-di-tert-butyl-2,2′-bipyridine	from $(CO)_3Re(N_2C_{10}H_6(C_4H_9\text{-}t)_2)C{\equiv}C\text{-}Si(CH_3)_3$ and $Cu(O_2CCH_3)_2$ (pyridine, 60 °C, 24 h); yield: 45% after chromatography air-stable, red rod-shaped crystals from acetone ^1H NMR (acetone-d_6): 1.4 (s, CH_3); 7.6 (dd), 8.5 (d), 8.8 (d) $^{13}C\{^1H\}$ NMR ($CDCl_3$): 30 (CH_3), 35 (**C**C_3), 117 (Re**C**≡C); 119, 123 ($N_2C_{10}H_6$); 124 (ReC≡**C**); 152, 155, 162 ($N_2C_{10}H_6$); 198, 207 (CO) IR (Nujol): 1883, 1903 (ν(CO)); 1981 (ν(C≡C)), 2003 (ν(CO)) UV (CH_2Cl_2): λ_{max} (ε) = 330 (15140), 362 (8510), 458 (3010); see also "Further information" CV (CH_3CN/0.1 M $[N(C_4H_9\text{-}n)_4]PF_6$, glassy carbon electrode):

References on pp. 38/9

Table 2 (continued)

No.	compound	method of preparation (yield) properties and remarks
*6 (continued)		$E_{1/2}$ (red.) = −1.62, $E_{p,a}$ = 0.72, 1.33, 1.84 (vs. $(C_5H_5)_2Fe^{I/0}$) FAB MS: $[M - n\ CO]^+$ (n = 0 to 2) [45]
7	$Li_4[(CO)_3Re(C(O)CH_3)_2-C(O)(CH_2)_nC(O)-Re(CO)_3(C(O)CH_3)_2]$ (n = 6, 7)	II (by employing 4 equivalents CH_3Li) IR: 1539 (ν(C=O)); 1863, 1952 (ν(CO)) successive treatment with $AlCl_3$ and MBF_4 yielded $M[\{(CO)_3Re(C(CH_3)O)_2-C(O)-\}_2(CH_2)_{n/2}Al]$ ($M = [N(P(C_6H_5)_3)_2]$) (see Table 12, Nos. 4, 5; pp. 268/9) [7]
compounds with 8 CO groups		
8	$(CO)_4Re(P(CH_3)_2C_6H_5)-(CH_2)_4-Re(CO)_4P(CH_3)_2C_6H_5$	III (56%); the intermediate diacyl complex No. 11 could also be isolated m.p. 92 to 98°C ^{1}H NMR ($CDCl_3$): 0.86, 1.60 (br m's, CH_2CH_2); 1.84 (d, CH_3; J(P,H) = 9), 7.40 (br, C_6H_5) IR (CH_2Cl_2): 1920, 1969, 1978, 2071 (ν(CO)) [24]
9	$(CO)_4Re(P(C_6H_5)_2CH_3)-(CH_2)_4-Re(CO)_4P(C_6H_5)_2CH_3$	III (60%) m.p. 148 to 154°C ^{1}H NMR ($CDCl_3$): 0.50, 1.49 (br m's, CH_2CH_2); 2.02 (d, CH_3; J(P,H) = 9), 7.33 (br, C_6H_5) IR (CH_2Cl_2): 1922, 1968, 1978, 2072 (ν(CO)) [24]
10	$(CO)_4Re(P(C_6H_5)_3)-(CH_2)_4-Re(CO)_4P(C_6H_5)_3$	III (68% (after refluxing for 42 h)) m.p. 154 to 159°C ^{1}H NMR ($CDCl_3$): 0.40 to 0.43 (br m, $ReCH_2$), 1.39 (br m, CC_2H_4C), 7.28 (br s, C_6H_5) IR (CH_2Cl_2): 1922, 1969, 1980, 2073 (ν(CO)) [24]
11	$(CO)_4Re(P(CH_3)_2C_6H_5)-C(O)(CH_2)_4C(O)-Re(CO)_4P(CH_3)_2C_6H_5$	could be isolated during the thermal reaction of No. 17 with $P(CH_3)_2C_6H_5$ according to Method III (no details) [24]

Table 2 (continued)

No.	compound	method of preparation (yield) properties and remarks
12	$(CO)_5Re-CH_2C[Re(CO)_3(^4D)]=CHC_6H_5$ ^4D = 1,10-phenanthroline or $4\text{-}CH_3C_6H_4N=CHCH=NC_6H_4CH_3\text{-}4$ (structure: $^4D-Re(CO)_3$; C_6H_5; $Re(CO)_5$)	obtained by low-temperature visible-light irradiation of $(CO)_5Re-Re(CO)_3(^4D)$ in the presence of phenylallene IR: 1875, 1890, 2010, 2118 (ν(CO)) [28]
13	$Li_2[(CO)_4Re(C(O)CH_3)-C(O)(CH_2)_nC(O)-Re(CO)_4C(O)CH_3]$ (n = 6, 7)	II (with 2 equivalents CH_3Li) IR: 1577 (ν(C=O)); 1917, 1947, 2067 (ν(CO)) [7]
	compounds with 10 CO groups	
*14	$(CO)_5Re-C_2H_4-Re(CO)_5$	IV (in dimethyl ether at −35 °C; yield: 60%) [23]; V (at −40 °C; yield: 86%) [10, 11] light yellow solid [11, 23]; m.p. 138 °C (dec.) [6, 9, 23] ^{1}H NMR ($CDCl_3$): 2.02 [9]; 2.05 [23]; 2.08 [11]; (acetone-d_6): 2.10 [19] $^{13}C\{^1H\}$ NMR ($CDCl_3$): 5.9 (CH_2); 187.0, 187.3 (CO) [23] (see also [9]) IR (KBr): 526, 585, 609 (ν(Re-CO)); (CH_2Cl_2): 1970, 2009, 2111 (ν(CO)) [23] (similar in [19]); (KBr): 2825, 2905, 2930 (ν(CH)) [11] FD MS: $[M - Re(CO)_5]^+$ [23]
*15	$(CO)_5Re-CH_2C(CH_3)H-Re(CO)_5$ (structure: CH_3; H_c; $Re(CO)_5$; H_b; $(CO)_5Re$; H_a)	V (at −50 to −45 °C; a yellow precipitate formed; yield: 76% [11]); also when treating $[(\pi\text{-}CH_2=CHCH_3)Re(CO)_5]^+$ with $[(CO)_4Os]^{2-}$ [25] air-stable for several days, but decomposed in CH_2Cl_2 into $(CO)_5ReCl$ and $(CO)_{10}Re_2$ [11] ^{1}H NMR (CD_2Cl_2): 2.01 (d, CH_3), 2.04 (dd, H_c), 2.41 (m, H_b), 2.83 (dd, H_a); J(H-a,b) = 1.8, J(H-a,c) = 11.4, J(H-b,c) = 12.8, J(H_b,CH_3) = 7.3 $^{13}C\{^1H\}$ NMR: 21.85, 22.01, 36.09; 188.24, 188.56 (CO) IR (C_6H_{12}): 1943, 1966, 1984, 1999, 2013, 2040, 2095, 2123 (ν(CO)); (KBr): 2835, 2925, 2950 (ν(CH)) MS: $[C_3H_mRe(CO)_n]^+$ (m = 6, 5, 3; n = 5 to 0), $[(CO)_nRe_2]^+$ (n = 10 to 0) [11]

References on pp. 38/9

Table 2 (continued)

No.	compound	method of preparation (yield) properties and remarks
*16	$(CO)_5Re-(CH_2)_3-Re(CO)_5$	IV (in THF, yield: 49% [24]; in dimethyl ether at −60°C, yield: 35% [23]) white solid from $CHCl_3$/hexane; m.p. 118 to 121°C [24], 119°C (dec.) [23] ^1H NMR ($CDCl_3$): 1.02, 2.12 (br m's, 2 and 1 CH_2) [24] $^{13}C\{^1H\}$ NMR ($CDCl_3$): 0.20 ($ReCH_2$), 47.00 (C**C**H_2C); 181.55, 186.14 (CO) [24]; (CD_2Cl_2): 0.5, 50.3; 182.3, 186.6 [23] IR (KBr): 528, 585, 610 (ν(Re-CO)) [23]; ($CHCl_3$): 1974, 2012, 2118 (ν(CO)) [24] (nearly identical in [23]) FD MS: $[M]^+$, $[M - C_3H_6]^+$ [24] rapidly decomposed in air in solvents such as THF, ether, alcohols [24]
17	$(CO)_5Re-(CH_2)_4-Re(CO)_5$	VI (42% [42]; 78% [24]); precipitated after concentration and cooling to 15 °C [24] white crystals; m.p. 171 to 172°C [24, 42] ^1H NMR ($CDCl_3$): 1.01 (br, CH_2), 1.78 (br m, $ReCH_2$) [24]; same values in [42] $^{13}C\{^1H\}$ NMR ($CDCl_3$): 9.67 ($ReCH_2$), 49.68 (C**C$_2$**H_4C); 181.54, 186.07 (CO) [24, 42] IR ($CHCl_3$): 1976, 2011, 2121 (ν(CO)) [24] MS: $[M]^+$, $[M - n\,CO]^+$, $[(CO)_{10}Re_2]^+$ [24, 42] rapidly decomposes in air in solvents such as THF, alcohols, ether [24] treatment with X_2 (X = Br, I) in CH_2Cl_2 yielded $(CO)_5ReX$ and $X(CH_2)_4X$ treatment with 2 equivalents $P(CH_3)_n(C_6H_5)_{3-n}$ (n = 2, 1, 0) yielded Nos. 8 to 11 [24]
*18	$(CO)_5Re-(CH_2)_5-Re(CO)_5$	IV (49%) [32]; VI (58%) [42] off-white solid [42]; m.p. 97.5°C [32]; but: m.p. 141 to 143 °C [42] ^1H NMR ($CDCl_3$): 0.98 (t), 1.30 (br s), 1.76 (br) $^{13}C\{^1H\}$ NMR ($CDCl_3$): −9.2, 37.8, 40.3; 181.4, 186.1 [42] (similar in [32])

Table 2 (continued)

No.	compound	method of preparation (yield) properties and remarks
		IR (KBr): 587, 612 (Re-CO); ($CHCl_3$): 1977, 2008, 2123 (ν(CO)) [42] (similar in [32]) MS: $[M - n\ CO]^+$ (n = 0 to 10 except 4), $[M - (CH_2)_5 - n\ CO]^+$ (n = 1 to 10) [42] FD MS: $[M]^+$ [32]
19	$(CO)_5Re-(CH_2)_6-Re(CO)_5$	VI (62%) off-white solid; m.p. 162 to 164°C ^1H NMR ($CDCl_3$): 0.91 (t), 1.26 (br s), 1.70 (br) $^{13}C\{^1H\}$ NMR ($CDCl_3$): −9.1, 37.1, 39.1; 181.5, 186.0 (CO) IR ($CHCl_3$): 1978, 2009, 2122 (ν(CO)) MS: $[M - n\ CO]^+$ (n = 0 to 10 except 1), $[M - (CH_2)_6 - n\ CO]^+$ (n = 0 to 10) [42]
20	$(CO)_5Re-(CH_2)_7-Re(CO)_5$	VI (69%) off-white solid; m.p. 122 to 127°C ^1H NMR ($CDCl_3$): 0.96 (t), 1.28 (br s), 1.76 (br) $^{13}C\{^1H\}$ NMR ($CDCl_3$): −9.1, 29.6, 37.3, 39.2; 181.5, 186.1 (CO) IR ($CHCl_3$): 1977, 2008, 2122 (ν(CO)) MS: $[M - n\ CO]^+$ (n = 0 to 10 except 2), $[M - (CH_2)_7 - n\ CO]^+$ (n = 0 to 10) [42]
21	$(CO)_5Re-(CH_2)_8-Re(CO)_5$	VI (54%) off-white solid; m.p. 151 to 155°C ^1H NMR ($CDCl_3$): 0.99 (t), 1.27 (br s), 1.77 (br) $^{13}C\{^1H\}$ NMR ($CDCl_3$): −9.1, 29.7, 37.5, 39.1; 181.5, 186.0 (CO) IR ($CHCl_3$): 1978, 2008, 2122 (ν(CO)) MS: $[M - n\ CO]^+$ (n = 0 to 10 except 3), $[M - (CH_2)_8 - n\ CO]^+$ (n = 0 to 10) [42]
22	$(CO)_5Re-(CH_2)_9-Re(CO)_5$	from $Br(CH_2)_9Br$ and $Na[(CO)_5Re]$ (THF, 8 h; evaporation and recrystallization; yield: 40%) off-white solid from CH_2Cl_2; m.p. 104 to 106°C ^1H NMR ($CDCl_3$): 0.96 (t), 1.27 (br s), 1.78 (br) $^{13}C\{^1H\}$ NMR ($CDCl_3$): −9.1, 29.2, 29.8, 37.4, 39.2; 181.5, 186.0 (CO) IR ($CHCl_3$): 1978, 2007, 2123 (ν(CO)) MS: $[M - n\ CO]^+$ (n = 0 to 10 except 3), $[M - (CH_2)_9 - n\ CO]^+$ (n = 0 to 10) [42]

References on pp. 38/9

Table 2 (continued)

No.	compound	method of preparation (yield) properties and remarks
23	$(CO)_5Re-(CH_2)_{10}-Re(CO)_5$	IV (53%) [32]; VI (71%) [42] off-white solid [42]; m.p. 91.5 °C [32], but: m.p. 138 to 140 °C [42] ^1H NMR ($CDCl_3$): 0.96 (t), 1.27 (br s), 1.77 (br) [42] (identical in [32]) $^{13}C\{^1H\}$ NMR ($CDCl_3$): −9.0, 29.1, 29.8, 37.5, 39.2; 181.5, 186.0 [42] (nearly identical in [32]) IR (KBr): 587, 609 (Re-CO) [32]; ($CHCl_3$): 1978, 2007, 2122 (ν(CO)) [42] (similar in [32]) MS: $[M - n\,CO]^+$ (n = 0 to 10 except 1, 2), $[M - (CH_2)_{10} - n\,CO]^+$ (n = 0 to 10) [42] FD MS: $[M]^+$ [32]
*24	$(CO)_5Re-CH_2(CHCH_2OCH_2CH\text{-cyclo})CH_2-Re(CO)_5$ 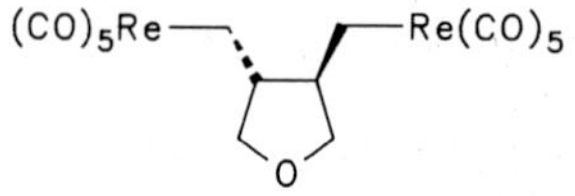 [structure: tetrahydrofuran ring with $(CO)_5Re$–CH_2 and CH_2–$Re(CO)_5$ substituents on 3,4-positions]	IV (64.2%) colorless solid; m.p. 104 °C ^1H NMR ($CDCl_3$): ABMM'XY system with 0.75, 1.47 (dd, $ReCH_2$, AB); 1.93 to 2.06 (m, CH, MM'); 3.42, 4.20 (dd, OCH_2, XY); $^2J(A,B)$ = 12.1, $^3J(A,M)$ = 10.64, $^3J(B,M)$ = 2.99, $^2J(X,Y)$ = 8.11, $^3J(M,X)$ = 7.85, $^3J(M,Y)$ = 6.71 $^{13}C\{^1H\}$ NMR ($CDCl_3$): −7.61 ($ReCH_2$), 59.43 (CH), 77.57 (OCH_2); 180.52, 185.39 (CO) IR (KBr): 585, 615 (Re-CO); (CH_2Cl_2): 1979, 2012, 2049, 2124 (ν(CO)) FD MS: $[M]^+$ $(CF_3SO_2)_2O$ did not induce ring cleavage [33]
25	$cyclo\text{-}C_6H_{10}[CH_2Re(CO)_5]_2\text{-}1,4$  [structure: cyclohexane chair with two $CH_2Re(CO)_5$ groups]	IV (67%) m.p. 159.8 °C ^1H NMR ($CDCl_3$): 0.78 to 0.85, 0.88 to 0.96, 1.02 to 1.05, 1.11 to 1.27, 1.35 to 1.37, 1.47 to 1.52, 1.68 to 1.74; all signals are m's from 2 H $^{13}C\{^1H\}$ NMR ($CDCl_3$): −0.59 ($ReCH_2$), 33.86 (CH_2CH_2), 42.91 (CH); 181.38, 186.27 (CO) IR (pentane): 1983, 2012, 2123 (ν(CO)) EI MS: $[M]^+$ [40]

References on pp. 38/9

Table 2 (continued)

No.	compound	method of preparation (yield) properties and remarks
*26	cyclo-$C_6H_{10}[CH_2Re(CO)_5]_2$-1,3 $(CO)_5Re$ … $Re(CO)_5$	IV (72%) ^{1}H NMR ($CDCl_3$): 0.44 to 0.58 (1 H), 0.63 to 0.78 (2 H), 0.83 to 0.99 (4 H), 1.11 to 1.30 (1 H), 1.38 to 1.56 (2 H), 1.65 to 1.75 (3 H), 1.81 to 1.86 (1 H); all signals are m's $^{13}C\{^1H\}$ NMR ($CDCl_3$): 1.81 ($ReCH_2$), 26.91 ($CHCH_2\mathbf{C}H_2$), 38.07 ($CH\mathbf{C}H_2CH_2$), 45.52 (CH), 51.97 ($CH\mathbf{C}H_2CH$); 181.57, 186.33 (CO) IR (pentane): 1983, 2012, 2123 (ν(CO)) EI MS: $[M]^+$ [40]
*27	trans-$(CO)_5Re-CH_2CH{=}CHCH_2-Re(CO)_5$ $(CO)_5Re$ H H $Re(CO)_5$	according to Method V when employing $[\pi\text{-}CH_2{=}CHCH{=}CH_2Re(CO)_5]^+$ (at −40°C); main product $(C_5H_6O)Re_2(CO)_9$ (see pp. 160/1) converted at room temperature to No. 27; separated by chromatography with CH_2Cl_2 yellow solid IR (KBr): 1923, 1960, 1993, 2030, 2120 (ν(CO)) [12]
28	$C_6H_4[C_2H_4Re(CO)_5]_2$-1,4 $(CO)_5Re$ … $Re(CO)_5$	IV (54%) colorless product; m.p. 131°C ^{1}H NMR (CD_2Cl_2): 1.18 to 1.25 (m, $ReCH_2$), 2.95 to 3.02 (m, $C_6H_4\mathbf{CH_2}$), 7.13 (s, C_6H_4) IR (CH_2Cl_2): 1977, 2010, 2125 (ν(CO)) EI MS: $[M]^+$ poorly soluble in organic solvents [36]
29	$C_6H_4[C_3H_6Re(CO)_5]_2$-1,4 $(CO)_5Re$ … $Re(CO)_5$	IV (61%) colorless product; m.p. 124°C ^{1}H NMR ($CDCl_3$): 0.97 to 1.04 (m, $ReCH_2$), 1.98 to 2.11 ($-CH_2-$), 2.53 to 2.59 (m, $C_6H_4\mathbf{CH_2}$), 7.10 (s, C_6H_4) $^{13}C\{^1H\}$ NMR ($CDCl_3$): −9.54 ($ReCH_2$), 41.29 ($-CH_2-$), 43.58 ($\mathbf{C}H_2C_6$); 128.28, 139.70 (C_6); 181.31, 185.76 (CO) IR (pentane): 1985, 2012, 2124 (ν(CO)) FD MS: $[M]^+$ [36]

References on pp. 38/9

Table 2 (continued)

No.	compound	method of preparation (yield) properties and remarks
30	$C_6H_4[C_2H_4Re(CO)_5]_2$-1,3 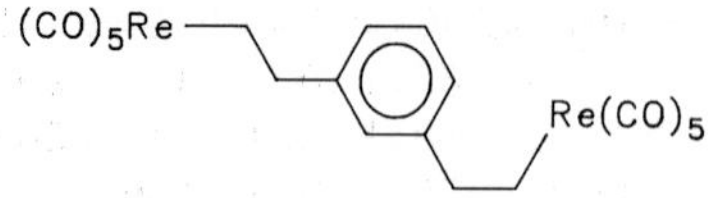	IV (69%) colorless product; m.p. 119 °C ^{1}H NMR ($CDCl_3$): 1.11 to 1.18 (m, $ReCH_2$), 2.89 to 2.96 (m, C_6H_4**CH_2**), 6.99 to 7.12 (m, C_6H_4) ^{13}C {^{1}H} NMR (CD_2Cl_2): −6.99 ($ReCH_2$), 45.12 (**C**H_2C_6); 125.11, 127.32, 128.57, 149.46 (C_6); 181.87, 186.21 (CO) IR (CH_2Cl_2): 1977, 2011, 2125 (ν(CO)) FD MS: $[M]^+$ [36]
31	$C_6H_4[C_3H_6Re(CO)_5]_2$-1,3 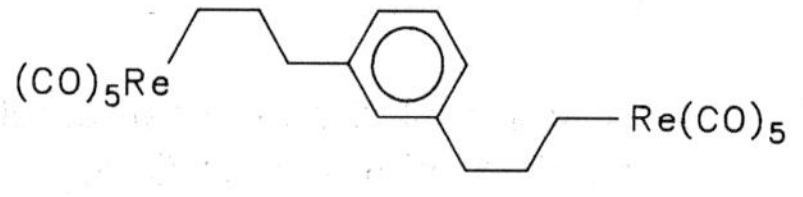	IV (77%) colorless product; m.p. 136 °C ^{1}H NMR ($CDCl_3$): 0.98 to 1.19 (m, $ReCH_2$), 1.95 to 2.34 (CH_2**CH_2**CH_2), 2.57 to 2.76 (m, C_6H_4**CH_2**), 6.98 to 7.35 (m, C_6H_4) ^{13}C {^{1}H} NMR ($CDCl_3$): −9.50 ($ReCH_2$), 41.11 (-CH_2-), 43.83 (**C**H_2C_6); 125.65, 128.00, 128.62, 142.42 (C_6); 181.23, 185.77 (CO) IR (pentane): 1985, 2012, 2124 (ν(CO)) FD MS: $[M]^+$ [36]
32	$C_6H_4[C_2H_4Re(CO)_5]_2$-1,2 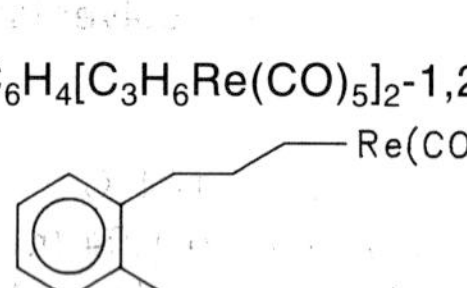	IV (58%) colorless product; m.p. 107 °C ^{1}H NMR ($CDCl_3$): 1.06 to 1.15 (m, $ReCH_2$), 2.91 to 2.98 (m, C_6H_4**CH_2**), 6.98 to 7.09 (m, C_6H_4) ^{13}C {^{1}H} NMR (CD_2Cl_2): −6.94 ($ReCH_2$), 42.14 (**C**H_2C_6); 126.13, 128.99, 146.44 (C_6); 181.89, 186.32 (CO) IR (CH_2Cl_2): 1976, 2011, 2125 (ν(CO)) FD MS: $[M]^+$ [36]
33	$C_6H_4[C_3H_6Re(CO)_5]_2$-1,2	IV (76%) colorless product; m.p. 151 °C ^{1}H NMR ($CDCl_3$): 0.97 to 1.05 (m, $ReCH_2$), 1.90 to 2.03 (CH_2**CH_2**CH_2), 2.53 to 2.59 (m, C_6H_4**CH_2**), 7.02 to 7.11 (m, C_6H_4) ^{13}C {^{1}H} NMR ($CDCl_3$): −9.11 ($ReCH_2$), 40.98 (-CH_2-), 41.08 (**C**H_2C_6); 125.61, 129.22, 140.22 (C_6); 181.22, 185.76 (CO) IR (pentane): 1985, 2013, 2124 (ν(CO)) FD MS: $[M]^+$ [36]

References on pp. 38/9

Table 2 (continued)

No.	compound	method of preparation (yield) properties and remarks
34	$[(CO)_5Cr{=}C(OCH_3)C^*HCH_2CH_2Re(CO)_5]_2$ 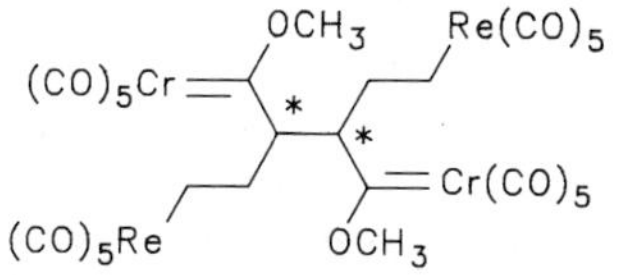	$[-CH_2(CH_3O)C{=}Cr(CO)_5]_2$ was doubly deprotonated with n-C_4H_9Li; subsequent addition of $[\pi\text{-}C_2H_4Re(CO)_5]BF_4$ in THF and chromatographic separation with pentane/CH_2Cl_2 gave the product; yield: 32% (2 diastereomers) yellow powder ^1H NMR (CD_2Cl_2): 0.9, 1.26 ($ReCH_2$); 1.9, 2.1 ($C^*HC\mathbf{H_2}$); 3.3, 3.65 (m's, C^*H); 4.73, 4.78 (s's, OCH_3) $^{13}C\{^1H\}$ NMR ($CDCl_3$): −13.6, −12.07 ($ReCH_2$); 41.1, 41.6 ($C^*H\mathbf{C}H_2$); 67.97, 68.39 (OCH_3); 70.82, 78.26 (C^*H); 180.1, 181.3, 185.2, 185.36 (ReCO); 216.7, 216.85, 223.3 (CrCO); 360, 364.6 (Cr=C) IR (Nujol): 1946, 1985, 2011, 2061, 2127 (ν(CO)) [43]
	OCH3, Re(CO)5, (CO)5Cr, *, *, Cr(CO)5, (CO)5Re, OCH3	
*35	$[(CO)_5ReC_2H_4C{\cdots}O]_2Fe_2(CO)_6$ 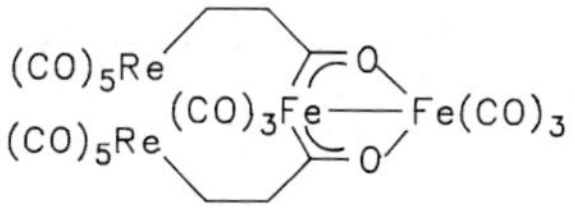	VII (when employing $[(CO)_8Fe_2]^{2-}$; yield: 61%) red crystals from CH_2Cl_2 ^1H NMR (C_6D_6): 0.88 (t, $ReCH_2$), 3.20 (t, $C(O)CH_2$) IR (Nujol): 1955, 1963, 1974, 1985, 1994, 1998, 2017, 2046, 2074, 2132 (ν(CO)) [20]
	(CO)5Re, O, (CO)3Fe, Fe(CO)3, (CO)5Re, O	
*36	$(CO)_6Fe_2[SC_2H_4Re(CO)_5]_2$	VII (when employing $[(CO)_6Fe_2S_2]^{2-}$; 2 isomers) [44]
	syn isomer 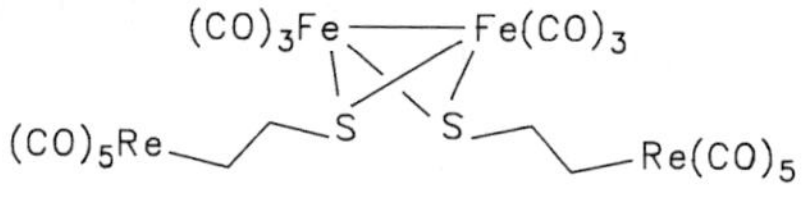 (CO)3Fe, Fe(CO)3, S, S, (CO)5Re, Re(CO)5	yield: 51.7% red solid; m.p. 119 to 121 °C ^1H NMR ($CDCl_3$): 1.24 ($ReCH_2$), 1.92 (SCH_2) $^{13}C\{^1H\}$ NMR ($CDCl_3$): −6.33 ($ReCH_2$), 49.32 (SCH_2); 184.12 (ReCO), 210.22 (FeCO) [44]
	anti isomer 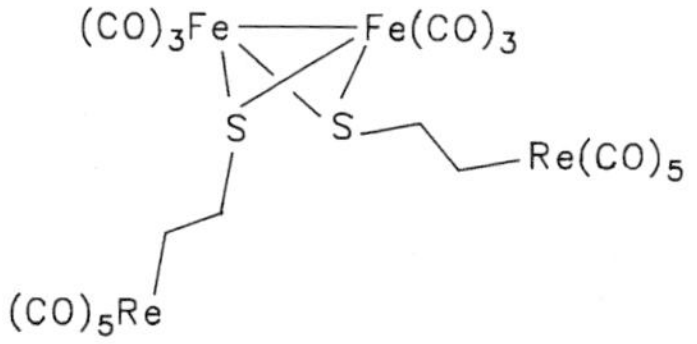 (CO)3Fe, Fe(CO)3, S, S, Re(CO)5, (CO)5Re	yield: 16.3% ruby red crystals; m.p. 117 °C ^1H NMR ($CDCl_3$): 1.04, 1.25 ($ReCH_2$); 2.60, 2.88 (SCH_2) $^{13}C\{^1H\}$ NMR ($CDCl_3$): −7.12, −6.76 ($ReCH_2$); 35.83, 50.39 (SCH_2); 180.22, 180.37, 184.19, 184.28 (ReCO); 209.43 (FeCO) [44]

References on pp. 38/9

Table 2 (continued)

No.	compound	method of preparation (yield) properties and remarks
37	$(CO)_6Fe_2[TeC_2H_4Re(CO)_5]_2$ 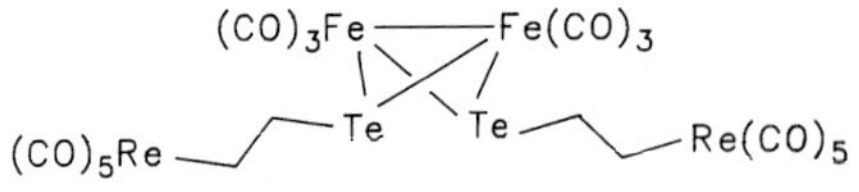	VII (when employing $[(CO)_6Fe_2Te_2]^{2-}$; yield: 70.9%) red solid; dec. 55 °C ^{1}H NMR ($CDCl_3$): 1.43 ($ReCH_2$), 3.54 ($TeCH_2$) $^{13}C\{^1H\}$ NMR ($CDCl_3$): −6.49, 52.76 ($ReCH_2$, $TeCH_2$); 176.0, 182.98 (ReCO); 215.28 (FeCO) IR (CH_2Cl_2): 1952, 1983, 2003, 2013, 2035, 2077, 2129 (ν(CO)) loses C_2H_4 in CH_2Cl_2 at room temperature to give $(CO)_6Fe_2[TeRe(CO)_5]_2$ [44]
38	$C_{30}H_{20}Fe_2O_8[Re(CO)_5]_2$ 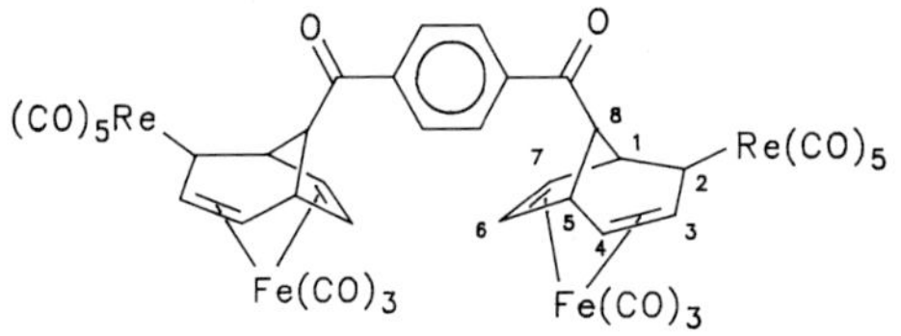	by adding $[(CO)_6Fe_2(C_8H_8C(O)C_6H_4\text{-}C(O)C_8H_8)]^{2+}$ to $Na[(CO)_5Re]$ in THF between −70 °C and room temperature; evaporation, extraction into CH_2Cl_2, and filtration over silica; yield: 30% pale yellow solid; dec. above 137 °C ^{1}H NMR ($CDCl_3$): 2.84 (m, 2 H-1), 3.08 (t, 2 H-4), 3.41 (m, H-2,2′,5,5′), 3.43 (s, H-8,8′), 3.58 (m, H-6,6′), 3.95 (m, H-7,7′), 4.27 (dt, H-3,3′), 7.98 (s, C_6H_4); J(H-4,H-3,5) = 7.3, J(H-3,3′,H-4,4′) = 7.2, J(H-3,3′, H-2,2′) = 2.3, ^{4}J(H-3,3′,H-5,5′) = 2.3 IR (Nujol): 1682 (C=O); 1954, 1988, 2006, 2025, 2055, 2126 (ν(CO)) poorly soluble in common solvents [39]
39	cis-$[(CO)_5ReC_2H_4]_2Os(CO)_4$ 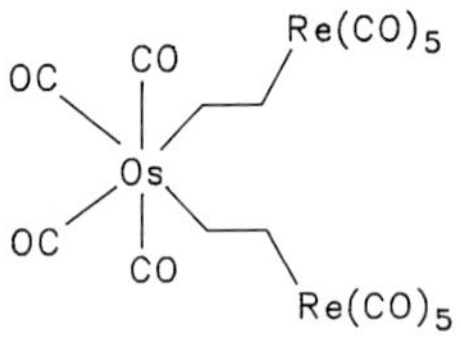	VII (when employing $[(CO)_4Os]^{2-}$; yield: 65%); No. 14 was also formed [18] ^{1}H NMR (CD_2Cl_2): 1.91 to 2.08 (14 lines) [18]; 1.97 ($OsCH_2$), 2.03 ($ReCH_2$); J_{gem} = −12.3, $^3J(CH_2,CH_2)$ = 14.7 and 4.2; spectrum unchanged between 25 and 75 °C [25] $^{13}C\{^1H\}$ NMR (CD_2Cl_2): 4.45, 12.04 (C_2H_4); 172.44 (OsCO), 181.90, 187.26 (ReCO) IR (pentane): 1983, 2009, 2033, 2110, 2119 (ν(CO)) MS: $[M]^+$, $[M - (CO)_5ReC_2H_4]^+$, and CO loss peaks [18] CI MS: $[M - H - n\,CO]^+$ (n = 0 to 4) [25] slowly decomposes in CD_2Cl_2 [25]

References on pp. 38/9

Table 2 (continued)

No.	compound	method of preparation (yield) properties and remarks
40	$[(CO)_5ReCH_2CH_2Os(CO)_4-]_2Os(CO)_4$ 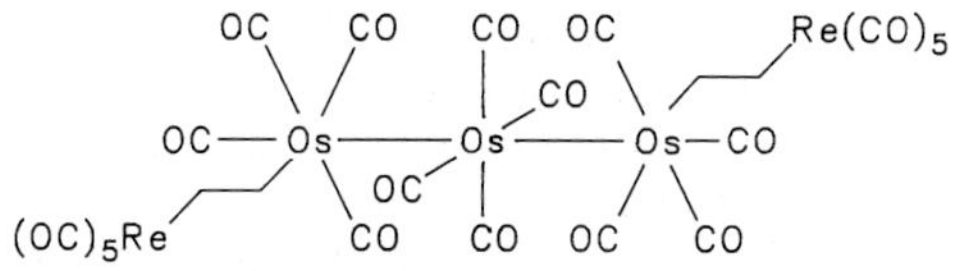 (suggested structure)	VII (when employing $[(CO)_{12}Os_3]^{2-}$; yield: 23%) yellow powder ^1H NMR (CD_2Cl_2): 1.99 (m, $OsCH_2$), 2.37 (m, $ReCH_2$); J_{gem} = −11.9 at both CH_2 groups, $^3J(CH_2,CH_2)$ = 14.6 and 4.2 IR (pentane): 1984, 2012, 2024, 2033, 2054, 2093, 2121, 2134 (ν(CO)) decomposed in $(CO)_{12}Os_3$, C_2H_4, and No. 14 in solution at room temperature [25]
41	$(CO)_5ReCH_2C[Re(CO)_5]{=}CHC_6H_5$ 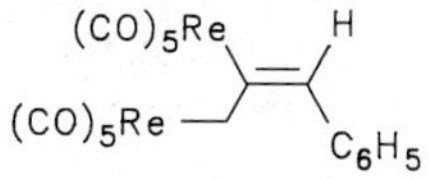	by irradiating $(CO)_5Re-Re(CO)_3(^4D)$ (^4D = $CH_3C_6H_4N{=}CHCH{=}NC_6H_4CH_3$ or 1,10-phenanthroline) and phenylallene (toluene, −33°C) IR: 2143 (highest value for ν(CO)) rearranged into $(\mu\text{-}CH_2CCHC_6H_5)Re_2(CO)_9$ (see p. 130) upon raising the temperature [28]
42	cyclo-$C_4H_2O_2[Re(CO)_5]_2$ 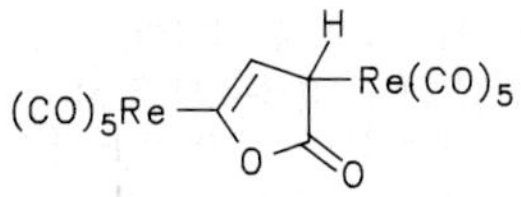	VIII (when employing fumaric acid dichloride); yield: 75% (2 isomers in the ratio 2:1) yellow solid; m.p. 92 to 95°C (dec.) ^1H NMR (CD_2Cl_2): 3.24, 5.93 (main isomer); 3.47, 6.05 (minor isomer); all signals: d's; J = 2 IR (CH_2Cl_2): 1617 (ν(C=C)), 1710 (ν(C=O)); 1991, 2033, 2136 (ν(CO)) [41]
43	cyclo-$C_5H_4O_2[Re(CO)_5]_2$ 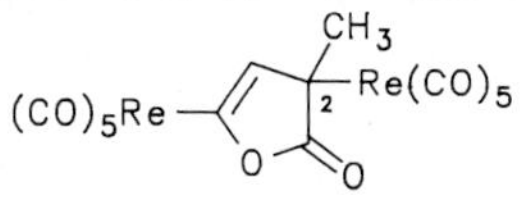	VIII (when employing mesaconic acid dichloride); yield: 49% (2 isomers in the ratio 2:1) ^1H NMR (CD_2Cl_2): 1.94 (CH_3), 6.06 (main isomer); 1.86 (CH_3), 6.19 (minor isomer) $^{13}C\{^1H\}$ NMR (CD_2Cl_2): 27.8 (CH_3), 29.0 (C-2); 142.2, 143.1 (C=C); 163.0 (C=O); 181.3, 181.7, 182.4, 192.6 (CO) IR (CH_2Cl_2): 1707 (ν(C=O)); 1999, 2032, 2068, 2133, 2143 (ν(CO)) [41]
44	$C_{60}[Re(CO)_5]_2$	by irradiating a mixture of $(CO)_{10}Re_2$ and C_{60} in benzene; CO did not suppress the reaction; yield: >90% (by IR); also

Table 2 (continued)

No.	compound	method of preparation (yield) properties and remarks
44 (continued)		by treating $((C_6H_5)_3C)Re(CO)_4$ with C_{60} under a CO atmosphere (with excess $((C_6H_5)_3C)Re(CO)_4$ likely higher-nuclear derivatives also formed) grayish brown, ESR-silent solution IR (C_6H_6): 1993, 2036, 2130, 2134 (ν(CO)) decomposed in solution within 1 d, recovering $(CO)_{10}Re_2$ and C_{60} via a first-order reaction with $k = (4.4 \pm 0.3) \times 10^{-3}\ min^{-1}$ decomposed in the presence of CCl_4 to $(CO)_5ReCl$ and C_{60} [37]
*45	$(cyclo\text{-}C_4O_2)[Re(CO)_5]_2$ 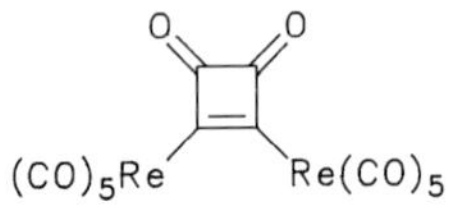	by reacting $Na[(CO)_5Re]$ with squaric acid dichloride; yield: 66% crude, ochre powder; recrystallization from $CHCl_3$ gave lemon yellow crystals; m.p. 165 °C IR (KBr): 1445, 1697, 1729, 1735, 1784; 1995, 2018, 2055, 2083, 2137, 2145 [15]
*46	$(CO)_5Re-C(=CH_2)CH(C_6H_5)CH(C_6H_5)C(=CH_2)-Re(CO)_5$ 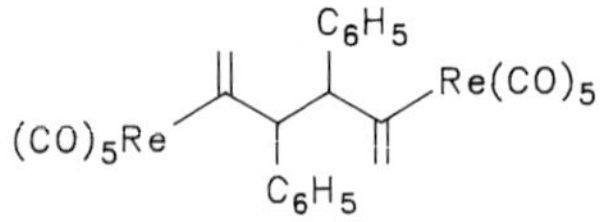	by irradiating a mixture of $(CO)_{10}Re_2$ and $CH_2=C=CHC_6H_5$ (hexane, −25 °C, 75 min); workup by column chromatography; yield: 11% colorless cubes ^1H NMR (CD_2Cl_2): 4.48, 5.41, 6.42 (s's); 7.32 (m, C_6H_5); at −40 °C splitting of the singlets: 4.32, 4.55, 4.92, 5.43, 5.48, 5.90, 6.26, 6.54 IR (n-hexane): 1975, 1995, 2012, 2030, 2125 (ν(CO)) [22]
*47	trans-$(CO)_5Re-CH=C(CO_2CH_3)-Re(CO)_5$ 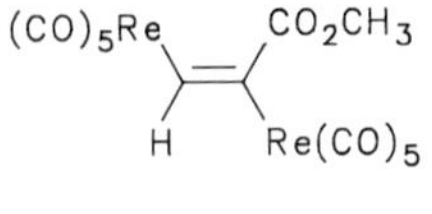	IX (when starting from $(CO)_4ReO=(CH_3O)CC[Re(CO)_5]=CH$-cyclo, 7 d; yield: 11%); also obtained at 103 °C (yield: 14%), but at this temperature two other compounds also formed colorless solid ^1H NMR ($CDCl_3$): 3.67 (CH_3), 7.65 (CH) IR (CH_2Cl_2): 1685 (C=O); (hexane): 1983, 2025, 2046, 2062, 2124 (ν(CO)) [30]

References on pp. 38/9

Table 2 (continued)

No.	compound	method of preparation (yield) properties and remarks
*48	cis-$(CO)_5Re-C(CO_2C_2H_5)=C(CO_2C_2H_5)-Re(CO)_5$ 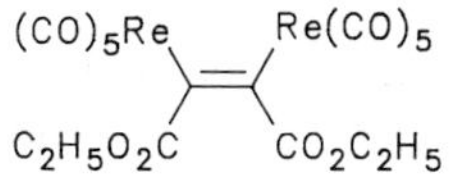 $(CO)_5Re$ $Re(CO)_5$ $C_2H_5O_2C$ $CO_2C_2H_5$	IX (when starting from $(CO)_4ReO=C(OC_2H_5)$-$C[=C(CO_2C_2H_5)Re(CO)_5]$-cyclo, 12 h; yield: 93%) colorless crystals ^1H NMR ($CDCl_3$): 1.26 (t, CH_3), 4.06 (q, CH_2); $^3J(H,H)$ = 7.2 IR (hexane): 1692 (C=O); 1984, 1994, 2005, 2030, 2043, 2067, 2129, 2140 (ν(CO)) [38]
49	$C_6H_2F_6[Re(CO)_5]_2$ (mixture of isomers) $C_6H_2F_6$=hexafluroro-bicyclo[2.2.0]hexadiyl	from hexafluoro bicyclo[2.2.0]hexa-2,5-diene and $(CO)_5ReH$ (closed tube, 25°C, 16 h) nonvolatile, orange solid from THF/hexane; m.p. 115 to 116°C too insoluble for molecular weight studies and ^{19}F NMR IR (Nujol): 664, 729, 741, 785, 859, 942, 969, 1101, 1156, 1201, 1219, 1241, 1369, 1381; 1939, 1978, 1987, 2023, 2042, 2051, 2072, 2079, 2145, 2176 (ν(CO)) [3]
50	$C_8F_{11}H[Re(CO)_5]_2$ $\beta-CF_3$ $\varepsilon,\gamma-F_2$ H $CF_3-\delta$ $(CO)_5Re$ $\alpha-CF_3$ $Re(CO)_5$	by reacting $CF_3C\equiv CCF_3$ with $Na[(CO)_5Re]$ (THF, −78°C to room temperature); yield: 9.5% (by-products were $(CO)_5ReC(CF_3)=C=CF_2$ and $(CO)_{10}Re_2$) [5] (compound also mentioned in [1] but with another suggested structure) air-stable, yellow solid; m.p. 138 to 138.5°C [5] ^1H NMR: 7.05 (J(H,F-γ) = 2.9, J(H,F-δ) = 2.0, J(H,F-ε) = 7.3) [5] ^{19}F NMR: 43.39 (F_α), 50.29 (F_β), 61.34 (F_γ), 62.84 (F_δ), 67.72 (F_ε); J(F-α,β) = 1.8, J(F-α,γ) = 4.0, J(F-γ,δ) = 3.7, J(F-γ,ε) = 176.0, J(F-δ,ε) = 1.7 [5] IR (Nujol): 641, 668, 690, 748, 830, 852, 882, 937, 1018, 1183, 1116, 1122, 1132, 1165, 1216, 1261, 1273, 1325, 1559, 1619 [5]; (C_6H_{12}): 2012, 2040, 2050, 2070, 2148 (ν(CO)) [4, 5] molecular weight (benzene, osmometry): 986 (calc. 958) [5] MS: $[M - n\,CO]^+$ (n = 0 to 10) [5]

References on pp. 38/9

Table 2 (continued)

No.	compound	method of preparation (yield) properties and remarks
51	$C_6H_4[Re(CO)_5]_2$-1,3	VI (under reduced pressure, at 105 °C; yield: 48.6%) sublimable, white crystalline product; m.p. 63.5 to 64.5 °C (note: 133.5 to 134.5 °C also given) IR: 1984, 2004, 2015, 2053, 2120 (ν(CO)) [2]
*52	$C_6H_4[Re(CO)_5]_2$-1,4	VI (in diglyme; yield: 65%) [2]; VI (in methyl isobutyl ketone; yield: 56%) [27] white microcrystals [27]; dec. >200 °C without melting [2], m.p. 210 to 215 °C [27] ^{1}H NMR (CD_2Cl_2): 7.26 ^{13}C {^{1}H} NMR ($CDCl_3$): 134.3 (ReC), 146.1 (C_6H_4); 181.5, 183.4 (CO) [27] IR: 1975, 2008, 2022, 2075, 2150 (ν(CO)) [2] (similar in [27]) MS: $[M - n\,CO]^+$ (n = 0 to 10), $[M - Re(CO)_5]^+$ etc. [27]
53	$[(CO)_5Re-C_6H_4Fe(C_5H_5)-Re(CO)_5]PF_6$ 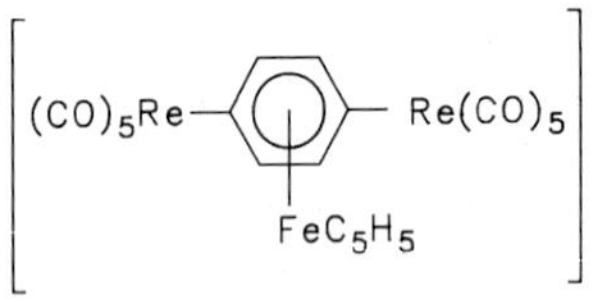 [structure: $(CO)_5Re$–C_6H_4(–FeC_5H_5)–$Re(CO)_5$]$^+$	by treating $[(\pi\text{-}1,4\text{-}Cl_2C_6H_4)FeC_5H_5]PF_6$ with $Na[(CO)_5Re]$ (ratio 2:1) in THF; workup by evaporation and extraction with CH_2Cl_2 ^{1}H NMR (CD_2Cl_2): 4.83 (C_5H_5), 6.04 (C_6H_4) ^{13}C {^{1}H} NMR (CD_2Cl_2): 74.7 (C_5), 95.6 (C-1,4), 101.2 (C-2,3,5,6); 179.3, 181.8 (CO) [35]
*54	$(CO)_5Re-C(O)C(O)-Re(CO)_5$ [structure: $(CO)_5Re$–C(=O)–C(=O)–$Re(CO)_5$] 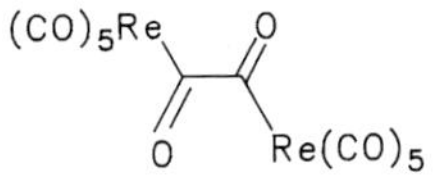	X (in ether; yield: >50%) deep purple, brick-like crystals from CH_2Cl_2 ^{13}C {^{1}H} NMR: 182.6, 184.1 (CO); 253.3 (C=O) IR: 1562, 1624, 1652 (C=O); 2002, 2035, 2125 (ν(CO)) UV: λ_{max} = 384, 533 (MLCT) DCI MS: $[Re_2(CO)_{12}H]^+$, $[Re(CO)_6]^+$ compound should be stored at <20 °C [13]
55	$(CO)_5Re-C(O)(CH_2)_4C(O)-Re(CO)_5$	X (48% [42]; 60% [24]) m.p. 144 to 147 °C [24, 42] ^{1}H NMR ($CDCl_3$): 1.40 (t, $-CH_2CH_2-$), 2.69 (t, CH_2CO) [42] (similar in [24])

References on pp. 38/9

Table 2 (continued)

No.	compound	method of preparation (yield) properties and remarks
		$^{13}C\{^1H\}$ NMR ($CDCl_3$): 23.0, 70.4; 181.6, 183.2 (CO); 248.1 (C(O)) [42] IR ($CHCl_3$): 1605 (ν(C=O)); 2021, 2062, 2132 (ν(CO)) [42] (similar in [24]) thermolysis (toluene, reflux) gave No. 17 [24, 42]
56	$(CO)_5Re-C(O)(CH_2)_5C(O)-Re(CO)_5$	X (46%) m.p. 117 to 120 °C ^1H NMR ($CDCl_3$): 1.23 (complex); 1.43, 2.69 (t's) $^{13}C\{^1H\}$ NMR ($CDCl_3$): 23.5, 28.2; 70.5; 181.2, 183.2 (CO); 247.7 (C(O)) IR ($CHCl_3$): 1605 (ν(C=O)); 2021, 2062, 2132 (ν(CO)) thermolysis gave No. 18 [42]
57	$(CO)_5Re-C(O)(CH_2)_6C(O)-Re(CO)_5$	X (52% [42]; 62% [7]) m.p. 121 to 126 °C [7]; 129 to 132 °C [42] ^1H NMR ($CDCl_3$): 1.22, 1.44 (complex); 2.69 (t) $^{13}C\{^1H\}$ NMR ($CDCl_3$): 23.5, 28.2; 70.6; 181.2, 183.3 (CO); 247.0 (C(O)) IR ($CHCl_3$): 1605, 2021, 2062, 2132 (ν(CO)) thermolysis gave No. 19 [42] treatment with 2 or 4 equivalents CH_3Li gave Nos. 13 and 7, respectively (for the case n = 6) [7]
58	$(CO)_5Re-C(O)(CH_2)_7C(O)-Re(CO)_5$	X (53% [42]; 67% [7]) m.p. 96 to 104 °C [42]; 90 to 104 °C [7] ^1H NMR ($CDCl_3$): 1.22, 1.43 (complex); 2.68 (t) $^{13}C\{^1H\}$ NMR ($CDCl_3$): 23.5, 28.6, 29.3; 70.7; 181.2, 183.3 (CO); 247.7 (C(O)) IR ($CHCl_3$): 1605, 2020, 2062, 2132 (ν(CO)) thermolysis gave No. 20 [42] treatment with 2 or 4 equivalents CH_3Li gave Nos. 13 and 7, respectively (for the case n = 7) [7]
59	$(CO)_5Re-C(O)(CH_2)_8C(O)-Re(CO)_5$	X (58%) m.p. 122 to 125 °C ^1H NMR ($CDCl_3$): 1.22, 1.44 (complex); 2.69 (t)

References on pp. 38/9

Table 2 (continued)

No.	compound	method of preparation (yield) properties and remarks
59 (continued)		$^{13}C\{^1H\}$ NMR ($CDCl_3$): 23.6, 28.8, 29.2; 70.7; 181.2, 183.3 (CO); 247.2 (C(O)) IR ($CHCl_3$): 1605, 2014, 2062, 2132 (ν(CO)) thermolysis gave No. 21 [42]
60	$(CO)_5Re-C(O)(CH_2)_{10}C(O)-Re(CO)_5$	X (50%) m.p. 106 to 110°C ^1H NMR ($CDCl_3$): 1.22, 1.43 (complex); 2.69 (t) $^{13}C\{^1H\}$ NMR ($CDCl_3$): 23.6, 28.8, 29.2, 29.3; 70.7; 181.2, 183.3 (CO); 247.2 (C(O)) IR ($CHCl_3$): 1605, 2019, 2062, 2132 (ν(CO)) thermolysis gave No. 23 [42]
61	$C_6H_4[C(O)Re(CO)_5]_2$-1,3 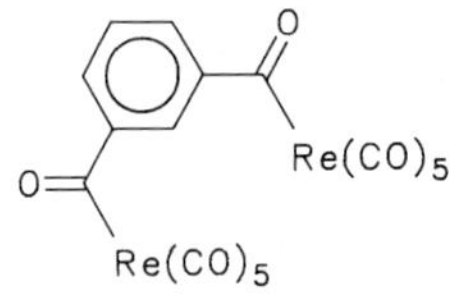 O Re(CO)5 O= Re(CO)5	X (71%) light yellow crystals; m.p. 105°C (dec.) IR: 1576 (ν(C=O)); 1980, 2008, 2020, 2078, 2140 (ν(CO)) thermolysis yielded No. 51 [2]
62	$C_6H_4[C(O)Re(CO)_5]_2$-1,4 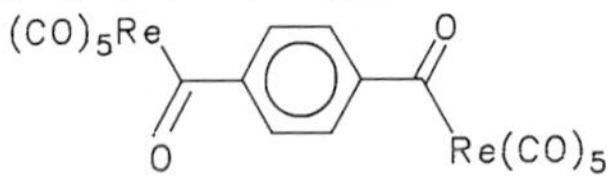 (CO)5Re O O Re(CO)5	X (96% [2], 44% [27]); also by reacting No. 52 with CO (30 atm, yield: 20%) or synthesis gas (40 atm, 70°C, THF) [27] light yellow crystals; m.p. >140°C (dec.) [2]; but: m.p. 210 to 218°C [27] IR: 1578; 1980, 2004, 2040, 2062, 2137 [2]; (CH_2Cl_2): 2002, 2018, 2060, 2133 (ν(CO)) [27] insoluble in common organic solvents [2] MS: $[M - 2\ CO]^+$ [27] thermolysis gave No. 52 [2, 27] treatment with X_2 (X = Br, I) gave $C_6H_4(CO_2H)_2$ (by subsequent hydrolysis) and $(CO)_5ReX$ [27] did not react with $BH_3 \cdot O(C_2H_5)_2$ [27]
*63	$(CO)_5Re-C{\equiv}C-Re(CO)_5$	for preparation see "Further information" pale yellow solid; m.p. ca. 125°C (dec.) [14] $^{13}C\{^1H\}$ NMR (CD_2Cl_2): 94.4, 180 [14] IR (CH_2Cl_2): 1974, 2032, 2073, 2135 [14] (similar in [21]); (Nujol): 1920, 1962, 2003, 2035, 2068, 2135 (ν(CO)) [14]

References on pp. 38/9

Table 2 (continued)

No.	compound	method of preparation (yield) properties and remarks
		Raman (CH_2Cl_2): 2002 (C≡C); 2073, 2154 [14] UV (THF): λ_{max} (ε) = 223 (68000; $12e_g \rightarrow 17a_{2u}$), 242 (50000; $17a_{1g} \rightarrow 17a_{2u}$ and $14e_u$), 319 (12000; $13e_u \rightarrow 13e_g$ and $14e_g$) [21] CV (CH_2Cl_2): $E_{p,a}$ = 1.12, 1.35 (vs. Ag); reduction not observed [21] MS: $[M - n\,CO]^+$ (n = 0 to 10) [14]
compounds of the type $RNCRe(CO)_4(\mu\text{-}\eta^{1:1}\text{-}{}^2L)Re(CO)_4CNR$		
64	$[t\text{-}C_4H_9NCRe(CO)_4-C(O)NHCH_2-]_2$ $(CO)_4ReCNC_4H_9$-t; NH−C_2H_4−NH; t−$C_4H_9NCRe(CO)_4$	XI (20%) ^1H NMR: 1.57 (s, CH_3), 3.24 (t, CH_2), 6.00 (br, NH) IR (Nujol): 1540, 1548 (ν(C=O)); 2191 (ν(CN)); 3288, 3350 (ν(NH)); (CH_2Cl_2): 1962, 1996, 2090 (ν(CO)) [17]
65	$[t\text{-}C_4H_9NCRe(CO)_4-C(O)NH(CH_2)_3-]_2$ $(CO)_4ReCNC_4H_9$-t; $HN(CH_2)_6NH$; t−$C_4H_9NCRe(CO)_4$	XI (22%) ^1H NMR: 1.15 (m), 1.42 (m), 1.57 (s, CH_3), 3.16 (t), 5.99 (br, NH) IR (Nujol): 1538, 1545 (ν(C=O)); 2192 (ν(CN)), 3318 (ν(NH)); (CH_2Cl_2): 1928, 1962, 1997, 2090 (ν(CO)) [17]
66	$[t\text{-}C_4H_9NCRe(CO)_4-C(O)N(CH_2-)_2]_2$ $(CO)_4ReCNC_4H_9$-t; N, N; t−$C_4H_9NCRe(CO)_4$	XI (15%) ^1H NMR: 1.54 (s, CH_3), 3.33 (CH_2) IR (Nujol): 1508 (ν(C=O)); (CH_2Cl_2): 1956, 1985, 2008, 2085 (ν(CO)); (Nujol): 2195 (ν(CN)) [17]

*Further information:

$[O(-CH_2Re(CO)(NO)N_3C_6H_{15})_2]I_2$ (Table **2**, No. **1**). Attempts to crystallize $[(CO)Re(NO)(CH_3)N_3C_6H_{15}]I$ from aqueous solution under an O_2 atmosphere led to crystals of No. 1. The compound could also be obtained, when the same starting salt was treated with NaI in an H_2O/acetone (1:1) mixture at 50 °C for 1 h or with 1 equivalent I_2 for 3 d. In either case the product separated with ca. 40% yield.

The salt crystallizes in the monoclinic space group $P2_1-C_2^2$ (No. 4) with a = 7.997(7), b = 11.23(1), c = 16.38(2) Å, β = 103.63(7)°; Z = 2 formula units per unit cell, D_{calc} = 2.43 g/cm³. The structure of the dication is illustrated in **Fig. 4**. The triazacyclononane ring assumes a λλλ conformation. There are two weak intramolecular NH···O(bridge) interactions. Both Re atoms are R-configurated [16].

References on pp. 38/9

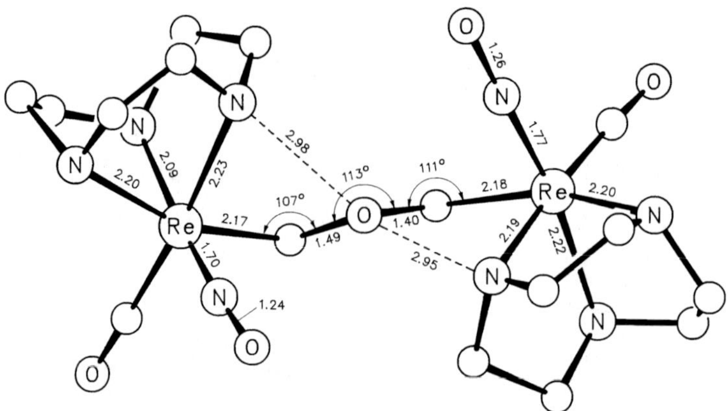

Fig. 4. Molecular structure of $[O(-CH_2Re(CO)(NO)N_3C_6H_{15})_2]^{2+}$ [16].

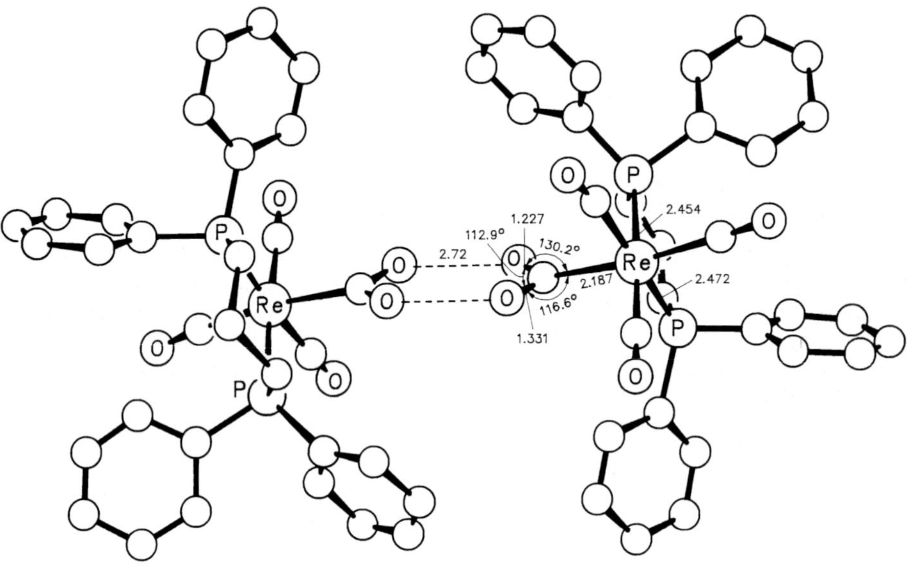

Fig. 5. Molecular structure of $[(CO)_3Re((C_6H_5)_2P(CH_2)_3P(C_6H_5)_2)C(O)OD]_2$ [34].

$[(CO)_3Re((C_6H_5)_2P(CH_2)_3P(C_6H_5)_2)C(O)OD]_2 \cdot C_6D_6$ (Table **2**, No. **2**) in the solid state exists as a hydrogen-bonded dimer. An X-ray study established the orthorhombic space group Pbca$-D_{2h}^{15}$ (No. 61) with a = 16.208(2), b = 20.028(2), c = 19.716(2) Å, Z = 4 dimer molecules per unit cell, D_{calc} = 1.60 g/cm^3. The dimer structure is shown in **Fig. 5** [34].

$(C_{44}H_{28}N_4)Sn[C{\equiv}Re(CO)_3]_2 \cdot 2\,C_6H_4Cl_2$-1,2 (Table **2**, No. **3**) crystallizes in the monoclinic space group $P2_1/n-C_{2h}^5$ (No. 14) with a = 10.44(2), b = 13.48(1), c = 20.19(2) Å, β = 110.8(1)°; Z = 2 molecules per unit cell, D_{calc} = 2.02 g/cm^3. The molecular structure is displayed in **Fig. 6**. The molecule is centrosymmetric, and the Sn-C-Re moiety is bent.

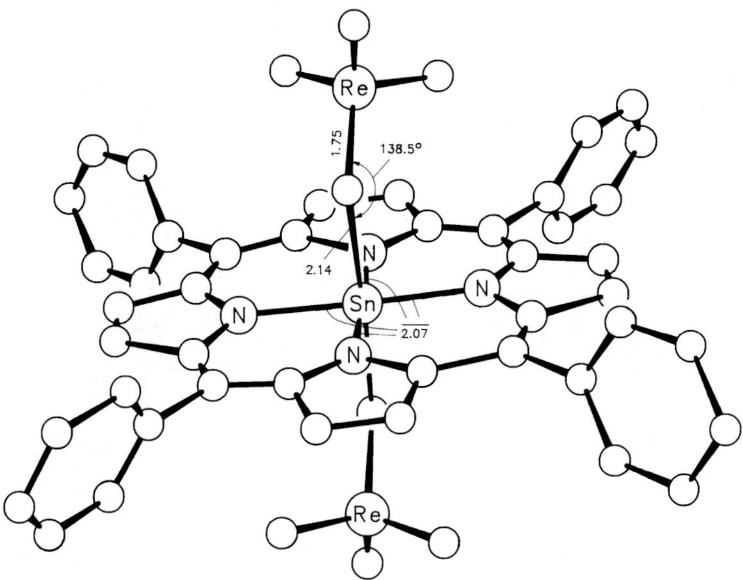

Fig. 6. Molecular structure of $(C_{44}H_{28}N_4)Sn[C{\equiv}Re(CO)_3]_2$ [8].

Heating at 180 °C released the carbide C atoms, and $[(CO)_3Re]_2SnN_4C_{44}H_{28}$ (see "Organorhenium Compounds" 5, 1994, p. 170) formed in the presence of excess $(CO)_{10}Re_2$ [8].

$\mathbf{((CO)_3Re(N_2C_{10}H_8)C{\equiv}N)_2Ru(N_2C_{10}H_6(CO_2H)_2)N_2C_{10}H_6(CO_2)_2 \cdot 10\ H_2O}$ (Table **2**, No. **5**) precipitates from aqueous solution as a neutral solid only at pH 2.5 to 3.5 by adding dilute $HClO_4$ or CF_3SO_3H. In neutral or alkaline solution (pH > 4) the carboxyl groups are deprotonated, and the complex is dianionic: $\mathbf{[((CO)_3Re(N_2C_{10}H_8)C{\equiv}N)_2Ru(N_2C_{10}H_6(CO_2)_2)_2]^{2-}}$. Otherwise, in strong acidic solutions (pH < 2) the compound is dicationic: $\mathbf{[((CO)_3Re(N_2C_{10}H_8)C{\equiv}N)_2Ru(N_2C_{10}H_6(CO_2H)_2)_2]^{2+}}$.

UV spectrum (alkaline C_2H_5OH, dianionic form): λ_{max} (ε) = 243, 302 (sh), 308 (60100) (all $\pi \rightarrow \pi^*$); 366 (16900), 500 (12600) ($Re^I \rightarrow {}^4D$ and $Ru^{II} \rightarrow {}^4D$ charge transfer) nm; the lowest energy band is solvent-sensitive (the absorption maximum is red-shifted from 492 to 500 nm in the order from CH_3OH to H_2O). Room-temperature emission spectrum: λ_{max} = 661 (τ < 40 ns) nm in C_2H_5OH, 670 nm in DMF, 707 nm in H_2O (assigned to Ru $\rightarrow$ ^4D); the emission is independent of the excitation wavelength over the range 350 to 600 nm. Emission at 77 K (CH_3OH/C_2H_5OH): λ_{max} = 655, 695 nm. The excitation spectrum for the Ru-based chromophore is nearly identical with the absorption spectrum; there is no emission from the Re-based chromophore, indicating efficient energy transfer from the peripheral Re units to the central Ru-based chromophore.

Difference absorption spectra were recorded during oxidation with Ce^{4+} salts in H_2SO_4. Isosbestic points were obtained at 358 and 607 nm, indicating a selective oxidation at the Ru^{II} atom. When the $Re^ICN{-}Ru^{III}{-}NCRe^I$ excited state was generated by 530-nm laser pulse excitation, transient difference absorption spectroscopy revealed excitation exclusively at Ru. The transient absorption decayed with a lifetime of ca. 35 ns [31].

$\mathbf{(CO)_3Re(N_2C_{10}H_6(C_4H_9\text{-}t)_2){-}C{\equiv}CC{\equiv}C{-}Re(CO)_3N_2C_{10}H_6(C_4H_9\text{-}t)_2 \cdot 2\ O{=}C(CH_3)_2}$ (Table **2**, No. **6**). The UV spectrum shows an intense lowest energy band in the region 430 to 495 nm (ace-

References on pp. 38/9

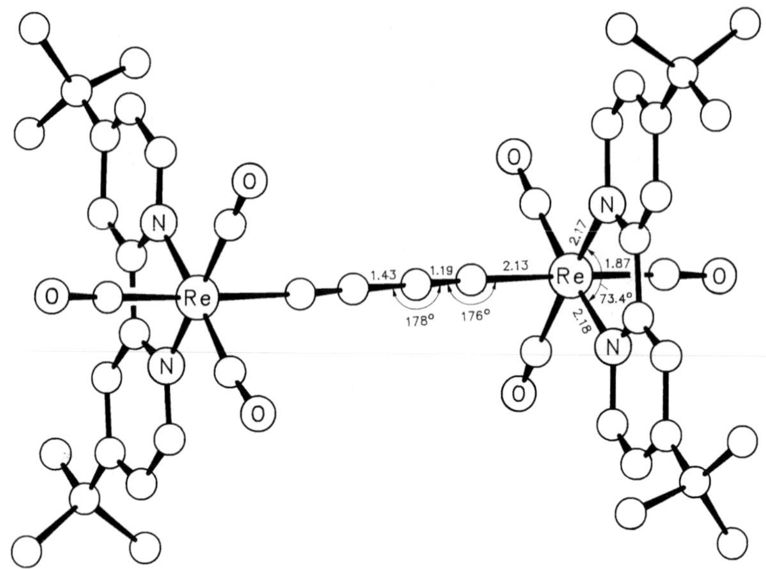

Fig. 7. Molecular structure of
$(CO)_3Re(N_2C_{10}H_6(C_4H_9\text{-}t)_2)-C{\equiv}CC{\equiv}C-Re(CO)_3N_2C_{10}H_6(C_4H_9\text{-}t)_2$ [45].

tone: 450, THF: 480, ether: 492, CH_3OH/C_2H_5OH (1:4): 432 nm). Excitation at $\lambda = 350$ to 400 nm resulted in long-lived luminescence at ca. 660 to 690 nm (acetone: 690, CH_2Cl_2: 620, THF: 630, ether: 660, CH_3OH/C_2H_5OH (1:4): 690 nm). However, upon excitation at $\lambda > 450$ nm a lower lying emission at ca. 750 nm appeared.

No. 6 crystallizes in the monoclinic space group $P2_1/a-C^5_{2h}$ (No. 14) with a = 11.490(9), b = 20.914(3), c = 12.375(5) Å, $\beta = 113.58(3)°$; Z = 2 formula units per unit cell, D_{calc} = 1.513 g/cm^3. The structure of the molecule is depicted in **Fig. 7**. The OC-Re-C_4-Re-CO unit is essentially linear. A separation of 8.1 Å is observed between the Re atoms [45].

$(CO)_5Re-C_2H_4-Re(CO)_5$ (Table **2**, No. **14**) was obtained with 40% yield by reacting $[((CO)_5ReCH{=}CH_2)Re(CO)_5]BF_4$ with $Na[(CO)_5Re]$ in THF [19]. In contrast, its formation by treating $[\pi\text{-}C_2H_4Re(CO)_5]^+$ with $Na[(CO)_5Re]$ at −50 °C was only briefly mentioned [10]. Several other reactions, sometimes unexpectedly, also produced $(CO)_5Re-C_2H_4-Re(CO)_5$ including the following examples: treatment of $Na_2[(CO)_4Os]$ with $[\pi\text{-}C_2H_4Re(CO)_5]^+$ in THF (main product was No. 39) [18]; decomposition of $(CO)_5ReC_2H_4Ru(C_5H_5)(CO)_2$ or $(CO)_4$-$Os[Os(CO)_4C_2H_4Re(CO)_5]_2$ (No. 40) in solution [25]; reaction of $[\pi\text{-}C_2H_4Re(CO)_5]^+$ with $Na_2[(CO)_4Ru]$ [25]; treating a mixture of $Na_3[(CO)_3Ir]$ and $[\pi\text{-}C_2H_4Re(CO)_5]^+$ with $[N(P(C_6H_5)_3)_2]Cl$ between −30 and 0 °C (No. 14 was NMR-spectroscopically detected) [26]; reaction of $(CO)_5ReC_2H_4Mn(CO)_5$ or $(CO)_5ReC_2H_4M(C_5H_5)(CO)_3$ (M = Mo, W) with $Na[(CO)_5Re]$ [6, 9, 11] (yield: 84% in the case of M = W) [9]; and reaction of $[C_5H_5W(C_2H_4)(CO)_3]^+$ with 2 equivalents $Na[(CO)_5Re]$ (yield: 71%) [9].

At −90 °C the compound crystallizes in the monoclinic lattice with a = 9.827(1), b = 6.536(2), c = 12.652(2) Å, $\beta = 86.15(1)°$; space group $P2_1/c-C^5_{2h}$ (No. 14); Z = 2 molecules per unit cell, D_{calc} = 2.78 g/cm^3 [11]. Similar room-temperature lattice data are available [9]. The molecule is centrosymmetric. An insight into the molecular geometry with bond parameters from the low-temperature measurement is given in **Fig. 8** [11].

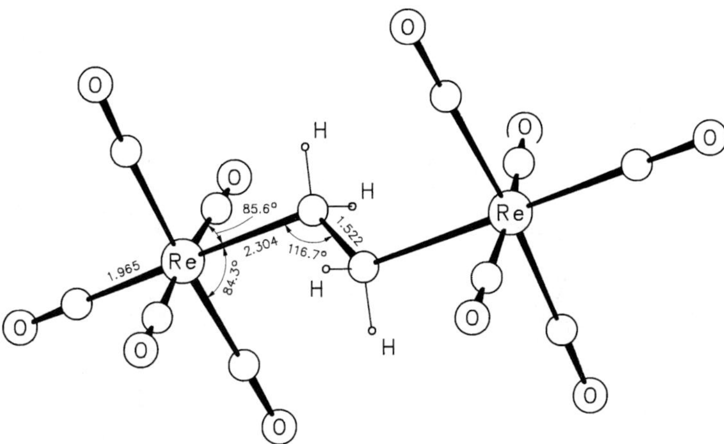

Fig. 8. Molecular structure of $(CO)_5Re-C_2H_4-Re(CO)_5$ [11].

Exposure to air in CD_2Cl_2 separated $[((CO)_5ReO)_3C]ReO_4 \cdot (CO)_{12}Re_4(\mu_3\text{-}OH)_4$ [26].

$(CO)_5Re-CH_2C(CH_3)H-Re(CO)_5$ (Table **2**, No. **15**). A variation of the standard preparation method involved warming the reaction mixture to room temperature followed by pouring the solution into H_2O. The resulting precipitate was filtered off and washed with H_2O. It contained No. 15 and $(CO)_{14}Re_3H$ as a by-product with yields up to 10%.

Exposure of the title complex to air for 2 months yielded a mixture of $(CO)_{14}Re_3H$, $(CO)_{10}Re_2$, and an unknown by-product [11].

$(CO)_5Re-(CH_2)_3-Re(CO)_5$ (Table **2**, No. **16**) can in principle exist in 9 rotameric forms if only staggered conformations are considered. Solution NMR data indicate a clear preference of the conformer with C_{2v} symmetry (see Formula I). The following parameters were measured: $\delta(X) = 1.0247$, $\delta(A) = 2.2375$ ppm; $^2J(XX') = {}^2J(X''X''') = -11.71$, $^2J(AA') = -13.15$, $^3J(AX') = {}^3J(A'X) = {}^3J(AX''') = {}^3J(A'X'') = 4.57$, $^3J(AX) = {}^3J(A'X') = {}^3J(AX'') = {}^3J(A'X''') = 12.01$ Hz. The conformers shown in Formulas II and III each contribute only 3.1% to the mixture [23].

$[Re] = Re(CO)_5$

I II III IV

$(CO)_5Re-(CH_2)_5-Re(CO)_5$ (Table **2**, No. **18**). There is a hindered ligand movement around the C_α-C_β bond. Coupling constants obtained from simulated ^1H NMR spectra indicate a trans arrangement of Re and X substituents in the fragment $ReCH_2CH_2X$ (see Formula IV): $\delta(A) = 0.9486$, $\delta(C) = 1.2882$, $\delta(B) = 1.7853$ ppm; $^2J(AA') = -11.683$, $^2J(CC') = -10.700$, $^2J(BB') = -11.000$, $^3J(AB) = 4.500$, $^3J(AB') = 11.700$, $^3J(BC) = 7.286$ Hz [32].

$(CO)_5Re-CH_2(cyclo\text{-}CHCH_2OCH_2CH)CH_2-Re(CO)_5$ (Table **2**, No. **24**) crystallizes in the orthorhombic space group $Pbcn-D_{2h}^{14}$ (No. 60) with a = 12.269(2), b = 13.015(3), c =

References on pp. 38/9

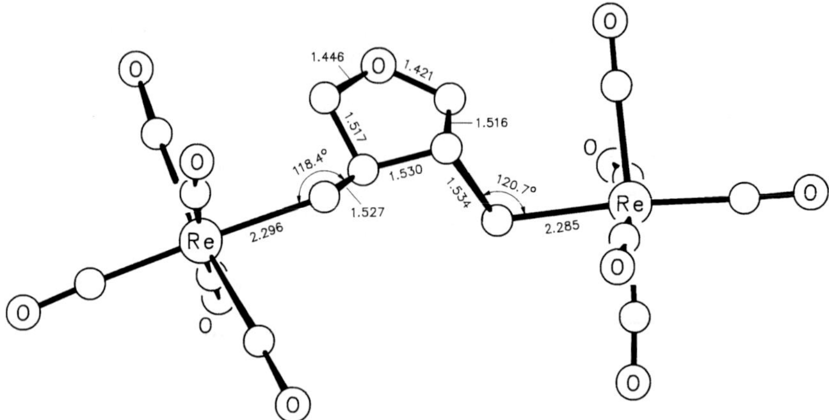

Fig. 9. Molecular structure of $(CO)_5Re-CH_2(cyclo\text{-}CHCH_2OCH_2CH)CH_2-Re(CO)_5$ [33].

24.994(5) Å; Z = 8 molecules per unit cell, D_{calc} = 2.499 g/cm^3. The molecular structure is illustrated in **Fig. 9** [33].

cyclo-$C_6H_{10}[CH_2Re(CO)_5]_2$-1,3 (Table **2**, No. **26**) crystallizes at −100 °C in the triclinic space group $P\bar{1}-C_i^1$ (No. 2) with a = 6.642(1), b = 13.00(2), c = 13.186(3) Å, α = 80.81(3)°, β = 80.44(2)°, γ = 80.46(3)°; Z = 2 molecules per unit cell, D_{calc} = 2.308 g/cm^3. The molecular structure is illustrated in **Fig. 10** [40].

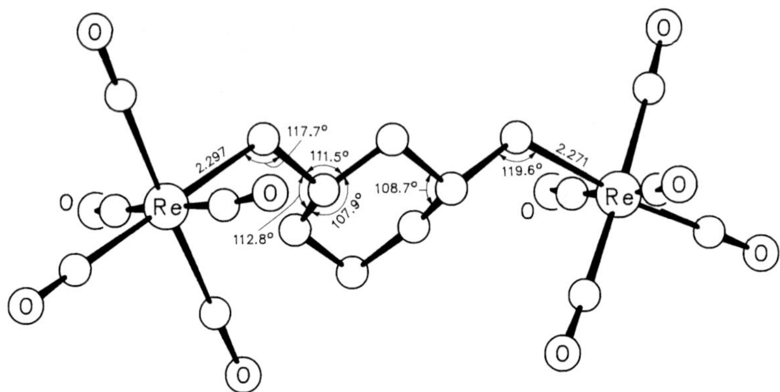

Fig. 10. Molecular structure of cyclo-$C_6H_{10}[CH_2Re(CO)_5]_2$-1,3 [40].

trans-$(CO)_5Re-CH_2CH{=}CHCH_2-Re(CO)_5$ (Table **2**, No. **27**). A low-temperature (−90 °C) crystal structure determination revealed a triclinic lattice with a = 6.126(3), b = 6.601(3), c = 12.440(6) Å, α = 83.44(4)°, β = 88.85(4)°, γ = 61.47(3)°; space group $P\bar{1}-C_i^1$ (No. 2), Z = 1 molecule per unit cell. The molecule shown in **Fig. 11** is centrosymmetric. The C atoms of the bridging ligand lie in a plane, whereas the C=C-C-Re moiety assumes a dihedral angle of 109° [12].

References on pp. 38/9

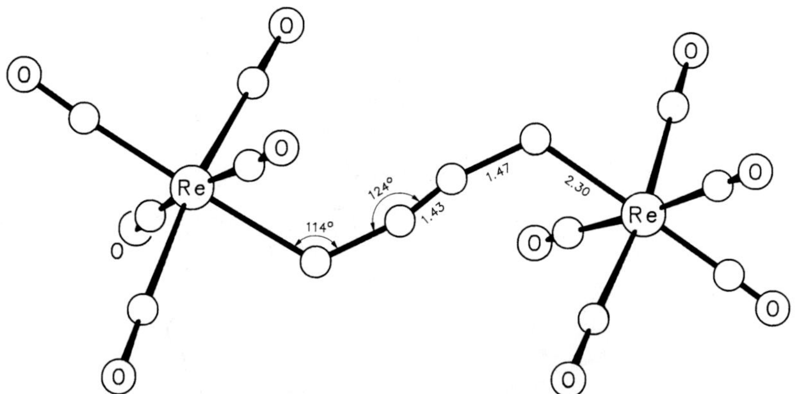

Fig. 11. Molecular structure of trans-$(CO)_5Re-CH_2CH=CHCH_2-Re(CO)_5$ [12].

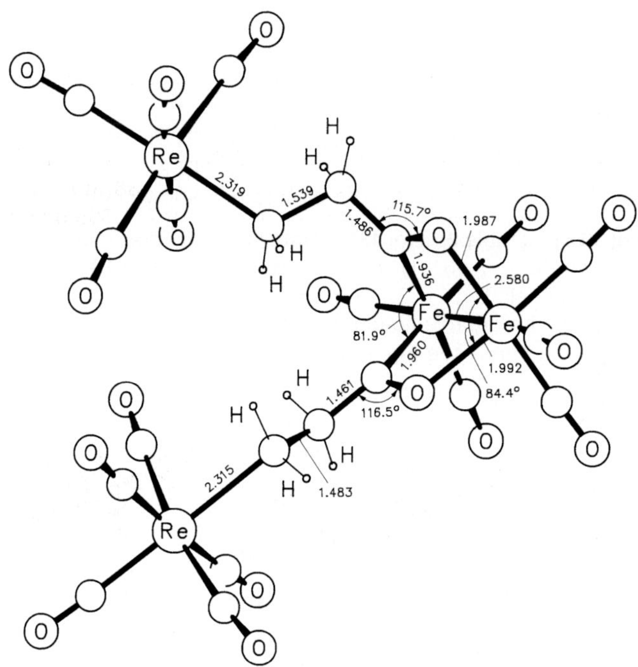

Fig. 12. Molecular structure of $[(CO)_5ReC_2H_4C{\cdots}O]_2Fe_2(CO)_6$ [20].

$[(CO)_5ReC_2H_4C{\cdots}O]_2Fe_2(CO)_6$ (Table **2**, No. **35**) crystallizes in the orthorhombic space group $Pna2_1-C_{2v}^9$ (No. 33) with a = 19.148(4), b = 22.498(6), c = 6.991(1) Å; Z = 4 molecules per unit cell, D_{calc} = 2.303 g/cm^3. The structure is illustrated in **Fig. 12** [20].

$(CO)_6Fe_2[SC_2H_4Re(CO)_5]_2$ (Table **2**, No. **36**). The IR spectrum of a mixture of the syn and anti form exhibits the following ν(CO) bands (CH_2Cl_2): 1980, 2015, 2025, 2064, 2130 (ν(CO)) cm^{-1}.

References on pp. 38/9

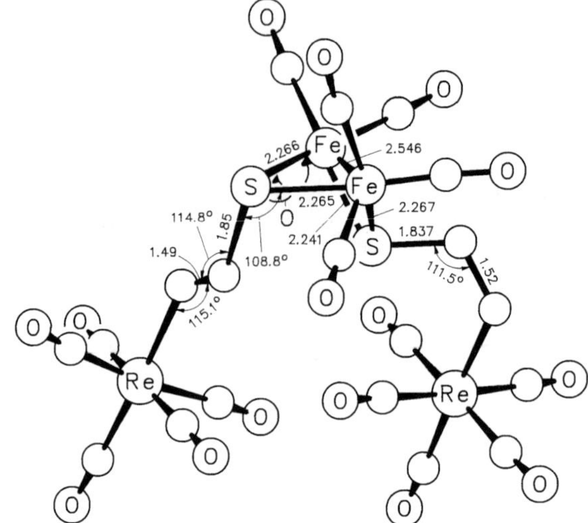

Fig. 13. Molecular structure of anti-$(CO)_6Fe_2[SC_2H_4Re(CO)_5]_2$ [44].

The anti isomer crystallizes in the triclinic space group $P\bar{1}-C_i^1$ (No. 2) with a = 11.058(2), b = 11.079(3), c = 14.033(3) Å, α = 75.99(2)°, β = 69.40(2)°, γ = 67.66(2)°; Z = 2 molecules per unit cell, D_{calc} = 2.259 g/cm^3. The structure is shown in **Fig. 13** [44].

(cyclo-C_4O_2)[Re(CO)$_5$]$_2$ (Table **2**, No. **45**) crystallizes in the monoclinic space group $P2_1/a-C_{2h}^5$ (No. 14) with a = 7.031(1), b = 11.532(1), c = 11.418(1) Å, β = 105.54(1)°; Z = 2 molecules per unit cell, D_{calc} = 2.727 g/cm^3. It is worth to note that the equatorial CO groups are bent towards the organic ligand and that the C=C bond is lengthened compared with a normal C-C double bond, see **Fig. 14** [15].

$(CO)_5Re-C(=CH_2)CH(C_6H_5)CH(C_6H_5)C(=CH_2)-Re(CO)_5$ (Table **2**, No. **46**) crystallizes in the monoclinic space group $P2_1/n-C_{2h}^5$ (No. 14) with a = 11.792(2), b = 11.686(3), c =

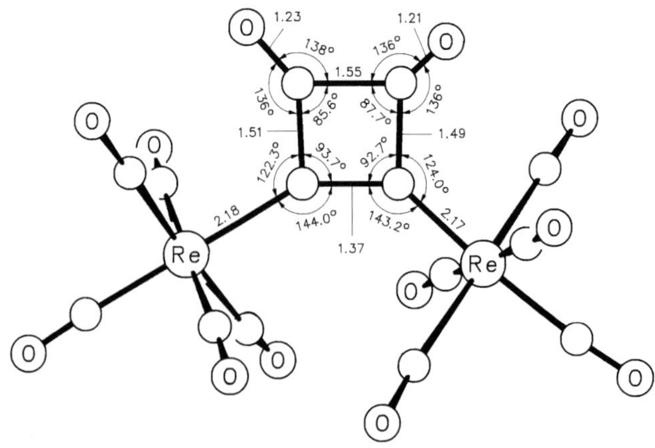

Fig. 14. Molecular structure of (cyclo-C_4O_2)[Re(CO)$_5$]$_2$ [15].

References on pp. 38/9

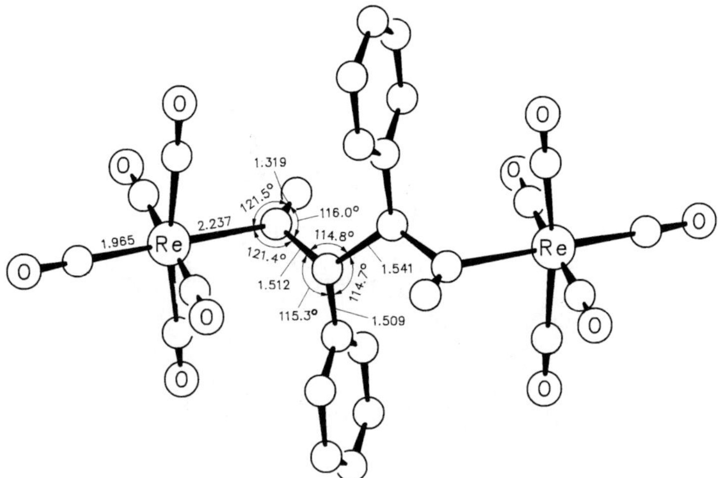

Fig. 15. Molecular structure of $(CO)_5Re-C(=CH_2)CH(C_6H_5)CH(C_6H_5)C(=CH_2)-Re(CO)_5$ [22].

11.247(2) Å, β = 114.58(2)°; Z = 4 molecules per unit cell, D_{calc} = 2.08 g/cm³. The geometry of the molecule is illustrated in **Fig. 15**. The complex assumes a meso configuration.

At −40°C three conformational rotamers exist in solution. The free energy of rotation around the central C-C single bond was estimated to be 52.3 ± 1 kJ/mol [22].

trans-$(CO)_5Re-CH=C(CO_2CH_3)-Re(CO)_5$ (Table **2**, No. **47**). Single crystals are triclinic with a = 10.730(2), b = 11.533(2), c = 9.544(2) Å, α = 114.14(1)°, β = 114.93(2)°, γ = 90.68(2)°; space group $P\bar{1}-C_i^1$ (No. 2), Z = 2 molecules per unit cell, D_{calc} = 2.57 g/cm³. The molecular structure is depicted in **Fig. 16** [30].

Thermolysis in refluxing hexane yielded $(CO)_4ReO{=}(CH_3O)CC[Re(CO)_5]{=}CH$-cyclo within 30 min [30].

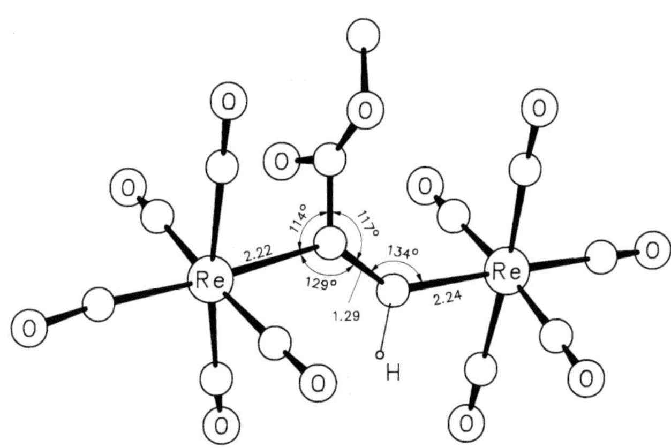

Fig. 16. Molecular structure of trans-$(CO)_5Re-CH=C(CO_2CH_3)-Re(CO)_5$ [30].

References on pp. 38/9

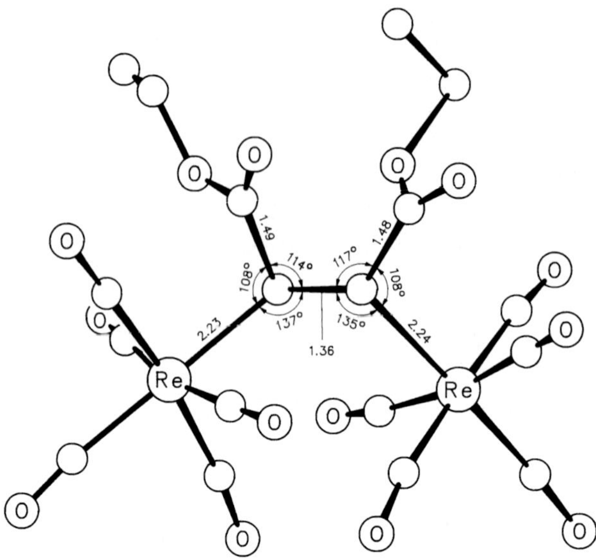

Fig. 17. Molecular structure of cis-$(CO)_5Re-C(CO_2C_2H_5)=C(CO_2C_2H_5)-Re(CO)_5$ [38].

cis-$(CO)_5Re-C(CO_2C_2H_5)=C(CO_2C_2H_5)-Re(CO)_5$ (Table **2**, No. **48**) crystallizes in the orthorhombic space group Pbca$-D_{2h}^{15}$ (No. 61) with a = 14.275(2), b = 33.00(1), c = 10.114(2) Å; Z = 8 molecules per unit cell, D_{calc} = 2.29 g/cm^3. The structure is shown in **Fig. 17**. The ReC=CRe grouping deviates only slightly from planarity.

Thermolysis reformed the starting product [38].

$C_6H_4[Re(CO)_5]_2$-1,4 (Table **2**, No. **52**). Treatment with CO (30 atm) formed $(CO)_5Re-C(O)C_6H_4C(O)-Re(CO)_5$ (No. 62), while treatment with synthesis gas (40 atm) not only gave No. 62, but also $C_6H_4(CH_2OH)$-1,4, and an unidentified aldehyde species.

The reaction with SO_2 in a sealed tube gave 1,4-$[(CO)_5Re-SO_2]_2C_6H_4$ by insertion into the Re-C bond, while treatment with Br_2 provided $C_6H_4Br_2$ and $(CO)_5ReBr$ [27].

$(CO)_5Re-C(O)C(O)-Re(CO)_5$ (Table **2**, No. **54**) crystallizes in the monoclinic space group $P2_1/n-C_{2h}^5$ (No. 14) with a = 6.527(4), b = 12.392(6), c = 10.192(6) Å, β = 95.69(8)°; Z = 2 molecules per unit cell, D_{calc} = 2.87 g/cm^3. The structure is shown in **Fig. 18**. The C_2O_2 moiety is strictly planar, and the $Re(CO)_5$ "substituents" are mutually trans and eclipsed with respect to each other and staggered with respect to the dionyl moiety.

Solid-state decomposition gave $(CO)_{10}Re_2$ and CO (under 2 atm CO, 60% of the complex degraded after 16 d) in an ESR-silent process. The decomposition is inhibited by CO, so that under 40 atm of CO the compound remains inert.

Catalytic hydrogenation of the carbonyl groups using a cationic Rh-phosphane complex was attempted but was not successful. The compound similarly neither reacted with AlH_3 nor BH_3 (except for slow decomposition into $(CO)_{10}Re_2$). In contrast, a stoichiometric treatment with $LiAlH_4$ actually yielded C_1, C_2, and C_4 products along with $(CO)_{10}Re_2$. Reaction with ylides such as $(C_6H_5)_3P=CHC_4H_9$-t did not initiate a Wittig reaction. Treatment with 4 equivalents $(CH_3)_3CCH_2Li$ gave $Li_4[\{mer\text{-}(CO)_3Re(C(O)CH_2C(CH_3)_3)_2\}_2(C(O)C(O))]$ [13].

References on pp. 38/9

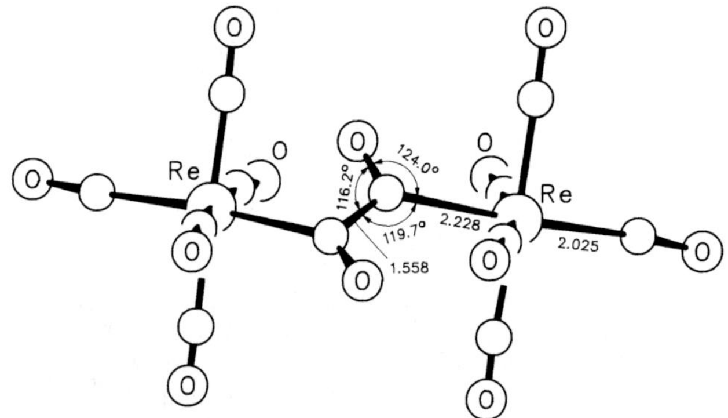

Fig. 18. Molecular structure of $(CO)_5Re-C(O)C(O)-Re(CO)_5$ [13].

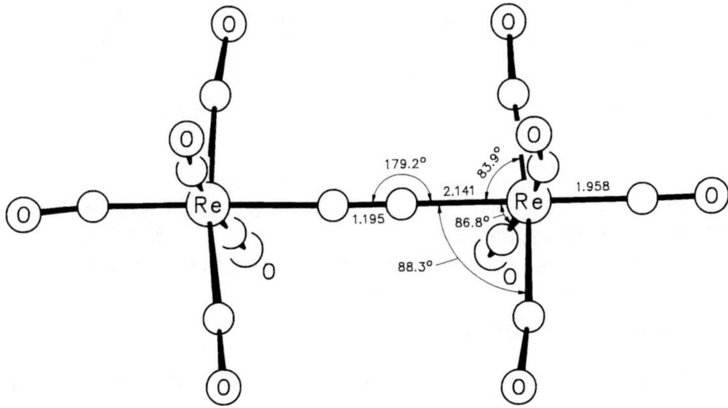

Fig. 19. Molecular structure of $(CO)_5Re-C{\equiv}C-Re(CO)_5$ [21].

$(CO)_5Re-C{\equiv}C-Re(CO)_5$ (Table **2**, No. **63**). $[((CO)_5ReC{\equiv}CH)Re(CO)_5]BF_4$ was treated with $NaOC_2H_5$ in THF at −10 °C. The solvent was removed after ca. 30 min and the residue extracted with CH_2Cl_2. Evaporation of the extract yielded 88% of the compound. However, the compound was not formed by reacting Na_2C_2 or Li_2C_2 with $(CO)_5ReFBF_3$ [14].

No. 63 crystallizes in the triclinic space group $P\bar{1}-C_i^1$ (No. 2) with a = 6.528(2), b = 6.516(2), c = 9.872(2) Å, α = 90.40(2)°, β = 96.77(3)°, γ = 98.62(3)°; Z = 1 molecule per unit cell. As can be seen in **Fig. 19**, the OC-Re-C≡C-Re-CO grouping is linear, and the equatorial CO groups of the $Re(CO)_5$ moieties are in an ideal eclipsed conformation. Furthermore, the equatorial CO groups are bent away from the axial CO ligand [21].

SCF-X_α calculations furnished the following ionization potentials: 8.17 ($13e_u$, HOMO), 9.50 ($17a_{1g}$), 10.11 ($12e_g$), 10.22 ($12e_u$), 10.25 ($16a_{2u}$), 10.37 ($3b_{1u}$, $3b_{2g}$), 12.90 ($11e_u$) eV. It could be shown that only a limited overlap (5 to 8%) exists between the C≡C π and Re π orbitals (d_{xz},d_{yz}). The acetylide system is thus nearly unperturbed, since there is only little Re-acetylide π interaction. Likewise, there is no Re → C≡C π backbonding interaction. Transition state calculations suggest that the first ionization should arise from the $13e_u$ C≡C π orbital, for which a high negative charge density has been calculated [21].

References on pp. 38/9

Protonation with $HBF_4 \cdot O(C_2H_5)_2$, HCl, or $HOSO_2F$ reformed $[((CO)_5ReC{\equiv}CH)Re(CO)_5]BF_4$ [14]. Treatment with $C_2H_4Pt(P(C_6H_5)_3)_2$, $(CO)_8Co_2$, or $(CO)_9Fe_2$ produced the complexes $[\mu\text{-}(CO)Pt(P(C_6H_5)_3)C{\equiv}C]Re_2(CO)_8P(C_6H_5)_3$ (see p. 98), $[\mu\text{-}(CO)_6Co_2(C{\equiv}C)]Re_2(CO)_8$ (see p. 161), or $[\mu\text{-}(CO)_4Fe(C{\equiv}C)]Re_2(CO)_9$ (see p. 139) [29].

References:

[1] Green, M.; Mayne, N.; Stone, F. G. A. (Chem. Commun. **1966** 755/6).
[2] Nesmeyanov, A. N.; Anisimov, K. N.; Kolobova, N. E.; Ioganson, A. A. (Dokl. Akad. Nauk SSSR **175** [1967] 358/60; Dokl. Chem. [Engl. Transl.] **172/177** [1967] 627/5).
[3] Cook, D. J.; Green, M.; Mayne, N.; Stone, F. G. A. (J. Chem. Soc. A **1968** 1771/5).
[4] Dalton, J.; Paul, I.; Stone, F. G. A. (J. Chem. Soc. A **1968** 1212/4).
[5] Goodfellow, R. J.; Green, M.; Mayne, N.; Rest, A. J.; Stone, F. G. A. (J. Chem. Soc. A **1968** 177/80).
[6] Beck, W.; Olgemöller, B. (J. Organomet. Chem. **127** [1977] C 45/C 47).
[7] Hobbs, D. T.; Lukehart, C. M. (Inorg. Chem. **18** [1979] 1297/301).
[8] Noda, I.; Kato, S.; Mizuta, M.; Yasuoka, N.; Kasai, N. (Angew. Chem. **91** [1979] 85/6; Angew. Chem. Int. Ed. Engl. **18** [1979] 83).
[9] Olgemöller, B.; Beck, W. (Chem. Ber. **114** [1981] 867/76).
[10] Raab, K.; Olgemöller, B.; Schloter, K.; Beck, W. (J. Organomet. Chem. **214** [1981] 81/6).

[11] Raab, K.; Nagel, U.; Beck, W. (Z. Naturforsch. **38b** [1983] 1466/76).
[12] Beck, W.; Raab, K.; Nagel, U.; Sacher, W. (Angew. Chem. **97** [1985] 498/9; Angew. Chem. Int. Ed. Engl. **24** [1985] 505).
[13] de Boer, E. J. M.; de With, J.; Meijboom, N.; Orpen, A. G. (Organometallics **4** [1985] 259/64).
[14] Appel, M.; Heidrich, M.; Beck, W. (Chem. Ber. **120** [1987] 1087/9).
[15] Beck, W.; Schweiger, M.; Müller, G. (Chem. Ber. **120** [1987] 889/93).
[16] Pomp, C.; Duddeck, H.; Wieghardt, K.; Nuber, B.; Weiss, J. (Angew. Chem. **99** [1987] 927/9; Angew. Chem. Int. Ed. Engl. **26** [1987] 924).
[17] Steil, P.; Nagel, U.; Beck, W. (J. Organomet. Chem. **339** [1988] 111/24).
[18] Beck, W.; Niemer, B.; Wagner, B. (Angew. Chem. **101** [1989] 1699/701; Angew. Chem. Int. Ed. Engl. **28** [1989] 1705/6).
[19] Steil, P.; Beck, W.; Stone, F. G. A. (J. Organomet. Chem. **368** [1989] 77/81).
[20] Breimair, J.; Robl, C.; Beck, W. (Chem. Ber. **123** [1990] 1661/3).

[21] Heidrich, J.; Steimann, M.; Appel, M.; Beck, W.; Phillips, J. R.; Trogler, W. C. (Organometallics **9** [1990] 1296/300).
[22] Kreiter, C. G.; Michels, W.; Exner, R. (Z. Naturforsch. **45b** [1990] 793/802).
[23] Lindner, E.; Pabel, M.; Eichele, K. (J. Organomet. Chem. **386** [1990] 187/94).
[24] Mapolie, S. F.; Moss, J. R. (J. Chem. Soc. Dalton Trans. **1990** 299/305).
[25] Breimair, J.; Niemer, B.; Raab, K.; Beck, W. (Chem. Ber. **124** [1991] 1059/63).
[26] Breimair, J.; Robl, C.; Beck, W. (J. Organomet. Chem. **411** [1991] 395/404).
[27] Mapolie, S. F.; Moss, J. R. (Polyhedron **10** [1991] 717/23; see also: Errata, Polyhedron **10** [1991] 1965).
[28] van Houwelingen, T.; Stufkens, D. J.; Oskam, A. (Coord. Chem. Rev. **111** [1991] 325/30).
[29] Weidmann, T.; Weinrich, V.; Wagner, B.; Robl, C.; Beck, W. (Chem. Ber. **124** [1991] 1363/8).
[30] Adams, R. D.; Chen, L.; Wu, W. (Organometallics **12** [1993] 1257/65).

[31] Kalyanasundaram, K.; Grätzel, M.; Nazeeruddin, Md. K. (Inorg. Chem. **31** [1992] 5243/53).
[32] Lindner, E.; Pabel, M.; Fawzi, R.; Mayer, H. A.; Wurst, K. (J. Organomet. Chem. **435** [1992] 109/21).
[33] Lindner, E.; Pabel, M.; Fawzi, R.; Steimann, M. (J. Organomet. Chem. **441** [1992] 63/74).
[34] Mandal, S. K.; Ho, D. M.; Orchin, M. (J. Organomet. Chem. **439** [1992] 53/64).
[35] Niemer, B.; Weidmann, T.; Beck, W. (Z. Naturforsch. **47b** [1992] 509/16).
[36] Lindner, E.; Wassing, W.; Fawzi, R.; Steimann, M. (Z. Naturforsch. **48b** [1993] 1651/60).
[37] Zhang, S.; Brown, T. L.; Du, Y.; Shapley, J. R. (J. Am. Chem. Soc. **115** [1993] 6705/9).
[38] Adams, R. D.; Chen, L. (Organometallics **13** [1994] 1264/71).
[39] Hüffer, S.; Wieser, M.; Polborn, K.; Beck, W. (J. Organomet. Chem. **481** [1994] 45/55).
[40] Lindner, E.; Wassing, W.; Fawzi, R.; Steimann, M. (J. Organomet. Chem. **472** [1994] 221/7).

[41] Weinrich, V.; Robl, C.; Beck, W. (J. Organomet. Chem. **484** [1994] 233/7).
[42] Andersen, J.-A. M.; Moss, J. R. (Polyhedron **14** [1995] 1881/93).
[43] Geisbauer, A.; Mihan, S.; Beck, W. (J. Organomet. Chem. **501** [1995] 61/6).
[44] Hüffer, S.; Polborn, K.; Beck, W. (Organometallics **14** [1995] 953/8).
[45] Yam, V. W.-W.; Lau, V. C.-Y.; Cheung, K.-K. (Organometallics **15** [1996] 1740/4).

2.2.1.2.3 Compounds of the Types $(CO)_5Re(\mu\text{-}^2L\text{-}^2D)Re(CO)_4$ and $(CO)_4Re(\mu\text{-}^2L\text{-}^4D)Re(CO)_4$ and Compounds Originating by Formal CO Substitution

The compounds dealt with in this section possess $(CO)_nRe$ groups (n = 4, 5) separated by a bridging $^2L\text{-}^2D$ or $^2L\text{-}^4D$ ligand (some compounds contain CO substitution groups). There are no Re-Re bonds.

Several compounds were obtained by special procedures, but generally the following methods were employed:

Method I: Treatment of $(CO)_9Re_2NCCH_3$ with the electron-poor alkynes $RC{\equiv}CCO_2CH_3$ (R = H, CH_3) and $C_2H_5O_2CC{\equiv}CCO_2C_2H_5$ in hexane (reflux, 2 h). The products precipitated on cooling or were separated by preparative TLC [10, 11]. The reaction with $HC{\equiv}CCO_2CH_3$ was also conducted in CH_2Cl_2 (reflux, 24 h) [18]. The mechanism was investigated (see below).

Method II: Treatment of $(CO)_4ReO{=}(CH_3O)CC[Re(CO)_5]{=}CH$-cyclo (No. 2) with $(CH_3)_3NO$ in CH_2Cl_2 [18] or CH_3CN [12] promotes substitution of one CO group at the $Re(CO)_5$ unit giving either No. 5 or No. 6.

Method III: Treatment of $(CO)_4ReO{=}(CH_3O)CC[Re(CO)_4NCCH_3]{=}CH$-cyclo (No. 6) with $HC{\equiv}CCO_2CH_3$ under a CO atmosphere (CH_2Cl_2, reflux, 4 h). The mixture was separated by preparative TLC using CH_2Cl_2/hexane (1 : 1) [12].

Method IV: Photolysis of $(CO)_4ReO{=}C(OC_2H_5)C[{=}C(CO_2C_2H_5)Re(CO)_5]$-cyclo (No. 4) at room temperature for 10 min. The mixture was separated by preparative TLC using CH_2Cl_2/hexane (1 : 2) [15].

Method V: Reaction of $(CO)_5ReCH_3$
a. with 1 equivalent $(CO)_4ReC_{14}H_7O_2$ (octane, reflux, 11 h). Subsequent column-chromatographic workup eluted a deep red and a green-black band,

References on pp. 52/3

followed by a thin red band and a fraction containing the starting compound. Another chromatographic separation of the first two fractions yielded the two isomers in pure form. However, when anthraquinone was combined with 1 equivalent $(CO)_5ReCH_3$, mainly $(CO)_4ReC_{14}H_7O_2$ was obtained and only minor amounts of the bimetallic product were formed [2];

b. with $(CO)_4Re(P(C_6H_4R\text{-}4)_2)(OC)C_6H_3R\text{-}2$ (R = H, F) (toluene, reflux, 3 to 5 h). Column-chromatographic separation of the mixture was followed by recrystallization from hexane/CH_2Cl_2 or ether/hexane [3].

Method VI: Thermolysis of $(CO)_8Re_2(P(C_6H_5)_3)_2$ (xylene, 230 °C, 14 d). Separation of the mixture by preparative HPLC with $CHCl_3$/hexane (3:1) gave several fractions; each of them was crystallized by vapor diffusion [8].

Investigations on the mechanism of the reaction described under "Method I" were undertaken: Neither the yield nor the reaction rate were significantly affected by the presence or absence of light, by the presence of radical scavengers such as air, duroquinone, or 2,6-$(t\text{-}C_4H_9)_2C_6H_3OH$, or by varying the solvent polarity. Moreover, treating a 1:1 mixture of $(CO)_9Re_2NCCH_3$ and $(^{13}CO)_9Re_2NCCH_3$ with $HC{\equiv}CCO_2CH_3$ did not yield the crossover products $(CO)_4ReO{=}(CH_3O)CC[Re(^{13}CO)_5]{=}CH$-cyclo or $(^{13}CO)_4ReO{=}(CH_3O)CC[Re(CO)_5]{=}CH$-cyclo. These results stronlgy support an intramolecular reaction path [10, 11] (see also [15]).

Table 3
Compounds of the Types $(CO)_5Re(\mu\text{-}^2L\text{-}^2D)Re(CO)_4$ and $(CO)_4Re(\mu\text{-}^2L\text{-}^4D)Re(CO)_4$ and Compounds Originating by Formal CO Substitution.
An asterisk indicates further information at the end of the table.
For explanations, abbreviations, and units see p. X.

No.	compound	method of preparation (yield) properties and remarks
compounds with ^2L-^2D ligands		
*1	$(\mu\text{-}C_7H_4O_2)Re_2(CO)(P(CH_3)_3)_4(NO)_2(CH_3)C(O)CH_3$	for preparation see "Further information" yellow-orange solid; m.p. 158 to 159 °C (dec.) ^1H NMR (acetone-d_6): −0.69 (t, $ReCH_3$; J = 10.7); 1.34, 1.86 ($P(CH_3)_3$; J(P,H) + J(P′,H) = 7.6 and 8.4); 2.35 (s, 3 H); 6.06, 6.21, 6.38, 6.92 (m's, 1 H) $^{13}C\{^1H\}$ NMR (THF-d_8, 12 °C): −9.0; 14.6, 16.4 (J(P,C) + J(P′,C) = 26 and 27); 50.8, 111.0, 122.6, 124.8, 127.6, 140.8, 205.7, 219.5, 238.1 $^{31}P\{^1H\}$ NMR (acetone-d_6, 15 °C): −36, −26 (s's) IR (Nujol): 1510, 1545; 1650, 1692 (ν(NO)); 1986 (ν(CO)) [5]
*2	$(CO)_4ReO{=}(CH_3O)CC[Re(CO)_5]{=}CH$-cyclo	I (in hexane, yield: 81% [10, 11]; 60% yield in CH_2Cl_2 [18]); also with 80% yield by pyrolyzing (hexane, 30 min) trans-$(CO)_5ReCH{=}C(CO_2CH_3)Re(CO)_5$ [11] pale yellow crystals from CH_2Cl_2 [10, 11, 18]; m.p. 151 to 155 °C [18]

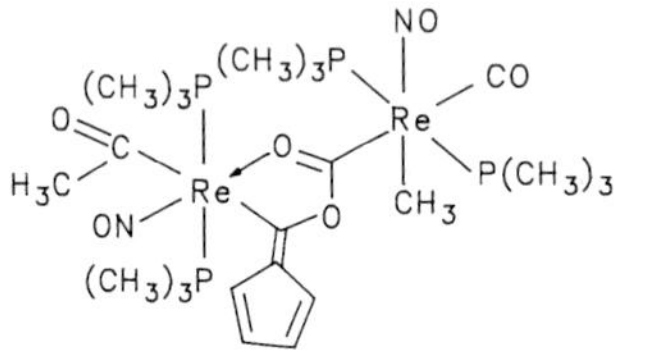

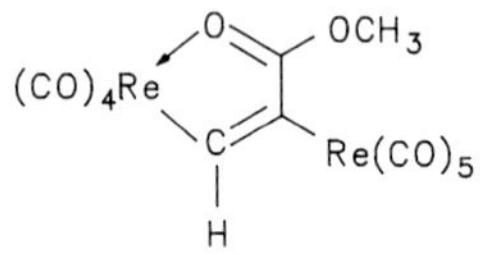

References on pp. 52/3

Table 3 (continued)

No.	compound	method of preparation (yield) properties and remarks
		^1H NMR ($CDCl_3$): 3.90 (CH_3), 10.42 (CH) [10, 11, 18] $^{13}C\{^1H\}$ NMR ($CDCl_3$): 54.73 (OCH_3); 121.05, 127.69 (HC=C); 182.01, 182.88, 189.36, 193.02, 193.33 (CO); 222.46 (CO_2) [18] IR (n-hexane): 1541 (ν(C=O)); 1937, 1982, 1986, 1991, 2024, 2090, 2134 (ν(CO)) [10, 11]; nearly identical in [18] MS: $[M - n\ CO]^+$ (n = 0 to 9) [10, 11, 18]
3	$(CO)_4ReO{=}(CH_3O)CC[Re(CO)_5]{=}CCH_3$-cyclo 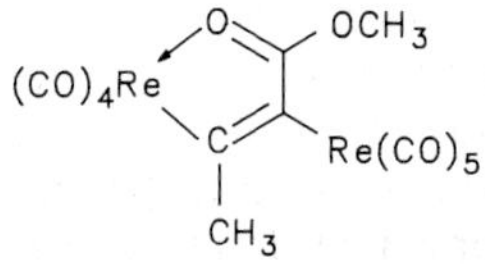	I (10%) [10, 11]; by-product was $(CO)_{10}Re_2$ [11] colorless solid [11] ^1H NMR ($CDCl_3$): 2.83 (CCH_3), 3.86 (OCH_3) IR (n-hexane): 1533 (C=O); 1935, 1982, 1990, 2021, 2087, 2132 (ν(CO)) [10, 11]
*4	$(CO)_4ReO{=}C(OC_2H_5)C[{=}C(CO_2C_2H_5)Re(CO)_5]$-cyclo 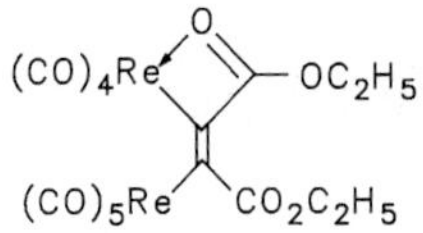	I (79%) [10, 11]; by-product: $(C_4(CO_2C_2H_5)_4)Re_2(CO)_7$ (see p. 150) [11, 15]; note that the compound was first thought to have a five-membered ring structure analogous to Nos. 2 and 3 [10, 11], until a crystal structure determination established the actual configuration [15] ^1H NMR ($CDCl_3$): 1.29, 1.31 (t's, CH_3); 4.16, 4.17 (q's, CH_2; J(H,H) = 7.2) IR (CH_2Cl_2): 1526, 1685 (br, ν(C=O)); (hexane): 1932, 1986, 1992, 1999, 2028, 2036, 2097, 2139 (ν(CO)) [10, 11]
*5	$(CO)_4ReO{=}(CH_3O)CC[Re(CO)_4N(CH_3)_3]{=}CH$-cyclo 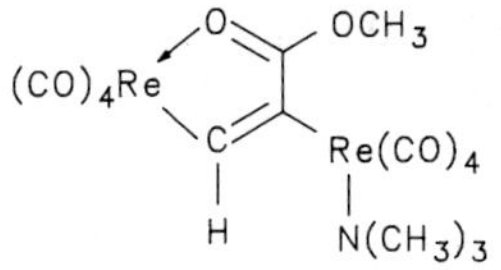	II (in CH_2Cl_2; yield: 38%); separation by preparative TLC with petroleum ether/acetone (10:3) lime green crystals; m.p. 129 to 132°C ^1H NMR ($CDCl_3$): 2.85 (s, NCH_3), 3.92 (s, OCH_3), 10.56 (CH) $^{13}C\{^1H\}$ NMR ($CDCl_3$): 54.21 (OCH_3), 59.98 (NCH_3); 189.87, 190.16, 191.27, 191.98, 193.67 (CO); 217.11 ($\mathbf{C}O_2CH_3$) IR (THF): 1924, 1969, 1982, 2081, 2092 (ν(CO)) FAB MS: $[M - n\ CO]^+$ (n = 0 to 6) [18]
*6	$(CO)_4ReO{=}(CH_3O)CC[Re(CO)_4NCCH_3]{=}CH$-cyclo 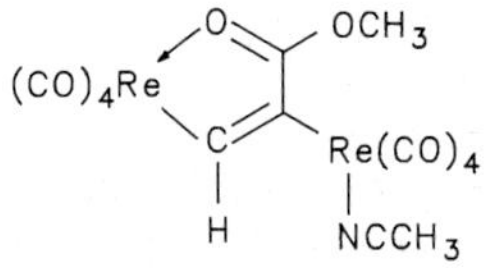	II (in CH_3CN, 5 min; yield: 71%) pale yellow crystals ^1H NMR (C_6D_6): 0.25 (CCH_3), 3.35 (OCH_3), 10.52 (=CH) IR (CH_2Cl_2): 1539 (br); 1914, 1936, 1996, 2084, 2098 (ν(CO))

References on pp. 52/3

Table 3 (continued)

No.	compound	method of preparation (yield) properties and remarks
*6 (continued)		MS: $[M - n\ CO]^+$ (n = 0 to 9), $[M - 9\ CO - CH_3CN]^+$ [12]
*7	$(CO)_4ReO{=}(CH_3O)CC[CH{=}C(Re(CO)_5)CO_2CH_3]{=}CH$-cyclo	III (11%; main product was the isomer mixture No. 11) ^1H NMR ($CDCl_3$): 3.73, 3.93 (OCH_3); 7.66, 9.62 (d's, CH; $^4J(H,H)$ = 1.5) IR (n-hexane): 1632, 1683, 1731; 1947, 1985, 1994, 2016, 2025, 2068, 2097, 2135 (ν(CO)) [12]
*8	$(CO)_4ReO{=}(C_2H_5O)CC[Re(CO)_5]{=}C(CO_2C_2H_5)$-cyclo	IV (43%) colorless solid ^1H NMR ($CDCl_3$): 1.33, 1.40 (t's, CH_3); 4.23, 4.38 (q's, CH_2; J(H,H) = 7.2) IR (n-hexane): 1536, 1701, 1942, 1986, 1995, 2019, 2029, 2068, 2096, 2137 (ν(CO)) MS: $[M - n\ CO]^+$ (n = 0 to 7) [15]
*9	$(CO)_4ReO{=}(CH_3O)CCH{=}C[C(O)Re(CO)_5]$-cyclo	by carbonylating (ca. 80 atm, 103°C, 11 h) No. 2 or $(CO)_4ReO{=}(CH_3O)CCH{=}C[Re(CO)_5]$-cyclo; yield: 8% and 22%, respectively; small amounts of $(CO)_{10}Re_2$ were also obtained orange solid ^1H NMR ($CDCl_3$): 3.92 (CH_3), 6.23 (=CH) IR (hexane): 1575; 1951, 1960, 1984, 2004, 2020, 2027, 2047, 2065, 2102, 2134 (ν(CO)) thermolysis in refluxing heptane reformed $(CO)_4ReO{=}(CH_3O)CCH{=}C[Re(CO)_5]$-cyclo [11]
	compounds with ^2L-^4D ligands	
*10	$(\mu\text{-}C_2H_5O_2CC{=}CCO_2C_2H_5)Re_2(CO)_8$	IV (32%), also by pyrolyzing No. 4 (heptane, reflux, 5 h, yield: 11%) or by photolyzing No. 8 for 30 min; yield: 62% orange solid ^1H NMR ($CDCl_3$): 1.44 (t, CH_3), 4.42 (q, CH_2; J = 7.2) IR (hexane): 1535, 1948, 1986, 1997, 2090 (ν(CO)) MS: $[M - n\ CO]^+$ (n = 0 to 5) [15]
*11	$(CO)_4ReO{=}(CH_3O)CC[CH{=}C(CH{=}C(CO_2CH_3)Re(CO)_4)C(O)OCH_3]{=}CH$-cyclo	III (49%); also in the absence of CO at room temperature (yield: 57%) or reflux temperature (yield: 36%); in these cases

Table 3 (continued)

No.	compound	method of preparation (yield) properties and remarks
		$(CO)_4ReC_{12}H_{11}O_6$ (see Formula III, p. 51) and $[CH{=}C(CO_2CH_3)CH{=}C(CO_2CH_3)]Re_2(CO)_7$ (see p. 150) were also obtained [12]

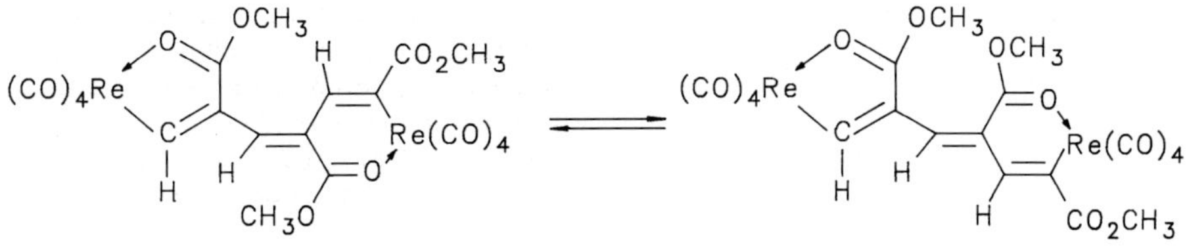

I (isomer A) II (isomer B)

No.	compound	method of preparation (yield) properties and remarks
	isomer A (see Formula I)	could be crystallized from the upper part of the eluted yellow band by fractional crystallization ^1H NMR ($CDCl_3$): 3.77, 3.93, 4.03 (s's, OCH_3); 7.28, 7.67, 10.63 (s's, CH) IR (n-hexane): 1590, 1633, 1731; 1953, 1983, 2002, 2097, 2101 (ν(CO)) [12]
	isomer B (see Formula II)	not isolated in pure form ^1H NMR ($CDCl_3$): 3.77, 3.78, 3.98 (CH_3); 6.68, 7.62, 10.00 (s's, CH) [12]
12	$(\mu\text{-}C_{14}H_6O_2)Re_2(CO)_8$	Va (two isomers formed) [2]
	isomer A 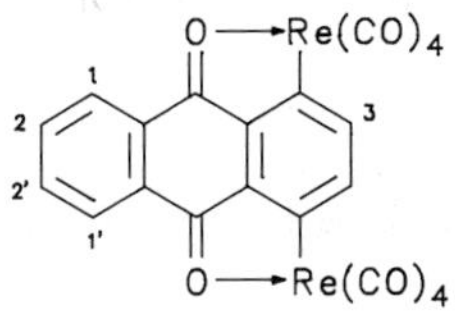	yield: 21% green-black crystals from hexane ^1H NMR (CD_2Cl_2): 7.84 (H-2), 8.26 (H-1; J(H-1,2) = 6, J(H-1′,2) = J(H-1,2′) = 3) IR (C_6H_{12}): 1947, 1999, 2092 (ν(CO)) MS: $[M - n\,CO]^+$ (n = 0 to 8) more soluble than isomer B [2]
	isomer B 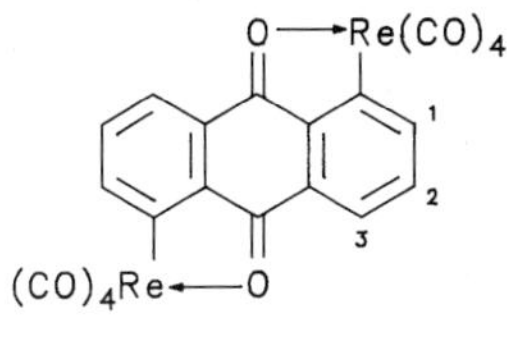	yield: 30% dark microcrystals from hexane/CH_2Cl_2 ^1H NMR (CD_2Cl_2): 7.54 (t, H-2), 7.90 (d, H-3), 8.35 (d, H-1; J(H-1,2) = J(H-2,3) = 7.5, J(H-1,3) < 1.5) IR (C_6H_{12}): 1946, 1999, 2094 (ν(CO)) MS: $[M - n\,CO]^+$ (n = 0 to 8) [2]
*13	$(\mu\text{-}C_{10}H_{20}S_2)Re_2(CO)_8$ 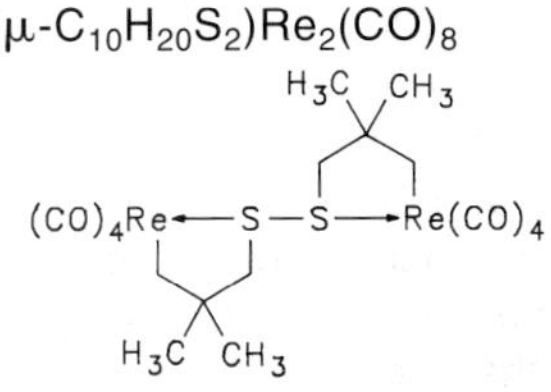	by photolyzing $C_5H_{10}S(CO)_4ReSCH_2C(CH_3)_2CH_2Re(CO)_4$-cyclo (see "Organorhenium Compounds" 5, 1994, p. 384) (CD_2Cl_2, 6 h) followed by chromatographic separation (silica, hexane/CH_2Cl_2 (9:1)); yield: 18%; main product was $(CO)_4Re(SC_5H_{10})Cl$

References on pp. 52/3

Table 3 (continued)

No.	compound	method of preparation (yield) properties and remarks
*13 (continued)		^1H NMR (CD_2Cl_2): 1.22, 1.52, 3.33 (br s's, 12:4:4 H) IR (n-hexane): 1950, 1989, 2082 (ν(CO)) [9]
14	(μ-$C_{24}H_7F_9N_4$)$Re_2(CO)_8$ 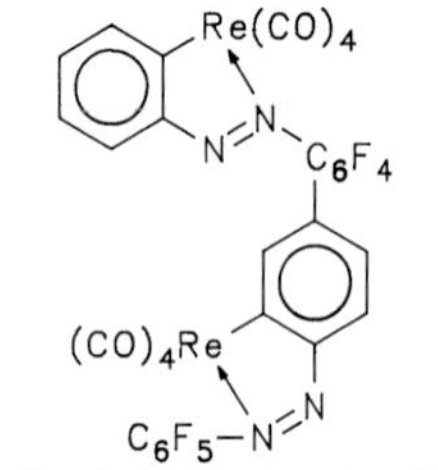 (tentative structure)	by treating $(CO)_5ReCH_3$ with a slight excess of $C_6F_5N{=}NC_6H_5$ (heptane, reflux, 51 h); chromatographic separation with ether/petroleum ether mixtures gave orange $(CO)_4ReC_6H_4N{=}NC_6F_5$ and the deep blue title product; yield: 2.5% sublimes at 100°C/0.1 Torr IR (C_6H_{12}): 1938, 1954, 1976, 2002, 2026, 2058 (ν(CO)) MS: $[M - n\ CO]^+$ (n = 0 to 8) [1]
15	(μ-$C_7H_3OP(C_6H_5)_2$)$Re_2(CO)_8$ 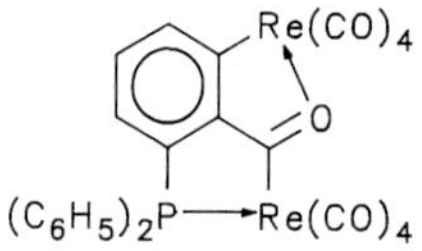	Vb (85%); also obtained by starting with $(CO)_4ReP(C_6H_5)_2C_6H_4$-2 (in methyl cyclohexane, 6 h; yield: 10%) yellow solid IR (CCl_4): 1452 (ν(C=O)); (C_6H_{12}): 1932, 1976, 1983, 1999, 2016, 2086, 2097 (ν(CO)) MS: $[M - n\ CO]^+$ (n = 0 to 9) [3]
16	(μ-$C_7H_2FOP(C_6H_4F\text{-}4)_2$)$Re_2(CO)_8$ 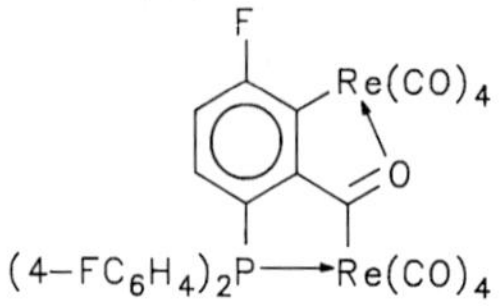	Vb (80%) yellow solid ^1H NMR ($CDCl_3$): 7.4 (m) ^{19}F NMR (ext. standard CF_3CO_2H): 1.18 (q, C_6H_2F), 28.1 (q, C_6H_4F; J ca. 5) IR (CCl_4): 1455 (ν(C=O)); (C_6H_{12}): 1948, 1979, 1989, 2002, 2009, 2090, 2101 (ν(CO)) EI MS: $[M - n\ CO]^+$ (n = 0 to 9) [3]
17	$[C_6H_4(P(C_6H_5)_2)CHRe(CO)_4]_2O$ 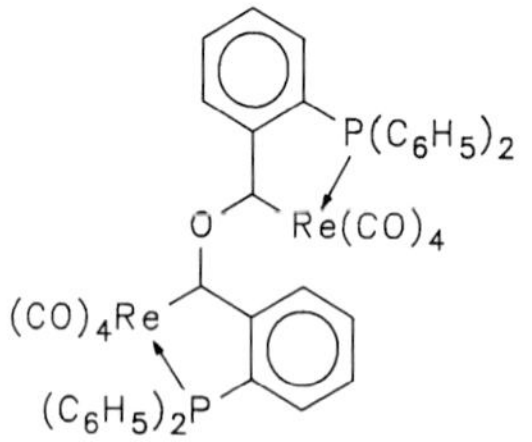	by reacting $(CO)_4ReC(OR)(H)C_6H_4P(C_6H_5)_2$-cyclo (R = H [4, 7], $Si(CH_3)_2C_4H_9$-t [7]) with ca. 0.06 equivalent CF_3SO_3H in acetone/H_2O; yield: 28 to 36% [4, 7] in the former, 84% in the latter case [7] yellow crystals from THF/CH_3CN; m.p. 212 to 216°C (dec.) [7] ^1H NMR (CD_2Cl_2): 6.21 (s, OCH); 6.92 to 7.05, 7.05 to 7.70, 7.84 to 7.95 (m's, 4:22:2 H) $^{13}C\{^1H\}$ NMR (THF-d_8): 75.8 (ReCH); 126.0, 127.9, 129.5, 129.7, 129.8, 131.3, 131.4, 132.1, 132.3, 133.7, 133.8, 134.9, 135.9, 165.8; 189.8, 191.4, 191.9, 193.4 (CO) IR (THF): 1939, 1980, 1992, 2073 (ν(CO))

References on pp. 52/3

Table 3 (continued)

No.	compound	method of preparation (yield) properties and remarks
		MS: $[M]^+$, $[M - n\,CO]^+$ (n = 0, 2, 6, 7), $[M - n\,CO - O]^+$ (n = 3 to 8) [7]
*18	$(\mu\text{-}(C_6H_5)_2PC_6H_3C{=}O)Re(CO)_4Re(CO)_3P(C_6H_5)_3$ 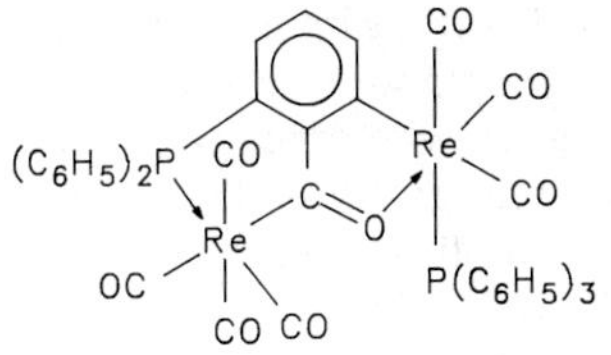	VI (32.4%) [8]; also in the presence of Ga_2I_3 at 150 °C [6] yellow solid [6, 8]; m.p. 219 °C [8] $^{31}P\{^1H\}$ NMR ($CDCl_3$): 20.85 (s, $P(C_6H_5)_3$), 28.42 (s, $P(C_6H_5)_2$) IR (KBr): 1453 (ν(C=O)); ($CHCl_3$): 1885, 1935, 1972, 1995, 2010, 2090 (ν(CO)) [8]
19	$(\mu\text{-}(C_6H_5)_2PC_6H_3C{=}O)Re(CO)_3(P(C_6H_5)_3)Re(CO)_3P(C_6H_5)_3$ isomer A 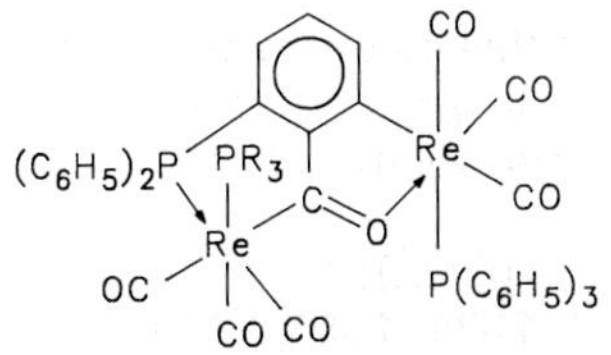 ($R = C_6H_5$)	VI (5.4%) yellow solid; m.p. 224 °C $^{31}P\{^1H\}$ NMR ($CDCl_3$): 3.82 (d, $P(C_6H_5)_3$-ax; $^2J(P,P)$ = 24.8), 19.38 (s, $P(C_6H_5)_3$), 30.95 (d, $P(C_6H_5)_2$; $^2J(P,P)$ = 24.9) IR (KBr): 1451 (ν(C=O)); ($CHCl_3$): 1800, 1822, 1885, 1922, 1958, 2008, 2030, 2090 (ν(CO)) [8]
	isomer B 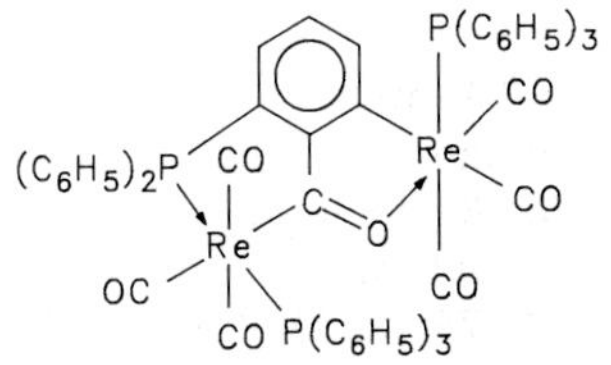	VI (2.7%) $^{31}P\{^1H\}$ NMR ($CDCl_3$): 14.99 (d, $P(C_6H_5)_3$-eq; $^2J(P,P)$ = 119.1), 19.55 (s, $P(C_6H_5)_3$), 32.94 (d, $P(C_6H_5)_2$; $^2J(P,P)$ = 118.75) [8]
20	$(\mu\text{-}(C_6H_5)_2PC_6H_3C{=}O)Re(CO)_4Re(CO)_2(P(C_6H_5)_3)_2$ 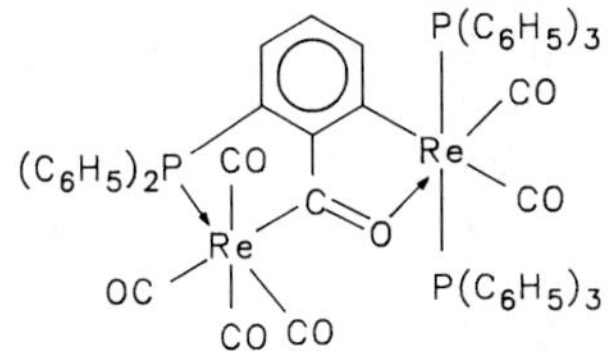	VI (1.4%) orange solid; m.p. 205 °C $^{31}P\{^1H\}$ NMR ($CDCl_3$): 26.61 (s, $P(C_6H_5)_3$), 27.72 (s, $P(C_6H_5)_2$) IR (KBr): 1454 (ν(C=O)); ($CHCl_3$): 1835, 1895, 1917, 1967, 1995, 2000, 2090 (ν(CO)) [8]

*Further information:

$(\mu\text{-}C_7H_4O_2)Re_2(CO)(P(CH_3)_3)_4(NO)_2(CH_3)C(O)CH_3$ (Table **3**, No. **1**) was obtained by heating $\sigma\text{-}C_5H_5Re(CO)(NO)(P(CH_3)_3)_2CH_3$ in benzene under ca. 70 atm of CO for 4 h. Subsequent evaporation followed by chromatographic separation on silica with ether yielded two yellow fractions. Evaporation gave an analytically pure product with 18% yield.

 References on pp. 52/3

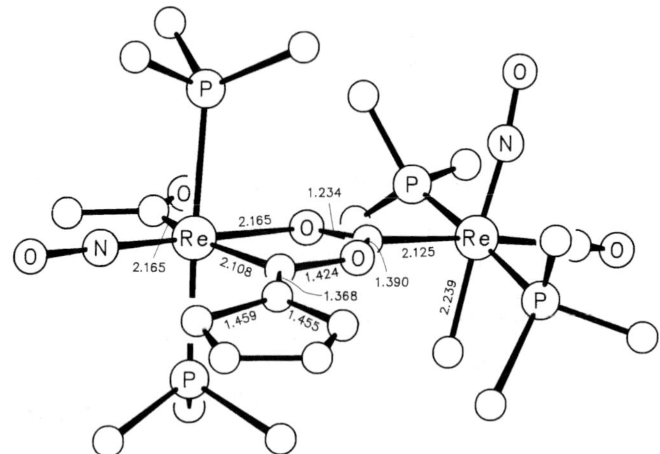

Fig. 20. Molecular structure of $(\mu\text{-}C_7H_4O_2)Re_2(CO)(P(CH_3)_3)_4(NO)_2(CH_3)C(O)CH_3$ [5].

No. 1 crystallizes in the triclinic space group $P1-C_1^1$ (No. 1) with a = 18.328(6), b = 23.907(7), c = 8.751(3) Å, α = 95.91(2)°, β = 101.98(2)°, γ = 111.33(2)°; Z = 4 molecules per unit cell, D_{calc} = 1.83 g/cm³. The molecular structure is shown in **Fig. 20** [5].

$(CO)_4ReO{=}(CH_3O)CC[Re(CO)_5]{=}CH$-cyclo (Table **3**, No. **2**) crystallizes in the monoclinic space group $P2_1/n-C_{2h}^5$ (No. 14) with a = 6.644(1), b = 30.005(8), c = 10.036(2) Å, (caution: 18.036(2) [11]), β = 96.09(1)°; Z = 8 molecules per unit cell, D_{calc} = 2.63 g/cm³. There are two symmetry-independent molecules in the asymmetric unit; both are structurally similar. The structure and selected bond parameters of one of these molecules are shown in **Fig. 21** [10, 11].

Room-temperature carbonylation (77 atm) exclusively gave trans-$(CO)_5ReCH{=}C(CO_2CH_3)Re(CO)_5$, whereas carbonylation at 103 °C not only produced this compound, but also $(CO)_4ReO{=}(CH_3O)CCH{=}C[Re(CO)_5]$-cyclo (see "Organorhenium Compounds" 5, 1994, pp. 427/8) and No. 9. Photolysis under a CO atmosphere (1 atm) yielded $(CO)_8Re_2(\mu\text{-}C{=}C\text{-}$

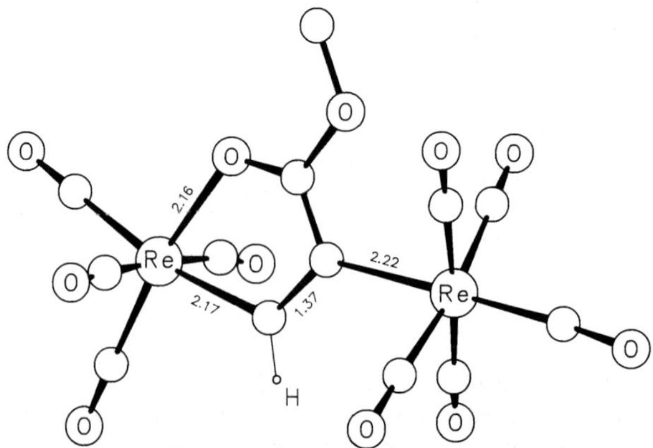

Fig. 21. Molecular structure of $(CO)_4ReO{=}(CH_3O)CC[Re(CO)_5]{=}CH$-cyclo [10, 11].

References on pp. 52/3

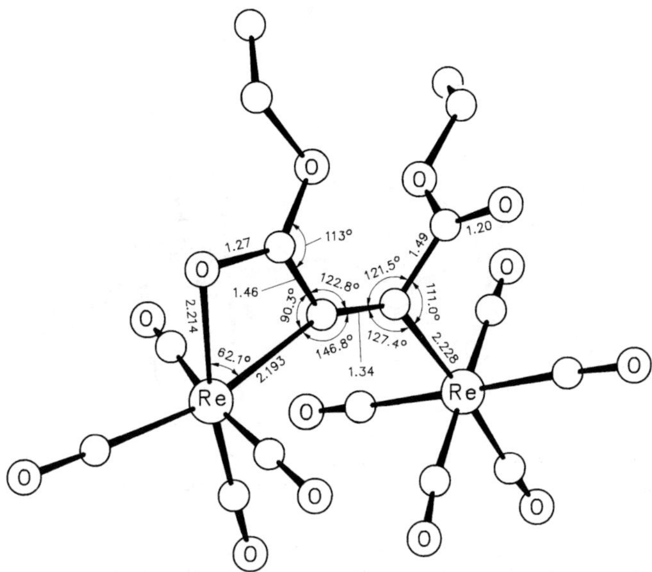

Fig. 22. Molecular structure of $(CO)_4ReO{=}C(OC_2H_5)C[{=}C(CO_2C_2H_5)Re(CO)_5]$-cyclo [15].

$HC(OCH_3){=}O)$ and $(CO)_4ReO{=}(CH_3O)CCH{=}C[Re(CO)_5]$-cyclo (see "Organorhenium Compounds" 5, 1994, pp. 382/3 and 427/8) [11]. The $(CH_3)_3NO$-promoted decarbonylation in CH_3CN yielded No. 6 [12], whereas in CH_2Cl_2 No. 5 was formed [18]. The compound did not react with $HC{\equiv}CCO_2CH_3$ below 98 °C [12].

$(CO)_4ReO{=}C(OC_2H_5)C[{=}C(CO_2C_2H_5)Re(CO)_5]$-cyclo (Table **3**, No. **4**) crystallizes in the monoclinic space group $P2_1/c-C_{2h}^5$ (No. 14) with a = 10.292(2), b = 11.731(2), c = 18.704(4) Å, β = 100.14(2)°; Z = 4 molecules per unit cell, D_{calc} = 2.37 g/cm^3. The molecular structure is shown in **Fig. 22**.

The compound is poorly soluble in hexane. During silica gel chromatography it decomposed to $(CO)_5ReC(CO_2C_2H_5){=}CHCO_2C_2H_5$. Brief UV irradiation at room temperature yielded a mixture of Nos. 8 and 10; in contrast, No. 10 was exclusively obtained when pyrolyzing the title compound. Room-temperature carbonylation under ca. 50 atm, however, gave cis-$(CO)_5Re-C(CO_2C_2H_5){=}C(CO_2C_2H_5)-Re(CO)_5$ (this reaction could be nearly completely reversed by heating the product in hexane). The reaction with a gaseous HCl/CO mixture gave $(CO)_5ReCl$ and $(CO)_5ReC(CO_2C_2H_5){=}CHCO_2C_2H_5$ [15].

$(CO)_4ReO{=}(CH_3O)CC[Re(CO)_4N(CH_3)_3]{=}CH$-cyclo (Table **3**, No. **5**). Single crystals are monoclinic with a = 7.670(3), b = 18.448(3), c = 15.200(7) Å, β = 108.72(4)°; space group $P2_1/c-C_{2h}^5$ (No. 14); Z = 4 molecules per unit cell, D_{calc} = 2.41 g/cm^3. The structure of the molecule is very similar to that of No. 6; the $N(CH_3)_3$ ligand replaces one equatorial CO group of the $Re(CO)_5$ fragment [18].

$(CO)_4ReO{=}(CH_3O)CC[Re(CO)_4NCCH_3]{=}CH$-cyclo (Table **3**, No. **6**) crystallizes in the triclinic space group $P\bar{1}-C_i^1$ (No. 2) with a = 9.272(1), b = 13.122(2), c = 8.509(1) Å, α = 106.46(1)°, β = 95.78(1)°, γ = 69.85(1)°; Z = 2 molecules per unit cell, D_{calc} = 2.57 g/cm^3. The molecule is structurally very similar to its parent compound. The CH_3CN ligand is coordinated to the Re atom in a position cis to the rhenacycle (see **Fig. 23**) [14].

References on pp. 52/3

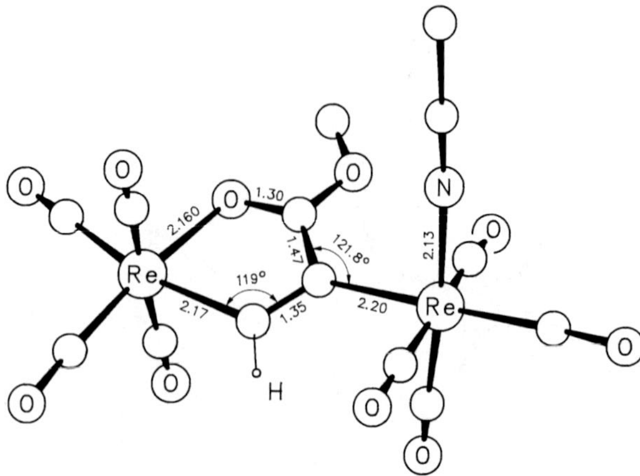

Fig. 23. Molecular structure of $(CO)_4ReO{=}(CH_3O)CC[Re(CO)_4NCCH_3]{=}CH$-cyclo [14].

The compound is unstable on silica gel and sensitive to moisture. Treatment with $HC{\equiv}CCO_2CH_3$ in refluxing CH_2Cl_2 under a CO atmosphere gave No. 7 and the isomer mixture of No. 11 by head-to-tail coupling of two or three alkynes, respectively, whereas in the absence of CO the isomer mixture formed with a higher yield (other products were $[CH{=}C(CO_2CH_3)CH{=}C(CO_2CH_3)]Re_2(CO)_7$ (see p. 150) and $(CO)_4ReC_{12}H_{11}O_6$ (see Formula III, p. 51)) [12]. Treatment with a series of aryl isothiocyanates, $4\text{-}RC_6H_4N{=}C{=}S$, in refluxing CH_2Cl_2 (R = H, Cl [14], CH_3 [13, 14]) yielded $(CO)_4ReO{=}(CH_3O)C(HC{=})C(C(S){=}C\text{-}C_6H_4CH_3\text{-}4)Re(CO)_4$ by insertion into one of the Re-C bonds (see "Organorhenium Compounds" 5, 1994, p. 391) [13, 14]. Treatment with CS_2 spontaneously gave $(CO)_4Re\text{-}O{=}(CH_3O)C(HC{=})CCS_2Re(CO)_4$ (see "Organorhenium Compounds" 5, 1994, p. 393) [16, 17].

$(CO)_4ReO{=}(CH_3O)CC[CH{=}C(Re(CO)_5)CO_2CH_3]{=}CH$-cyclo (Table **3**, No. **7**) crystallizes in the triclinic space group $P\bar{1}-C_i^1$ (No. 2) with a = 13.792(2), b = 14.009(2), c = 6.2734(6) Å, α = 91.47(1)°, β = 92.95(1)°, γ = 114.764(8)°; Z = 2 molecules per unit cell, D_{calc} = 2.40 g/cm^3. The structure of the molecule is illustrated in **Fig. 24** [12].

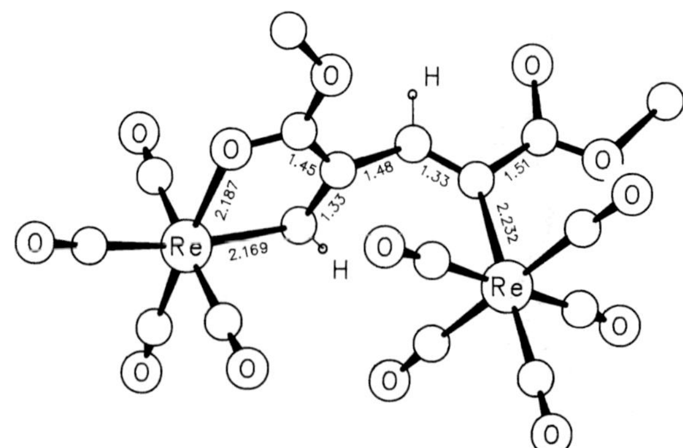

Fig. 24. Molecular structure of $(CO)_4ReO{=}(CH_3O)CC[CH{=}C(Re(CO)_5)CO_2CH_3]{=}CH$-cyclo [12].

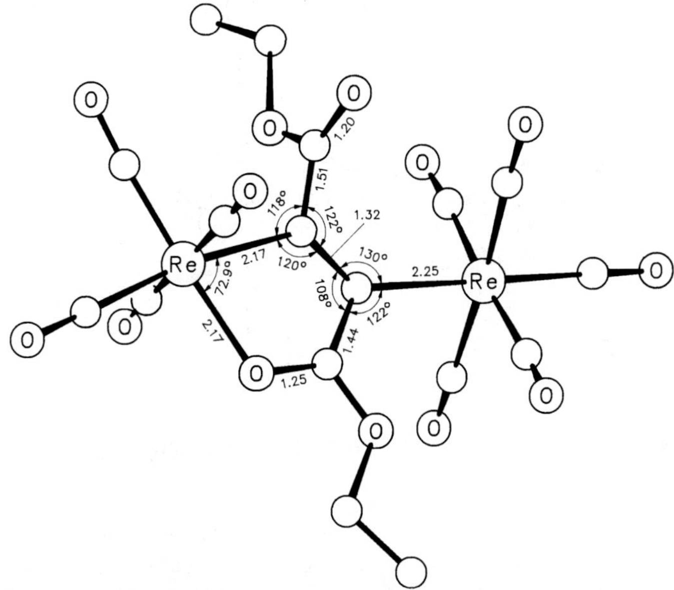

Fig. 25. Molecular structure of $(CO)_4ReO{=}(C_2H_5O)CC[Re(CO)_5]{=}C(CO_2C_2H_5)$-cyclo [15].

Dissolution in CH_3CN in the presence of $(CH_3)_3NO$ and $HC{\equiv}CCO_2CH_3$ gave the rhenacycle $[CH{=}C(CO_2CH_3)CH{=}C(CO_2CH_3)]Re_2(CO)_7$ (see p. 150) and an isomer mixture of No. 11. The yield of the former product increased when leaving the alkyne out [12].

$(CO)_4ReO{=}(C_2H_5O)CC[Re(CO)_5]{=}C(CO_2C_2H_5)$-cyclo (Table **3**, No. **8**) crystallizes in the orthorhombic space group $Fdd2-C_{2v}^{19}$ (No. 43) with a = 24.498(3), b = 35.102(5), c = 10.681(1) Å; Z = 16 molecules per unit cell, D_{calc} = 2.30 g/cm³. A view of the molecular structure is given in **Fig. 25** [15].

Room-temperature photolysis yielded No. 10 [15].

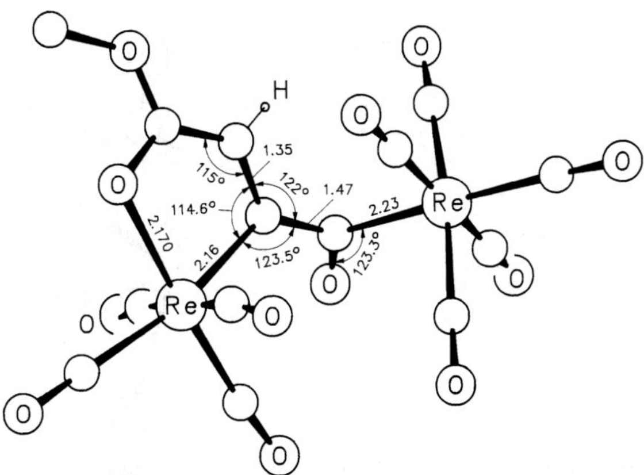

Fig. 26. Molecular structure of $(CO)_4ReO{=}(CH_3O)CCH{=}C[C(O)Re(CO)_5]$-cyclo [11].

$(CO)_4ReO{=}(CH_3O)CCH{=}C[C(O)Re(CO)_5]$-cyclo (Table **3**, No. **9**). Single crystals have a triclinic lattice with a = 9.894(1), b = 12.434(2), c = 8.748(2) Å, α = 108.98(1)°, β = 113.593(9)°, γ = 77.69(1)°; space group $P\bar{1}-C_i^1$ (No. 2), Z = 2 molecules per unit cell, D_{calc} = 2.63 g/cm^3. The molecular structure is illustrated in **Fig. 26** [11].

$(\mu\text{-}C_2H_5O_2CC{=}CCO_2C_2H_5)Re_2(CO)_8$ (Table **3**, No. **10**) crystallizes in the triclinic space group $P\bar{1}-C_i^1$ (No. 2) with a = 7.838(3), b = 9.830(3), c = 7.047(2) Å, α = 106.93(2)°, β = 94.84(3)°, γ = 84.28(3)°; Z = 2 molecules per unit cell, D_{calc} = 2.47 g/cm^3. The structure of the molecule is shown in **Fig. 27**. It contains a center of symmetry, which is why the metallacyclic rings lie in a common plane [15].

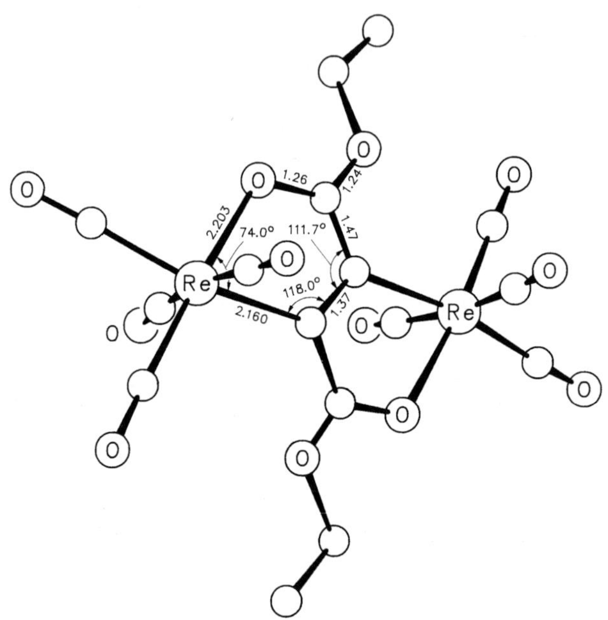

Fig. 27. Molecular structure of $(\mu\text{-}C_2H_5O_2CC{=}CCO_2C_2H_5)Re_2(CO)_8$ [15].

$(CO)_4ReO{=}(CH_3O)CC[CH{=}C(CH{=}C(CO_2CH_3)Re(CO)_4)C(O)OCH_3]{=}CH$-cyclo (Table **3**, No. **11**). The isomer mixture was also formed by decarbonylation of No. 7 with $(CH_3)_3NO$ in CH_3CN followed by adding $HC{\equiv}CO_2CH_3$ in refluxing CH_2Cl_2. Yield: 33%.

In solution the compound exists as a mixture of two isomers. Dissolution of the pure isomer A in benzene resulted in continuing conversion into isomer B. The ratio A:B became 73:27, 56:44, 52:48, and 49:51 after 1, 3, 6, and 9 days, respectively. Otherwise, a solution enriched with isomer B slowly transformed into isomer A until the A:B equilibrium ratio 49:51 was reached.

Single crystals of isomer A are triclinic with a = 10.7738(8), b = 13.179(2), c = 9.024(1) Å, α = 96.52(1)°, β = 101.467(8)°, γ = 84.604(9)°; space group $P\bar{1}-C_i^1$ (No. 2), Z = 2 molecules per unit cell, D_{calc} = 2.27 g/cm^3. The geometry of the molecule is illustrated in **Fig. 28**.

References on pp. 52/3

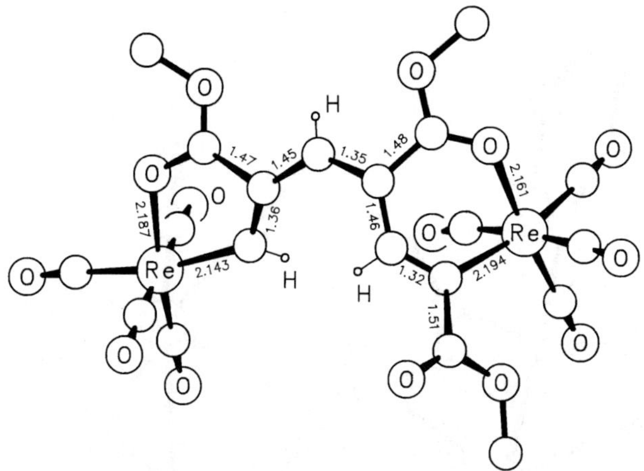

Fig. 28. Molecular structure of
$(CO)_4ReO{=}(CH_3O)CC[CH{=}C(CH{=}C(CO_2CH_3)Re(CO)_4)C(O)OCH_3]{=}CH$-cyclo (isomer A) [12].

Thermolysis of the pure isomer A in hot heptane provided $(CO)_4ReC_{12}H_{11}O_6$ (see Formula III) and a small amount of $(CO)_{10}Re_2$ [12].

$(CO)_4Re$ O OCH_3
CH_3O_2C
CO_2CH_3

III

$(\mu$-$C_{10}H_{20}S_2)Re_2(CO)_8$ (Table **3**, No. **13**) crystallizes in the triclinic space group $P\bar{1}-C_i^1$ (No. 2) with a = 8.938(2), b = 9.594(3), c = 7.549(2) Å, α = 100.10(2)°, β = 93.44(2)°,

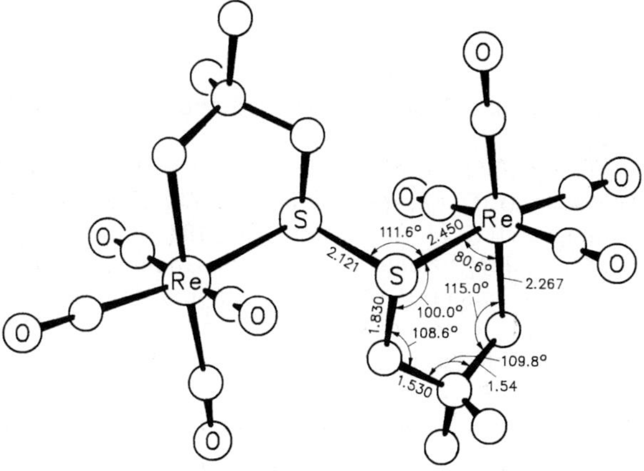

Fig. 29. Molecular structure of $(\mu$-$C_{10}H_{20}S_2)Re_2(CO)_8$ [9].

References on pp. 52/3

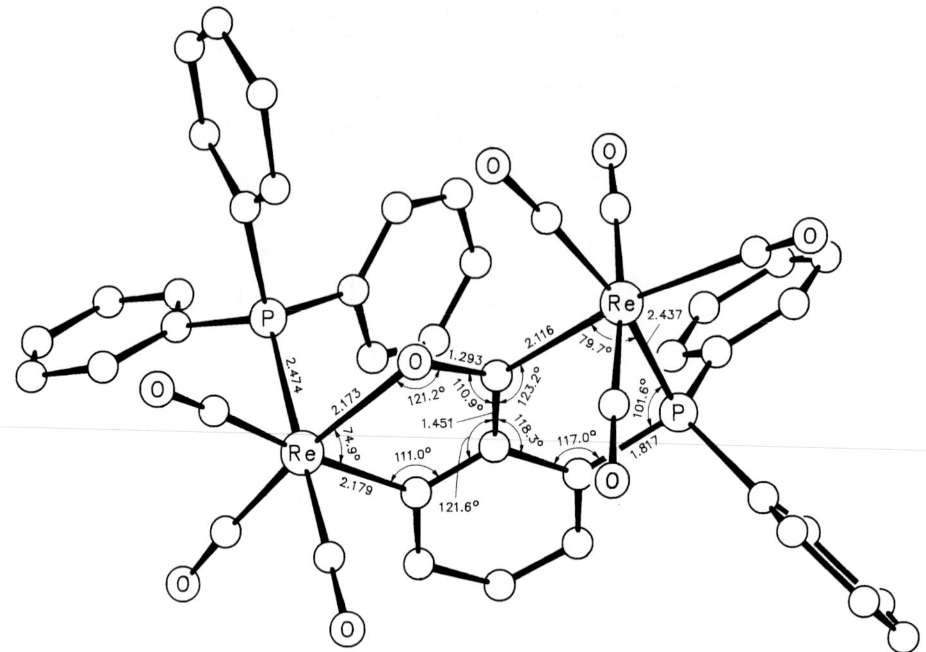

Fig. 30. Molecular structure of (μ-$(C_6H_5)_2PC_6H_3C{=}O)Re(CO)_4Re(CO)_3P(C_6H_5)_3$ [6].

$\gamma = 66.60(2)°$; Z = 1 molecule per unit cell, $D_{calc} = 2.27$ g/cm^3. The molecular centrosymmetric structure is illustrated in **Fig. 29**. The geometry around S is pyramidal [9].

(μ-$(C_6H_5)_2PC_6H_3C{=}O)Re(CO)_4Re(CO)_3P(C_6H_5)_3$ (Table **3**, No. **18**). Single crystals have a monoclinic lattice with a = 16.800(2), b = 12.077(3), c = 23.048(3) Å, $\beta = 92.82(1)°$; Z = 4 molecules per unit cell, $D_{calc} = 1.591$ g/cm^3 [8]. Another study by the same authors gave a = 15.265(2), b = 13.506(5), c = 19.869(2) Å, $\beta = 93.54(9)°$ [6]. The molecular geometry is shown in **Fig. 30** [6].

References:

[1] Bruce, M. I.; Goodall, B. L.; Sheppard, G. L.; Stone, F. G. A. (J. Chem. Soc. Dalton Trans. **1975** 591/5).
[2] McKinney, R. J.; Firestein, G.; Kaesz, H. D. (Inorg. Chem. **14** [1975] 2057/61).
[3] McKinney, R. J.; Kaesz, H. D. (J. Am. Chem. Soc. **97** [1975] 3066/72).
[4] Vaughn, G. D.; Gladysz, J. A. (J. Am. Chem. Soc. **103** [1981] 5608/9).
[5] Casey, C. P.; O'Connor, J. M.; Haller, K. J. (J. Am. Chem. Soc. **107** [1985] 3172/7).
[6] Haupt, H.-J.; Flörke, U.; Balsaa, P. (Acta Crystallogr. C **41** [1985] 1307/9).
[7] Vaughn, G. D.; Strouse, C. E.; Gladysz, J. A. (J. Am. Chem. Soc. **108** [1986] 1462/73).
[8] Haupt, H.-J.; Balsaa, P.; Flörke, U. (Inorg. Chem. **27** [1988] 280/6).
[9] Adams, R. D.; Belinski, J. A.; Schierlmann, J. (J. Am. Chem. Soc. **113** [1991] 9004/6).
[10] Adams, R. D.; Chen, L.; Wu, W. (Organometallics **11** [1992] 3505/7).
[11] Adams, R. D.; Chen, L.; Wu, W. (Organometallics **12** [1993] 1257/65).
[12] Adams, R. D.; Chen, L.; Wu, W. (Organometallics **12** [1993] 1623/8).

[13] Adams, R. D.; Chen, L.; Wu, W. (Organometallics **12** [1993] 2404/5).
[14] Adams, R. D.; Chen, L.; Wu, W. (Organometallics **12** [1993] 3812/8).
[15] Adams, R. D.; Chen, L. (Organometallics **13** [1994] 1264/71).
[16] Adams, R. D.; Chen, L.; Wu, W. (Organometallics **13** [1994] 1257/63).
[17] Adams, R. D.; Chen, L.; Wu, W. (Angew. Chem. **106** [1994] 591/2; Angew. Chem. Int. Ed. Engl. **33** [1994] 568/9).
[18] Bruce, M. I.; Low, P. J.; Skelton, B. W.; White, A. H. (J. Organomet. Chem. **464** [1994] 191/5).

2.2.1.2.4 Compounds with Rhenium-Heterometal Bonds-Bridging CO Groups

The structures of the compounds are illustrated by Formulas I to VII.

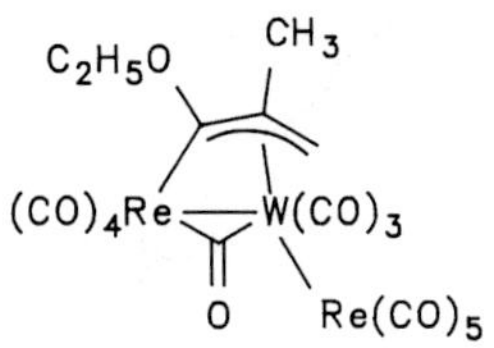

$R = C_6H_4CH_3-4$

I 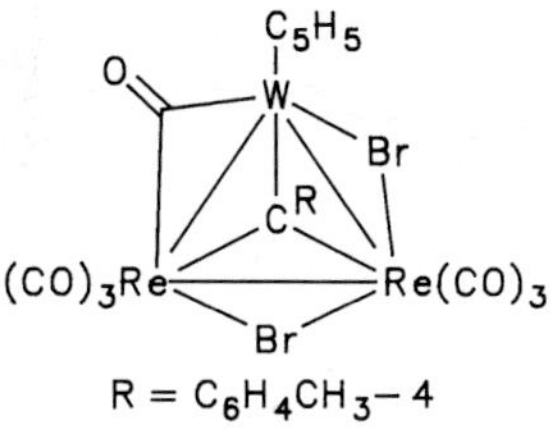 II 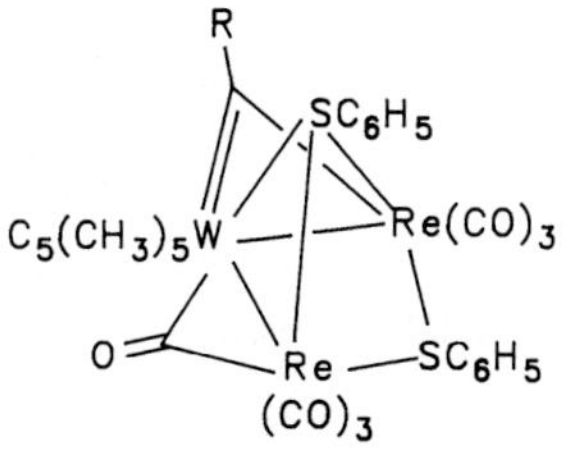 III

[$C_2H_5OCC(CH_3)CH_2W(CO)_3CO]Re_2(CO)_9$ (see Formula I) was produced by photochemically reacting $(CO)_{10}Re_2$ with 1.25 equivalents CH_2=$(CH_3)C(C_2H_5O)C$=$W(CO)_5$ (pentane, −50°C, 70 min). Subsequent chromatographic workup (silica, pentane/CH_2Cl_2 mixtures) separated the title compound from (μ-$\eta^{3:1}$-$C_2H_5OCC(CH_3)CH_2)Re_2(CO)_8$ (see p. 130) and $(CO)_8W_2$-(μ-$\eta^{3:1}$-$C_2H_5OCC(CH_3)$=CH_2). Purification by HPLC with pentane/CH_2Cl_2 (9:1) at −10°C eventually yielded 0.9% of orange crystals.

^1H NMR spectrum (C_6D_6): δ = 1.22 (t, CH_2**CH_3**; J = 6.9 Hz), 1.91 (d, CH_2⋯; J = 4.4 Hz), 2.18 (s, CH_3C⋯), 2.58 (d, ⋯CH_2); 4.00, 4.27 (dq's, OCH_2; J = 6.9 and 8.9 Hz, respectively) ppm. ^{13}C NMR spectrum (CD_2Cl_2, 10°C): δ = 16.96 (q, CH_2CH_3), 24.27 (q, CH_3), 44.12 (t, CH_2), 74.55 (t, CH_2; J(C,H) = 127, 129, 156, 141 Hz, respectively), 97.45 (s); 182.13, 187.37, 187.89, 190.51, 191.02, 193.65 (ReCO); 199.7 (s); 214.7 (WCO; J(W,C) = 119 Hz) ppm (the WCO signal broadened on cooling to −70°C). IR spectrum (n-hexane): 1811, 1952, 1962, 1977, 1987, 1998, 2019, 2048, 2080, 2118 (ν(CO)) cm^{-1}.

The compound crystallizes in the monoclinic space group $P2_1/n-C_{2h}^5$ (No. 14) with a = 8.559(1), b = 16.170(2), c = 18.191(2) Å, β = 102.68°; Z = 4 molecules per unit cell, D_{calc} = 2.831 g/cm^3. The molecular structure is shown in **Fig. 31** [2].

[μ-$C_5H_5W(CC_6H_4CH_3$-4)(CO)Br]$Re_2(CO)_6$(μ-Br) (see Formula II) was obtained by reacting equimolar amounts of $(CO)_6Re_2(OC_4H_8)_2(\mu\text{-}Br)_2$ and $C_5H_5W(\equiv CC_6H_4CH_3)(CO)_2$ in refluxing hexane for 1 h. The solvent was removed, and the residue was taken up in ether/CH_2Cl_2 (1:1), passed through an alumina column, and concentrated; subsequent addition of petroleum ether precipitated the compound which was recrystallized from hexane/CH_2Cl_2 (3:1) at −20°C. Yield: 10%. Orange crystals; m.p. 160°C. When the reaction was conducted with a 5-fold excess of the tungsten compound, $(CO)_3Re(\mu\text{-}CO)W_2(C_5H_5)_2(\mu\text{-}Br)(\mu\text{-}CC_6H_4\text{-}CH_3\text{-}4)(\mu_3\text{-}CC_6H_4CH_3\text{-}4)$ (see "Organorhenium Compounds" 2, 1989, p. 324) was also formed in addition to the title compound.

References on p. 57

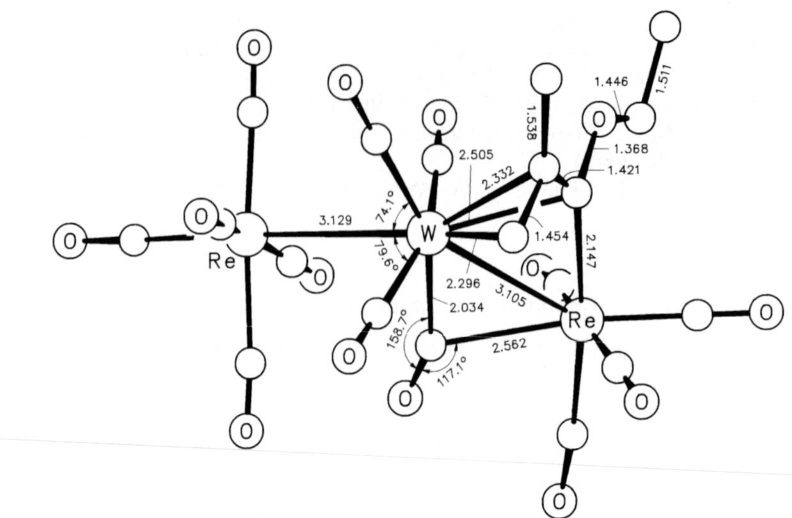

Fig. 31. Molecular structure of $[C_2H_5OCC(CH_3)CH_2W(CO)_3CO]Re_2(CO)_9$ [2].

^{1}H NMR spectrum (CD_2Cl_2): δ = 2.39 (s, CH_3), 5.89 (s, C_5H_5), 7.16 (AB, C_6H_4; J = 8 Hz) ppm. ^{13}C {^{1}H} NMR spectrum ($CDCl_3$): δ = 21.4 (CH_3), 96.7 (C_5H_5); 121.9, 128.8, 138.1 (C_6H_4); 158.5 (C_{ipso}); 188.2, 188.7, 188.8, 191.0, 191.4, 191.8 (CO); 237.9 (μ-CO); 312.3 (μ_3-C) ppm. IR spectrum (CH_2Cl_2): 1814, 1929, 1967, 2034, 2052 (ν(CO)) cm^{-1} [1].

$[\mu\text{-}(C_5(CH_3)_5)W(=CR)(SC_6H_5)CO]Re_2(CO)_6(\mu\text{-}SC_6H_5)\cdot 0.5\ CH_2Cl_2$ (see Formula III). These compounds were obtained by treating $[\mu\text{-}(C_5(CH_3)_5)W(CO)_2C{\equiv}CR']Re_2(CO)_6(\mu\text{-}CO)$ (R′ = C_6H_9-cyclo, $C(CH_3){=}CH_2$; see pp. 125/6) with excess C_6H_5SH in refluxing toluene for 5 min. The products were separated by preparative TLC with CH_2Cl_2/hexane (1:1) [5].

The compounds were characterized as follows (for abbreviations and units see p. X):

R = $CH{=}C(CH_2)_5$: Yield: 31%. – Orange crystals.
^{1}H NMR ($CDCl_3$): 1.40 to 1.71 (m, CH_2); 1.87 ($(CH_3)_5$); 2.03 to 2.24 (m, CH_2), 6.18 (s, CH=C), 7.07 (t; J(H,H) = 7.4), 7.21 to 7.25 and 7.46 to 7.54 (m's, C_6H_5). – ^{13}C {^{1}H} NMR ($CDCl_3$): 9.5 (CH_3); 26.3, 27.3, 29.2, 30.9, 35.8 (CH_2); 53.3 (CH=**C**$(CH_2)_5$), 106.9 (C_5); 126.2, 126.4, 127.9, 129.6, 129.7, 132.0, 137.8 (C_6H_5); 141.2 (**C**H=$C(CH_2)_5$; J(W,C) = 30); 142.3 (C_6H_5); 190.4, 190.6, 190.7, 192.3, 196.5, 196.7 (ReCO); 243.8 (WCO; J(W,C) = 151), 302.7 (W=**C**CH=; J(W,C) = 215). – IR (CH_2Cl_2): 1802, 1910, 1948, 2017, 2032 (ν(CO)). – FAB MS: $[M]^+$.
Single-crystal data: triclinic; space group $P\bar{1}-C_i^1$ (No. 2); a = 10.149(1), b = 12.151(4), c = 17.442(2) Å, α = 84.79(2)°, β = 76.40(1)°, γ = 70.39(2)°; Z = 2 molecules per unit cell; D_{calc} = 2.116 g/cm^3.
The molecular structure is illustrated in **Fig. 32** [5].

R = $CH{=}C(CH_3)_2$: Yield: 36%.
^{1}H NMR ($CDCl_3$): 1.70 (CH_3), 1.88 ($(CH_3)_5$), 1.88 (CH_3), 6.30 (s, C**H**=$C(CH_3)_2$); 7.07 (t; J = 7.4), 7.21 to 7.25 (t), 7.46 to 7.54 (C_6H_5). – IR (C_6H_{12}): 1810, 1912, 1934, 1959, 2022, 2037 (ν(CO)). – FAB MS: $[M]^+$ [5].

References on p. 57

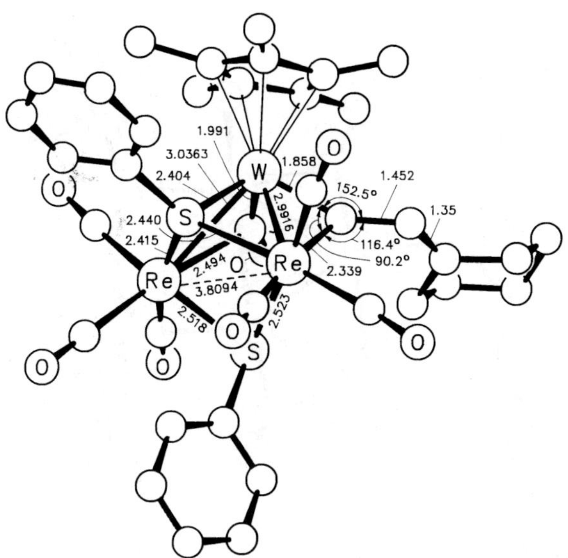

Fig. 32. Molecular structure of
[μ-($C_5(CH_3)_5$)W(=CCH=C(CH_2)$_5$)(SC_6H_5)CO]$Re_2(CO)_6$(μ-SC_6H_5) [5].

[μ-((C_6H_5)$_2$PCH$_2$P(C_6H_5)$_2$)$_2$Pt$_2$(CO)$_2$]Re$_2$(CO)$_6$ · 2 O=C(CH$_3$)$_2$ (see Formula IV) was obtained by treating $Pt_2[(C_6H_5)_2PCH_2P(C_6H_5)_2]_2Cl_2$ with Na[(CO)$_5$Re] in THF for several days. The reaction was shown by NMR spectroscopy to proceed via at least two observable intermediates (see "Organorhenium Compounds" 5, 1994, p. 395).

IV V VI

^{1}H NMR spectrum (acetone-d$_6$): δ = 4.78 (m, CH$_2$) ppm. ^{13}C {^{1}H} NMR spectrum (acetone-d$_6$): δ = 204.8, 208.3 (m's, ReCO); 216.1 (CO-μ; J(Pt,C) = 747, J(P,C) = 28 Hz) ppm. ^{31}P {^{1}H} NMR spectrum (acetone-d$_6$): δ = 10.2 (m, ReP), 44.2 (m, PtP; ^{1}J(Pt,P) = 3317, ^{2}J(Pt,P) = 310 Hz) ppm. IR spectrum (Nujol): 1821, 1904, 1935, 1998 (ν(CO)) cm^{-1}.

The compound crystallizes in the triclinic space group $P\bar{1}-C_i^1$ (No. 2) with a = 11.503(1), b = 14.069(3), c = 11.455(1) Å, α = 104.31(2)°, β = 119.10(9)°, γ = 91.53(2)°; Z = 1 molecule per unit cell. The molecule is centrosymmetric; therefore, the central metal core is strictly planar. It has a Pt-Pt single bond but no Re-Re bond (compare with **Fig. 33**) [4].

[μ-(C$_6$H$_{11}$)$_3$PPt(CO)$_2$]Re$_2$(CO)$_8$ (see Formula V) rapidly formed when (CO)$_{10}$Re$_2$ was combined with 1 equivalent (π-C$_2$H$_4$)$_2$PtP(C$_6$H$_{11}$)$_3$ in CH$_2$Cl$_2$. The compound was not isolated,

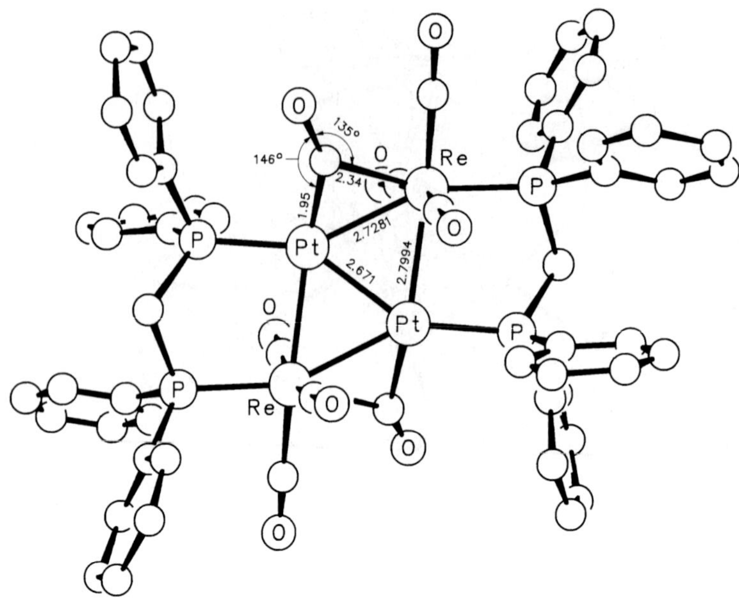

Fig. 33. Molecular structure of $[\mu\text{-}((C_6H_5)_2PCH_2P(C_6H_5)_2)_2Pt_2(CO)_2]Re_2(CO)_6$ [4].

since $[\mu\text{-}(CO)_4Pt_2(P(C_6H_{11})_3)_2]Re_2(CO)_6$ (see p. 161) formed with additional $(\pi\text{-}C_2H_4)_2\text{-}PtP(C_6H_{11})_3$.

IR spectrum (CH_2Cl_2): 1804, 1856 $(\nu(CO)_\mu)$ cm^{-1} [3].

$[\mu\text{-}((C_6H_{11})_3P)_2Pt_2(CO)_2(P(C_6H_5)_2)H]Re_2(CO)_6$ (see Formula VI) was obtained when a CH_2Cl_2 solution containing $(CO)_8Re_2(\mu\text{-H})(\mu\text{-}P(C_6H_5)_2)$ and 2.8 equivalents $(\pi\text{-}C_2H_4)_2PtP(C_6H_{11})_3$ was allowed to stand for 1 week. Subsequent addition of hexane and recrystallization of the precipitate from CH_2Cl_2/hexane gave orange-red plates with 49.3% yield. The reaction proceeds via $(CO)_8Re_2(\mu\text{-H})(\mu\text{-}P(C_6H_5)_2)PtP(C_6H_{11})_3$ ("Organorhenium Compounds" 5, 1994, p. 318) as shown spectroscopically.

^1H NMR spectrum (CD_2Cl_2): $\delta = -11.2$ (ReHPt; J(Pt,H) = 762, J(P,H) = 60 and 15 Hz) ppm. $^{31}P\{^1H\}$ NMR spectrum (CD_2Cl_2): $\delta = 171.0$ $(P(C_6H_{11})_3$; $^1J(Pt,P) = 3714$ and 4574, $^2J(Pt,P) = 168$ and 342, J(P,P) = 17, $J(P,P_\mu) = 6$ Hz), 248 $(P_\mu$; J(Pt,P) = 2389 Hz) ppm. IR spectrum (CH_2Cl_2): 1785, 1837, 1925, 1956, 1992, 2033 $(\nu(CO))$ cm^{-1}.

$P(C_6H_{11})_3$ H Pt O R_2 P $(CO)_3Re$ $Re(CO)_3$ Pt O O $P(C_6H_{11})_3$

$(R = C_3H_7\text{–}n)$

VII

References on p. 57

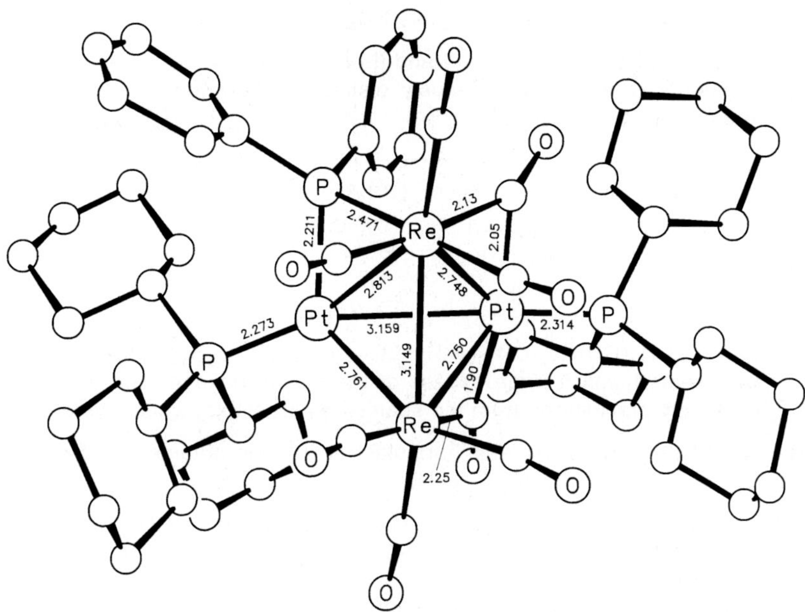

Fig. 34. Molecular structure of $[\mu\text{-}((C_6H_{11})_3P)_2Pt_2(CO)_2(P(C_6H_5)_2)H]Re_2(CO)_6$ [3].

Single crystals are monoclinic with a = 15.215(9), b = 19.633(9), c = 19.446(8) Å, β = 92.90(4)°; space group $P2_1/n-C^5_{2h}$ (No. 14); Z = 4 molecules per unit cell, D_{calc} = 1.984 g/cm³. The structure is illustrated in **Fig. 34** [3].

$[\mu\text{-}((C_6H_{11})_3P)_2Pt_2(CO)_3H]Re_2(CO)_6(\mu\text{-}P(C_3H_7\text{-}n)_2)$ (see Formula VII). This composition was tentatively assigned to a product obtained by reacting $(CO)_9Re_2P(C_3H_7\text{-}n)_2H$ with excess $(\pi\text{-}C_2H_4)_2PtP(C_6H_{11})_3$ in CD_2Cl_2. $(CO)_8Re_2(\mu\text{-}CO)(\mu\text{-}H)(\mu\text{-}P(C_3H_7\text{-}n)_2)PtP(C_6H_{11})_3$ (see "Organorhenium Compounds" 5, 1994, p. 395) was initially formed, but over a period of hours the title species could also be spectroscopically observed.

^{1}H NMR spectrum (CD_2Cl_2): δ = −11.0 (q; ^{1}J(Pt,H) = 759, ^{2}J(Pt,H) = 44, J(P,H) = 15 Hz) ppm. ^{31}P {^{1}H} NMR spectrum (CD_2Cl_2): δ = 73 ($P(C_6H_{11})_3$), 93 (P_μ) ppm; ^{1}J(Pt,PC_6H_{11}) = 4370, J(Pt,P_μ) = 52, ^{2}J(Pt,P) = 252 and 6, J(P,P) = 6 Hz [3].

References:

[1] Carriedo, G. A.; Jeffery, J. C.; Stone, F. G. A. (J. Chem. Soc. Dalton Trans. **1984** 1597/603).

[2] Kreiter, C. G.; Schufft, S.; Heckmann, G. (J. Organomet. Chem. **424** [1992] 163/72).

[3] Powell, J.; Brewer, J. C.; Gulia, G.; Sawyer, J. F. (J. Chem. Soc. Dalton Trans. **1992** 2503/16).

[4] Xiao, J.; Vittal, J. J.; Puddephatt, R. J. (J. Chem. Soc. Chem. Commun. **1993** 167/9).

[5] Peng, J.-J.; Peng, S.-M.; Lee, G.-H.; Chi, Y. (Organometallics **14** [1995] 626/33).

2.2.1.2.5 Compounds of the Type (μ-^{2}L)$Re_2(CO)_8$(μ-H)

The compounds depicted in Formulas I and II have either the structure a or b. Spectroscopic measurements cannot unambiguously distinguish between the structures. These complexes were prepared by Method I.

Ia Ib IIa IIb

Method I: (μ-$\eta^{2:1}$-CH_2=CH)$Re_2(CO)_8$(μ-H) [2] or (μ-$\eta^{2:1}$-$C_6H_5C{\equiv}C$)$Re_2(CO)_8$(μ-H) [1] was treated with a large excess of $P(CH_3)_3$ in hexane. The products instantaneously precipitated from the solution as air-stable, white crystalline solids.

(μ-$(CH_3)_3PCHCH_2$)$Re_2(CO)_8$(μ-H) (see Formulas Ia, Ib) was obtained by Method I when starting from (μ-$\eta^{2:1}$-CH_2=CH)$Re_2(CO)_8$(μ-H).

^{1}H NMR spectrum (toluene-d_8, 0°C): δ = −15.75 (dd, ReHRe; J(P,H) = 6.6 Hz), 0.29 (d, $P(CH_3)_3$; J(P,H) = 11.9 Hz), 0.71 (td, H-3), 1.26 (ddd, H-2), 1.55 (m, H-1) ppm; J(H-1,3) = 9.9, J(H-2,3) = 14.7, J(P,H-3) = 14.7, J(H-1,2) = 4.9, J(P,H-2) = 8.5, J(H-1,μ-H) = 2.7, J(P,H-1) = 4.4 Hz. IR spectrum (toluene): 1920, 1927, 1949, 1965, 1982, 1988, 2065, 2089 (ν(CO)) cm^{-1}. FD mass spectrum: $[M]^+$.

The compound slowly decomposes in toluene into the starting complex. With excess $P(CH_3)_3$, $(CO)_8Re_2(P(CH_3)_3)_2$ was obtained [2].

(μ-$(CH_3)_3PC{=}CC_6H_5$)$Re_2(CO)_8$(μ-H) (see Formulas IIa, IIb) was obtained by Method I when starting with (μ-$\eta^{2:1}$-$C_6H_5C{\equiv}C$)$Re_2(CO)_8$(μ-H). The compound is stable in benzene or toluene in the presence of $P(CH_3)_3$, but slowly decomposes in the absence of $P(CH_3)_3$.

^{1}H NMR spectrum (CD_2Cl_2, 0°C): δ = −15.39 (d, ReH; J(P,H) = 4.6 Hz), 1.35 (d, $P(CH_3)_3$; J(P,H) = 12.0 Hz); 6.79, 7.03, 7.25 (d, t, and t, C_6H_5) ppm. IR spectrum (toluene): 1920, 1940, 1974, 1986, 1993, 2067, 2091 (ν(CO)) cm^{-1}. FD mass spectrum: $[M]^+$ [1].

III IV

(μ-$CH_3C(H)CN(CH_3)_2$)$Re_2(CO)_8$(μ-H) (see Formula III) was obtained by exposing [μ-CH_3-$C((CH_3)_2N)C$=]$Re_2(CO)_8$ (see p. 123) to an H_2 atmosphere (10 atm, hexane, 70°C, 70 min). Separation of the mixture by preparative TLC (silica, hexane/CH_2Cl_2 (4:1)) gave the compound with 13% yield. The yield was lowered when the reaction time was lengthened. Another product was (μ-$CH_3CH{=}CHN(CH_3)_2$)$Re_2(CO)_8$ (see p. 117).

^{1}H NMR spectrum ($CDCl_3$): δ = −15.29 (ReH), 2.00 (d, CHC**H$_3$**), 3.51 (br s, NCH_3), 3.57 (q, **C**HCH_3; ^{3}J(H,H) = 6.6 Hz) ppm. IR spectrum (n-hexane): 1932, 1067, 1989, 2040, 2086 (ν(CO)) cm^{-1}.

Single-crystal X-ray diffraction revealed a monoclinic lattice with a = 9.384(2), b = 18.786(4), c = 10.188(1) Å, β = 100.98(1)°; space group $P2_1/n-C_{2h}^5$ (No. 14); Z = 4 mole-

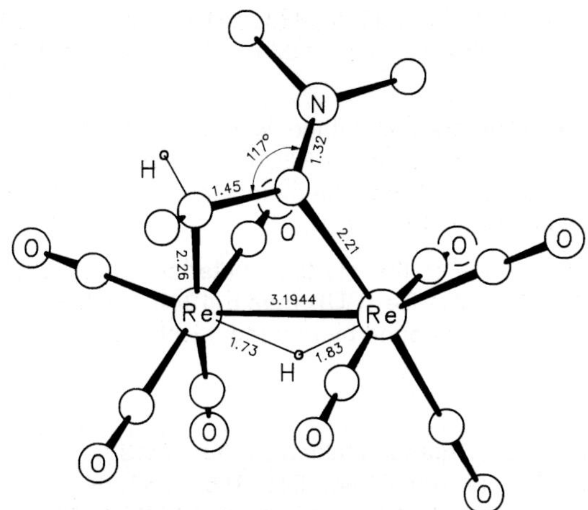

Fig. 35. Molecular structure of (μ-$CH_3C(H)CN(CH_3)_2$)$Re_2(CO)_8$(μ-H) [3].

cules per unit cell. The molecular structure is illustrated in **Fig. 35**. The N atom has a planar geometry, and the C-N bond has multiple bond character.

Thermolysis at 60 °C for 6 h initiated isomerization to (μ-$CH_3CH{=}CHN(CH_3)_2$)$Re_2(CO)_8$; by-products were not observed. The intramolecular pathway was confirmed by carrying out a crossover experiment using a mixture of (μ-$CH_3C(H)CN(CH_3)_2$)$Re_2(CO)_8$(μ-H) and (μ-$CH_3C(D)CN(CH_3)_2$)$Re_2(CO)_8$(μ-D). The product contained either two H or two D atoms [3].

(μ-$CH_2{=}CCN(CH_3)_2$)$Re_2(CO)_8$(μ-H) (see Formula IV) was obtained with only 0.5% yield by heating [μ-$CH_3C((CH_3)_2N)C{=}$]$Re_2(CO)_8$ (see p. 123) in C_6H_{12} for 2 h. Colorless solid.

^1H NMR spectrum ($CDCl_3$): δ = −15.80 (ReH); 3.42, 3.60 (s, NCH_3); 5.06, 5.25 (d's, CH_2; J = 3.0 Hz) ppm. $^{13}C\{^1H\}$ NMR spectrum ($CDCl_3$): δ = 41.3, 50.1 (NCH_3); 108.1(=CH_2);

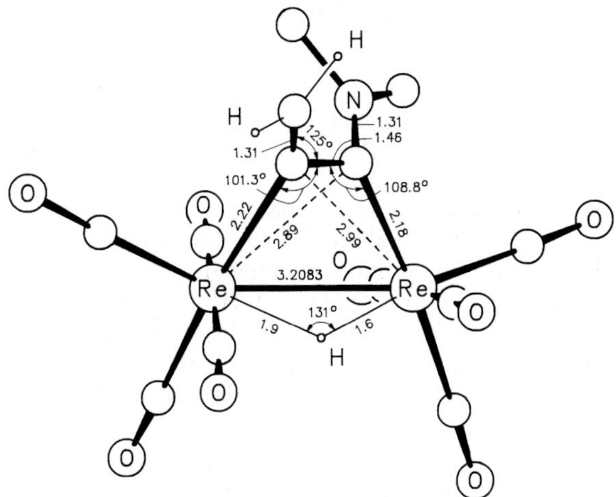

Fig. 36. Molecular structure of (μ-$CH_2{=}CCN(CH_3)_2$)$Re_2(CO)_8$(μ-H) [4].

References on p. 60

150.4 (C=**C**CN); 184.2, 185.3, 185.5, 185.9, 186.0, 186.1, 186.4, 187.9 (CO); 212.7 (Re=CN) ppm. IR spectrum (C_6H_{12}): 1522 (ν(CN)); 1959, 1966, 1979, 1992, 1998, 2012, 2082, 2105 (ν(CO)) cm^{-1}.

Single crystals have a monoclinic lattice with a = 9.268(2), b = 18.652(4), c = 10.276(2) Å, β = 101.24(1)°; space group $P2_1/n-C^5_{2h}$ (No. 14), Z = 4 molecules per unit cell, D_{calc} = 2.59 g/cm^3. **Fig. 36** illustrates the molecular structure. Within the bridging ligand the C-C-C-N torsion angle amounts to 78(2)°.

Thermolysis gave a mixture of the starting material, (μ-$CH_2C((CH_3)_2N)C$=)$Re_2(CO)_7$-(μ-H), and (μ-CH_2=$CH((CH_3)_2N)C$=)$Re_2(CO)_8$ (see p. 134). The heptacarbonyl product could be converted back to the title compound by treating it with CO at atmospheric pressure [4].

References:

[1] Nubel, P. O.; Brown, T. L. (Organometallics **3** [1984] 29/32).
[2] Nubel, P. O.; Brown, T. L. (J. Am. Chem. Soc. **106** [1984] 644/52).
[3] Adams, R. D.; Chen, G.; Yin, J. (Organometallics **10** [1991] 2087/8).
[4] Adams, R. D.; Chen, G.; Chi, Y.; Wu, W.; Yin, J. (Organometallics **11** [1992] 1480/6).

2.2.2 Compounds with σ,π-Bonded ^{2}L Ligands

2.2.2.1 Compounds with Alkenyl Ligands

2.2.2.1.1 Compounds of the Type (μ-$\eta^{2:1}$-CR_2=CR)$Re_2(CO)_8$(μ-H) and Their CO Substitution Derivatives

General. Structure. All compounds have one σ,π-bonded alkenyl ligand which bridges two Re atoms connected by a single bond. There is also a bridging H atom. Formula Ia illustrates the molecular geometry; however, the simplified representation in Formula Ib suffices for all practical purposes (note that in compounds No. 1 to 12 some CO groups are replaced by Re-bonded ^{2}D or ^{4}D ligands).

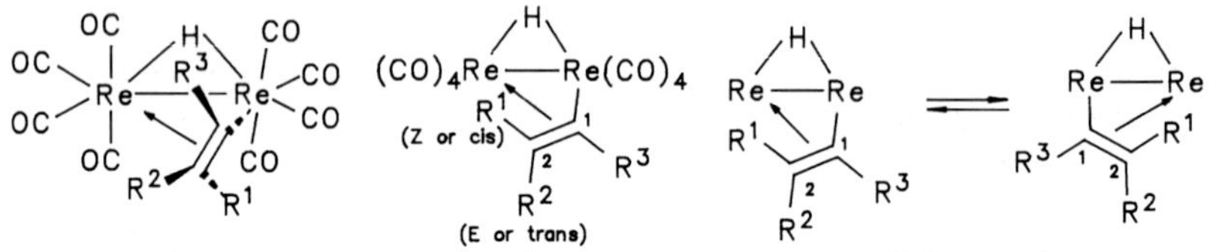

Ia Ib Scheme 1

Isomerism. Formula Ib contains several stereochemical descriptors used in the table. In most isolated compounds, R^1 (the substituent cis to the σ-bonded Re atom) and R^3 is H. Otherwise, R^2 (the substituent trans to the σ-bonded Re atom) represents a widely varying alkyl moiety. The opposite isomers, i.e. those with R^1 = alkyl and R^2 = H, could also be identified in several instances (at least in solution at low temperatures). When such isomerism occurs, Table 4 refers to these compounds as trans (R^1 = H) and cis (R^2 = H) isomers. The cis isomers in almost all cases appear to be thermodynamically unstable and transform to the respective trans isomers.

In only a few compounds (Nos. 26 to 33) is R^3 different from H. This is usually due to the presence of a σ,π-coordinated, cyclic alkene or polyalkene ligand; Nos. 26 and 27, however, are isomers of No. 17.

Preparation. The compounds were prepared by the following methods:

Method I: Photochemical reaction of $(CO)_8Re_2(\mu\text{-}R_2P(CH_2)_nPR_2)$ (n = 1, 2; R = CH_3, C_6H_5) with an alkene CH_2=CHR (R = H, CH_3) in toluene at room temperature (time of irradiation given in the table). Separation of the mixture by preparative TLC on silica with CH_2Cl_2/hexane.

In compounds with $R_2PCH_2PR_2$ the bridging diphosphane is retained, whereas in compounds with $R_2PC_2H_4PR_2$ (R = CH_3, C_6H_5) the diphosphane rearranges to a chelating, but no longer bridging ligand [12].

Method II: Photochemical reaction of

a. $(CO)_9Re_2PR_3$ (R = CH_3, C_6H_5)

b. $(CO)_8Re_2(PR_3)_2$ (R = CH_3, C_6H_5)

c. a 1:1 mixture of $(CO)_8Re_2(P(CH_3)_3)_2$ and $(CO)_8Re_2(P(C_6H_5)_3)_2$

d. a 1:1 mixture of $(CO)_8Re_2(PR_3)_2$ (R = CH_3 or C_6H_5) and $(CO)_{10}Re_2$

with C_2H_4 in toluene at room temperature. The resulting mixture contained compounds of the type $(\mu\text{-}\eta^{2:1}\text{-}CH_2\text{=}CH)Re_2(CO)_n(PR_3)_{8-n}(\mu\text{-}H)$ (n = 6 to 8); it was separated by preparative TLC on silica with hexane/CH_2Cl_2 mixtures [12].

Method III: Photochemical reaction of $(CO)_{10}Re_2$ with

a. an open-chain alkene of the type CH_2=CHR (R = H, CH_3, C_2H_5, C_4H_9-n, C_6H_5, CH=CH_2, C(CH_3)=CH_2) or cis-RCH=CHR (R = CH_3) in hexane or toluene [2 to 4, 8 to 10, 14].

b. a substituted allene (C_6H_5CH=C=CH_2 or $(CH_3)_2$C=C=CR_2 with R = H, CH_3) in hexane at −30 °C for 1 h [19, 24].

c. a 100-fold excess of a cylic alkene (C_6H_{10}, C_5H_6, C_6H_8, C_7H_8, C_8H_8, 6,6-dimethyl fulvene) in hexane at −20 °C [4, 11].

The reactions generally yielded multi-component mixtures (see the chapter "Chemical Behavior of $(CO)_{10}Re_2$" for the complete product pattern in "Organorhenium Compounds" 6). The mixtures were separated by column chromatography, preparative TLC, or HPLC.

Method IV: Photochemical treatment of $(\mu\text{-}\eta^{2:1}\text{-}CH_2\text{=}CH)Re_2(CO)_8(\mu\text{-}H)$ (No. 13) with methyl vinyl ether for 5 h. The products were only characterized by NMR spectroscopy [10].

Method V: Thermal treatment of $(\mu\text{-}\eta^{2:1}\text{-}CH_2\text{=}CH)Re_2(CO)_8(\mu\text{-}H)$ (No. 13) with the respective terminal alkene CH_2=CHR (R = C_4H_9-n, C_6H_5) [3, 9, 10].

NMR Spectroscopy. In cases where both cis (R^2 = R^3 = H) and trans (R^1 = R^3 = H) isomers exist, they can usually be distinguished by the magnitude of the vicinal coupling constant of the RCH=CH fragment: J(H-1,2) which for the trans isomers is in the range 14 to 16 Hz and for the cis isomers in the range 10 to 12 Hz [9, 12].

^{1}H and ^{13}C NMR spectroscopy at variable temperatures demonstrated a fluxional behavior. The changes in the spectra can be rationalized by a rapid σ-π interchange of the bridging alkenyl ligand (see Scheme 1). In some cases T_{coal} was determined and the activation energy calculated; their values are given in the table [2, 9, 10, 12, 14].

Chemical Behavior. The title compounds can be considered "stabilized $(CO)_8Re_2$ fragments", because the alkenyl ligand is only weakly bonded. Treatment with a large number of nucleophiles under mild conditions readily eliminates the alkene, producing other octacar-

References on p. 78

bonyl dirhenium complexes with good yield. The complex most frequently studied in this respect is (μ-$\eta^{2:1}$-n-C_4H_9CH=CH)$Re_2(CO)_8$(μ-H) (No. 19). Even several heterometal-containing clusters could be built by starting from this compound. The weak bonding of the alkenyl ligand becomes also obvious by the ease, with which the ligand can be exchanged by other alkenes or alkynes. Various studies including deuteration experiments relating to these exchange reactions have been undertaken (see "Further information").

Table 4
Compounds of the Type (μ-$\eta^{2:1}$-R^1R^2C=CR^3)$Re_2(CO)_8$(μ-H)
and Their CO Substitution Derivatives.
An asterisk indicates further information at the end of the table.
For explanations, abbreviations, and units see p. X.

No.	compound	method of preparation (yield) properties and remarks
	compounds of the type (μ-$\eta^{2:1}$-R^1CR^2=CR^3)$Re_2(CO)_6$(^4D)(μ-H)	
1	(μ-$\eta^{2:1}$-CH_2=CH)$Re_2(CO)_6$(($CH_3)_2PCH_2P(CH_3)_2$)(μ-H) $(CH_3)_2P$, $P(CH_3)_2$, H, $(CO)_3Re$—$Re(CO)_3$	I (48 h irradiation; yield: 40%) yellow solid ^{1}H NMR ($CDCl_3$): −14.17 (t; J(P,H) = 10.3); 1.20 (m, 1 H); 1.64, 1.71 (t's, CH_3; J(P,H) = 3.0 and 3.6, respectively); 1.78 (m, 1 H), 4.63 (dd, 1 H; J = 14.5, 3.3), 5.67 (m, 1 H), 6.02 (dd, 1 H; J = 11.0, 3.3) $^{13}C\{^1H\}$ NMR (CD_2Cl_2): 189.2 (s), 189.7 (t; J(P,C) + J(P′,C) = 67.0), 193.2 (s); all peaks 2 CO; (−90 °C): 187.5, 187.7, 190.8, 191.6, 192.9, 194.9 (d's, CO; J(P,C) = 54.5, 7.0, 7.5, 6.9, 61.8, 1.0, respectively); T_{coal} = −55 °C, thus $\Delta G^{\neq}$ ca. 41.0 kJ/mol IR (toluene): 1903, 1920, 1957, 2003, 2042 (ν(CO)) UV (CH_2Cl_2): λ_{max} (ε) = 297 (6800) FD MS: $[M]^+$ [12]
2	(μ-$\eta^{2:1}$-CH_3CH=CH)$Re_2(CO)_6$(($CH_3)_2PCH_2P(CH_3)_2$)(μ-H)	I (48 h irradiation gave cis and trans isomers in the ratio 1:1; they could not be separated from each other; combined yield: 30%) yellow solids [12]
	trans isomer $(CH_3)_2P$, $P(CH_3)_2$, H, $(CO)_3Re$—$Re(CO)_3$, 2, 1, CH_3	^{1}H NMR ($CDCl_3$): −13.76 (t; J(P,H) = 10.7); 1.20 (m, 1 H); 1.64, 1.71 (t's, PCH_3; J(P,H) = 3.3, 3.6, respectively); 1.80 (m, 1 H), 2.02 (d, CH_3; J(CH_3,H) = 4.5), 5.07 (dt, H-1), 5.50 (dq, H-2); J(P,H-1) = 10.6, ^{3}J(CH,CH) = 15.1 IR (toluene): 1902, 1919, 1952, 2001, 2037 (ν(CO)) FD MS: $[M]^+$ [12]

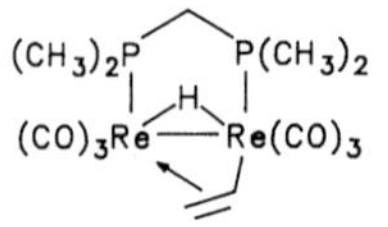

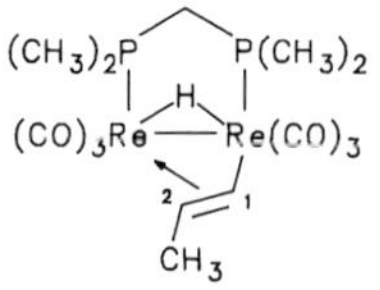

Table 4 (continued)

No.	compound	method of preparation (yield) properties and remarks
	cis isomer 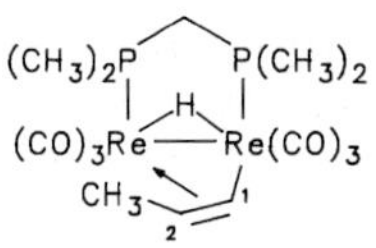	^1H NMR ($CDCl_3$): −13.53 (t, ReH; J(P,H) = 10.4); 2.34 (d, CH_3; J = 6.1), 5.23 (dt, H-1; J(H,H) = 11.6, J(P,H) = 14.9), 7.12 (dq, H-2) did not convert to the trans isomer upon standing at room temperature for 3 d [12]
3	(μ-$\eta^{2:1}$-CH_2=CH)$Re_2(CO)_6$(($CH_3)_2PC_2H_4P(CH_3)_2$)(μ-H) 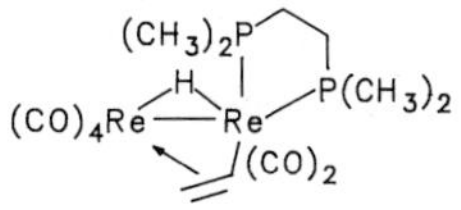	I (24 h irradiation; yield: 45%) yellow solid ^1H NMR ($CDCl_3$): −15.0 (t, ReH; J(P,H) = 15.0); 1.38, 1.70, 1.87, 1.88 (d's, CH_3; J(P,H) = 8.0, 8.1, 9.4, 9.4, respectively); 3.47 (t, 1 H; J(P,H) = 13.1, J(H,H) = 13.7), 4.90 (d, 1 H; J(H,H) = 9.5), 6.79 (dd, 1 H) $^{13}C\{^1H\}$ NMR ($CDCl_3$): 183.6, 185,5, 189.5, 190.6 (s's); 194.9 (t; J = 7.4), 196.5 (dd; J = 50.8, 9.5) IR (toluene): 1880, 1931, 1944, 1961, 1975, 2010, 2075 (ν(CO)) FD MS: $[M]^+$ [12]
4	(μ-$\eta^{2:1}$-CH_3CH=CH)$Re_2(CO)_6$(($CH_3)_2PC_2H_4P(CH_3)_2$)(μ-H) 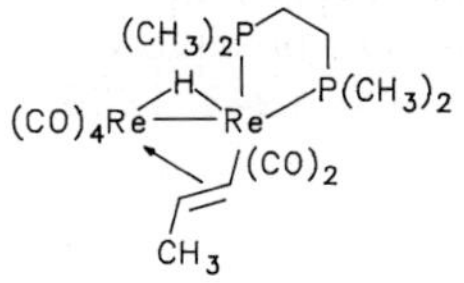	I (24 h irradiation; yield: 30%); no cis isomer detected as in the case of No. 2 yellow solid ^1H NMR ($CDCl_3$): −14.76 (t, ReH; J(P,H) = 14.9); 1.38, 1.66 (d's, CH_3; J(H,H) = 8.1); 1.87 (m, 6 H), 1.93 (d, CH_3; J(H,H) = 5.4), 4.14 (ddq, 1 H; J(H,H) = 13.3, 5.4, J(P,H) = 12.5), 6.38 (d, 1 H) IR (toluene): 1856, 1877, 1930, 1944, 1958, 1976, 2020, 2071 (ν(CO)) FD MS: $[M]^+$ [12]
5	(μ-$\eta^{2:1}$-CH_2=CH)$Re_2(CO)_6$(($C_6H_5)_2PC_2H_4P(C_6H_5)_2$)($\mu$-H) 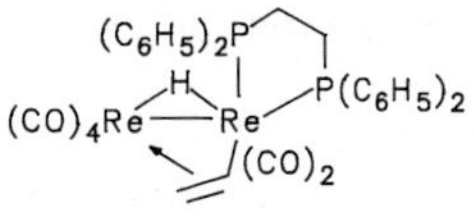	I (9 h irradiation; yield: 50%) yellow solid ^1H NMR ($CDCl_3$): −14.11 (t; J(P,H) = 12.8); 2.54, 2.77 (m's, CH_2); 3.66 (dd; J(H,H) = 13.9, J(P,H) = 10.2), 5.03 (d; J(H,H) = 9.5), 6.86 (dd); 7.24 to 7.78 (m, C_6H_5) $^{13}C\{^1H\}$ NMR ($CDCl_3$): 184.8, 185.9, 187.7, 190.2 (s's); 194.1 (d; J = 5.9), 194.9 (dd; J = 53.4, 10.0) IR (toluene): 1892, 1931, 1955, 1968, 1978, 2026, 2078 (ν(CO)) FD MS: $[M]^+$ [12]

References on p. 78

Table 4 (continued)

No.	compound	method of preparation (yield) properties and remarks
6	$(\mu\text{-}\eta^{2:1}\text{-}CH_3CH{=}CH)Re_2(CO)_6((C_6H_5)_2PC_2H_4P(C_6H_5)_2)(\mu\text{-}H)$ 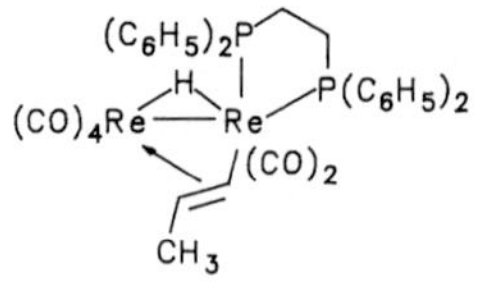	I (9 h irradiation; yield: 38%); in contrast to No. 2, only the trans isomer was identified yellow solid ^{1}H NMR ($CDCl_3$): −13.93 (t; J(P,H) = 11.6); 1.27 (d, CH_3; J = 5.3); 2.55 to 2.85 (m, CH_2); 4.16 (ddq, H-1; J(H,H) = 14.0, 5.3, J(P,H) = 10.6), 6.40 (d, H-2); 7.30 to 7.68 (m, C_6H_5) IR (toluene): 1888, 1930, 1967, 1979, 2026, 2077 (ν(CO)) FD MS: $[M]^+$ [12]

compounds of the type $(\mu\text{-}\eta^{2:1}\text{-}CH_2{=}CH)Re_2(CO)_6(^2D)_2(\mu\text{-}H)$

No.	compound	method of preparation (yield) properties and remarks
7	$(\mu\text{-}\eta^{2:1}\text{-}CH_2{=}CH)Re_2(CO)_6(P(CH_3)_3)_2(\mu\text{-}H)$ 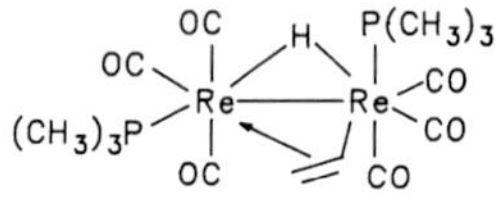	IIa (3%); IIb (66%); IIc (17%); IId (7.6%) ^{1}H NMR ($CDCl_3$): −14.01 (dd, ReH; J(P,H) = 19.0, 11.1); 4.04 (dd; J(H,H) = 15.8, 3.2), 5.80 (dd; J(H,H) = 9.6, 3.2), 6.21 (ddd; J(H,H) = 15.8, 9.6; J(P,H) = 11.0); several other peaks due to unidentified side products were also observed IR (toluene): 1892, 1920, 1943, 1999, 2085 (ν(CO)) FD MS: $[M]^+$ [12]
8	$(\mu\text{-}\eta^{2:1}\text{-}CH_2{=}CH)Re_2(CO)_6(P(C_6H_5)_3)(P(CH_3)_3)(\mu\text{-}H)$	IIc (19%) ^{1}H NMR ($CDCl_3$): −12.42 (dd; J(P,H) = 25.6, 12.1) FD MS: $[M]^+$ [12]
9	$(\mu\text{-}\eta^{2:1}\text{-}CH_2{=}CH)Re_2(CO)_6(P(C_6H_5)_3)_2(\mu\text{-}H)$ 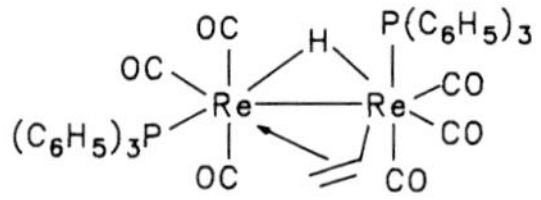	IIa (24%); IIb (65%); IIc (24%); IId (17%) ^{1}H NMR ($CDCl_3$): −11.99 (dd; J(P,H) = 21.8, 11.2); 6.59 to 6.65 (m, 2 H), 6.80 (t, 1 H; J = 6.8); 7.16 to 7.55 (m, C_6H_5); several other peaks due to minor products were also observed $^{13}C\{^1H\}$ NMR ($CDCl_3$): 183.4, 183.9 (d's; J(P,C) = 7.3 and 5.9); 185.1 (s); 192.2 (d, C trans to P), 193.9 (d), 194.2 (t; J(P,C) = 57.0, 9.0, 6.1, respectively) IR (toluene): 1903, 1925, 1963, 1995, 2014, 2096 (ν(CO)) FD MS: $[M]^+$

References on p. 78

Table 4 (continued)

No.	compound	method of preparation (yield) properties and remarks
		photolysis gave but-1-ene and trans-but-2-ene [12]
compounds of the type (μ-$\eta^{2:1}$-R^1CR^2=CR^3)$Re_2(CO)_7(^2D)(\mu$-H)		
*10	(μ-$\eta^{2:1}$-CH_2=CH)$Re_2(CO)_7(P(CH_3)_3)(\mu$-H) (mixture of isomers)	IIa (60%); IIb (17%); IIc (6.7%); IId (46%, 57%, or 63%, depending on the ratio of the starting compounds) [12]; see also "Further information"
	IIa, IIb, IIc, IId	
	isomer A (see Formula IIa)	yield: 15% ^{1}H NMR ($CDCl_3$): −14.75 (d; J(P,H) = 18.8); 4.12 (dd; J(H,H) = 15.6, 2.2), 5.77 (dd; J(H,H) = 11.2, 2.2), 6.58 (dd; J(H,H) = 15.6, 11.2) [12]
	isomer B (see Formula IIb)	yield: 18% ^{1}H NMR ($CDCl_3$): −14.54 (d; J(P,H) = 13.6); 3.93 (dd; J(H,H) = 16.7, 3.3), 5.80 (dd; J(H,H) = 10.7, 3.3), 7.41 (dd; J(H,H) = 16.7, 10.7) [12]
	isomer C (see Formula IIc)	yield: 18% ^{1}H NMR ($CDCl_3$): −14.54 (d; J(P,H) = 13.6); 3.88 (dd; J(H,H) = 16.7, 3.3), 5.80 (dd; J(H,H) = 10.7, 3.3), 7.41 (dd; J(H,H) = 16.7, 10.7) [12]
	isomer D (see Formula IId)	yield: 13% ^{1}H NMR ($CDCl_3$): −14.08 (d; J(P,H) = 8.6); 4.61 (dd; J(H,H) = 17.2, 3.1), 6.23 (dd; J(H,H) = 11.2, 3.1), 6.60 (dd; J(H,H) = 17.2, 11.2) [12]
*11	(μ-$\eta^{2:1}$-CH_2=CH)$Re_2(CO)_7(P(C_6H_5)_3)(\mu$-H) (mixture of isomers)	IIa (44%); IIb (15%); IIc (8.4%); IId (55%) [12]; see also "Further information"
	isomer A (main isomer)	^{1}H NMR (CD_2Cl_2): −13.55 (d; J(P,H) = 8.5); 4.55, 6.18, 6.26 (dd's); 7.48 (m, C_6H_5); J(H-1,2) = 16.7, J(H-1,3) = 11.0, J(H-2,3) = 4.0, J(P,H-1) = 11; (−70 °C): −13.82 and −13.75 with J(P,H) = 8.4 and 7.1, respectively, ratio 5:4; T_{coal} = −30°C, i.e. $\Delta G^{\neq}$ = 51 kJ/mol; also: 4.55 peak splits into 4.35 (dd; J(H-1,2) = 16.9, J(H-2,3) = 3.0) and 4.61 (dd; J(H-1,2) = 17.3,

References on p. 78

Table 4 (continued)

No.	compound	method of preparation (yield) properties and remarks
*11	(continued)	$J(H\text{-}2,3) = 2.4$) in the ratio 5:4; $T_{coal} = -15\,°C$, i.e. $\Delta G^{\neq} = 51.5$ kJ/mol [10]
	isomer B	^{1}H NMR (CD_2Cl_2): −13.87 (d; J(P,H) = 16.6); 4.25, 5.93 (dd's); J(H-1,2) = 15.3, J(H-2,3) = 2, J(H-1,3) = 10.1 [10]
	isomer C	^{1}H NMR (CD_2Cl_2): −13.95 (d; J(P,H) = 11.5); 3.64 (m), 5.78 (dd), 6.71 (m); J(H-1,2) = 10.7, J(H-2,3) = 3.5 [10]
12	(μ-$\eta^{2:1}$-C_2H_5CH=CH)$Re_2(CO)_7(NC_5H_5)$(μ-H)	by irradiating No. 17 in the presence of pyridine, two isomers formed (no details) [3]

compounds of the type (μ-$\eta^{2:1}$-R^1CR^2=CR^3)$Re_2(CO)_8$(μ-H)

No.	compound	method of preparation (yield) properties and remarks
*13	(μ-$\eta^{2:1}$-CH_2=CH)$Re_2(CO)_8$(μ-H) 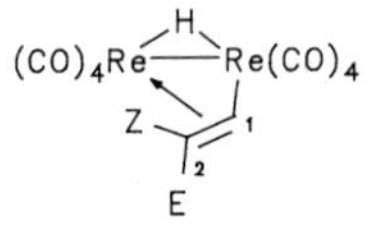	IIa (R = CH_3: 24%, C_6H_5: 20%) [12]; IId (16% for both R = CH_3 or C_6H_5; 41% when employing a reactant ratio of 2:3 and R = CH_3) [12]; IIIa [3, 4, 9, 14] (at −50 °C, 80 min; yield: 40.5% [4, 14]; 70 to 80% [9]; main product, side products were Nos. 17 and 26 (ca. 10 to 15% each) [3, 9]) pale yellow crystals [3, 4, 9, 10, 14]; air-stable [9]; sublimes at 70 °C/0.1 mm Hg [9, 10] ^{1}H NMR ($CDCl_3$): −14.72 (s, ReH); 4.46 (dd, H-2Z), 6.26 (dd, H-2E), 7.18 (dd, H-1); J(H-1,2Z) = 17.2, J(H-1,2E) = 10.9, J(H-E,Z) = 2.9 [3, 9, 10]; unchanged from −83 to +25 °C [9] (similar in [4, 14]) ^{13}C NMR (C_6D_6): 81.8 (t; J(C,H) = 158), 129.3 (d; J(C,H) = 138) [14]; 181.9, 183.0, 183.8, 185.1 (CO) [9] (δ(CO) also in [14]) IR (n-hexane): 1970, 1978, 1988, 1998, 2021, 2085, 2117 (ν(CO)) [3, 9, 10] (nearly identical in [14]) UV (C_6H_{12}): λ_{max} (ε) = 285 (sh), 328 (3200) [10] MS: $[M]^+$ [10]
14	(μ-$\eta^{2:1}$-CH_3OCH=CH)$Re_2(CO)_8$(μ-H)	IV (main product) ^{1}H NMR (CD_2Cl_2): −14.19 (s, ReH), 3.69 (s, OCH_3); 5.10, 6.91 (d's, OCH=CH; J = 13.6) [10]
15	(μ-$\eta^{2:1}$-n-C_4H_9OCH=CH)$Re_2(CO)_8$(μ-H)	from No. 19 and excess n-butyl vinyl ether (heptane, 50 h); isolated by TLC with hexane air-stable, colorless oil

Table 4 (continued)

No.	compound	method of preparation (yield) properties and remarks
		^{1}H NMR ($CDCl_3$): −14.21 (s, ReH), 0.96 (t, CH_3; J = 13.6); 1.44, 1.72 (m's, CH_2); 3.86 (t, OCH_2); 5.04, 6.85 (d's, H-1 and H-2) IR (n-hexane): 1964, 1970, 1994, 2005, 2014, 2083, 2112 (ν(CO)) EI MS: $[M]^+$ reaction with $P(C_6H_5)_3$ gave $(CO)_8Re_2(P(C_6H_5)_3)_2$ [9]
*16	(μ-$\eta^{2:1}$-CH_3CH=CH)$Re_2(CO)_8$(μ-H) trans isomer $(CO)_4Re$—(μ-H)—$Re(CO)_4$, bridging C(1)H=C(2)H–CH_3	IIIa (>90% [3], 70 to 80% [8, 9, 10]); isolated with CH_2Cl_2/hexane (1:4) and by subsequent sublimation (0.1 mm Hg, 50°C) [8, 10] pale yellow oil [9, 10]; air-stable over several hours but decomposed within several weeks even under an inert atmosphere [9] ^{1}H NMR ($CDCl_3$): −14.42 (s, ReH), 2.20 (d, CH_3), 5.30 (dq, H-2), 6.51 (d, H-1); J(H-1,2) = 15.9, J(H-2,CH_3) = 5.4 [3, 9, 10] IR (n-hexane): 1967, 1975, 1979, 1994, 2017, 2083, 2114 (ν(CO)) [3, 8 to 10] EI MS: $[M]^+$ [3, 9, 10]
	cis isomer $(CO)_4Re$—(μ-H)—$Re(CO)_4$, bridging C(1)H=C(2)H–CH_3 (cis)	IIIa (when immediately cooled after photolysis); not isolated ^{1}H NMR (CD_2Cl_2, −10°C): −14.26 (s, ReH); 2.25 (d, CH_3), 6.82 (d, H-1); 7.37 (dq, H-2); J(H-1,2) = 11.9, J(CH_3,H-2) = 6.5 isomerized to the trans isomer with a half-life time of ca. 10 min [9, 10]
*17	(μ-$\eta^{2:1}$-C_2H_5CH=CH)$Re_2(CO)_8$(μ-H) trans isomer $(CO)_4Re$—(μ-H)—$Re(CO)_4$, bridging C(1)H=C(2)H–C_2H_5	IIIa (>90% [3], 70 to 80% [9, 10]); yield: only 5.3% when performing the reaction at −50 to −20°C (other products were $C_2H_4Re_2(CO)_9$ and $(CO)_{14}Re_3H$) [4, 14]; IIIa (when employing trans-but-2-ene (!); only product) [3, 9] pale yellow oil at room temperature [4, 9, 10, 14]; air-stable for several hours, but even the pure complex very slowly decomposed under an inert atmosphere [9] ^{1}H NMR ($CDCl_3$): −14.45 (s, ReH), 1.14 (t, CH_3), 2.29 (qd, CH_2), 5.36 (dt, H-2), 6.55 (d, H-1); J(H-1,2) = 16.1, J(H-2,CH_2) = 5.2, J(CH_2,CH_3) = 7.3 [3, 9, 10] (similar in [4]); decoupling of the CH_3 group and cooling caused the δ(CH_2) signal to split (T_{coal} = −55°C, $\Delta G^{\neq}$ = 41.8 kJ/mol) [9]

References on p. 78

Table 4 (continued)

No.	compound	method of preparation (yield) properties and remarks
*17 (continued)		^{13}C NMR (C_6D_6): 182.3, 183.4, 184.4, 185.4 (CO) [9] IR (n-hexane): 1967, 1976, 1979, 1995, 2017, 2083, 2114 (ν(CO)) [3, 9, 10] UV (C_6H_{12}): λ_{max} (ε) = 284 (sh), 328 (3000) [10] EI MS: $[M]^+$ [10]
	cis isomer 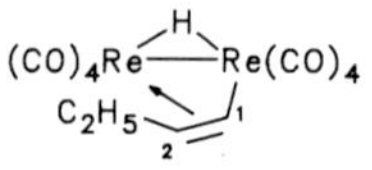	IIIa (when immediately cooled after irradiation); not isolated [9] ^1H NMR (CD_2Cl_2, −10 °C): −14.32 (s, ReH), 1.17 (t, CH_3), 2.22 (qui, CH_2), 6.82 (d, H-1), 7.22 (dt, H-2); J(H-1,2) = 11.8, J(CH_3,CH_2) = 7.2, J(CH_2,H-2) = 7.2 [9, 10] completely isomerized to the trans isomer with a half-life time of ca. 10 min [9]
18	(μ-$\eta^{2:1}$-$CH_3OC_2H_4CH$=CH)$Re_2(CO)_8$(μ-H)	
	trans isomer 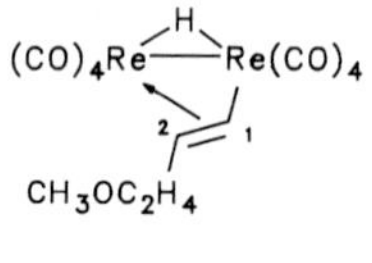	IV (minor product) ^1H NMR (CD_2Cl_2): −14.40 (s, ReH), 2.51 (q, CH**CH_2**; J(H-2,CH_2) = 6), 3.45 (s, OCH_3), 3.65 (t, OCH_2; J(CH_2,CH_2) = 6 to 6.5), 5.29 (dt, H-2; J(H-1,2) = 16.1), 6.71 (d, H-1) [10]
	cis isomer 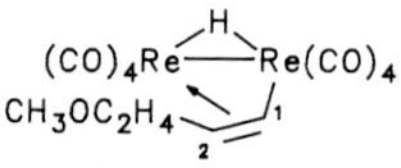	IV (^1H NMR-spectroscopically observed as intermediate; see "Further information" to No. 13) ^1H NMR (CD_2Cl_2, −20 °C): −14.35 (s); peak diminished upon warming with concomitant increase in the intensity of the trans isomer [10]
*19	(μ-$\eta^{2:1}$-n-C_4H_9CH=CH)$Re_2(CO)_8$(μ-H)	IIIa (>90%) [3, 9]; V (quantitative) [3] pale yellow oil, air-stable even in solution for several hours, but the pure compound decomposed within several weeks even under an inert atmosphere [9] ^1H NMR ($CDCl_3$): −14.44 (s, ReH), 0.96 (t, CH_3), 1.4 to 1.6 (m, C_2H_4), 2.19 (td, **CH_2**CH), 5.27 (dt, H-2), 6.51 (d, H-1); J(H-1,2) = 16.0, J(H-2,CH_2) = 5.8, J(CH_2,CH_3) = 7.3 [3, 9] IR (n-hexane): 1967, 1976, 1979, 1994, 2017, 2083, 2114 (ν(CO)) [3, 9] EI MS: $[M]^+$ [9]
20	(μ-$\eta^{2:1}$-C_6H_5CH=CH)$Re_2(CO)_8$(μ-H)	
	trans isomer 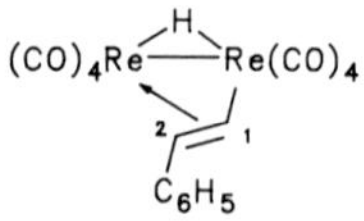	IIIa (38.7%) [4, 14]; V (quantitative) [9, 10] pale yellow oil [9, 10]; but: yellow needles [4, 14] ^1H NMR (CD_2Cl_2): −14.16 (s, ReH), 6.22 (d, H-1 or H-2); 7.28 to 7.34, 7.41 to 7.43 (m's, C_6H_5);

References on p. 78

Table 4 (continued)

No.	compound	method of preparation (yield) properties and remarks
		7.53 (d, H-2 or H-1; J(H-1,2) = 16.6) [9, 10]; similar in [4, 14] ^{13}C NMR: 110.7, 117.6 (C-2, C-1; J(C,H) = 156, 135); 127.6, 130.5, 130.7 (C_6H_5; J(C,H) = 154, 160, 160, respectively), 143.1 (C_{ipso}); 183.6, 185.3, 186.1, 187.4 (CO) [14] IR (n-hexane): 1969, 1974, 1988, 1996, 2017, 2083, 2113 (ν(CO)) [9, 10]; very similar in [14] EI MS: $[M]^+$ [9, 10]
	cis isomer 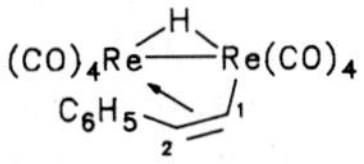 $(CO)_4Re$—H—$Re(CO)_4$, C_6H_5, 1, 2	IIIa (at −50 to −20 °C; yield: 12.8%) [4, 14] light yellow needles [4, 14] ^{1}H NMR (CD_2Cl_2): −13.88 (s), 7.00 (d, CH=CH), 7.23, 7.34 (t, H_γ and H_β of C_6H_5), 7.61 (d, H_α of C_6H_5; J(H-α,β) = J(H-β,γ) = 7.5), 8.76 (d, CH=CH; J(H,H) = 12.3) [10]; see also [4, 14] ^{13}C NMR: 112.3, 120.1 (C-2, C-1; J(C,H) = 155, 133); 129.7, 130.9, 131.9 (C_6H_5; J(C,H) = 162, 161, 159, respectively); 140.7 (C_{ipso}); 184.7, 185.4, 186.8 (CO) [14] IR (n-hexane): 1964, 1978, 1991, 2005, 2015, 2080, 2110 (ν(CO)) [14] isomerized to the trans isomer in the dark with $t_{1/2}$ of ca. 2.5 h (CH_2Cl_2) [10, footnote No. 30] cis and trans isomers at room temperature form an equilibrium with trans:cis = 8:1 [14]
21	(μ-$\eta^{2:1}$-CH_2=CHCH=CH)$Re_2(CO)_8$(μ-H) 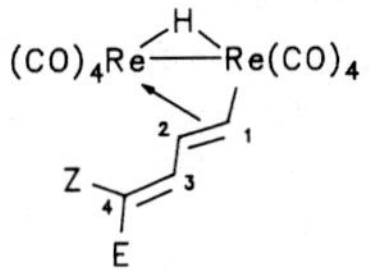 $(CO)_4Re$—H—$Re(CO)_4$, 1, 2, 3, 4, Z, E	IIIa (yield after 20 min: 61% at −35 °C, 37% at −85 °C) pale yellow needles ^{1}H NMR (toluene-d_8, 0 °C): −14.43 (ReH); 4.77, 4.83, 5.36, 5.56, 6.14; J(H-1,2) = 15.0, J(H-2,3) = 9.0, J(H-3,4E) = 9.0, J(H-3,4Z) = 16.0, J(H-4E,4Z) = 1.0 ^{13}C {^{1}H} NMR (toluene-d_8, 0 °C): 111.17, 117.51, 120.49, 144.01 (C-2,4,1,3; J(C,H) = 155, 160, 138, 155, respectively); 182.13, 183.42, 184.63, 185.45 (CO, all 2 C); unchanged at −80 °C IR (n-hexane): 1967, 1975, 1985, 1995, 2015, 2080, 2110 (ν(CO)) rearranged into (μ-$\eta^{2:2}$-C_4H_6)$Re_2(CO)_8$ (see p. 141) in polar solvents [2]

References on p. 78

Table 4 (continued)

No.	compound	method of preparation (yield) properties and remarks
22	(μ-$\eta^{2:1}$-CH_2=C(CH_3)CH=CH)$Re_2(CO)_8$(μ-H) 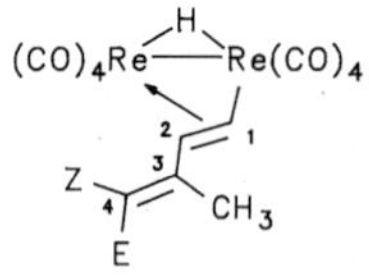	IIIa (at −80 °C; yield: 42.3%) [14]; IIIb (starting from $(CH_3)_2C$=C=CH_2; yield: 7%) [24]; also by photolyzing (μ-$(CH_3)_2$CCHCH)$Re_2(CO)_9$ (see p. 133) (hexane, 2 h; yield: 90%) [5, 16] low-melting [5, 16], light yellow crystals [5, 14, 16]; yellow needles [24] ^{1}H NMR (CD_2Cl_2): −14.26 (ReH), 1.75 (CH_3); 5.25, 5.34 (H-4Z,4E; J = 0.6); 6.02, 6.70 (H-2, H-1; J = 16.6) [14]; −14.2 (s); 1.78 (CH_3); 5.35, 6.06, 6.13, 6.74 (d's, H-4Z,2,4E,1; J = 3.2, 16.5, 3.2, 16.5, respectively) [24]; in solution the compound exists as two rotameric species in the ratio 2:1; minor isomer: −14.02, 1.07; 5.28, 6.70 (d's; J = 12.2); 7.50 (m, CH_2); major isomer: −14.28, 1.67; 6.00, 6.68 (d's; J = 16.4); 7.50 (m, CH_2) [5, 16] ^{13}C NMR: 16.7 (CH_3); 115.1, 115.7, 118.6 (C-1,2,4; J(C,H) = 127, 135, 152, 160, respectively); 146.7 (C-3); 182.9, 184.3, 185.2, 186.2 (CO) [14] IR (n-hexane): 1967, 1979, 1994, 2015, 2020, 2080, 2110 [14] (similar in [5, 16]); 1970, 1976, 1984, 1992, 2013, 2089, 2124 (ν(CO)) [24] thermolysis gave (μ-$\eta^{2:2}$-CH_2=CHC(CH_3)=CH_2)$Re_2(CO)_8$ [24]
23	(μ-$\eta^{2:1}$-$(CH_3)_2$C=CHCH=CH)$Re_2(CO)_8$(μ-H) 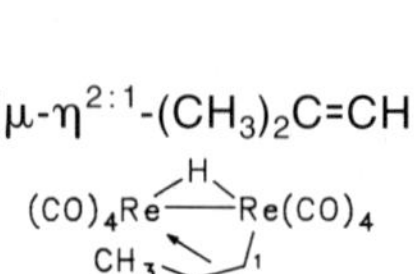	IIIb (starting from $(CH_3)_2C$=C=$CHCH_3$; yield: 5%) yellow needles from pentane ^{1}H NMR (CD_2Cl_2): −14.18 (s); 1.89, 1.92 (s's, CH_3-4); 5.60 (d, H-3), 6.18 (dd, H-2), 6.58 (d, H-1); J(H-1,2) = 16.0, J(H-2,3) = 9.7 IR (n-hexane): 1968, 1975, 1982, 1993, 2013, 2086, 2125 (ν(CO)) thermolysis yielded (μ-$\eta^{2:2}$-CH_2=CHCH=C$(CH_3)_2$)$Re_2(CO)_8$ [24]
24	(μ-$\eta^{2:1}$-$(CH_3)_2$C=CH(CH_3)C=CH)$Re_2(CO)_8$(μ-H) 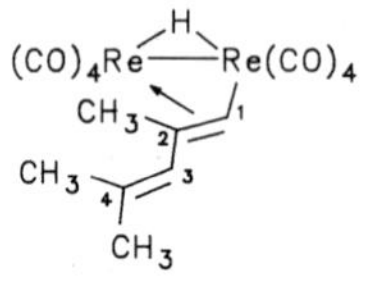	IIIb (starting from $(CH_3)_2C$=C=C$(CH_3)_2$; yield: 13%) yellow crystals from pentane ^{1}H NMR (CD_2Cl_2): −13.56 (ReH); 1.75, 1.94 (d's, CH_3-4; J = 0.6); 2.44 (CH_3-2), 6.11 (br s, H-3), 6.52 (H-1)

Table 4 (continued)

No.	compound	method of preparation (yield) properties and remarks
		IR (n-hexane): 1966, 1978, 1982, 1991, 2009, 2083, 2118 (ν(CO)) thermolysis yielded (μ-$\eta^{2:2}$-CH_2=C(CH_3)CH=C(CH_3)$_2$)$Re_2(CO)_8$ [24]
25	(μ-$\eta^{2:1}$-C_6H_5CH=C=CH)$Re_2(CO)_8$(μ-H) 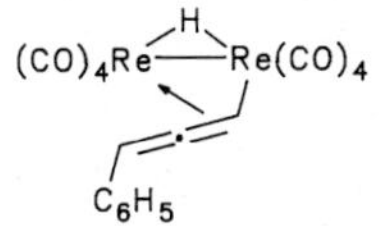	IIIb (starting from C_6H_5CH=C=CH_2; yield: 12%) yellow crystals from pentane ^{1}H NMR: −14.16 (ReH); 6.22, 7.52 (d's, CH=; J = 16.5); 7.51 (m, C_6H_5) IR (n-hexane): 1963, 1971, 1982, 1989, 2024, 2088, 2117 (ν(CO)) thermolysis gave (μ-CH_2=C=CHC_6H_5)$Re_2(CO)_8$ (see p. 131) [19]
26	(μ-$\eta^{2:1}$-CH_2=CC_2H_5)$Re_2(CO)_8$(μ-H) 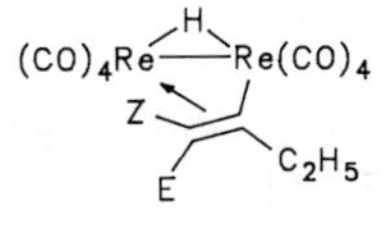	IIIa (when using C_2H_4 (!), yield: 10 to 15%; main product was No. 13, other by-product: No. 17 (cis and trans isomer)) [3, 9]; also by the photochemical reaction of No. 13 with C_2H_4 [3, 10] (main product after short irradiation times) [10] ^{1}H NMR ($CDCl_3$): −14.42 (s, ReH), 1.31 (t, CH_3), 2.88 (br q, CH_2); 4.06, 5.76 (d's, H-E, H-Z); J(H-E,Z) = 2.2, J(CH_2,CH_3) = 7.4 [3, 9]; decoupling of the CH_3 group causes broadening and, on cooling to −40 °C, splitting of the CH_2 resonance: δ = 2.50, 3.06 (d's; J_{gem} = 11.3) (T_{coal} = 1 °C, $\Delta G^{\neq}$ = 53.1 kJ/mol) [9] isomerized to No. 17 in the dark; $t_{1/2}$ ca. 10 h slowly converted to No. 13 under C_2H_4 [9]
27	(μ-$\eta^{2:1}$-CH_3CH=CCH_3)$Re_2(CO)_8$(μ-H) 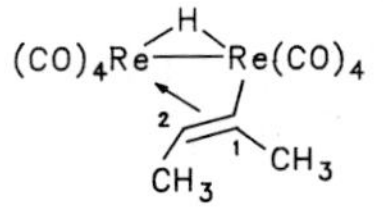	IIIa (ca. 20% (by ^{1}H NMR)) [3, 9]; also from No. 16 and cis-but-2-ene within 2 d, yield: 50% [9] ^{1}H NMR (CD_2Cl_2): −13.99 (s, ReH), 2.04 (d, CH_3-2), 2.99 (s, CH_3-1), 5.71 (q, CH); J(CH,CH_3) = 5.8 [3, 9] IR (n-hexane): 1966, 1978, 1992, 2016, 2083, 2113 (ν(CO)) photochemically isomerized to No. 17 [3, 9]

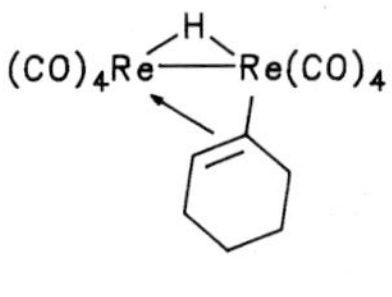

III 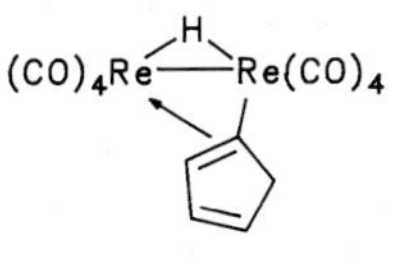 IVa 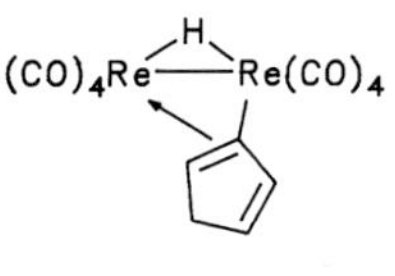 IVb 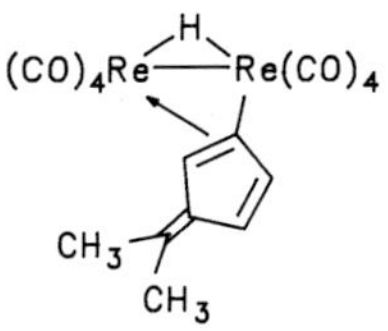 V

References on p. 78

Table 4 (continued)

No.	compound	method of preparation (yield) properties and remarks
28	(μ-$\eta^{2:1}$-cyclo-C_6H_9)$Re_2(CO)_8$(μ-H) (see Formula III)	IIIc (41.2%; only product of the reaction) pale yellow needles ^{1}H NMR (CD_2Cl_2): −13.72 (s); 1.67, 2.52, 3.45, 5.63 (m's, 4, 2, 2, and 1 H) [4]
29	(μ-$\eta^{2:1}$-C_5H_5)$Re_2(CO)_8$(μ-H)	IIIc; 2 isomers were obtained both isomers form colorless needles [4]
	isomer A (see Formula IVa)	yield: 5% ^{1}H NMR (CD_2Cl_2): −13.79 (s); 3.97 (m, CH_2); 6.15, 6.71, 6.79 (m's, CH) [4]
	isomer B (see Formula IVb)	yield: 42% ^{1}H NMR (CD_2Cl_2): −13.87 (s); 3.47 (m, CH_2); 5.47, 6.28, 6.99 (m's, CH) [4]
30	(μ-$\eta^{2:1}$-C_8H_9)$Re_2(CO)_8$(μ-H) (see Formula V)	IIIc (7.4%) yellow spherical crystals ^{1}H NMR (CD_2Cl_2): −13.73 (s); 2.23, 2.34 (s's, CH_3); 5.63 (d); 6.43, 6.96 (dd's, CH) [4]

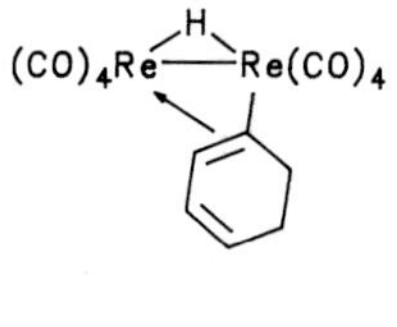

VIa 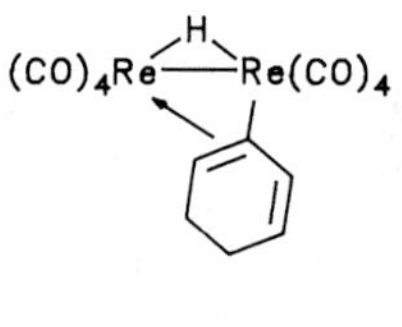 VIb 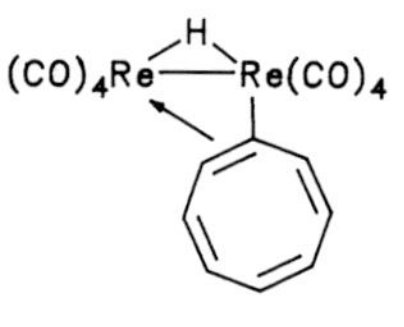 VII 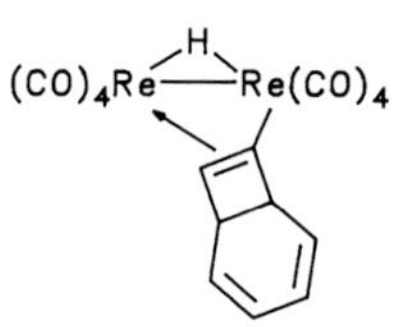 VIII

No.	compound	method of preparation (yield) properties and remarks
31	(μ-$\eta^{2:1}$-C_6H_7)$Re_2(CO)_8$(μ-H)	IIIc; two isomers obtained both isomers: pale yellow needles [4]
	isomer A (see Formula VIa)	yield: 28.2% ^{1}H NMR (CD_2Cl_2): −14.02 (s); 1.96 (m), 3.44 (t), 5.75 (dd); 5.85, 6.19 (dtt's) [4]
	isomer B (see Formula VIb)	yield: 30.3% ^{1}H NMR (CD_2Cl_2): −13.90 (s); 2.05, 2.42 (dt's, CH_2), 5.26 (s), 5.62 (dt), 6.71 (d) [4]
*32	(μ-$\eta^{2:1}$-C_7H_7)$Re_2(CO)_8$(μ-H)	IIIc; 2 isomers were obtained [4, 11] IR (n-hexane, each isomer): 1967, 1975, 1978, 1993, 2012, 2080, 2110 (ν(CO)) [11]
	isomer A 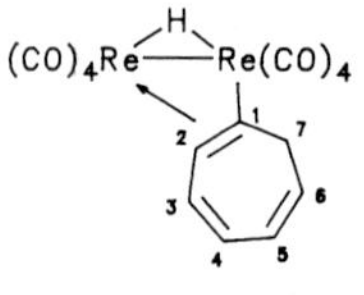	yield: 12.5% [4, 11] yellow, rhombic crystals [4, 11] ^{1}H NMR (CD_2Cl_2): −13.85 (s); 3.23 (d, CH_2), 5.53 (d), 5.94 (dt), 6.36 (dd), 6.69 (s) [4]; (toluene-d_8, −3°C): −13.86 (s, ReH); 3.01 (d, CH_2; J_{gem} = 12); 5.34 (d, H-2; J(H-2,3) =

References on p. 78

Table 4 (continued)

No.	compound	method of preparation (yield) properties and remarks
		3.8), 5.53 (dt, H-6; J(H-6,7) = 6.8), 5.95 (dd, H-5; J(H-5,6) = 9.5), 6.03 (dd, H-3; J(H-3,4) = 10.7), 6.24 (dd, H-4; J(H-4,5) = 5.0) [11]; (−98 °C): −14.02 (s, ReH); 2.40, 3.37 (dd's, CH_2; J_{gem} ca. 12); 5.11 (d, H-2), 5.50 (dt, H-6), 5.67 (dd, H-3), 5.94 (dd, H-5), 6.08 (dd, H-4) [11] $^{13}C\{^1H\}$ NMR (CD_2Cl_2, −20 °C): 50.30 (CH_2), 103.72 (C-2); 129.07, 132.14, 133.41, 136.61 (C-3 to C-6); 142.64 (C-1); 183.26, 184.91, 185.39, 186.71 (CO) [11]
	isomer B $(CO)_4Re$—H—$Re(CO)_4$ (bridging cycloheptadienyl, C-1 to C-7)	yield: 31.3% [4, 11] pale yellow crystals [4, 11] ^1H NMR (C_6D_6): −13.99 (s); 3.78 (t, 2 H), 5.43 (dt), 5.89 (dd), 6.23 (dd), 7.32 (d, 2 H) [4]; (toluene-d_8, −83 °C): −14.09 (s, ReH); 0.97, 2.13 (ddd's, CH_2; J_{gem} = 12.0), 3.55 (dd, H-1; J(H-1,7) = 5.5, J(H-1,7′) = 7.6), 5.28 (dt, H-6; J(H-6,7) = 7.5, J(H-6,7′) = 5.5), 5.78 (dd, H-4; J(H-4,5) = 5.3), 6.19 (dd, H-5; J(H-5,6) = 9.2), 7.33 (d, H-3; J(H-3,4) = 10.3); (77 °C): −13.95 (s, ReH); 2.01 (t, CH_2; J_{gem} ca. 12), 3.88 (t, H-1; J(H-1,7) = J(H-1,7′) = 6.9), 3.91 (dd, H-4), 5.44 (dt, H-6; J(H-6,7) = J(H-6,7′) = 6.5), 6.22 (dd, H-5), 7.33 (d, H-3); other coupling constants the same as at −83 °C [11] $^{13}C\{^1H\}$ NMR (CD_2Cl_2, −75 °C): 38.93 (CH_2), 82.34 (C-1), 124.34 (C-4); 127.47, 128.13 (C-5,6); 133.02 (C-2), 148.66 (C-3); 180.92, 183.13, 183.35, 183.75, 185.04, 185.57, 186.18, 187.76 (CO); (25 °C): 39.92 (CH_2), 84.36 (C-1), 125.45 (C-4), 128.50 (C-5,6; J(C,H) = 128, 162, 155, 155, respectively); 134.53 (s, C-2), 149.25 (C-3; J(C,H) = 159); 182.62, 184.78, 185.33, 186.77 (CO) [11]
33	$(\mu-\eta^{2:1}-C_8H_7)Re_2(CO)_8(\mu-H)$	IIIc (starting from cyclooctatetraene); two isomers both isomers form pale yellow crystals [4]
	isomer A (see Formula VII)	yield: 10.3% ^1H NMR (CD_2Cl_2): −13.9 (s); 5.30 (d), 5.39 (s), 6.03 (d), 6.07 (s), 6.36 (dd), 7.12 (d) thermally equilibrated with isomer B [4]

References on p. 78

Table 4 (continued)

No.	compound	method of preparation (yield) properties and remarks
33 (continued)		
	isomer B (see Formula VIII)	yield: 5.8% ^{1}H NMR (C_6D_6): −14.09 (s); 3.04, 3.26 (t's); 5.29 (m), 5.31 (d); 5.63, 5.73, 5.78 (m's, 1 H each) [4]

* Further information:

(μ-$\eta^{2:1}$-CH_2=CH)$Re_2(CO)_7(P(CH_3)_3)$(μ-H) (Table **4**, No. **10**) was obtained as a mixture of 4 nonseparable isomers (total yield: 60%); however, the components could be NMR-spectroscopically identified through 4 independent sets of protons.

IR spectrum (n-hexane, mixture): 1940, 1965, 1977, 1987, 1996, 2019, 2034, 2084, 2115 (ν(CO)) cm^{-1}. FD mass spectrum (mixture): $[M]^+$ [12].

(μ-$\eta^{2:1}$-CH_2=CH)$Re_2(CO)_7(P(C_6H_5)_3)$(μ-H) (Table **4**, No. **11**) was obtained via Method II as a mixture of 3 isomers in the ratio 11:3:2 [12]. The mixture was also formed by treating (μ-$\eta^{2:1}$-CH_2=CH)$Re_2(CO)_8$(μ-H) with $P(C_6H_5)_3$ (by-product: $(CO)_4Re(P(C_6H_5)_3)H$) [10].

^{1}H NMR spectrum ($CDCl_3$, mixture): δ = −14.01, −13.92, −13.61 (d's, ReH; J(P,H) = 11.1, 16.7, 8.4 Hz, respectively); 3.59 (m); 4.51 (dd; J(H,H) = 15.4, 4.5 Hz); 5.74 (dd; J(H,H) = 10.0, 2.5 Hz); 6.16 (m); 7.44 to 7.51 (m) ppm [12]. IR spectrum (pentane, mixture): 1928, 1948, 1957, 1968, 1979, 1990, 2019, 2038, 2055, 2083, 2088, 2107 (ν(CO)) cm^{-1} [10] (similar in [12]). FD mass spectrum: $[M]^+$ [12], $[M - n\ CO]^+$ (n = 0 to 4) [10].

(μ-$\eta^{2:1}$-CH_2=CH)$Re_2(CO)_8$(μ-H) (Table **4**, No. **13**) was also obtained by treating (μ-$\eta^{2:1}$-RCH=CH)$Re_2(CO)_8$(μ-H) (R = CH_3 [3, 9, 10], C_2H_5 [3, 9], n-C_4H_9 [3]) or (μ-$\eta^{2:1}$-CH_2=CC_2H_5)-$Re_2(CO)_8$(μ-H) [9] with C_2H_4 at atmospheric pressure (toluene, 25 to 40 °C, 1 d, repeatedly degassing and resaturating the solution). **(μ-$\eta^{2:1}$-CD_2=CD)$Re_2(CO)_8$(μ-D)** could be analogously prepared from (μ-$\eta^{2:1}$-CH_3CH=CH)$Re_2(CO)_8$(μ-H) and C_2D_4 (1 atm, 40 °C, 2 × 10 h) [9, 10]. Small amounts of No. 13 were also formed by the photochemical reaction of $(CO)_{10}Re_2$ with acetylene (0.8%) [6] or by photolyzing π-$C_2H_4Re_2(CO)_9$ at −20 °C [14].

Exposure of No. 13 to an H_2 atmosphere (toluene, 1 atm) produced $(CO)_8Re_2$(μ-H)$_2$; prolonged (several weeks) exposure to H_2 led to $(CO)_{12}Re_3$(μ-H)$_3$ via $(CO)_8Re_2$(μ-H)$_2$ [9].

Reactions with nucleophiles usually proceed via liberation of the organic ligand: While dissolution in wet THF (2% H_2O) slowly led to $(CO)_{12}Re_4$(μ_3-OH)$_4$, exposure to an H_2S atmosphere (1 atm) produced mainly $(CO)_8Re_2$(μ-SH)$_2$. Treatment with ^{2}D = CH_3CN, pyridine, or $P(OR)_3$ (R = CH_3, C_6H_5) readily yielded compounds of the type 1,2-$(CO)_8$-$Re_2(^2D\text{-eq})_2$. A kinetics study of the reaction with pyridine revealed $k_1 = 4.6 \times 10^{-4}\ s^{-1}$, $k_{-1}/k_2 = 1.7 \times 10^{-3}$ M (at 28.5 °C). The same reaction of (μ-$\eta^{2:1}$-CD_2=CD)$Re_2(CO)_8$(μ-D) gave $k_{obs} = 4.4 \times 10^{-4}\ s^{-1}$; thus there is no isotope effect. Treatment with $P(C_6H_5)_3$ in hexane gave a ca. 4:1 mixture of 1,2-eq,ax and 1,2-diax $(CO)_8Re_2(P(C_6H_5)_3)_2$; likewise, treatment with $P(C_4H_9\text{-n})_3$ in toluene yielded $(CO)_8Re_2(P(C_4H_9\text{-n-})_3\text{-ax})_2$ via the 1,2-eq,ax isomer. In contrast, the reaction with $P(CH_3)_3$ in hexane initially gave the dipolar addition product (μ-$(CH_3)_3PCHCH_2$)$Re_2(CO)_8$(μ-H) (see p. 58), whereas the reaction in toluene led to $(CO)_8Re_2(P(CH_3)_3\text{-eq})_2$. Similarly, $(CO)_{10}Re_2$ was slowly formed in the presence of CO [9].

References on p. 78

Thermal reactions with several terminal alkenes, RCH=CH_2 (R = CH_3 [10], C_2H_5 [9], n-C_4H_9 [9], C_6H_5 [9, 10]), led to the derivatives (μ-$\eta^{2:1}$-RCH=CH)$Re_2(CO)_8$(μ-H). Treatment with buta-1,3-diene (1 atm, 40 h) quantitatively gave (μ-$\eta^{2:2}$-C_4H_6)$Re_2(CO)_8$ [9].

Photolysis in inert solvents (hexane, toluene) slowly yielded $(CO)_{12}Re_4(\mu_3\text{-}OH)_4$ (probably due to traces of moisture). However, photolysis in a ^{13}CO atmosphere for 15 to 30 min exchanged several CO groups, providing a mixture of (μ-$\eta^{2:1}$-CH_2=CH)$Re_2(^{13}CO)_n$-$(CO)_{8-n}$(μ-H) (with n = 1 to 4). ^{13}CO could not be incorporated without irradiation. Extended photolysis under CO for at least 1 d eventually produced $(CO)_{10}Re_2$. In contrast, irradiation in the presence of $P(C_6H_5)_3$ yielded (μ-$\eta^{2:1}$-CH_2=CH)$Re_2(CO)_7(P(C_6H_5)_3)$(μ-H) and $(CO)_4$-$Re(P(C_6H_5)_3)H$ within 45 min; disubstitution was almost never observed [10].

The photochemical reaction with ethene (1 atm, saturated solution) at room temperature produced Nos. 17 (both isomers) and 26 [3, 10, 14]. No. 26 clearly predominated (ratio 9:2:1 (by ^1H NMR)), when the mixture was only briefly irradiated; however, even after 5 h it still was the dominating product. Traces of $(CO)_{12}Re_4(\mu_3\text{-}OH)_4$ were immediately detected, yet after 24 h ca. 40% of No. 13 were converted to this tetranuclear cluster. But-1-ene and trans-hex-3-ene were concurrently formed (C_3 or C_5 products were not detected), but their formation rate was slow (1 to 3 turnovers per day). These hydrocarbons were more rapidly formed when the photolysis was carried out at 50°C [10]. In addition to Nos. 17 and 26, a small amount of (μ-$\eta^{2:2}$-C_4H_6)$Re_2(CO)_8$ was also observed [14]. Moreover, when C_2H_4 was bubbled through the solution, in situ IR spectroscopy accounted for the formation of compounds of the type π-$C_2H_4Re_2$(μ-$\eta^{2:1}$-C_4H_7)$(CO)_7$(μ-H) (see p. 103) after 45 min of photolysis; the combined yield of these products reached 15% [10].

Photochemical treatment with propene in hexane or toluene yielded (μ-$\eta^{2:1}$-CH_3-CH=CH)$Re_2(CO)_8$(μ-H) (cis and trans isomer) in nearly equal amounts. Pent-1-ene (other isomers were almost negligible) was also detected in the mixture [10].

The photo-initiated reaction with methyl vinyl ether for 5 h yielded a mixture consisting of the title product, No. 14, and No. 18 (trans isomer) in the ratio 10:8:3 (by ^1H NMR). Evidence for the intermediate presence of (μ-$\eta^{2:1}$-cis-$CH_3OC_2H_4$CH=CH)$Re_2(CO)_8$(μ-H) was provided by ^1H NMR spectroscopy when the mixture was cooled to −20°C immediately after irradiation. Signals ascribed to this intermediate diminished within 30 min upon warming to 25°C, while signals due to the trans isomer of No. 18 grew simultaneously [10].

The photochemical reaction of (μ-$\eta^{2:1}$-CD_2=CD)$Re_2(CO)_8$(μ-D) with propene in hexane yielded H/D exchange products in addition to (μ-$\eta^{2:1}$-CH_3CH=CH)$Re_2(CO)_8$(μ-H). The main initial product is (μ-$\eta^{2:1}$-CD_2=CD)$Re_2(CO)_8$(μ-H). After a 45 min irradiation period, the mixture contained No. 13 with a D_4, D_3H, D_2H_2, DH_3, and H_4 isotope distribution of 14, 36, 36, 11, and 3%, respectively (H/D exchange neither occurred under thermal conditions nor upon irradiation in the absence of propene). Photolysis of (μ-$\eta^{2:1}$-CD_2=CD)$Re_2(CO)_8$(μ-D) with ethene similarly caused H/D exchange, such that after only 15 min D_0 to D_4 species could mass-spectroscopically be detected (the hydride position is more labile than any vinyl position). Subsequently, the partially deuterated species ($\eta^{2:1}$-$C_4H_nD_{7-n}$)$Re_2(CO)_8$(μ-H/D) (Nos. 17 and 26) were also formed (with mainly 3 and 4 D atoms incorporated) [10].

Irradiation with $(CO)_{10}Re_2$ at −20°C gave (μ-$\eta^{2:1}$-C_2H_3)$Re_3(CO)_{13}$ and $(CO)_{14}Re_3H$ [14].

cis- and **trans-(μ-$\eta^{2:1}$-CH_3CH=CH)$Re_2(CO)_8$(μ-H)** (Table **4**, No. **16**) were simultaneously obtained in almost equal amounts, when No. 13 was irradiated with propene for 90 min. However, a thermal reaction selectively yielded the trans isomer. Photolysis of the trans isomer in toluene rapidly (30 min) gave a stationary cis-trans isomer mixture, where the trans:cis ratio was 1.4:1 at 5 and 25°C; however, it became 5.5:1 at 45°C [10].

References on p. 78

Addition of C_2H_4 to this photostationary cis-trans isomer mixture, while continuously irradiating, slowly produced trans-pent-2-ene (other isomers were negligible) and No. 13 [10]. The thermal reaction of No. 16 with C_2H_4 or C_2D_4 (1 atm, repeatedly saturating and degassing the solution) also gave $(\mu\text{-}\eta^{2:1}\text{-}CH_2\text{=}CH)Re_2(CO)_8(\mu\text{-}H)$ or $(\mu\text{-}\eta^{2:1}\text{-}CD_2\text{=}CD)Re_2(CO)_8(\mu\text{-}D)$, respectively, but did not simultaneously yield pentene [9, 10]. Irradiation in the presence of propene slowly produced hexene. At room temperature cis- and trans-hex-2-ene both predominated, whereas at 45 °C trans-hex-2-ene exclusively prevailed [10]. Upon vigorously bubbling propene through the solution, additional weak ν(CO) bands were observed in an IR spectrum taken on the solution; these bands were assigned to the species $(\pi\text{-}CH_3CH\text{=}CH_2)Re_2(\mu\text{-}\eta^{2:1}\text{-}CH\text{=}CHCH_3)(CO)_7(\mu\text{-}H)$ (see p. 104) [10].

Extensive photolysis of the pure compound produced $(CO)_{12}Re_4(\mu_3\text{-}OH)_4$; half of the initial starting material was consumed after 20 h. Photolysis under a ^{13}CO atmosphere (1 atm) for only 15 to 30 min provided a mixture of the ^{13}CO exchange products $(\mu\text{-}\eta^{2:1}\text{-}CH_3\text{-}CH\text{=}CH)Re_2(^{13}CO)_n(CO)_{8-n}(\mu\text{-}H)$ (n = 1 to 4). Thermal ^{13}CO incorporation was not observed [10]. However, treatment with $P(CH_3)_3$ or $P(C_6H_5)_3$ under thermal conditions readily yielded $(CO)_8Re_2(P(CH_3)_3\text{-eq})_2$ or $(CO)_8Re_2(P(C_6H_5)_3\text{-ax})_2$ [9].

Although No. 16 did not react with trans-but-2-ene, treatment with cis-but-2-ene gave $(\mu\text{-}\eta^{2:1}\text{-}CH_3CH\text{=}CCH_3)Re_2(CO)_8(\mu\text{-}H)$ [9]. Likewise, treatment with terminal alkynes, HC≡CR (R = C_6H_5, $C_6H_4CH_3$-4), yielded compounds of the type $(\mu\text{-}\eta^{2:1}\text{-}RC{\equiv}C)Re_2(CO)_8(\mu\text{-}H)$ [8].

cis- and **trans-$(\mu\text{-}\eta^{2:1}\text{-}C_2H_5CH\text{=}CH)Re_2(CO)_8(\mu\text{-}H)$** (Table **4**, No. **17**) were detected as byproducts in a mixture obtained by irradiating $(\mu\text{-}\eta^{2:1}\text{-}CH_2\text{=}CH)Re_2(CO)_8(\mu\text{-}H)$ in the presence of C_2H_4 [3, 10]. The reaction mainly yielded $(\mu\text{-}\eta^{2:1}\text{-}CH_2\text{=}CC_2H_5)Re_2(CO)_8(\mu\text{-}H)$ (No. 26); however, this compound slowly isomerized in the dark to the title compounds [9]. These three compounds were also obtained, each with ca. 10 to 15% yield (along with No. 13), by photochemically reacting $(CO)_{10}Re_2$ with ethene [3, 9]. Otherwise, the isomer mixture also formed by photolyzing $(\mu\text{-}\eta^{2:1}\text{-}CH_3CH\text{=}CCH_3)Re_2(CO)_8(\mu\text{-}H)$ [3].

Photolysis of the pure trans isomer produced a steady-state cis-trans isomer mixture (cis:trans = 1:1.7) within 30 min [10]. Photolysis in the presence of 1 equivalent pyridine, however, gave 2 isomers of $(\eta^{2:1}\text{-}C_2H_5CH\text{=}CH)Re_2(CO)_7(NC_5H_5)(\mu\text{-}H)$ (No. 12) [3].

Exposure to an H_2 atmosphere (1 atm) slowly led to $(CO)_8Re_2(\mu\text{-}H)_2$. Treatment with acrylonitrile provided $(CO)_8Re_2(NCCH\text{=}CH_2\text{-eq})_2$, and treatment with $P(C_6H_5)_3$ yielded $(CO)_8Re_2(P(C_6H_5)_3)_2$ [9]. Repeated treatment with C_2H_4 (1 atm) gave No. 13 [3, 9]; likewise, interaction with $HC{\equiv}CC_6H_5$ gave $(\eta^{2:1}\text{-}C_6H_5C{\equiv}C)Re_2(CO)_8(\mu\text{-}H)$ [3]. Stirring with excess 3,3-dimethyl-cyclopropene in hexane separated $(\mu\text{-}(CH_3)_2CCHCH)Re_2(CO)_8$ (see p. 132) [5, 16].

$(\mu\text{-}\eta^{2:1}\text{-n-}C_4H_9CH\text{=}CH)Re_2(CO)_8(\mu\text{-}H)$ (Table **4**, No. **19**). Solutions containing No. 19 in several instances were prepared via Method IIIa. However, the compound was not isolated but immediately used for subsequent reactions [7, 15, 17, 18, 22].

Treatment with various nucleophiles removed the organic ligand: Exposure to a CO atmosphere slowly gave $(CO)_{10}Re_2$ with a half-life time of ca. 3 to 4 d [3, 9] and treatment with 3,3-dimethyl thietane ($C_5H_{10}S$) gave 1,2-$(CO)_8Re_2(SC_5H_{10})_2$ [20]. Likewise, the reaction with pyridine yielded $(CO)_8Re_2(NC_5H_5\text{-eq})_2$ [3, 9] (a kinetic study revealed the first-order reaction constants $k_1 = 1.3\times10^{-3}\ s^{-1}$, $k_{-1}/k_2 = 2.8\times10^{-2}$ M at 28.5 °C [9]), while the reaction with $P(C_6H_5)_3$ yielded $(CO)_8Re_2(P(C_6H_5)_3)_2$ [9]. Treatment with diphosphanes in CH_2Cl_2 usually provided compounds of the type $(CO)_8Re_2(\mu\text{-}R_2P(CH_2)_nPR_2)$. These reactions were performed with $(CH_3)_2P(CH_2)_nP(CH_3)_2$ (n = 1, 2) [7], $(C_6H_5)_2P(CH_2)_nP(C_6H_5)_2$ (n = 1 [7, 9], 2 [7], 3 to 6 [15]), $(C_6H_{11})_2PCH_2P(C_6H_{11})_2$, and cis-$(C_6H_5)_2PCH\text{=}CHP(C_6H_5)_2$ [15]. Treatment

References on p. 78

with $(CH_3)_2POC_5H_4N$-2 at room temperature yielded $(CO)_8Re_2(\mu\text{-}P(CH_3)_2)(\mu\text{-}OC_5H_4N)$; NMR-monitoring established however at least 8 intermediate products [13, footnote No. 36]. Treatment with $((C_6H_5)_2P)_2CHC_5H_4N$-2 yielded $(CO)_8Re_2[((C_6H_5)_2P)_2CHC_5H_4N]$ and a red-brown, unidentified compound [18], while a treatment with $((CH_3)_2P)_3CH$ gave $(CO)_8Re_2$-$(\mu\text{-}((CH_3)_2P)_2CHP(CH_3)_2)$ [22]. Interaction with $(CH_3O)_3SiH$ in hot heptane produced $(CO)_8$-$Re_2(\mu\text{-}H)(\mu\text{-}(CH_3O)_2SiOCH_3)$ (see "Organorhenium Compounds" 5, 1994, p. 352) [23].

Repeated treatment with C_2H_4 (1 atm) gave $(\mu\text{-}\eta^{2:1}\text{-}CH_2{=}CH)Re_2(CO)_8(\mu\text{-}H)$ [3]. Interaction with vinyl butyl ether gave $(\mu\text{-}\eta^{2:1}\text{-n-}C_4H_9OCH{=}CH)Re_2(CO)_8(\mu\text{-}H)$ [9]. Treatment with excess $CH_3C{\equiv}CN(CH_3)_2$ at 25°C yielded $[\mu\text{-}CH_3C((CH_3)_2N)C{=}]Re_2(CO)_8$ (see p. 123), whereas at 50°C rhenacycles of the type $[C_4(CH_3)_2(N(CH_3)_2)_2]Re_2(CO)_7$ (^{4}L compounds, see Table 7, Nos. 1 to 3, p. 150) were concomitantly obtained [21].

Treatment with $C_5H_5M({\equiv}CR)(CO)_2$ (M = Mo, R = $C_6H_4CH_3$-4; M = W, R = CH_3 or $C_6H_4CH_3$-4) in hot THF yields clusters of the type $[\mu\text{-}(C_5H_5)_2M_2(CO)_3(CR)_2]Re_2(CO)_6$; in addition, the reaction with $C_5H_5W({\equiv}CCH_3)(CO)_2$ also yielded $[\mu\text{-}(C_5H_5)_2W_2(CO)_3(CCH_3)$-$(C{=}CH_2)]Re_2(CO)_6(\mu\text{-}H)$ (see pp. 156/7 for these products) [17].

$(\mu\text{-}\eta^{2:1}\text{-}C_7H_7)Re_2(CO)_8(\mu\text{-}H)$ (Table **4**, No. **32**). Two other compounds, namely $(\eta^5\text{-}C_7H_9)$-$Re(CO)_3$ and $(\eta^5\text{-}C_7H_7)Re(CO)_3$, were also obtained [4, 11]. These products preferably formed when the mixture was extensively irradiated [11]. The photochemical reaction of $(CO)_{10}Re_2$ with C_7H_8 has already been studied earlier [1]; however, any products found at that time were not mentioned in [4, 11].

^{1}H NMR spectroscopy at various temperatures demonstrated coalescence for the CH_2 protons at T_{coal} = −65°C and −48°C with isomers A and B, respectively. The corresponding activation barriers are $\Delta G^{\neq}$ (−65°C) = 39.9 ± 2 and $\Delta G^{\neq}$ (−48°C) = 48.0 ± 2 kJ/mol. The signals arising from the CH groups were not significantly affected by changing the temperature. Either isomer is chiral, thus exists as an enantiomeric pair. High-temperature NMR spectroscopy revealed that exo and endo protons of the CH_2 groups become equivalent; moreover, only four δ(CO) signals indicate an apparent high symmetry. There are two mechanistic pathways for enantiomerization: σ-π bond exchange or ring inversion. It could be

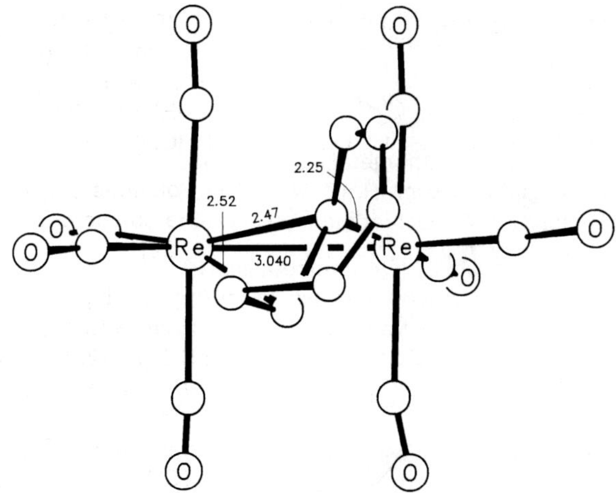

Fig. 37. Molecular structure of $(\mu\text{-}\eta^{2:1}\text{-}C_7H_7)Re_2(CO)_8(\mu\text{-}H)$ (isomer B) [11].

References on p. 78

shown that enantiomerization occurs through σ-π bond exchange rather than through ring inversion [11].

Isomer B crystallizes in the monoclinic space group C2/c$-C^6_{2h}$ (No. 15) with a = 9.743(3), b = 14.450(4), c = 48.72(2) Å, β = 90.46°; Z = 16 molecules per unit cell, D_{calc} = 2.66 g/cm^3. The structure of one of the two symmetry-independent molecules within the asymmetric unit is depicted in **Fig. 37**. It can be seen that the bridging ligand assumes a boat conformation [11] (for selected bond lengths, see also [2, footnotes No. 18, 19]).

Heating of either pure isomer always gave a mixture containing isomer A and isomer B in the ratio 0.03:0.97. The conversion of isomer A was monitored at 77°C, and a reaction rate constant of k = $0.038 \pm 0.004\ s^{-1}$ was measured, accounting for $\Delta G^{\neq}$ (77°C) = 95.7 ± 5 kJ/mol [11].

References:

[1] Davis, R.; Ojo, I. A. O. (J. Organomet. Chem. **110** [1976] C 39/C 41).
[2] Franzreb, K.-H.; Kreiter, C. G. (Z. Naturforsch. **37b** [1982] 1058/69).
[3] Nubel, P. O.; Brown, T. L. (J. Am. Chem. Soc. **104** [1982] 4955/7).
[4] Franzreb, K.-H.; Kreiter, C. G. (J. Organomet. Chem. **246** [1983] 189/95).
[5] Green, M.; Orpen, A. G.; Schaverien, C. J.; Williams, I. D. (J. Chem. Soc. Chem. Commun. **1983** 1399/401).
[6] Franzreb, K.-H.; Kreiter, C. G. (Z. Naturforsch. **39b** [1984] 81/5).
[7] Lee, K.-W.; Pennington, W. T.; Cordes, A. W.; Brown, T. L. (Organometallics **3** [1984] 404/13).
[8] Nubel, P. O.; Brown, T. L. (Organometallics **3** [1984] 29/32).
[9] Nubel, P. O.; Brown, T. L. (J. Am. Chem. Soc. **106** [1984] 644/52).
[10] Nubel, P. O.; Brown, T. L. (J. Am. Chem. Soc. **106** [1984] 3474/84).

[11] Kreiter, C. G.; Franzreb, K. H.; Michels, W.; Schubert, U.; Ackermann, K. (Z. Naturforsch. **40b** [1985] 1188/98).
[12] Lee, K.-W.; Brown, T. L. (Organometallics **4** [1985] 1030/6).
[13] Collum, B. D.; Klang, J. A.; Depue, R. T. (J. Am. Chem. Soc. **108** [1986] 2333/40).
[14] Kreiter, C. G.; Franzreb, K. H.; Sheldrick, W. S. (Z. Naturforsch. **41b** [1986] 904/14).
[15] Lee, K.-W.; Hanckel, J. M.; Brown, T. L. (J. Am. Chem. Soc. **108** [1986] 2266/73).
[16] Green, M. L.; Orpen, A. G.; Schaverien, C. J.; Williams, I. D. (J. Chem. Soc. Dalton Trans. **1987** 1313/8).
[17] Jeffery, J. C.; Parrott, M. J.; Pyell, U.; Stone, F. G. A. (J. Chem. Soc. Dalton Trans. **1988** 1121/9).
[18] Mattson, B. M.; Ito, L. N. (Organometallics **8** [1989] 391/5).
[19] Kreiter, C. G.; Michels, W.; Exner, R. (Z. Naturforsch. **45b** [1990] 793/802).
[20] Adams, R. D.; Belinski, J. A.; Schierlmann, J. (J. Am. Chem. Soc. **113** [1991] 9004/6).

[21] Adams, R. D.; Chen, G.; Yin, J. (Organometallics **10** [1991] 1278/82).
[22] Mague, J. T.; Lloyd, C. L. (Organometallics **11** [1992] 26/34).
[23] Adams, R. D.; Cortopassi, J. E.; Yamamoto, J. H. (Organometallics **12** [1993] 3036/41).
[24] Kreiter, C. G.; Michels, W.; Heeb, G. (J. Organomet. Chem. **502** [1995] 9/18).

2.2.2.1.2 Miscellaneous Compounds

(μ-$\eta^{2:1}$-C_6H_5CH=CC_6H_5)$Re_2(CO)_7$(μ-$P(C_6H_5)_2$) (see Formula I) was obtained by reacting equimolar quantities of $(CO)_8Re_2(\mu\text{-}H)(\mu\text{-}P(C_6H_5)_2)$ and diphenyl acetylene in refluxing

References on p. 81

xylene for 6 h. The solvent was removed, and the remaining solid was separated chromatographically with hexane/$CHCl_3$ (5:1). Yield: 20.9%. Orange solid.

^1H NMR spectrum ($CDCl_3$): δ = 5.11 (d; J(P,H) = 8.8 Hz), 7.40 (m, C_6H_5) ppm. ^{31}P {^1H} NMR spectrum ($CDCl_3$): δ = 60.74 (s) ppm. IR spectrum ($CHCl_3$): 1913, 1943, 1965, 1980, 2020, 2073 (ν(CO)) cm^{-1}.

$(C_6H_5)_2$ P; $(CO)_3Re$—$Re(CO)_4$; H; C_6H_5; C_6H_5 — I

⊖; $(CO)_4Re$—$Re(CO)_4$; ⊕ $N(CH_3)_2$ — II

$[(CO)_5Re$; H−3; 1; H−2; $Re(CO)_5]^+$ — III

SC_6H_5; $(CO)_3Re$—$Re(CO)_3$; $(CO)_2W$; $C_5(CH_3)_5$; H; C_6H_5 — IV

Single crystals are triclinic with a = 9.436(2), b = 11.774(3), c = 14.353(4) Å, α = 95.91(2)°, β = 99.57(2)°, γ = 95.71(2)°; space group $P\bar{1}-C_i^1$ (No. 2), Z = 2 molecules per unit cell, D_{calc} = 1.995 g/cm^3. The molecular structure is illustrated in **Fig. 38** [2].

(μ-η$^{2:1}$-CHCHCH=N(CH_3)$_2$)Re_2(CO)$_8$ (see Formula II). (μ-CH_2=CH((CH_3)$_2$N)C=)Re_2(CO)$_8$ (see p. 134) was heated in C_6D_6 for 35 h. Subsequent chromatographic workup separated only (μ-CHCHCHN(CH_3)$_2$)Re_2(CO)$_6$(μ-CO) (see p. 135); however, a ^1H NMR spectrum recorded prior to elution evidenced the presence of another species (the title complex), which rapidly decomposed during elution. When the thermolysis was conducted under a CO atmosphere, this labile compound almost exclusively formed.

^1H NMR spectrum (C_6D_6): δ = 1.37, 1.46 (s's, CH_3); 3.13 (dd, 1 H; J(H,H) = 11.6, 10.8 Hz), 5.12 (d, 1 H; J = 10.8 Hz), 7.11 (d, 1 H; J = 11.6 Hz) ppm [3].

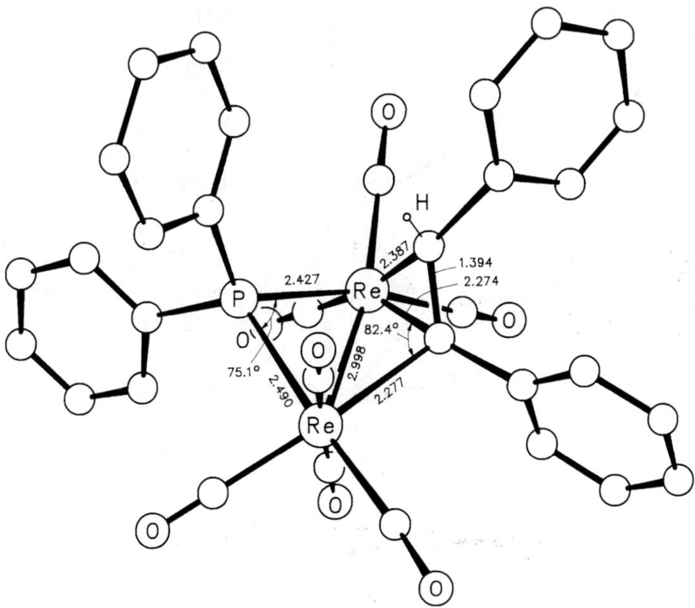

Fig. 38. Molecular structure of (μ-η$^{2:1}$-C_6H_5CH=CC_6H_5)Re_2(CO)$_7$(μ-P(C_6H_5)$_2$) [2].

References on p. 81

$[((CO)_5ReCH{=}CH_2)Re(CO)_5]BF_4$ (see Formula III) was obtained by reacting $(CO)_5ReFBF_3$ with the 10-fold excess of $(CH_3)_3SiCH{=}CH_2$ (CH_2Cl_2, room temperature, 72 h). The product precipitated as a yellow solid with 80% yield.

^{1}H NMR spectrum (CD_3NO_2): δ = 5.15 (dd, H-3), 5.63 (dd, H-2), 6.81 (dd, H-1) ppm; J(H-1,3) = 18.81 Hz, J(H-2,3) = 3.67 Hz, J(H-1,2) = 11.47 Hz. IR spectrum (KBr): 1535 (ν(C=C)); 1965, 1977, 2012, 2032, 2062, 2077, 2132, 2156, 2172 (ν(CO)) cm^{-1}.

Treatment with I^- in THF gave $(CO)_5ReI$ and $[\pi\text{-}C_2H_4Re(CO)_5]BF_4$. Treatment with $[OCH_3]^-$ was only studied by IR spectroscopy and revealed attack at CO. A reaction with $[(CO)_5Re]^-$ in THF furnished $(CO)_5Re{-}C_2H_4{-}Re(CO)_5$ and $(CO)_{10}Re_2$ [1].

$[\mu\text{-}(C_5(CH_3)_5)W(CO)_2CH{=}CC_6H_5]Re_2(CO)_6(\mu_3\text{-}SC_6H_5)$ (see Formula IV) was obtained by treating $[\mu\text{-}(C_5(CH_3)_5)W(CO)_2C{\equiv}CC_6H_5]Re_2(CO)_6$ (see p. 126) with excess thiophenol (toluene, reflux, 10 min). After evaporating the mixture, the residue was separated chromatographically with CH_2Cl_2/hexane (1:1). Yield: 32%. Red crystals.

^{1}H NMR spectrum ($CDCl_3$): δ = 2.06 (CH_3); 6.86 to 7.32 (m, C_6H_5); 8.06 (s, CH=) ppm. $^{13}C\{^1H\}$ NMR spectrum ($CDCl_3$): δ = 10.8 ($(CH_3)_5$), 106.2 (C_5); 124.2, 126.2, 127.6, 129.1, 130.4, 131.6, 135.8, 160.2 (C_6H_5); 168.0 (**C**H=CC_6H_5; J(W,C) = 62 Hz); 191.3, 197.8 (ReCO); 200.7 (=**C**C_6H_5); 216.7, 219.2 (WCO; J(W,C) = 170 and 165 Hz) ppm. IR spectrum (C_6H_{12}): 1897, 1937, 1955, 1985, 2009, 2036 (ν(CO)) cm^{-1}. FAB mass spectrum: $[M]^+$.

Single crystals with the composition **$[\mu\text{-}(C_5(CH_3)_5)W(CO)_2CH{=}CC_6H_5]Re_2(CO)_6(\mu_3\text{-}SC_6H_5)\cdot 0.5\,CH_2Cl_2$** have a monoclinic lattice with a = 20.017(4), b = 11.869(6), c = 29.622(5) Å, β = 98.50(2)°; space group I2/c$-C_{2h}^3$ (No. 12); Z = 8 formula units per unit cell, D_{calc} = 2.230 g/cm^3. The structure of the molecule is depicted in **Fig. 39**.

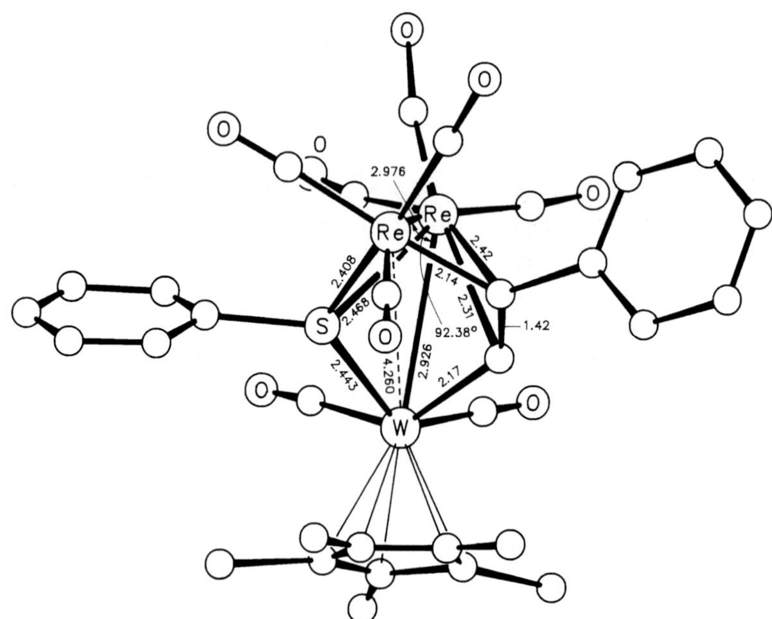

Fig. 39. Molecular structure of $[\mu\text{-}(C_5(CH_3)_5)W(CO)_2CH{=}CC_6H_5]Re_2(CO)_6(\mu_3\text{-}SC_6H_5)$ [4].

Thermolysis (toluene, 1 h) yielded [μ-$(C_5(CH_3)_5)W(SC_6H_5)(HC{=}CC_6H_5)CO]Re_2(CO)_6$ (see p. 136). The reaction could be partially reversed by heating this product under CO [4].

References:

[1] Steil, P.; Beck, W.; Stone, F. G. A. (J. Organomet. Chem. **368** [1989] 77/81).
[2] Haupt, H.-J.; Heinekamp, C.; Flörke, U. (Inorg. Chem. **29** [1990] 2955/63).
[3] Adams, R. D.; Chen, G.; Chi, Y.; Wu, W.; Yin, J. (Organometallics **11** [1992] 1480/6).
[4] Peng, J.-J.; Peng, S.-M.; Lee, G.-H.; Chi, Y. (Organometallics **14** [1995] 626/33).

2.2.2.2 Compounds with Alkynyl Ligands

2.2.2.2.1 Compounds of the Type $(\mu\text{-}\eta^{2:1}\text{-}RC{\equiv}C)Re_2(CO)_8(\mu\text{-}H)$ and Their CO Substitution Derivatives

General. Structure. This section presents compounds whose main structural feature is one σ,π-bonded alkyn-1-yl ligand bridging two Re atoms which are connected by a single bond. There is also a bridging hydride. The structure of most compounds is represented by Formulas I to IV. R stands for H, C_6H_5, $CH_3OC_6H_4$, or 17α-estradiol ($C_{18}H_{23}O_2$, see Formula V). ^2D and bridging ^4D ligands are said to coordinate at an apical site, i.e. perpendicular to the molecular plane formed by the Re atoms, the hydride, and the C≡C unit. This arrangement has been substantiated in all cases, where X-ray crystallography was performed.

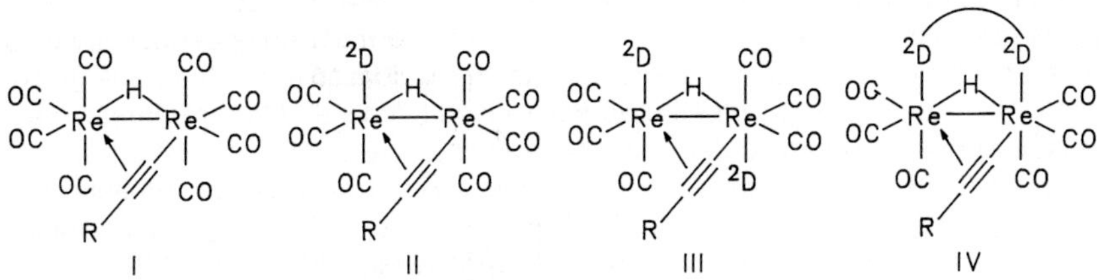

I II III IV

Preparation. The compounds were prepared by the following methods:

Method I: Treatment of $(\mu\text{-}\eta^{2:1}\text{-}RC{\equiv}C)Re_2(R'_2PCH_2PR'_2)(CO)_6(\mu\text{-}H)$ (R = H, C_6H_5; R' = CH_3, C_6H_5) with excess $(C_6H_5)_2PCH_2P(C_6H_5)_2$ (xylene, reflux, 6 h) followed by chromatographic separation (TLC; CH_2Cl_2/hexane/toluene (2:1:1)) [4].

Method II: Treatment of $(CO)_6Re_2(\mu\text{-}R_2PCH_2PR_2)(\mu\text{-}R'_2PCH_2PR'_2)$ (R, R' = CH_3, C_6H_5) with $HC{\equiv}CC_6H_5$ in $C_6H_4Cl_2$-1,3 at reflux temperature. The product was separated by preparative TLC with CH_2Cl_2/hexane/toluene (2:1:1) [4].

Method III: Photochemical treatment of $(CO)_8Re_2(\mu\text{-}R_2PCH_2PR_2)$ (R = CH_3, C_6H_5) with $HC{\equiv}CR'$ (R' = H, C_6H_5) in toluene for 12 h followed by chromatographic workup on preparative TLC plates (silica/CH_2Cl_2/hexane mixtures). The title complex could be separated from products of the types $(\mu\text{-}\eta^{2:1}\text{-}R'C{\equiv}C)$-$Re_2(\mu\text{-}C(R'){=}CH_2)(CO)_6(R_2PCH_2PR_2)$ and $R'C{\equiv}CRe_2(CO)_7(R_2PCH_2PR_2)(\mu\text{-}H)$ (see p. 100 and "Organorhenium Compounds" 5, 1994, pp. 260/2) [5].

Method IV: Photolysis of $C_6H_5C{\equiv}CRe_2(CO)_7(R_2PCH_2PR_2)(\mu\text{-}H)$ (R = CH_3, C_6H_5; see "Organorhenium Compounds" 5, 1994, pp. 260/2) in toluene (room temperature, 24 h) [5].

Method V: Treatment of $(\mu\text{-}\eta^{2:1}\text{-}RC{\equiv}C)Re_2(CO)_8(\mu\text{-}H)$ (R = 4-$CH_3OC_6H_4$, C_6H_5, $C_{18}H_{23}O_2$) with the respective ^2D ligand in CH_2Cl_2 [3, 7, 11].

References on p. 96

Method VI: Reaction of $(CO)_8Re_2(NCCH_3)_2$ with HC≡CR (R = C_6H_5, $C_{18}H_{23}O_2$ [9, 11], $C{\equiv}CSi(CH_3)_3Co_2(CO)_4((C_6H_5)_2PCH_2P(C_6H_5)_2)$ [16]) in refluxing CH_2Cl_2 for 5 to 15 h [9, 11] or light petroleum ether for ca. 6 h [16].

Method VII: Treatment of $(\mu\text{-}\eta^{2:1}\text{-}RC{\equiv}C)Re_2(CO)_7(NCCH_3)(\mu\text{-}H)$ (R = C_6H_5, $C_{18}H_{23}O_2$)
a. with the respective ^2D ligand in CH_2Cl_2 at room temperature [11];
b. with CO (1 atm) in benzene or toluene at ca. 70 °C [11].

Method VIII: Photochemical reaction of $(CO)_{10}Re_2$ with the terminal alkyne [2, 5, 7].

Method IX: Treatment of $(\mu\text{-}\eta^{2:1}\text{-}RCH{=}CH)Re_2(CO)_8(\mu\text{-}H)$ (R = CH_3, C_2H_5) with the respective alk-1-yne in hydrocarbon solution at room temperature [1, 3].

CO substitution reactions on $(\mu\text{-}\eta^{2:1}\text{-}RC{\equiv}C)Re_2(CO)_8(\mu\text{-}H)$ by ^2D reagents (see Method V) proceed under surprisingly mild conditions (compared with $(CO)_{10}Re_2$). In some cases even the monosubstitution products could not be isolated. This is said to be due to the labilization of the CO groups cis to the alkynyl ligand due to stabilization of the coordinatively unsaturated intermediate by the π electrons of the C≡C group. Actually, kinetic studies evidenced a CO dissociative mechanism. The alkynyl ligand, however, was in no case affected during these reactions, suggesting that C≡CR is much stronger bonded to the Re centers than RCH=CH in compounds of the type $(\mu\text{-}\eta^{2:1}\text{-}RCH{=}CH)Re_2(CO)_8(\mu\text{-}H)$ [3].

Fluxionality. The room-temperature ^{13}C NMR spectra display less resonances in the δ(CO) region than expected from the solid-state structure. This behavior can be rationalized in terms of a fluxional process that rapidly interchanges the σ and π bonds between the metal atoms (see Scheme 2a) [2 to 4, 9]. Variable-temperature NMR spectroscopy on several derivatives established activation parameters for the process; the values are given in the table (except that the $P(C_6H_5)_3$-substituted derivatives No. 18 and 23 are nonfluxional [11]). A closer look at the temperature-dependent changes in the spectra of some heptacarbonyl compounds (Nos. 16, 17) suggests an alternative scrambling process, namely initial dissociation of the ^2D ligand followed by rapid interconversion of the CO groups bonded to the $Re(CO)_3$ fragment and recoordination of the ^2D ligand (see Scheme 2b) [11].

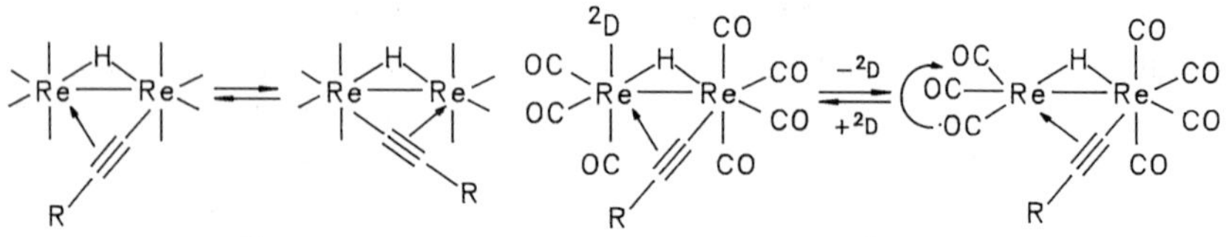

a b

Scheme 2

17α-Estradiol-Substituted Compounds. The alkynyl ligand is chiral; thus, the heptacarbonyl compounds No. 20 to 23 form as two diastereomers (see Formulas VIa, VIb, R* =

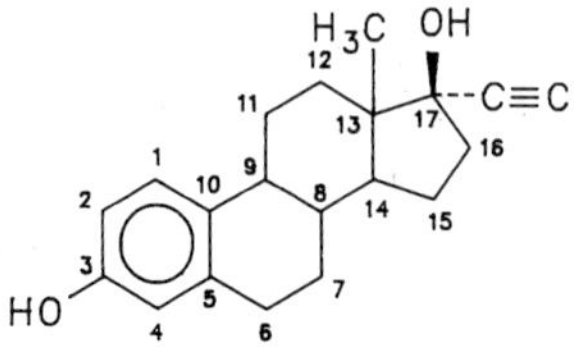

V ($C_{18}H_{23}O_2C{\equiv}C$) 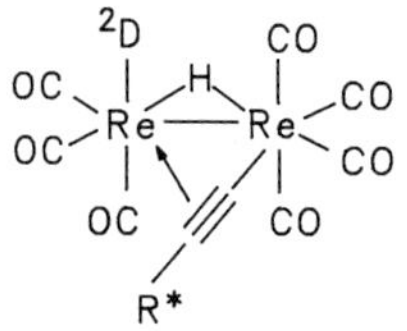 VIa 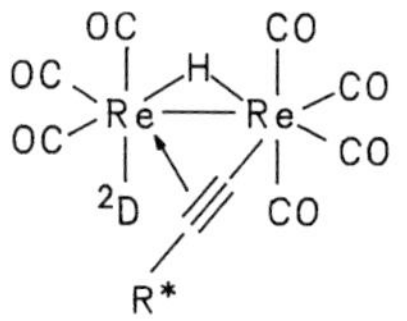 VIb

References on p. 96

$C_{18}H_{23}O_2$). As a consequence, some resonances in the ^{1}H and ^{13}C NMR spectra are doubled [9, 11].

The compounds were explored in form of an Re-bonded hormone to be used as a marker for biological systems. However, all complexes had a relative binding affinity (a measure for the ability to recognize their specific receptor) of zero [11].

Table 5
Compounds of the Type (μ-$\eta^{2:1}$-RC≡C)$Re_2(CO)_8$(μ-H) and Their CO Substitution Derivatives.
An asterisk indicates further information at the end of the table.
For explanations, abbreviations, and units see p. X.

No.	compound	method of preparation (yield) properties and remarks
	compounds with 4 CO groups	
1	(μ-$\eta^{2:1}$-HC≡C)$Re_2(CO)_4((C_6H_5)_2PCH_2P(C_6H_5)_2)_2$($\mu$-H)	I (70%) ^{1}H NMR (CD_2Cl_2): −11.89 (qui; J(P,H) = 7.1); 3.08, 3.98 (m's, CH_2); 4.11 (s, ≡CH); 7.20 to 7.42 (C_6H_5) ^{13}C {^{1}H} NMR (CD_2Cl_2): 196.5, 199.6 (CO); (−80°C): 195.1, 197.5, 198.1, 199.7 (equal-intensity s's); T_{coal} = −48°C, thus: $\Delta G^{\neq}$ = 43.1 kJ/mol IR (toluene): 1881, 1935 (ν(CO)) FD MS: $[M]^+$ [4]
2	(μ-$\eta^{2:1}$-C_6H_5C≡C)$Re_2(CO)_4((C_6H_5)_2PCH_2P(C_6H_5)_2)((CH_3)_2PCH_2P(CH_3)_2)$($\mu$-H)	I (76%); II (60%) yellow solid ^{1}H NMR (CD_2Cl_2): −12.03 (qui; J(P,H) = 7.5); 1.57, 1.76 (t's, CH_3; J(P,H) = 2.6); 1.66, 2.61, 3.07, 3.54 (q's, CH_2; J = 11.3, 11.0); 7.07 to 7.46 (m, C_6H_5) IR (toluene): 1878, 1929 (ν(CO)) ^{13}C {^{1}H} NMR (CD_2Cl_2): 197.3, 201.0 (CO) FD MS: $[M]^+$ [4]
3	(μ-$\eta^{2:1}$-C_6H_5C≡C)$Re_2(CO)_4((C_6H_5)_2PCH_2P(C_6H_5)_2)_2$($\mu$-H)	I (82%); II (65%) yellow solid ^{1}H NMR (CD_2Cl_2): −11.65 (qui; J(P,H) = 7.5); 3.15, 3.71 (m's, CH_2); 7.06 to 7.40 (m, C_6H_5); ($CDCl_3$): 3.06, 3.70 (m's, CH_2) ^{13}C {^{1}H} NMR (CD_2Cl_2): 196.8, 199.9 (CO) IR (toluene): 1882, 1936 (ν(CO)) FD MS: $[M]^+$ [4]

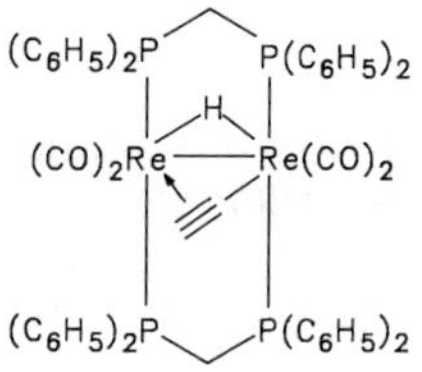

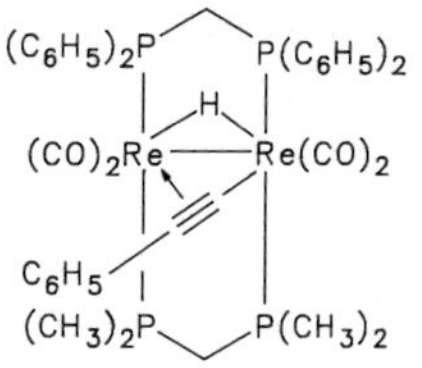

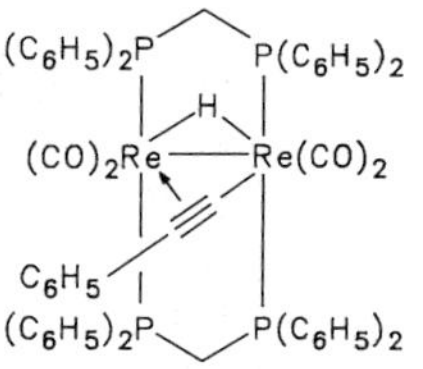

Table 5 (continued)

No.	compound	method of preparation (yield) properties and remarks

compound with 5 CO groups

4 $(\mu\text{-}\eta^{2:1}\text{-}C_6H_5C{\equiv}C)Re_2(CO)_5(P(C_6H_5)_3)((C_6H_5)_2PCH_2P(C_6H_5)_2)(\mu\text{-}H)$

(most likely structure)

by treating No. 9 with 4 equivalents $P(C_6H_5)_3$ (toluene, reflux, 3 h); yield: 70%
^1H NMR ($CDCl_3$): −11.64 (q, ReH; J(P,H) = 7.5); 2.87, 3.72 (q's; J(P,H) = 10.5)
$^{13}C\{^1H\}$ NMR ($CDCl_3$): 186.5, 190.5, 192.0 (d's; J(P,C) = 76.0, 9.0, 3.9, respectively), 194.5 (t; J(P,C) = 8.0), 199.7 (s); all peaks: 1 CO
IR (toluene): 1878, 1927, 1950, 2026 (ν(CO))
FD MS: $[M]^+$ [5]

compounds with 6 CO groups

5 $(\mu\text{-}\eta^{2:1}\text{-}HC{\equiv}C)Re_2(CO)_6((CH_3)_2PCH_2P(CH_3)_2)(\mu\text{-}H)$

III
^1H NMR ($CDCl_3$): −12.75 (t; J(P,H) = 10.0), 1.12 (q; J = 11.8), 1.64 (s), 1.85 (q; J = 11.9), 3.94 (s)
IR (toluene): 1922, 1933, 1955, 2013, 2045 (ν(CO))
MS: $[M]^+$ [5]

6 $(\mu\text{-}\eta^{2:1}\text{-}HC{\equiv}C)Re_2(CO)_6((C_6H_5)_2PCH_2P(C_6H_5)_2)(\mu\text{-}H)$

III
^1H NMR ($CDCl_3$): −12.20 (t; J(P,H) = 10.0); 2.42, 2.92 (q's; J = 12.1, 11.4); 4.04 (s); 7.09 to 7.35 (m C_6H_5)
$^{13}C\{^1H\}$ NMR (toluene-d_8): 186.6 (t; J(P,C) + J(P′,C) = 71.2); 188.9, 192.7 (s); all peaks: 2 CO; ($CDCl_3$, −77°C): 186.2, 187.7 (dd's; J(P,C) = 75.6 and 63.9); 188.0, 190.1, 191.7, 193.9 (s's, CO cis to P); T_{coal} = −46°C, thus: $\Delta G^{\neq}$ = 43.9 kJ/mol
IR (toluene): 1928, 1936, 1960, 2017, 2047 (ν(CO))
MS: $[M]^+$ [5]
treatment with $(C_6H_5)_2PCH_2P(C_6H_5)_2$ gave No. 1 [4]

*7 $(\mu\text{-}\eta^{2:1}\text{-}C_6H_5C{\equiv}C)Re_2(CO)_6(C_6H_5N{=}NC_6H_5)(\mu\text{-}H)$

by irradiating No. 25 in the presence of $N_2(C_6H_5)_2$ (toluene, 10°C, 5 h); chromatographic workup (alumina, benzene) gave a 50% yield
large prismatic, green-brown crystals
IR (KBr): 675, 690, 710, 750, 1400, 1478; 1905, 1913, 1948, 1960, 2015, 2040 (ν(CO)) [10]

References on p. 96

Table 5 (continued)

No.	compound	method of preparation (yield) properties and remarks
8	$(\mu\text{-}\eta^{2:1}\text{-}C_6H_5C{\equiv}C)Re_2(CO)_6((CH_3)_2PCH_2P(CH_3)_2)(\mu\text{-}H)$ 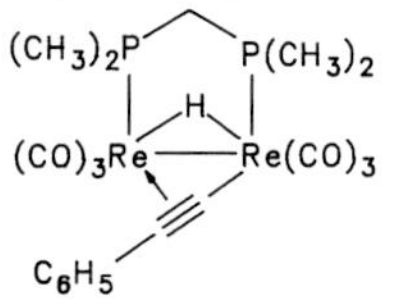	III (41%); IV (quantitative) ^{1}H NMR (CD_2Cl_2): −12.42 (t; J(P,H) = 10.3); 1.23, 1.55, 1.66 (t's), and 1.88 (q; J = 11.2, 3.6, 3.8, 11.2, respectively); 7.31 to 7.41 (m), 7.53 (dd; J = 1.6, 8.0) ^{13}C {^{1}H} NMR (toluene-d_8): 187.8 (t; J(P,C) + J(P′,C) = 72.1); 190.3, 193.0 (s's); all peaks: 2 CO IR (toluene): 1922, 1933, 1954, 2015, 2045 (ν(CO)) MS: $[M]^+$ [5] treatment with $(C_6H_5)_2PCH_2P(C_6H_5)_2$ gave No. 2 [4] photochemical reaction with $HC{\equiv}CC_6H_5$ mainly yielded $(\mu\text{-}\eta^{2:1}\text{-}C_6H_5C{\equiv}C)Re_2(\mu\text{-}\eta^{2:1}\text{-}C(C_6H_5){=}CH_2)(CO)_6\text{-}((CH_3)_2PCH_2P(CH_3)_2)$ (see p. 100) [5]
*9	$(\mu\text{-}\eta^{2:1}\text{-}C_6H_5C{\equiv}C)Re_2(CO)_6((C_6H_5)_2PCH_2P(C_6H_5)_2)(\mu\text{-}H)$ 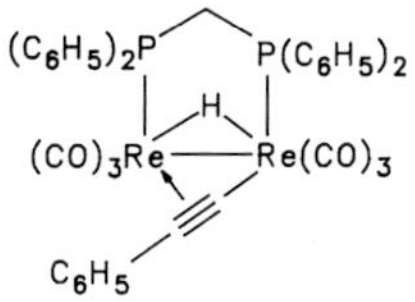	III (46%); IV (75%); see also "Further information" ^{1}H NMR ($CDCl_3$): −11.85 (t, μ-H; J(P,H) = 10.0); 2.57, 3.11 (q's, CH_2; J = 11.8 and 11.4); 6.85 to 7.85 (m, C_6H_5) ^{13}C {^{1}H} NMR (CD_2Cl_2): 187.4 (t; J(P,C) + J(P′,C) = 71.3); 189.8, 192.7 (s's); all peaks: 2 CO IR (toluene): 1928, 1936, 1960, 2019, 2046 (ν(CO)) MS: $[M]^+$ [5]
*10	$(\mu\text{-}\eta^{2:1}\text{-}C_6H_5C{\equiv}C)Re_2(CO)_6((C_6H_5)_2PCH(AuP(C_6H_5)_3)P(C_6H_5)_2)(\mu\text{-}H)$ 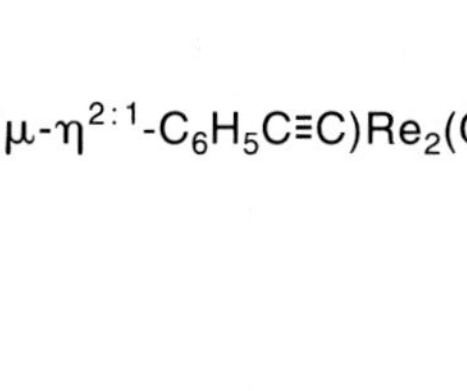	by successively treating No. 9 with CH_3Li and either $ClAuP(C_6H_5)_3$ or $[O(AuP(C_6H_5)_3)_3]BF_4$; yield: 45 or 95%, respectively ^{1}H NMR ($CDCl_3$): −9.91 (br s, μ-H); 4.15 (br s, CH), 6.78 to 8.09 (m, C_6H_5) ^{13}C {^{1}H} NMR ($CDCl_3$): 94.2, 102.0; 127.2 to 140.85 (m, C_6H_5); 187.41, 187.88, 188.21, 191.72 (CO) IR (THF): 1917, 1941, 1994, 2008, 2035 (ν(CO)) FAB MS: $[M - C_6H_5]^+$, $[M - AuP(C_6H_5)_3 - n\ CO]^+$ (n = 0 to 6) [13]
*11	$(\mu\text{-}\eta^{2:1}\text{-}C_6H_5C{\equiv}C)Re_2(CO)_6(NCCH_3)_2(\mu\text{-}H)$	V light yellow solid IR (KBr): 675, 750; 1870, 1885, 1900, 1925, 1940, 2000 (ν(CO)); 2025 (ν(C≡C)); 2910, 2940, 2955 MS: $[M]^+$ [7]
12	$(\mu\text{-}\eta^{2:1}\text{-}C_6H_5C{\equiv}C)Re_2(CO)_6(NC_5H_5)_2(\mu\text{-}H)$	V; also from No. 17 and excess C_5H_5N pale yellow needles, air-stable (even in solution) ^{1}H NMR (CD_2Cl_2): −7.40 (s, μ-H); 7.22 (m), 7.40 (t), 7.48 (t), 7.71 (m), 8.61 (dd)

References on p. 96

Table 5 (continued)

No.	compound	method of preparation (yield) properties and remarks
12 (continued)		IR (heptane): 1919, 1928, 2011, 2033 (ν(CO)) FD MS: $[M]^+$ [3]
13	$(\mu\text{-}\eta^{2:1}\text{-}C_6H_5C{\equiv}C)Re_2(CO)_6(P(C_4H_9\text{-}i)_3)_2(\mu\text{-}H)$	V (in hexane, 3 to 4 h); compound not isolated IR (n-hexane): 1927, 1935, 2012, 2032 (ν(CO)) [3]
14	$(\mu\text{-}\eta^{2:1}\text{-}C_6H_5C{\equiv}C)Re_2(CO)_6(P(C_6H_5)_3)_2(\mu\text{-}H)$	V pale yellow, air-stable solid; slowly decomposes in solution ^{1}H NMR (CD_2Cl_2): −11.28 (t, μ-H; J(P,H) = 7.9); 7.1 to 7.6 (m, C_6H_5) ^{13}C {^{1}H} NMR (CD_2Cl_2): 187.3 (d; J(P,C) = 70); 189.3, 194.1 (s's); all peaks: 2 CO IR (heptane): 1926, 1938, 2018, 2033 (ν(CO)) FD MS: $[M - CO]^+$ [3]
compounds with 7 CO groups		
15	$(\mu\text{-}\eta^{2:1}\text{-}C_6H_5C{\equiv}C)Re_2(CO)_7(NH_2C_3H_7\text{-}n)(\mu\text{-}H)$	V (within 2 h with quantitative yield) colorless crystals from CH_2Cl_2/hexane; m.p. 140 °C ^{1}H NMR (CD_2Cl_2): −11.36 (s, μ-H); 0.84 (t, CH_3), 1.37 (m, CH_2), 1.75 (s, NH_2), 2.66 (m, CH_2N); 7.39 (m, 3 H) and 7.53 (d, 2 H, C_6H_5) ^{13}C {^{1}H} NMR (CD_2Cl_2): 9.96 (CH_3), 25.99 (CH_2), 52.66 (CH_2N); 102.53, 104.87 (C≡C); 127.91, 128.07, 128.39, 131.15 (C_6H_5); 181.10, 181.60, 185.61, 186.13 (s's, 1 CO); 189.84 (m, 3 CO) IR (CH_2Cl_2): 1919, 1932, 1966, 1976, 2028, 2100 (ν(CO)) EI MS: $[M - H_2NC_3H_7 - n\,CO]^+$ (n = 0, 2 to 4, 6, 7) [11]
*16	$(\mu\text{-}\eta^{2:1}\text{-}C_6H_5C{\equiv}C)Re_2(CO)_7(NCCH_3)(\mu\text{-}H)$	V (within 1 h) [11]; VI (5 h, yield: 54%) [9, 11] colorless solid from CH_2Cl_2/pentane; m.p. 156 °C [11] ^{1}H NMR (CD_2Cl_2): −11.45 (s, μ-H); 2.20 (s, CH_3); 7.39, 7.45 (m's, 2 and 1 H, C_6H_5), 7.56 (dd, 2 H, C_6H_5) [9, 11] ^{13}C {^{1}H} NMR (CD_2Cl_2): 3.29 (CH_3); 98.85 (≡**C**C_6H_5; J(C,H) = 5), 101.78 (ReC≡; J(C,μ-H) = 4 in the ^{1}H-coupled spectrum); 119.63 (NC); 128.35, 128.71, 128.80, 131.75 (C_6H_5); 181.54, 182.75, 185.38, 187.27 (s's, 1 CO); 189.84 (m, 3 CO) [9, 11]; (−40 °C): the peak at δ = 189.84 splits into 189.50, 189.70, and 190.09 [11]; otherwise, on

References on p. 96

Table 5 (continued)

No.	compound	method of preparation (yield) properties and remarks
		raising the temperature the 181.5 and 185.4 peaks merge; $\Delta G^{\neq} = 65.7 \pm 1.7$ kJ/mol [11] IR (CH_2Cl_2): 1929, 1955, 1987, 2000, 2034, 2100 (ν(CO)) [9, 11] EI MS: $[M - CH_3CN - n\,CO]^+$ (n = 0 to 7) [9, 11]
17	$(\mu\text{-}\eta^{2:1}\text{-}C_6H_5C{\equiv}C)Re_2(CO)_7(NC_5H_5)(\mu\text{-}H)$	V (within 1.5 h) [3]; VIIa (quantitative) [11] light yellow, air-stable crystals [3, 11]; m.p. 180 °C [11] ^{1}H NMR (CD_2Cl_2): −10.42 (s, μ-H); 7.25 (dd, 2 H, C_5H_5N); 7.46 (dd + t, 3 H, C_6H_5); 7.73 (dd, 2 H, C_6H_5); 7.73 (m, 1 H, C_5H_5N); 8.47 (dd, 2 H, C_5H_5N) [11] (similar in [3]) $^{13}C\{^1H\}$ NMR (CD_2Cl_2): 103.21, 109.11 (C≡C); 126.13 (NC_5H_5); 128.20, 128.58, 129.09, 132.22 (C_6H_5); 138.84, 153.93 (NC_5H_5); 181.39, 182.90, 183.48, 187.28 (s's, CO); 193.16 (m, 3 CO) [11] IR (CH_2Cl_2): 1919, 1934, 1955, 2028, 2101 (ν(CO)) [11] (similar in [3]) FD and DCI MS: $[M]^+$ [3, 11] treatment with excess C_5H_5N gave No. 12 [3]
18	$(\mu\text{-}\eta^{2:1}\text{-}C_6H_5C{\equiv}C)Re_2(CO)_7(P(C_6H_5)_3)(\mu\text{-}H)$	V (intermediate in the reaction to give No. 14) [3]; VIIa (quantitative) [11] ^{1}H NMR (CD_2Cl_2): −12.02 [3] or −11.98 [11] (d, μ-H; J(P,H) = 8.5) [3, 11]; 7.44 (m, C_6H_5) [11] $^{13}C\{^1H\}$ NMR (CD_2Cl_2): 89.74 (ReC≡); 100.83 (d, ≡C; J(C,H) = 9.0); 128.35, 128.50, 128.67, 130.76, 131.97; 134.61 (d; J(P,C) = 10.9); 180.81, 181.63, 183.31, 185.46 (s's, 1 CO); 187.08, 188.42, 192.87 (d's, 1 CO; J(P,C) = 62.4, 7.9, 8.7, respectively); large number of δ(CO) peaks reveals nonfluxionality IR (CH_2Cl_2): 1927, 1961, 1996, 2009, 2029, 2104 (ν(CO)) [11]
19	$(\mu\text{-}\eta^{2:1}\text{-}4\text{-}CH_3OC_6H_4C{\equiv}C)Re_2(CO)_7(NC_5H_5)(\mu\text{-}H)$	V IR (toluene): 1923, 1934, 1960, 1987, 2005, 2030, 2101 (ν(CO)) [3]
20	$(\mu\text{-}\eta^{2:1}\text{-}C_{18}H_{23}O_2C{\equiv}C)Re_2(CO)_7(NH_2C_3H_7\text{-}n)(\mu\text{-}H)$ $C_{18}H_{23}O_2$ = estradiol (see Formula V)	V m.p. 135 °C ^{1}H NMR (CD_2Cl_2): −11.68 (s, μ-H); 0.79 (t, CH_3); 1.02 (s, CH_3-13); 1.35 (m, CH_2); 2.80 (m, CH_2N); 6.53 (d, H-4); 6.59 (dd, H-2); 7.17 (d, H-1)

References on p. 96

Table 5 (continued)

No.	compound	method of preparation (yield) properties and remarks
20 (continued)		$^{13}C\ \{^1H\}$ NMR (CD_2Cl_2): 13.65 (CH_3-13), 22.76 (C-15), 26.42 (C-11), 26.97 (C-7), 29.31 (C-6), 33.02 (C-12), 39.59 (C-8), 40.32 (C-16), 43.54 (C-9); 48.23, 48.43 (C-13,14); 85.20 (C-17); 100.09, 111.53 (C≡C); 112.38 (C-2), 114.90 (C-4), 126.11 (C-1), 131.75 (C-10); 137.80, 137.93 (C-5); 153.74 (C-3); n-C_3H_7: 9.98 (CH_3), 26.08 (CH_2), 51.92 (CH_2N); 179.89, 181.63, 185.31, 186.37 (all 1 CO); 191.57 (m, 3 CO) IR (CH_2Cl_2): 1914, 1930, 1961, 1976, 2005, 2026, 2099 (ν(CO)) [11]
21	(μ-$\eta^{2:1}$-$C_{18}H_{23}O_2C$≡C)$Re_2(CO)_7(NCCH_3)(\mu$-H) $C_{18}H_{23}O_2$ = estradiol (see Formula V)	V (within 1 h) [11]; VI (15 h, yield: 58%) [9, 11] colorless solid; m.p. 150°C (dec.) [11] ^1H NMR (CD_2Cl_2): −11.76, −11.75 (s, μ-H); 1.05 (CH_3-13); 2.14, 2.18 (CH_3CN); 2.83 (m, 2 H-6); 5.08, 5.38 (s, OH); 6.56 (d, H-4), 6.61 (dd, H-2), 7.20 (d, H-1) [9, 11] $^{13}C\ \{^1H\}$ NMR (CD_2Cl_2): 3.17, 3.38 (NC**C**H_3); 14.20, 14.32 (CH_3-13); 23.07, 23.26 (C-15); 26.98 (C-11); 27.52 (C-7); 29.78 (C-6); 33.32, 33.44 (C-12); 39.96 (C-8); 39.44, 40.34 (C-16); 44.13 (C-9); 49.00, 49.25, 49.57 (C-13,14); 84.65, 84.90 (C-17); 98.14, 98.55, 109.44, 109.84 (C≡C); 112.69 (C-2); 115.25 (C-4); 119.37 (CN); 126.67 (C-1); 132.78 (C-10); 138.42, 138.60 (C-5); 153.67 (C-3); 181.42, 182.65, 185.26, 187.27 (all 1 CO); 189.75 (m, 3 CO) [9] IR (CH_2Cl_2): 1925, 1953, 1984, 1999, 2033, 2099 (ν(CO)) [9] treatment with C_5H_5N, $P(C_6H_5)_3$, and CO gave Nos. 22, 23, and 27, respectively [11]
22	(μ-$\eta^{2:1}$-$C_{18}H_{23}O_2C$≡C)$Re_2(CO)_7(NC_5H_5)(\mu$-H) $C_{18}H_{23}O_2$ = estradiol (see Formula V)	V (within 10 min); VIIa (in benzene, yield: 85%) colorless solid; m.p. 130°C (dec.) ^1H NMR (CD_2Cl_2): −10.72 (s, μ-H); 1.13 (s, CH_3-13), 2.84 (m, 2 H-6), 5.32 (br s, OH), 6.56 (d, H-4), 6.62 (dd, H-2), 7.20 (m, 3 H, H-1 + 2 H of NC_5H_5); 7.72 (m), 8.67 (d), 8.74 (m); all 1 H of NC_5H_5 $^{13}C\ \{^1H\}$ NMR (CD_2Cl_2): 14.08, 14.37 (CH_3-13); 23.17 (C-15), 27.10 (C-11), 27.72 (C-7), 29.86 (C-6), 33.43 (C-12), 40.23 (C-16), 41.41 (C-8), 44.35 (C-9), 49.23 (C-13), 49.42 (C-14), 85.79 (C-17); 105.09, 112.52 (C≡C); 112.89 (C-2), 115.43 (C-4),

References on p. 96

Table 5 (continued)

No.	compound	method of preparation (yield) properties and remarks
		126.73 (C-1), 132.99 (C-10), 137.73 (C-5), 153.89 (C-3); 181.56, 182.95, 183.33, 187.38 (all 1 CO); 193.07 (m, 3 CO); C_5H_5N: 125.57, 125.81 (C-3,5); 138.46, 138.68 (C-4), 154.79, 155.30 (C-2,6) IR (CH_2Cl_2): 1917, 1927, 1953, 1990, 1993, 2028, 2099 (ν(CO)) EI MS: $[M - C_5H_5N - H_2O - n\,CO]^+$ (n = 0, 1) [11]
23	$(\mu\text{-}\eta^{2:1}\text{-}C_{18}H_{12}O_2C{\equiv}C)Re_2(CO)_7(P(C_6H_5)_3)(\mu\text{-}H)$ $C_{18}H_{23}O_2$ = estradiol (see Formula V)	V (within 10 min); VIIa (quantitative) white solid; m.p. 168 °C (dec.) ^{1}H NMR (CD_2Cl_2): −12.25 (d, μ-H; J(P,H) = 9); 1.02, 1.09 (s, CH_3-13); 2.86 (m, 2 H-6), 5.45 (br s, OH), 6.56 (d, H-4), 6.61 (dd, H-2), 7.18 (d, H-1), 7.43 (m, C_6H_5) $^{13}C\{^1H\}$ NMR (CD_2Cl_2): 14.24, 14.42 (CH_3-13); 22.82, 23.39 (C-15); 27.05 (C-11); 27.56, 27.74 (C-7); 29.78, 29.91 (C-6); 33.72, 33.93 (C-12); 39.94, 40.00 (C-8); 39.73, 40.59 (C-16); 44.05, 44.15 (C-9); 49.16 (C-13); 49.52, 49.55 (C-14); 84.51, 84.61 (C-17); 94.53, 95.35, 101.89, 102.13 (d's, C≡; J(P,C) = 6.4, 8.7, 1.6, 1.6, respectively); 112.75, 112.81 (C-2); 115.27, 115.28 (C-4); 126.64, 126.75 (C-1); 132.45 (C-10), 138.39 (C-5), 153.82 (C-3); 181.52, 181.86; 181.30, 181.59; 182.80, 183.27; 185.40, 185.82 (pairs of s's); 186.73, 186.86; 188.46, 188.98; 191.93, 192.32 (pairs of d's, CO; J(P,C) = 66.3, 62.3, 8.3, 6.5, 7.4, 8.7, respectively); $P(C_6H_5)_3$: 128.64, 130.77 (s's); 134.46 (d, C_{ortho}; J(P,C) = 11.1) IR (CH_2Cl_2): 1928, 1959, 1999, 2011, 2028, 2101 (ν(CO)) [11]
compounds with 8 CO groups		
24	$(\mu\text{-}\eta^{2:1}\text{-}HC{\equiv}C)Re_2(CO)_8(\mu\text{-}H)$	VIII [2, 5] (40.4%) [2] pale yellow needles from hexane [2] ^{1}H NMR (CD_2Cl_2): −13.30 (μ-H), 4.22 (CH≡) ^{13}C NMR (CD_2Cl_2, 10 °C): 67.06 (d; ^{1}J(C,H) = 238.4), 90.50 (d; ^{2}J(C,H) = 30.7); 181.50, 182.60, 185.15, (CO); (−33 °C): 180.12, 181.20, 181.51, 183.33, 183.75, 186.09 (2:1:2:1:1:1 CO); $\Delta G^{\neq}$ = 48.2, $\Delta H^{\neq}$ = 43.1 kJ/mol, $\Delta S^{\neq}$ = −21 J · mol^{-1} · K^{-1} IR: 1985, 1995, 2005, 2020, 2028, 2098 (ν(CO)) [2]

References on p. 96

Table 5 (continued)

No.	compound	method of preparation (yield) properties and remarks
*25	$(\mu\text{-}\eta^{2:1}\text{-}C_6H_5C{\equiv}C)Re_2(CO)_8(\mu\text{-}H)$	
		VIIb (2 h, quantitative yield) [11]; VIII [5, 7] (47%) [7]; IX (R = C_2H_5 [1], CH_3 [3]; yield: 70% (R = CH_3) [3]) cinnamon orange, air-stable (even in solution) crystals, but decomposed on silica [3, 7]; orange-brown [3]; but colorless crystals; m.p. 110°C [11] ^{1}H NMR (CD_2Cl_2): −13.01 (s, μ-H); 7.44 (m, H_{meta}, H_{para}); 7.57 (m, H_{ortho}) [1, 3] (see also [11]) ^{13}C NMR (CD_2Cl_2): 88.46, 89.07 (C≡C); 129.07, 129.21, 132.35 (C_6H_5) [11]; (C_6D_6): 181.6, 182.1 (s's); 184.6 (d; J = 3.5; collapses to an s upon broad-band ^{1}H decoupling) [3]; δ(CO) nearly identical in [11] IR (n-heptane): 1982, 2002, 2023, 2094, 2119 [1, 3]; (CH_2Cl_2): 1970, 1997, 2020, 2092, 2117 (ν(CO)) [11] FD MS: $[M]^+$ [1, 3, 7], $[Re_2]^+$ [1]
26	$(\mu\text{-}\eta^{2:1}\text{-}4\text{-}CH_3OC_6H_4C{\equiv}C)Re_2(CO)_8(\mu\text{-}H)$	
		IX (for R = CH_3) orange-brown crystals ^{1}H NMR (CD_2Cl_2): −13.00 (s, μ-H); 3.86 (s, OCH_3); 6.96, 7.51 (d's, C_6H_4; J = 8.6) IR (n-hexane): 1979, 2001, 2021, 2093, 2119 (ν(CO)) reaction with C_5H_5N gave No. 19; $k_{obs} = 4.9 \times 10^{-4}\ s^{-1}$ (toluene, 26°C, pseudo first-order conditions) [3]
27	$(\mu\text{-}\eta^{2:1}\text{-}C_{18}H_{23}O_2C{\equiv}C)Re_2(CO)_8(\mu\text{-}H)$	
	$C_{18}H_{23}O_2$ = estradiol (see Formula V)	VIIb (in benzene for 15 h; yield: 52%) colorless solid; m.p. 130°C (dec.) ^{1}H NMR (CD_2Cl_2): −13.36 (s, μ-H); 1.04 (s, CH_3-13), 2.82 (m, 2 H-6), 5.03 (s, OH), 6.55 (d, H-4), 6.61 (dd, H-2), 7.19 (d, H-1) ^{13}C {^{1}H} NMR (CD_2Cl_2): 14.25 (CH_3-13), 23.11 (C-15), 26.91 (C-11), 27.50 (C-7), 29.77 (C-6), 33.52 (C-12), 39.94 (C-16), 40.43 (C-8), 44.07 (C-9), 48.98 (C-13), 49.51 (C-14), 84.57 (C-17); 86.00, 99.65 (C≡C); 112.77 (C-2), 115.28 (C-4), 126.73 (C-1), 132.70 (C-10), 138.54 (C-5), 153.65 (C-3); 181.35, 181.58, 182.23, 184.82 (all 2 CO) IR (CH_2Cl_2): 1970, 1998, 2019, 2092, 2119 (ν(CO)) EI MS: $[M]^+$, $[M - H_2O - n\ CO]^+$ (n = 0 to 3) treatment with ^{2}D = $H_2NC_3H_7$-n, CH_3CN, C_5H_5N, $P(C_6H_5)_3$ gave Nos. 20 to 23 [11]

References on p. 96

Table 5 (continued)

No.	compound	method of preparation (yield) properties and remarks
*28	$[\mu\text{-}\eta^{2:1}\text{-}((C_6H_5)_2PCH_2P(C_6H_5)_2)Co_2(CO)_4((CH_3)_3SiC{\equiv}CC{\equiv}C)]Re_2(CO)_8(\mu\text{-}H)$	VI (two fractions were initially obtained, but the second fraction isomerized to No. 28 during recrystallization; total yield: 85%) deep red solid from CH_2Cl_2/CH_3OH ^1H NMR ($CDCl_3$): −12.88 (s, μ-H), 0.54 (s, $SiCH_3$); 3.63, 4.09 (m's, PCH_2P); 7.02 to 7.50 (m, C_6H_5) $^{13}C\{^1H\}$ NMR ($CDCl_3$): 2.09 ($SiCH_3$), 31.81 (t, CH_2P; J(P,C) = 16.92); 80.01, 86.77 (br, SiC≡C); 94.37, 96.86 (s's, $C{\equiv}CRe_2$); 127.9 to 137.2 (C_6H_5); 181.33, 182.06, 184.78 (ReCO); 202.35, 206.59 (CoCO) IR (C_6H_{12}): 1937, 1959, 1972, 1982, 1995, 2001, 2006, 2022, 2026, 2089, 2115 (ν(CO)) FAB MS: $[M - n\,CO]^+$ (n = 3 to 9) [16]

*Further information:

$(\mu\text{-}\eta^{2:1}\text{-}C_6H_5C{\equiv}C)Re_2(CO)_6(C_6H_5N{=}NC_6H_5)(\mu\text{-}H)$ (Table **5**, No. **7**) crystallizes in the triclinic space group $P\bar{1}-C_i^1$ (No. 2) with a = 9.892(2), b = 11.068(4), c = 14.473(4) Å, α = 96.56(2)°, β = 94.62(2)°, γ = 100.82(3)°; Z = 2 molecules per unit cell. The molecular structure is illustrated in **Fig. 40**. The angle between the planes formed by Re_2H and Re_2C is 157.9° [10].

$(\mu\text{-}\eta^{2:1}\text{-}C_6H_5C{\equiv}C)Re_2(CO)_6((C_6H_5)_2PCH_2P(C_6H_5)_2)(\mu\text{-}H)$ (Table **5**, No. **9**) was also obtained by heating $C_6H_5C{\equiv}CRe_2(CO)_7((C_6H_5)_2PCH_2P(C_6H_5)_2)(\mu\text{-}H)$ in toluene for 3 h. Yield: 85%. Either

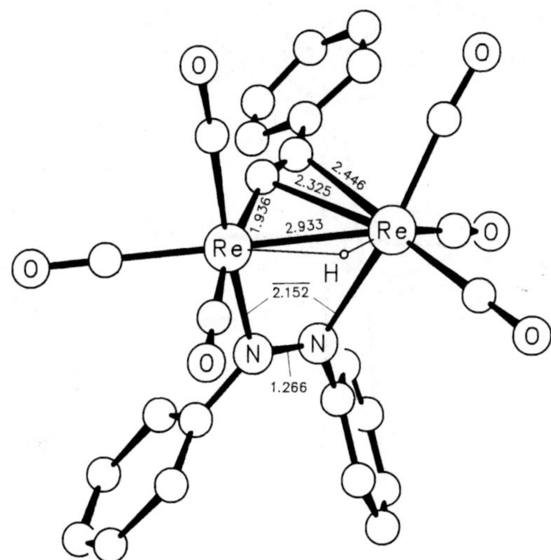

Fig. 40. Molecular structure of $(\mu\text{-}\eta^{2:1}\text{-}C_6H_5C{\equiv}C)Re_2(CO)_6(C_6H_5N{=}NC_6H_5)(\mu\text{-}H)$ [10].

References on p. 96

isomer (cis or trans) of the starting compound could be employed for the photochemical or thermal reaction. No. 9 was also accessible from $(CO)_8Re_2(\mu\text{-}(C_6H_5)_2PCH_2P(C_6H_5)_2)$ and $HC{\equiv}CC_6H_5$ in the presence of 3 equivalents $(CH_3)_3NO$ [5].

The complex did not react with CO, not even under pressure. In contrast, treatment with excess $P(C_6H_5)_3$ gave No. 4 [5], and treatment with excess $(C_6H_5)_2PCH_2P(C_6H_5)_2$ gave No. 3 [4]. Irradiation in the presence of excess $HC{\equiv}CR$ (R = H, C_6H_5) produced the μ-alkynyl-μ-alkenyl compounds $(\mu\text{-}\eta^{2:1}\text{-}C_6H_5C{\equiv}C)Re_2(\mu\text{-}\eta^{2:1}\text{-}C(R){=}CH_2)(CO)_5((C_6H_5)_2PCH_2P(C_6H_5)_2)$ (see p. 101). A study of the kinetics revealed the first-order reaction constant $k_{obs} = 2.40 \times 10^{-2}\ min^{-1}$ for the reaction with acetylene [5].

Successive treatment with CH_3Li and either $ClAuP(C_6H_5)_3$ or $[O(AuP(C_6H_5)_3)_3]BF_4$ gave No. 10. The reactions with $CH_3AuP(C_6H_5)_3$ and $CH_3Li/[C_8H_{12}Rh(\mu\text{-}Cl)]_2$ provided the cluster compounds $(\mu\text{-}\eta^{2:1}\text{-}C_6H_5C{\equiv}C)Re_2(CO)_6((C_6H_5)_2PCH_2P(C_6H_5)_2)(\mu\text{-}AuP(C_6H_5)_3)$ and $(\mu\text{-}\eta^{2:1}\text{-}C_6H_5C{\equiv}C)Re_2(CO)_6[\mu\text{-}(P(C_6H_5)_2)_2CHRh(C_8H_{12})H]$, respectively (see pp. 96/7) [13].

$(\mu\text{-}\eta^{2:1}\text{-}C_6H_5C{\equiv}C)Re_2(CO)_6((C_6H_5)_2PCH(AuP(C_6H_5)_3)P(C_6H_5)_2)(\mu\text{-}H)$ (Table **5**, No. **10**) crystallizes in the triclinic space group $P\bar{1}-C_i^1$ (No. 2) with a = 15.335(4), b = 14.420(6), c = 13.114(3) Å, α = 80.12(3)°, β = 71.29(3)°, γ = 87.74(3)°; Z = 2 molecules per unit cell, D_{calc} = 1.93 g/cm³. The molecular structure is depicted in **Fig. 41** [13].

$(\mu\text{-}\eta^{2:1}\text{-}C_6H_5C{\equiv}C)Re_2(CO)_6(NCCH_3)_2(\mu\text{-}H)$ (Table **5**, No. **11**) was also obtained when a reaction between $(\mu\text{-}\eta^{2:1}\text{-}C_6H_5C{\equiv}C)Re_2(CO)_8(\mu\text{-}H)$ and $(CO)_3Mo(NCCH_3)_3$ in boiling CH_3CN was attempted. After concentration and cooling, only No. 11 separated with 70% yield [7].

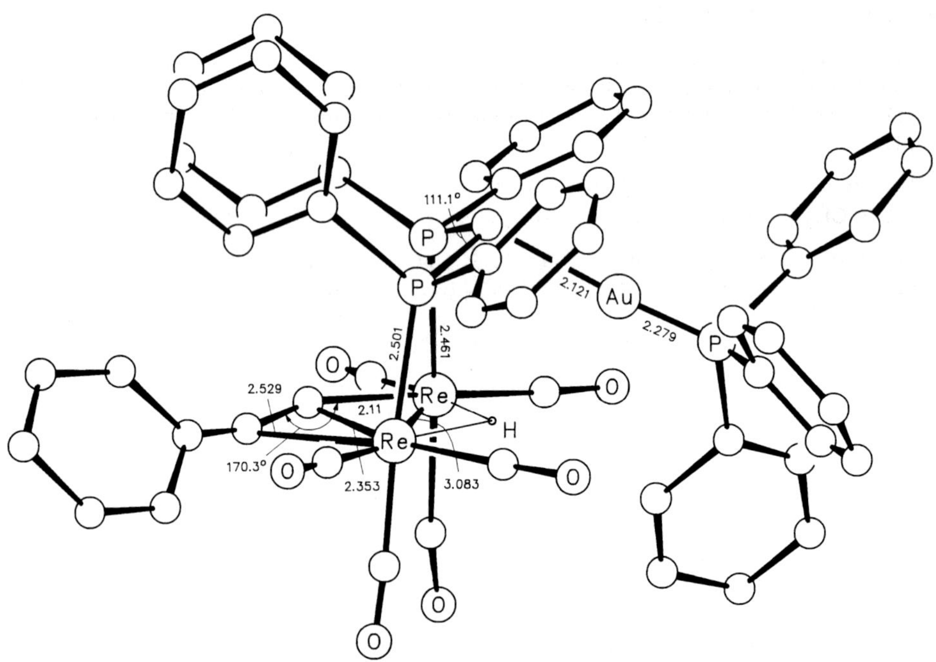

Fig. 41. Molecular structure of $(\mu\text{-}\eta^{2:1}\text{-}C_6H_5C{\equiv}C)Re_2(CO)_6((C_6H_5)_2PCH(AuP(C_6H_5)_3)P(C_6H_5)_2)(\mu\text{-}H)$ [13].

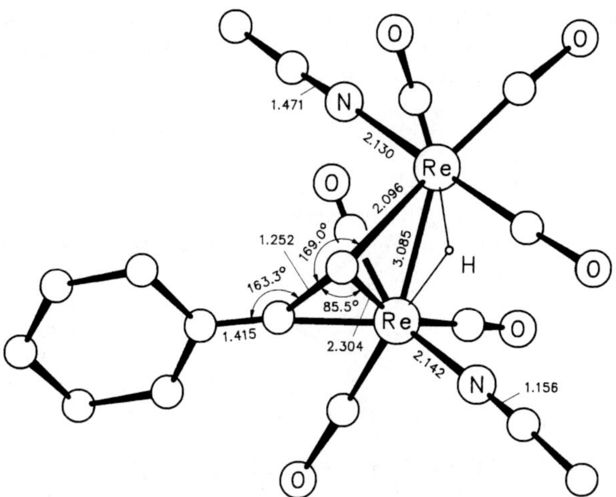

Fig. 42. Molecular structure of $(\mu\text{-}\eta^{2:1}\text{-}C_6H_5C{\equiv}C)Re_2(CO)_6(NCCH_3)_2(\mu\text{-}H)$ [7].

The compound crystallizes in the monoclinic space group $P2_1/n-C_{2h}^5$ (No. 14) with a = 16.557(5), b = 10.908(2), c = 22.594(6) Å (β not given); Z = 8 molecules per unit cell. The molecular structure is shown in **Fig. 42** [7].

Treatment with $(C_5H_5)_2Mo_2(CO)_6$ in boiling xylene for 10 h yielded the cluster $[(C_5H_5)_2\text{-}Mo_2(CO)_2(CH{=}CC_6H_5)(\mu\text{-}CO)_2]Re_2(CO)_6$ (see p. 158) [7].

$(\mu\text{-}\eta^{2:1}\text{-}C_6H_5C{\equiv}C)Re_2(CO)_7(NCCH_3)(\mu\text{-}H)$ (Table **5**, No. **16**). Single-crystal crystallography furnished the following lattice characteristics: monoclinic, a = 7.2462(7), b = 16.146(2), c = 17.060(3) Å, β = 101.13(1)°; space group $P2_1/n-C_{2h}^5$ (No. 14), Z = 4 molecules per unit cell, D_{calc} = 2.41 g/cm³. **Fig. 43** shows the molecular structure [9, 11].

Treatment with C_5H_5N, $P(C_6H_5)_3$, or CO (atmospheric pressure was sufficient) substituted the labile CH_3CN ligand, thereby providing No. 17, 18, or 25, respectively [11].

$(\mu\text{-}\eta^{2:1}\text{-}C_6H_5C{\equiv}C)Re_2(CO)_8(\mu\text{-}H)$ (Table **5**, No. **25**) crystallizes in the monoclinic space group B2/b (non-standard setting of $C2/c-C_{2h}^6$ (No. 15)) with a = 11.787(2), b = 17.594(3), c = 17.770(3) Å (β not given); Z = 8 molecules per unit cell. The molecular structure is illustrated in **Fig. 44**. The hydride ligand bridges asymmetrically [7].

The compound readily undergoes thermal CO substitutions with donor reagents. Treatment with $H_2NC_3H_7$-n quantitatively gave No. 15 [11]. Addition of CH_3CN yielded No. 16 within 1 h [11] and No. 11 when the mixture was heated [7]. The room-temperature reaction with pyridine initially yielded the monosubstitution product No. 17 (a kinetic study assuming first-order conditions revealed $k_{obs} = 4.9 \times 10^{-4}\ s^{-1}$ (toluene, 26 °C)); long-term exposure to C_5H_5N (10 d) caused exchange of a second CO group to give No. 12. Treatment with PR_3 (R = C_4H_9-i, C_6H_5) rapidly formed $(\mu\text{-}\eta^{2:1}\text{-}C_6H_5C{\equiv}C)Re_2(CO)_6(PR_3)_2(\mu\text{-}H)$ (in the case of R = C_6H_5, NMR-monitoring revealed the intermediate presence of No. 18). In contrast, treatment with $P(CH_3)_3$ separated a dipolar addition product with the composition $(\mu\text{-}(CH_3)_3PC{=}CC_6H_5)Re_2(CO)_8(\mu\text{-}H)$ (see p. 58). No reaction, however, was observed with C_2H_4 or excess $HC{\equiv}CC_6H_5$ at 50 °C [3]. Irradiation in the presence of $C_6H_5N{=}NC_6H_5$ yielded No. 7 [10]. Treatment with 4,5-(diphenylphosphino)cyclopent-4-en-1,3-dione gave the complex shown in Formula VII [15].

References on p. 96

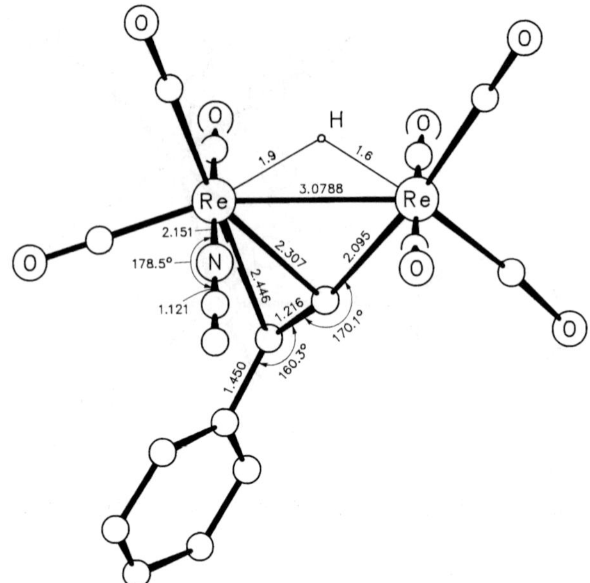

Fig. 43. Molecular structure of (μ-$\eta^{2:1}$-$C_6H_5C{\equiv}C$)$Re_2(CO)_7(NCCH_3)$(μ-H) [9, 11].

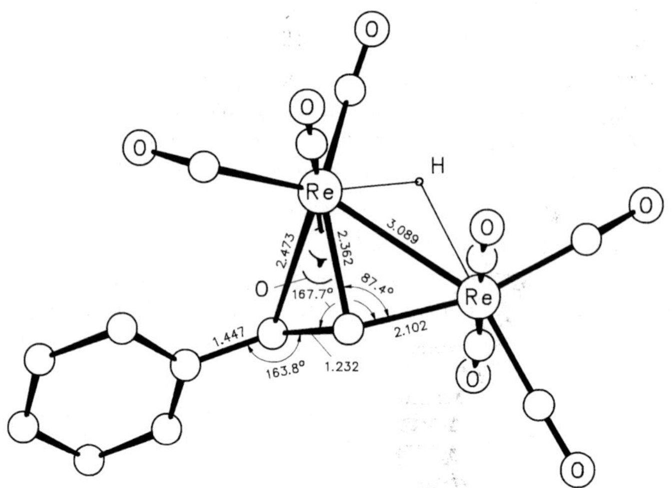

Fig. 44. Molecular structure of (μ-$\eta^{2:1}$-$C_6H_5C{\equiv}C$)$Re_2(CO)_8$(μ-H) [7].

Several clusters were also built from No. 25: Treatment with $(C_5H_5)_2Ni$ in refluxing xylene gave $(CO)_{10}Re_2$, $(C_5H_5)_3Ni_3(CO)_2$, [μ-$(C_5H_5)_2Ni_2(C_6H_5C{=}CCH{=}CHC_6H_5)$]$Re_2(CO)_6$ (see p. 118) [8, 12], and [μ-$(C_5H_5)_2Ni_2CCHC_6H_5$]$Re_2(CO)_6$(μ-CO) (see Formula VIII) [12]. The room-temperature reaction with $(CO)_8Co_2$ yielded only $(CO)_5ReC{\equiv}CC_6H_5 \cdot Co_2(CO)_6$ and $(CO)_5ReH$ [12, 14]. Interaction with $(CO)_{12}Fe_3$ in refluxing benzene gave [$(CO)_6Fe_2$-$(C{=}CHC_6H_5)CO$]$Re_2(CO)_8$ (see p. 2) [12]; however, in refluxing toluene $(CO)_6Re_2$(μ-CO)-

VII VIII IX

[μ-$CCHC_6H_5Fe_2(CO)_6$] (see Formula IX) [12] or (μ-$C_8H_6Fe(CO)_3$)$Re_2(CO)_6$ (see p. 160) [8] were obtained depending on the Re:Fe ratio (1.5:1 [12] or 1:1 [8]). The thermal reaction with $(C_5H_5)_2Mo_2(CO)_6$ in xylene gave [$(C_5H_5)_2Mo_2(CO)_2(CH{=}CC_6H_5)(\mu\text{-}CO)_2$]$Re_2(CO)_6$ (see p. 158) [6, 7], while treatment with $(CO)_3Mo(NCCH_3)_3$ in refluxing CH_3CN produced only No. 11 [7].

[μ-$\eta^{2:1}$-($(C_6H_5)_2PCH_2P(C_6H_5)_2$)$Co_2(CO)_4$($(CH_3)_3SiC{\equiv}CC{\equiv}C$)]$Re_2(CO)_8$($\mu$-H) (Table **5**, No. **28**) crystallizes in the orthorhombic space group Pccn−D_{2h}^{10} (No. 56) with a = 43.683(10), b = 17.804(3), c = 12.185(4) Å; Z = 8 molecules per unit cell, D_{calc} = 1.87 g/cm^3. The structure of the molecule is depicted in **Fig. 45**. The C≡CC≡C torsion angle is 153° [16].

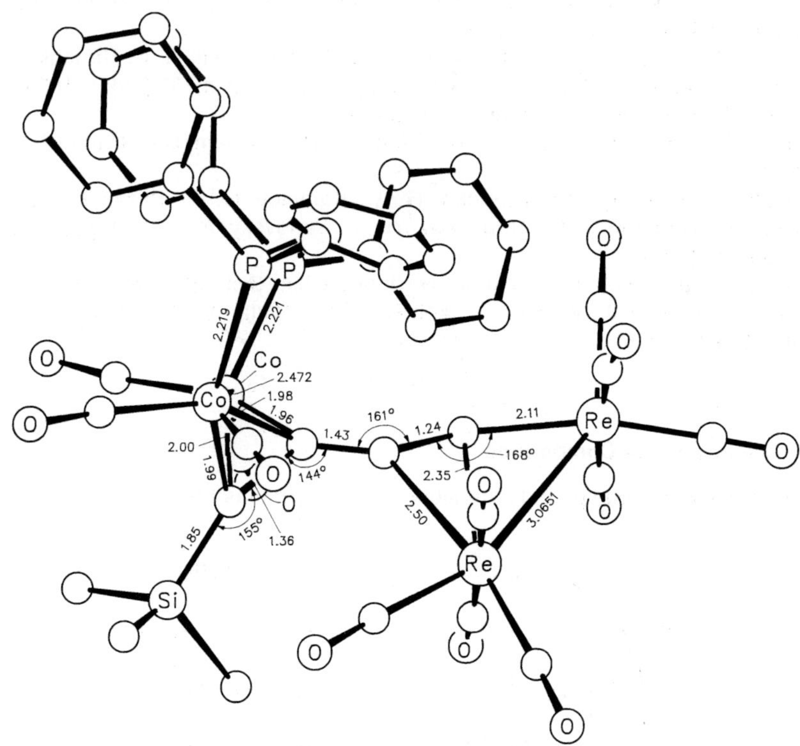

Fig. 45. Molecular structure of [μ-$\eta^{2:1}$-($(C_6H_5)_2PCH_2P(C_6H_5)_2$)$Co_2(CO)_4$($(CH_3)_3SiC{\equiv}CC{\equiv}C$)]$Re_2(CO)_8$($\mu$-H) [16].

References on p. 96

References:

[1] Nubel, P. O.; Brown, T. L. (J. Am. Chem. Soc. **104** [1982] 4955/7).
[2] Franzreb, K.-H.; Kreiter, C. G. (Z. Naturforsch. **39b** [1984] 81/5).
[3] Nubel, P. O.; Brown, T. L. (Organometallics **3** [1984] 29/32).
[4] Lee, K.-W.; Brown, T. L. (Organometallics **4** [1985] 1025/30).
[5] Lee, K.-W.; Pennington, W. T.; Cordes, A. W.; Brown, T. L. (J. Am. Chem. Soc. **107** [1985] 631/41).
[6] Shaposhnikova, A. D.; Stadnichenko, R. A.; Bel'skii, V. K.; Pasynskii, A. A. (Izv. Akad. Nauk SSSR Ser. Khim. **1987** 1913/4; Bull. Acad. Sci. USSR Div. Chem. Sci. [Engl. Transl.] **36** [1987] 1776/7).
[7] Shaposhnikova, A. D.; Stadnichenko, R. A.; Bel'skii, V. K.; Pasynskii, A. A. (Metalloorg. Khim. **1** [1988] 945/51; Organomet. Chem. USSR [Engl. Transl.] **1** [1988] 522/6).
[8] Shaposhnikova, A. D.; Kamalov, G. L.; Stadnichenko, R. A.; Pasynskii, A. A.; Eremenko, I. L.; Nefedov, S. E.; Struchkov, Yu. T.; Yanovsky, A. I. (J. Organomet. Chem. **405** [1991] 111/20).
[9] Top, S.; Gunn, M.; Jaouen, G.; Vaissermann, J.; Daran, J.-C.; Thornback, J. R. (J. Organomet. Chem. **414** [1991] C 22/C 27).
[10] Eremenko, I. L.; Pasynskii, A. A.; Nefedov, S. E.; Katugin, A. S.; Kolobkov, B. I.; Shaposhnikova, A. D.; Stadnichenko, R. A.; Yanovski, A. I.; Struchkov, Yu. T. (Zh. Neorg. Khim. **37** [1992] 574/82; Russ. J. Inorg. Chem. [Engl. Transl.] **37** [1992] 284/9).

[11] Top, S.; Gunn, M.; Jaouen, G.; Vaissermann, J.; Daran, J.-C.; McGlinchey, M. J. (Organometallics **11** [1992] 1201/9).
[12] Pasynskii, A. A.; Eremenko, I. L.; Nefedov, S. E.; Yanovskii, A. I.; Struchkov, Yu. T.; Shaposhnikova, A. D.; Stadnichenko, R. A. (Zh. Neorg. Khim. **38** [1993] 455/65; Russ. J. Inorg. Chem. [Engl. Transl.] **38** [1993] 423/32).
[13] Bruce, M. I.; Low, P. J.; Skelton, B. W.; White, A. H. (J. Chem. Soc. Dalton Trans. **1993** 3145/6).
[14] Shaposhnikova, A. D.; Stadnichenko, R. A.; Kamalov, G. L.; Pasynskii, A. A.; Eremenko, I. L.; Nefedov, S. E.; Struchkov, Yu. T.; Yanovsky, A. I. (J. Organomet. Chem. **453** [1993] 279/81).
[15] Xia, C.-G.; Bott, S. G.; Richmond, M. G. (Inorg. Chim. Acta **230** [1995] 45/50).
[16] Bruce, M. I.; Low, P. J.; Werth, A.; Skelton, B. W.; White, A. H. (J. Chem. Soc. Dalton Trans. **1996** 1551/66).

2.2.2.2.2 Miscellaneous Compounds

The structures of the compounds are illustrated in Formulas I to V.

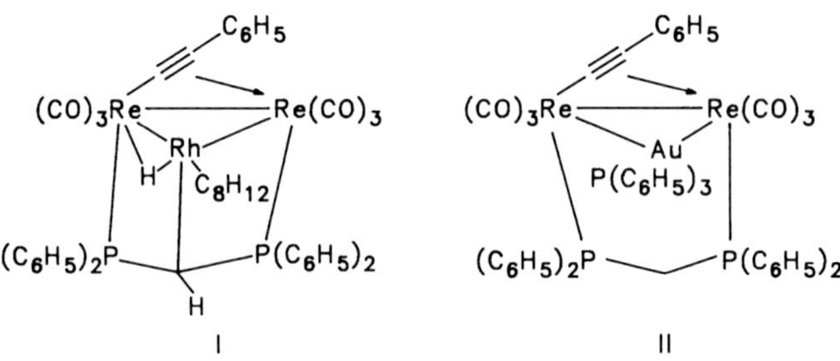

References on p. 100

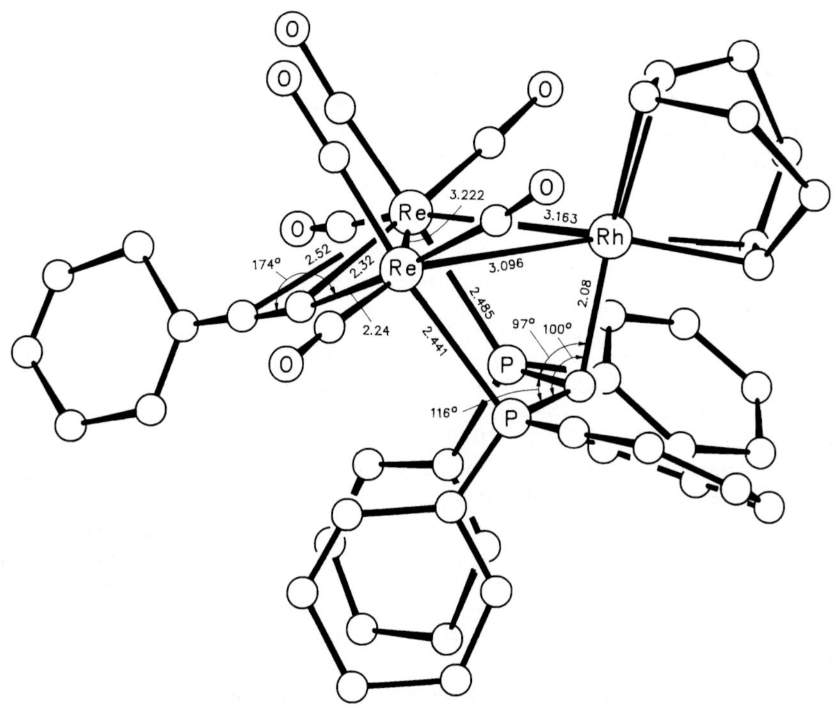

Fig. 46. Molecular structure of
$(\mu\text{-}\eta^{2:1}\text{-}C_6H_5C{\equiv}C)Re_2(CO)_6[\mu\text{-}(P(C_6H_5)_2)_2CHRh(C_8H_{12})H]$ [3].

$(\mu\text{-}\eta^{2:1}\text{-}C_6H_5C{\equiv}C)Re_2(CO)_6[\mu\text{-}(P(C_6H_5)_2)_2CHRh(C_8H_{12})H] \cdot 0.5\ CH_2Cl_2$ (see Formula I; C_8H_{12} = cycloocta-1,5-diene) was obtained by treating $(\mu\text{-}\eta^{2:1}\text{-}C_6H_5C{\equiv}C)Re_2(CO)_8(\mu\text{-}H)$ with CH_3Li and $[C_8H_{12}Rh(\mu\text{-}Cl)]_2$. Yield: 30%. The solid melts at 227 to 230 °C (dec.).

^{1}H NMR spectrum ($CDCl_3$): δ = −19.43 (d, RhH; J(Rh,H) = 13.5 Hz); 1.76, 2.21 (m's, C_8H_{12}); 3.45 (t, CHP_2; J(P,H) = 10.1 Hz); 4.27, 4.72 (s's, CH= of C_8H_{12}); 6.65, 6.82, 6.98, 7.44 to 7.58, 7.96 to 8.01 (m's, C_6H_5) ppm. IR spectrum (C_6H_{12}): 1920, 1931, 1938, 1955, 2004, 2017, 2041 (ν(CO)) cm^{-1}. FAB mass spectrum: $[M - n\ CO]^+$ (n = 0 to 6).

The cluster crystallizes in the orthorhombic space group Pbca$-D_{2h}^{15}$ (No. 61) with a = 23.470(8), b = 21.199(6), c = 18.814(5) Å; Z = 8 molecules per unit cell, D_{calc} = 1.82 g/cm^3. The molecular structure is illustrated in **Fig. 46** [3].

$(\mu\text{-}\eta^{2:1}\text{-}C_6H_5C{\equiv}C)Re_2(CO)_6((C_6H_5)_2PCH_2P(C_6H_5)_2)(\mu\text{-}AuP(C_6H_5)_3)$ (see Formula II) was obtained with 60% yield by treating $(\mu\text{-}\eta^{2:1}\text{-}C_6H_5C{\equiv}C)Re_2(CO)_8(\mu\text{-}H)$ with $CH_3AuP(C_6H_5)_3$ in refluxing THF. Bright yellow-green crystals; m.p. 248 to 253 °C (dec.).

^{1}H NMR spectrum ($CDCl_3$): δ = 2.53 to 2.64 and 4.18 to 4.27 (m's, CH_2); 6.92 to 7.56 (m, C_6H_5) ppm. IR spectrum (THF): 1903, 1914, 1935, 1987, 2018 (ν(CO)) cm^{-1}. FAB mass spectrum: $[M - n\ CO]^+$ (n = 0 to 6) [3].

$[((CO)_5ReC{\equiv}CH)Re(CO)_5]BF_4$ (see Formula III) was obtained by treating $(CO)_5ReFBF_3$ with either 3.3 equivalents $(CH_3)_3SiC{\equiv}CH$ or 0.5 equivalent $(CH_3)_3EC{\equiv}CE(CH_3)_3$ (E = Si, Sn)

References on p. 100

$\left[(CO)_5Re{-}C{\equiv}C{-}H \atop (CO)_5Re\right]^+$ (III)

$(CO)_4Re(P(C_6H_5)_3){-}C{\equiv}C{-}Pt(P(C_6H_5)_3)(CO)$, $(CO)_4Re$ (IV)

$(CO)_3Re$, $Re(CO)_3$, C_6H_5, H, W, O, $C_5(CH_3)_5$ (V)

III IV V

in CH_2Cl_2 at 0°C for 26 h or at room temperature for 0.5 h. Yields amounted to 73% or 30%. The compound also separated with nearly quantitative yield (93%) when $(CO)_5ReC{\equiv}CRe(CO)_5$ was treated with 1 equivalent $HBF_4 \cdot O(C_2H_5)_2$ in CH_2Cl_2.

^1H NMR spectrum (CD_3NO_2): δ = 4.55 ppm. $^{13}C\{^1H\}$ NMR spectrum (CD_3NO_2): δ = 57.6 (Re**C**≡CH), 82.4 (≡CH; J(C,H) = 134 Hz); 178.5, 179.8, 180.7, 181.5, 181.7, 182.4 (CO) ppm. IR spectrum (Nujol): 1870 (ν(C≡C)); 2012, 2038, 2055, 2110, 2115, 2125, 2165, 2170 (ν(CO)); 3155 (ν(CH)) cm^{-1}; (KBr): 1865; 1965, 2010, 2032, 2050, 2070, 2110, 2135, 2156, 2170; 3150 cm^{-1}. EI mass spectrum: $[M - HBF_4]^+$, $[M - HBF_4 - n\,CO]^+$ (n = 1 to 10).

Treatment with $NaOC_2H_5$ in THF gave $(CO)_5Re{-}C{\equiv}C{-}Re(CO)_5$ with good yield [1].

[μ-(CO)Pt(P(C₆H₅)₃)C≡C]Re₂(CO)₈P(C₆H₅)₃ (see Formula IV). π-$C_2H_4Pt(P(C_6H_5)_3)_2$ and $(CO)_5ReC{\equiv}CRe(CO)_5$ were combined in THF at −78°C. The mixture was stirred at this temperature for 1 h and then warmed to room temperature. Removal of the solvent left an oil; crystallization from CH_2Cl_2 gave a yellow, air-stable solid; dec.p. 165°C. Yield: 73%.

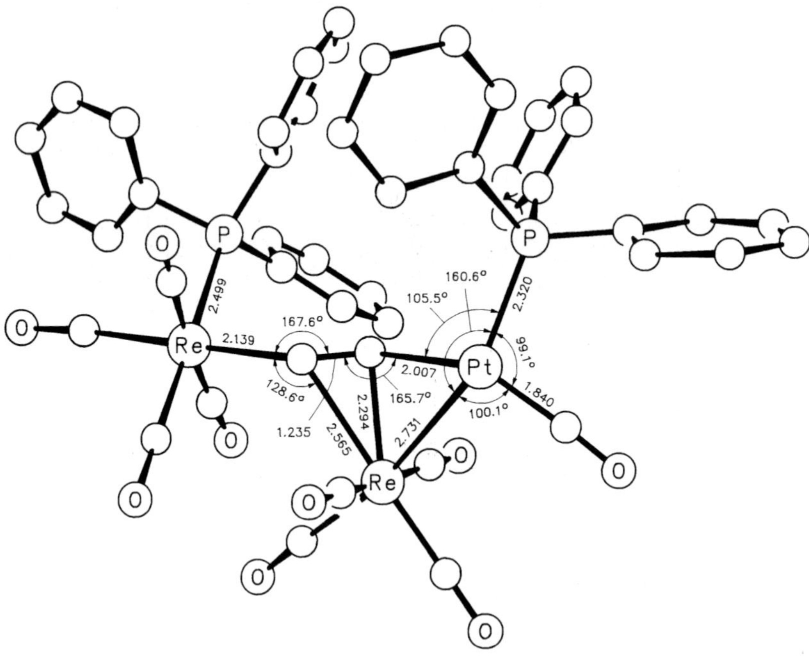

Fig. 47. Molecular structure of $[\mu\text{-}(CO)Pt(P(C_6H_5)_3)C{\equiv}C]Re_2(CO)_8P(C_6H_5)_3$ [2].

References on p. 100

$^{13}C\{^1H\}$ NMR spectrum (CD_2Cl_2): δ = 84.3 (d, ≡C; J(P,C) = 13 Hz), 148.6 (dd, ≡C; J(P,C) = 7 and 3 Hz); 185.2, 185.9, 186.6, 192.1, 193.0, 193.9, 199.1 (CO) ppm. $^{31}P\{^1H\}$ NMR spectrum (CD_2Cl_2): δ = 2.3 (d; $^5J(P,P)$ = 2 Hz), 2.3 (dd; $^4J(Pt,P)$ = 14, $^5J(P,P)$ = 2 Hz), 35.0 (d), 35.0 (dd; $^1J(Pt,P)$ = 2643 Hz) ppm. ^{195}Pt NMR (CD_2Cl_2): δ = −818 (dd; $^1J(Pt,P)$ = 2643, $^4J(Pt,P)$ = 14 Hz) ppm.

X-ray crystallography showed the compound to crystallize in the monoclinic space group $P2_1/c-C^5_{2h}$ (No. 14) with a = 15.758(4), b = 16.123(4), c = 17.693(5) Å, β = 95.45(2)°; Z = 4 molecules per unit cell, D_{calc} = 2.031 g/cm³. The structure is illustrated in **Fig. 47** [2].

[μ-(C$_5$(CH$_3$)$_5$)W(O)(C≡CC$_6$H$_5$)H]Re$_2$(CO)$_6$(μ-H) · 0.5 CHCl$_3$ (see Formula V) was obtained with 50% yield by reacting $(CO)_8Re_2(\mu_3\text{-}CCC_6H_5)W(C_5(CH_3)_5)(O)$ (see Formulas IVa, IVb, R = C_6H_5; p. 126) with H_2 under atmospheric pressure (heptane, 90°C, 40 min). $[\mu\text{-}(C_5(CH_3)_5)W(O)CH{=}CHC_6H_5]Re_2(CO)_8$ (see p. 117) and $(CO)_8Re_2(\mu\text{-}H)W(O)(C_5(CH_3)_5)$-($\mu$-$CHCH_2C_6H_5$) (see Formula IV, p. 118) were simultaneously obtained, but the title product was shown to be the initial product of the reaction. It forms a dark red solid.

^1H NMR spectrum ($CDCl_3$): δ = −14.08 (ReHRe), −7.44 (ReHW; J(W,H) = 135 Hz); 2.09 (s, $C_5(CH_3)_5$); 7.38, 7.53 (t's, 1 and 2 H; J = 7.2 Hz), 7.92 (d, 2 H; J = 8 Hz) ppm. IR spectrum (C_6H_{12}): 1932, 1938, 1954, 1971, 2017, 2052 (ν(CO)) cm^{-1}.

Single crystals have a triclinic lattice with a = 9.740(1), b = 10.533(1), c = 15.357(1) Å, α = 102.50(1)°, β = 93.50(1)°, γ = 114.76(1)°; space group $P\bar{1}-C^1_i$ (No. 2). The molecular structure is illustrated in **Fig. 48**. The C≡C bond is perpendicular to one W-Re edge, while the two bridging hydrides span the other W-Re edge and the Re-Re bond.

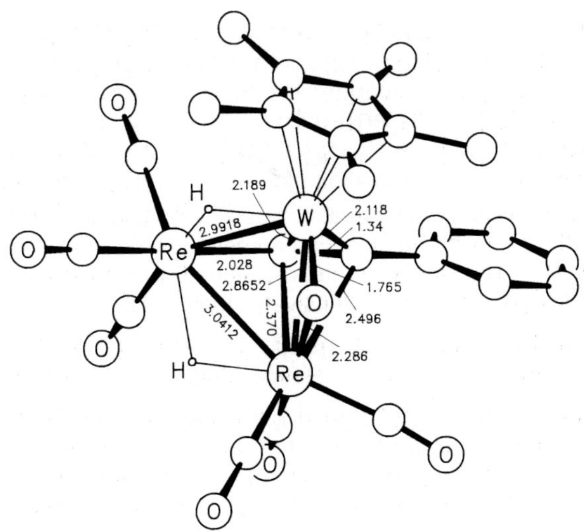

Fig. 48. Molecular structure of $[\mu\text{-}(C_5(CH_3)_5)W(O)(C{\equiv}CC_6H_5)H]Re_2(CO)_6(\mu\text{-}H)$ [4].

Treatment with CO in refluxing CH_2Cl_2 gave $[\mu\text{-}(C_5(CH_3)_5)W(O)CH{=}CHC_6H_5]Re_2(CO)_8$, which is already one of the coproducts of the preparation (above) [4].

References:

[1] Appel, M.; Heidrich, M.; Beck, W. (Chem. Ber. **120** [1987] 1087/9).
[2] Weidmann, T.; Weinrich, V.; Wagner, B.; Robl, C.; Beck, W. (Chem. Ber. **124** [1991] 1363/8).
[3] Bruce, M. I.; Low, P. J.; Skelton, B. W.; White, A. H. (J. Chem. Soc. Dalton Trans. **1993** 3145/6).
[4] Chi, Y.; Cheng, P.-S.; Wu, H.-L.; Hwang, D.-K.; Su, P.-C.; Peng, S.-M.; Lee, G. H. (J. Chem. Soc. Chem. Commun. **1994** 1839/40).

2.2.2.3 Compounds with Alkenyl and Alkynyl Ligands

This section describes several compounds each possessing bridging σ,π-Re bonded alkenyl and alkynyl ligands. Their structures are illustrated by Formulas I and II (assignment of Formula I is based on the spectroscopic properties only and thus tentative; in contrast, the structure of one of the Formula-II-type compounds was crystallographically established).

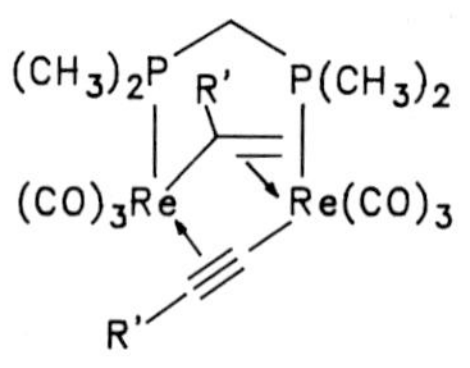

I 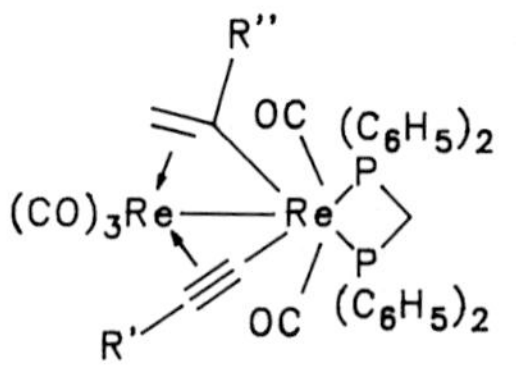 II

Preparation. The compounds were prepared as follows:

Method I: Photochemical reaction of $(CO)_8Re_2(\mu\text{-}R_2PCH_2PR_2)$ (R = CH_3, C_6H_5) with HC≡CR′ (R′ = H, C_6H_5) in toluene. Workup by preparative TLC (silica plates; CH_2Cl_2/hexane (3:1)) separated the title compounds from products of the types $R'C{\equiv}CRe_2(CO)_7(R_2PCH_2PR_2)(\mu\text{-}H)$ (see "Organorhenium Compounds" 5, 1994, pp. 260/2) and $(\mu\text{-}\eta^{2:1}\text{-}R'C{\equiv}C)Re_2(CO)_6(R_2PCH_2PR_2)(\mu\text{-}H)$ (see Table 5). The yields depend on the irradiation time and the bridging ligand.

The kinetics was investigated. It was shown that the reactions of $(CO)_8Re_2$-$(\mu\text{-}R_2PCH_2PR_2)$ with acetylenes proceed slower than the analogous reactions of $(CO)_{10}Re_2$. The kinetics of the reaction of $(CO)_8Re_2(\mu\text{-}(C_6H_5)PCH_2P(C_6H_5)_2)$ with $HC{\equiv}CC_6H_5$ has been extensively studied [1].

Method II: Photochemical reaction of $(\mu\text{-}\eta^{2:1}\text{-}R'C{\equiv}C)Re_2(CO)_6((C_6H_5)_2PCH_2P(C_6H_5)_2)$-$(\mu\text{-}H)$ with excess HC≡CR″ in toluene. Separation by preparative TLC [1].

$(\mu\text{-}\eta^{2:1}\text{-}HC{\equiv}C)Re_2(\mu\text{-}\eta^{2:1}\text{-}CH{=}CH_2)(CO)_6((CH_3)_2PCH_2P(CH_3)_2)$ (see Formula I, R′ = H) was obtained by Method I, but could not be isolated, because it rapidly decomposed on TLC plates. Nevertheless, the following ^1H NMR signals were recorded on the reaction mixture: δ = 5.45, 5.70 (d's; J = 10.9, 17.0 Hz, respectively); 6.30 (dd) ppm [1].

$(\mu\text{-}\eta^{2:1}\text{-}C_6H_5C{\equiv}C)Re_2(\mu\text{-}\eta^{2:1}\text{-}C(C_6H_5){=}CH_2)(CO)_6((CH_3)_2PCH_2P(CH_3)_2)$ (see Formula I, R′ = C_6H_5) was obtained by Method I with 19% yield. The compound was also prepared according to Method II inside an IR cell, yielding the complex as the dominant product (70%).

^1H NMR spectrum ($CDCl_3$): δ = 1.15 (t; J = 7.2 Hz); 1.68 to 1.97 (m), 4.67 (s); 7.07 to 7.28 (m); 7.38 (t; J = 7.4 Hz), 7.81 (d; J = 7.6 Hz) ppm. ^{13}C {^1H} NMR spectrum ($CDCl_3$):

Reference on p. 102

δ = 187.6, 189.1 (d's; J(P,C) = 65.7, 72.5 Hz); 192.9 (s); 195.5, 195.7, 196.2 (d's; J(P,C) = 7.8, 6.9, 8.2 Hz, respectively) ppm. IR spectrum (toluene): 1880, 1907, 1932, 1951, 1996, 2030 (ν(CO)) cm^{-1}. FAB mass spectrum: $[M]^+$ [1].

(μ-$\eta^{2:1}$-$C_6H_5C{\equiv}C$)Re_2(μ-$\eta^{2:1}$-CH=CH_2)$(CO)_5$(($C_6H_5)_2PCH_2P(C_6H_5)_2$) (see Formula II, R′ = C_6H_5, R″ = H) was obtained by Method II as a yellow, crystalline solid.

^1H NMR spectrum ($CDCl_3$): δ = 3.05 (d; J(H,H) = 15.7 Hz); 4.80, 5.00 (m's); 5.14 (dd; J(H,H) = 11.0, J(P,H) = 5.4 Hz); 6.85 to 7.9 (m, C_6H_5), 9.08 (dd; J(H,H) = 11.0, 15.7 Hz) ppm. ^{13}C {^1H} NMR spectrum ($CDCl_3$): δ = 193.0, 194.9, 197.3 199.1 (s's); 201.6 (dd; J(P,C) = 14.0, 32.4 Hz) ppm. IR spectrum (toluene): 1892, 1914, 1928, 1937, 1959, 2022 (ν(CO)) cm^{-1}. FD mass spectrum: $[M]^+$.

The compound crystallizes in the orthorhombic space group Pbca$-D_{2h}^{15}$ (No. 61) with a = 12.743(2), b = 15.974(2), c = 35.174(4) Å; Z = 8 molecules per unit cell, D_{calc} = 1.902 g/cm^3. The molecular structure is depicted in **Fig. 49** [1].

(μ-$\eta^{2:1}$-$C_6H_5C{\equiv}C$)Re_2(μ-$\eta^{2:1}$-C(C_6H_5)=CH_2)$(CO)_5$(($C_6H_5)_2PCH_2P(C_6H_5)_2$) (see Formula II, R′ = R″ = C_6H_5) was obtained by Method I with 25% yield or by Method II with 50% yield.

^1H NMR spectrum ($CDCl_3$): δ = 2.44 (s, 1 H); 4.87 to 4.97 (m, 3 H); 6.89 to 7.60 (m, C_6H_5) ppm. ^{13}C {^1H} NMR spectrum ($CDCl_3$): δ = 194.0, 195.2, 197.0, 198.8 (s's); 200.1 (dd; J(P,C) = 12.9, 33.7 Hz) ppm. IR spectrum (toluene): 1887, 1915, 1931, 1942, 1944, 2022 (ν(CO)) cm^{-1}. FD mass spectrum: $[M]^+$ [1].

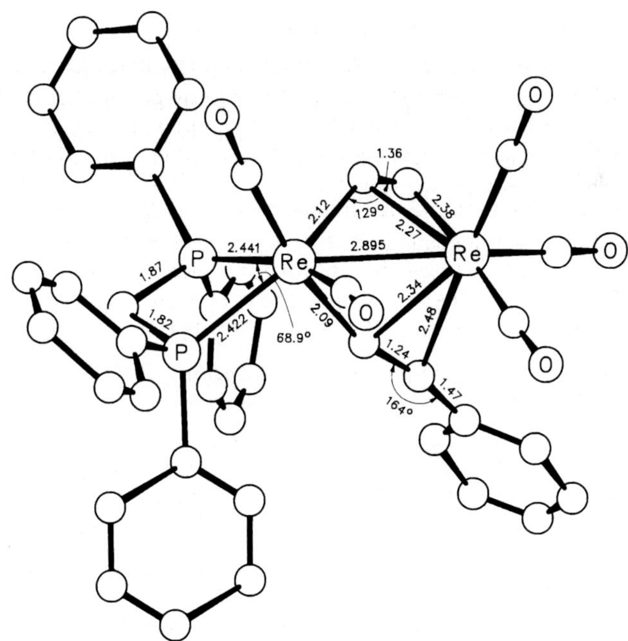

Fig. 49. Molecular structure of (μ-$\eta^{2:1}$-$C_6H_5C{\equiv}C$)Re_2(μ-$\eta^{2:1}$-CH=CH_2)$(CO)_5$(($C_6H_5)_2PCH_2P(C_6H_5)_2$) [1].

Reference on p. 102

Reference:

[1] Lee, K.-W.; Pennington, W. T.; Cordes, A. W.; Brown, T. L. (J. Am. Chem. Soc. **107** [1985] 631/41).

2.2.3 Compounds with π-Bonded ^2L Ligands

2.2.3.1 Compounds with Terminal ^2L Ligands

2.2.3.1.1 Compounds with Alkenes

(π-$C_4H_2O_3$)$Re_2(CO)_7(P(OCH_3)_3)_2$ and **(π-$C_4H_2O_3$)$Re_2(CO)_9$** ($C_4H_2O_3$ = maleic anhydride). Laser flash photolysis on toluene solutions containing $(CO)_8Re_2(P(OCH_3)_3)_2$ or $(CO)_{10}Re_2$ in the presence of $C_4H_2O_3$ initiates Re-Re bond cleavage and CO dissociation, thereby simultaneously generating the radical $(CO)_4Re^{\bullet}(^2D)$ and the CO loss species $(CO)_7Re_2(^2D)_2$ (^2D = $P(OCH_3)_3$ or CO). While the radicals react with maleic anhydride via an electron transfer reaction, the CO loss species form adducts with the organic ligand. The latter have never been separated; yet their formation was confirmed by transient absorption spectroscopy: Absorptions assigned to the radical species soon decayed, but after that a new absorption band appeared which was attributed to adduct formation. It was also demonstrated that $(C_4H_2O_3)Re_2(CO)_7(P(OCH_3)_3)_2$ is not as stable as $(C_4H_2O_3)Re_2(CO)_9$ [7].

π-$C_2H_4Re_2(CO)_9$ was one product obtained by the photochemical reaction of $(CO)_{10}Re_2$ with C_2H_4 (hexane, −50 to −20°C, workup by preparative PLC with hexane/CH_2Cl_2 mixtures, yield: 6.5%) [2, 5]. The compound was also observed when the reaction was conducted in Ar-diluted C_2H_4 matrices [8]. It forms colorless needles [5].

^{1}H NMR spectrum (CD_2Cl_2): δ = 3.05 (s) ppm. ^{13}C {^{1}H} NMR spectrum: δ = 31.6 (t; J(C,H) = 163 Hz); 178.8, 183.4, 183.6, 195.6 (CO, ratio 1:2:1:5) ppm [5]. IR spectrum (n-hexane): 1947, 1965, 1980, 1990, 2012, 2050, 2115 [5]; 1940, 1975, 2015.3, 2077.9, 2114.4 [8] (ν(CO)) cm^{-1}. Spectroscopic data suggest an equatorial C_2H_4 coordination.

Photolysis in hexane at −20°C gave (μ-$\eta^{2:1}$-CH_2=CH)$Re_2(CO)_8$(μ-H) [5]. Upon long-wavelength irradiation of the Ar matrix-embedded compound new weak IR bands were observed (1914, 2028, 2086 cm^{-1}) which were suggested to be part of the absorption spectrum of axially substituted π-$C_2H_4Re_2(CO)_9$ [8].

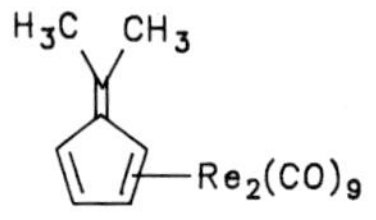

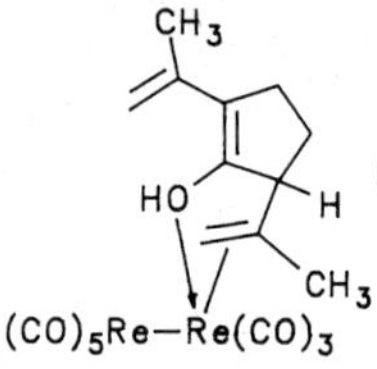

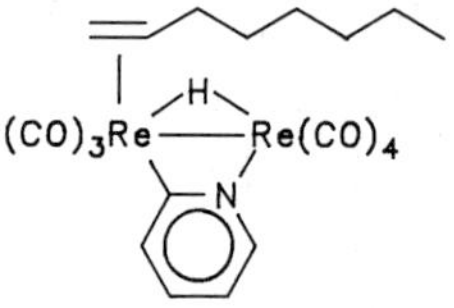

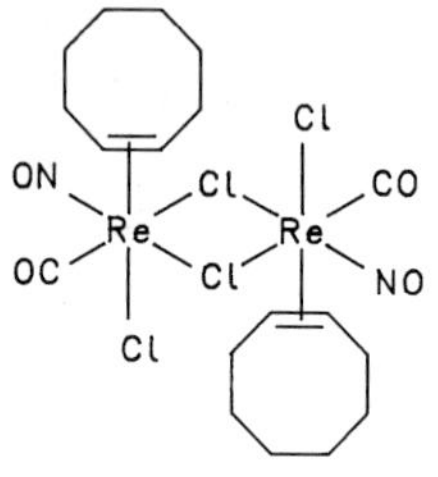

π-$C_8H_{10}Re_2(CO)_9$ (see Formula I) was obtained along with (μ-$\eta^{2:1}$-C_8H_9)$Re_2(CO)_8$(μ-H) and ($C_5H_4C(CH_3)$=CH_2)$Re(CO)_3$ by photochemically reacting $(CO)_{10}Re_2$ with a 100-fold excess of 6,6-dimethylfulvene in hexane at −20°C. Separation was achieved by preparative HPLC with hexane/CH_2Cl_2. Yield: 2%. Yellow, spherical crystals.

^{1}H NMR spectrum (CD_2Cl_2): δ = 2.03, 2.27 (s's, CH_3); 5.3 (s, 1 H); 5.56, 6.07, 6.72 (d's, 1 H) ppm [2].

References on p. 104

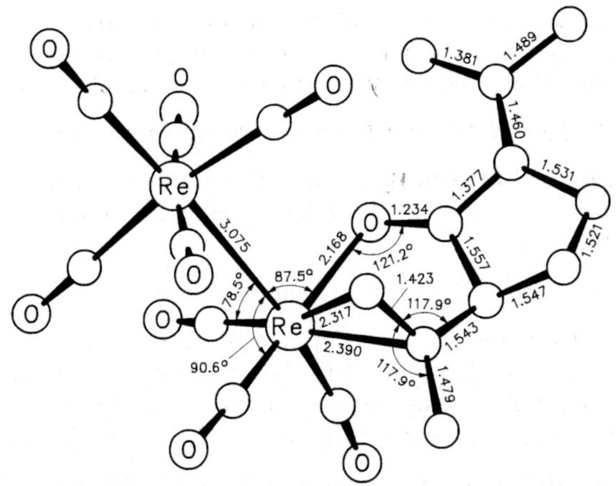

Fig. 50. Molecular structure of (π-$C_{11}H_{16}O$)$Re_2(CO)_8$ [9].

π-$C_8H_8Re_2(CO)_9$ (C_8H_8 = cyclooctatetraene) was obtained with 4.6% [2] yield by photochemically reacting $(CO)_{10}Re_2$ with cyclooctatetraene (hexane, −20 °C [2]; pentane, −35 °C, 10 to 180 min [6]). Separation by HPLC with hexane/CH_2Cl_2 yielded colorless needles [2].

^{1}H NMR spectrum (CD_2Cl_2): δ = 4.49 (s); 5.88 (d); 6.01 (s); 6.72 (d's; all 2 H) ppm. Mass spectrum: $[M]^+$ [2].

(π-$C_{11}H_{16}O$)$Re_2(CO)_8$ (see Formula II) was produced with low yield (0.5%) by the photochemical reaction of $(CO)_{10}Re_2$ with 1,1-dimethyl allene (hexane, −30 °C, 1 h). After evaporating the solvent, the mixture was crudely separated chromatographically (alumina, CH_2Cl_2/hexane). Each fraction was purified by HPLC and recrystallized from pentane.

Red rhomboids. IR spectrum (n-hexane): 1598; 1911, 1929, 1962, 1971, 1983, 1998, 2005, 2092 (ν(CO)) cm^{-1}.

The complex crystallizes in the orthorhombic space group $P2_12_12_1-D_2^4$ (No. 19) with a = 12.477(2), b = 14.537(1), c = 11.899(1) Å; Z = 4 molecules per unit cell. The molecular structure is illustrated in **Fig. 50** [9].

(π-C_8H_{16})$Re_2(CO)_7(\mu$-H)(μ-C_5H_4N) (C_8H_{16} = oct-1-ene, see Formula III) was obtained by irradiating an Ar-purged mixture of $(CO)_8Re_2(\mu$-H)(μ-C_5H_4N) in oct-1-ene for 1 h. The spectroscopically determined yield exceeded 50%.

^{1}H NMR spectrum (C_6D_6): δ = −13.70 (br s, μ-H, half-width ca. 9 Hz); 0.87 (t, CH_3); 1.17 (br s, 5 CH_2); 1.5 to 2.9 (br m, H_2C=); 3.7 to 4.8 (br m, CH=); 6.02 (td, H_γ or H_β); 6.47 (td, H_β or H_γ); 7.33 (dd, $H_{\beta'}$); 7.67 (dd, H_α) ppm. IR spectrum (toluene): 1928, 1945, 1978, 1993, 2035, 2088 (ν(CO)) cm^{-1}.

The compound rapidly decomposed in air. Treatment with CO (1 atm, benzene, room temperature) readily formed $(CO)_8Re_2(\mu$-H)(μ-C_5H_4N), whereas thermolysis (hot benzene) gave $(CO)_7Re_2(NC_5H_5)(\mu$-H)(μ-C_5H_4N) along with other insoluble, unidentified species. The complex showed no tendency to isomerize excess oct-1-ene to internal octenes at 25 °C [3].

π-$C_2H_4Re_2(\mu$-$\eta^{2:1}$-$C_4H_7)(CO)_7(\mu$-H) (C_4H_7 = CH=CHC_2H_5 and C(C_2H_5)=CH_2) was obtained by irradiating a toluene solution containing (μ-$\eta^{2:1}$-CH_2=CH)$Re_2(CO)_8(\mu$-H) for 45 min while

References on p. 104

bubbling C_2H_4 through the solution. In addition to the IR bands of the three isomers of $(\mu\text{-}\eta^{2:1}\text{-}C_4H_7)Re_2(CO)_8(\mu\text{-}H)$, weak bands were detected which were assigned to the CO-substituted title product. The combined yield of all these products amounted to 15%.

IR spectrum: 1920, 2044, 2105 (ν(CO)) cm^{-1} (disappeared upon exposure to CO) [4].

$(\pi\text{-}CH_3CH{=}CH_2)Re_2(\mu\text{-}\eta^{2:1}\text{-}CH{=}CHCH_3)(CO)_7(\mu\text{-}H)$ was obtained by irradiating a toluene solution containing $(\mu\text{-}\eta^{2:1}\text{-}CH_3CH{=}CH)Re_2(CO)_8(\mu\text{-}H)$ while bubbling propene through the solution. Yield (by IR): ca. 10%.

IR spectrum: 2038, 2105 (ν(CO)) cm^{-1} (disappeared upon exposure to CO) [4].

$(\pi\text{-}C_8H_{14})_2Re_2(CO)_2(NO)_2Cl_2(\mu\text{-}Cl)_2$ (see Formula IV) formed from $(CO)_4Re_2(NO)_2Cl_2$ and cyclooctene (ratio 1:1.5) in CCl_4 (reflux, 8 h). The mixture was filtered and concentrated; addition of hexane separated the crude product. Recrystallization from CCl_4/hexane gave a cream yellow, air-stable, crystalline powder (m.p. >300 °C) with 65% yield.

IR spectrum (Nujol): 250, 283, 326 (ν(ReCl)); (CCl_4): 1770 (ν(NO)), 2058 (ν(CO)) cm^{-1}. Molecular weight (determined osmometrically): M = 862 (calc. 851) g/mol.

Treatment with ^2D = pyridine, $P(OC_6H_5)_3$, or $As(C_6H_5)_3$ in CCl_4 or 1,2-dichloroethane at reflux temperature gave compounds of the type $(CO)Re(^2D)_2(NO)Cl_2$ [1].

References:

[1] Trovati, A.; Uguagliati, P.; Zingales, F. (Inorg. Chem. **10** [1971] 851/3).
[2] Franzreb, K.-H.; Kreiter, C. G. (J. Organomet. Chem. **246** [1983] 189/95).
[3] Nubel, P. O.; Wilson, S. C.; Brown, T. L. (Organometallics **2** [1983] 515/25).
[4] Nubel, P. O.; Brown, T. L. (J. Am. Chem. Soc. **106** [1984] 3474/84).
[5] Kreiter, C. G.; Franzreb, K. H.; Sheldrick, W. S. (Z. Naturforsch. **41b** [1986] 904/14).
[6] Caballero, C.; Oberdorfer, F.; Guggolz, E.; Nuber, B.; Korswagen, R. P.; Ziegler, M. L. (Bol. Soc. Qim. Peru **53** [1987] 135/49; C.A. **110** [1989] No. 213022).
[7] Rushman, P.; Brown, T. L. (J. Am. Chem. Soc. **109** [1987] 3632/9).
[8] Klotzbücher, W. E. (J. Mol. Struct. **174** [1988] 5/10).
[9] Kreiter, C. G.; Michels, W.; Heeb, G. (J. Organomet. Chem. **502** [1995] 9/18).

2.2.3.1.2 Compounds with Alkynes

2.2.3.1.2.1 Compounds of the Type $[(\pi\text{-}RC{\equiv}CR')Re_2(^1L)((C_6H_5)_2PCH_2P(C_6H_5)_2)_2X_3]^+$

General. Structure. This section presents several cations with a structure illustrated by Formula I, where ^1L represents CO or RNC (R = t-C_4H_9, 2,6-$(CH_3)_2C_6H_3$) and X stands for Cl, Br. The counteranions always are PF_6^- for ^1L = CO (in some cases BF_4^- salts were also separated) and $O_3SCF_3^-$ for ^1L = RNC.

Crystal-structure analyses performed on two of the cations revealed similar molecular geometries: The Re atoms possess different geometries: distorted octahedral and distorted square-pyramidal. The ^1L ligand and the π-bonded acetylene ligand are in an anti arrangement to one another. The C≡C vector is perpendicularly oriented with respect to the equatorial plane of the cation (defined by the Re atoms, X atoms, and ^1L). The bridgehead CH_2 groups of the diphosphanes are directed to that side of the molecule which contains the coordinated alkyne [1, 2]. The Re-Re connection is a triple bond with a $\sigma^2\pi^4\delta^2\delta^{*2}$ electron occupation [1].

References on p. 109

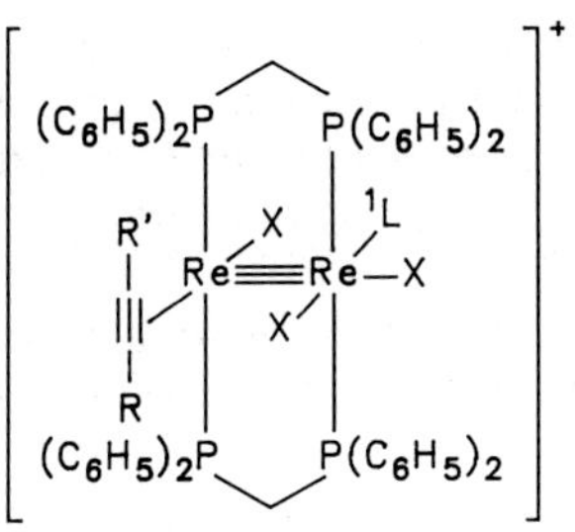

I

Preparation. All salts were prepared by treating $(CO)Re_2((C_6H_5)_2PCH_2P(C_6H_5)_2)_2X_4$ or $RNCRe_2((C_6H_5)_2PCH_2P(C_6H_5)_2)_2X_4$ (X = Cl, Br) with the respective alkyne RC≡CR′ in the presence of $TlPF_6$ or TlO_3SCF_3 in CH_2Cl_2 at room temperature (Method I). The respective mixture was filtered, and the filtrate was subsequently treated as follows:

a) Concentration and addition of ether to precipitate the product [1, 2].

b) Addition of $HBF_4 \cdot O(C_2H_5)_2$. After stirring the mixture for 20 min, the treatment continued as in Method Ia [1].

The reactions did not proceed in the absence of $TlPF_6$. An attempt to perform an analogous reaction with $C_6H_5C{\equiv}CC_6H_5$ failed [1].

Properties. The compounds are yellow-brown, diamagnetic solids. They are thermally very stable and air-stable as well [1]. The IR spectra of the PF_6 salts always display one band at ca. 830 to 845 cm^{-1} due to ν(PF) [1], those of the O_3SCF_3 salts display one band at 1250 to 1265 due to ν(SO) [2]. The conductivity of acetone solutions is between 90 and 115 $cm^2 \cdot \Omega^{-1} \cdot mol^{-1}$ [1, 2]. The $^{31}P\ \{^1H\}$ NMR spectra display two multiplets which are components of an AA′BB′ or ABCD pattern [1, 2].

Electrochemistry. Cyclic voltammograms (CV) were recorded in CH_2Cl_2/0.1 M $[N(C_4H_9\text{-}n)_4]PF_6$ solution at a Pt bead electrode. The data were referenced to the Ag/AgCl electrode. The salts with ^1L = CO showed a reversible one-electron reduction close to −0.5 V and an irreversible reduction between −1.2 and 1.6 V [1]. The salts with ^1L = RNC showed an irreversible reduction at ca. −0.8 V and a reversible one-electron oxidation at ca. 1.3 V [2].

Table 6
Compounds of the Type $[(\pi\text{-}RC{\equiv}CR')Re_2(^1L)((C_6H_5)_2PCH_2P(C_6H_5)_2)_2X_3]^+$.
An asterisk indicates further information at the end of the table.
For explanations, abbreviations, and units see p. X.

No.	compound	method of preparation (yield) properties and remarks
compounds with ^1L = CO		
*1	$[(\pi\text{-}HC{\equiv}CH)Re_2(CO)((C_6H_5)_2PCH_2P(C_6H_5)_2)_2Cl_3]PF_6$	
		Ia (84%) ^1H NMR (CD_2Cl_2): 10.4 (AA′BB′; CH≡); unchanged over the range −80 to 25 °C $^{13}C\ \{^1H\}$ NMR (^{13}CO-labeled): 143.2 (t, CH≡; J(P,C) = 10.4), 200.8 (t, CO; J(P,C) = 6.7)

References on p. 109

Table 6 (continued)

No.	compound	method of preparation (yield) properties and remarks
*1 (continued)		$^{31}P\{^1H\}$ NMR (CD_2Cl_2): −3.5, −1.5 IR (Nujol): 1944 (ν(CO)); ^{13}CO-labeled cation: 1919 ($\nu(^{13}CO)$) CV: $E_{p,c}$ = −1.26, $E_{1/2}$ (red.) = −0.48 [1]
2	[(π-$CH_3C{\equiv}CCH_3$)Re_2(CO)((C_6H_5)$_2PCH_2P$(C_6H_5)$_2$)$_2Cl_3$]X	
	X = PF_6	Ia (67%) IR (Nujol): 1936 (ν(CO)) CV: $E_{p,c}$ = −1.45, $E_{1/2}$ (red.) = −0.50 [1]
	X = BF_4	Ib $^{31}P\{^1H\}$ NMR (CD_2Cl_2): −1.4, 2.9 [1]
*3	[(π-$CH_3C{\equiv}CC_2H_5$)Re_2(CO)((C_6H_5)$_2PCH_2P$(C_6H_5)$_2$)$_2Cl_3$]X	
	X = PF_6	Ia (72%) IR (Nujol): 1937 (ν(CO)) CV: $E_{p,c}$ = −1.58, $E_{1/2}$ (red.) = −0.51 [1]
	X = BF_4	Ib (59%) ^1H NMR (CD_2Cl_2): 1.17 (t, $\mathbf{CH_3}CH_2$), 2.83 (s, CH_3); 2.75, 3.60 (m's, $\equiv CCH_2$); 5.32, 5.74 (m's, PCH_2); 7.0 to 8.1 (m, C_6H_5) $^{31}P\{^1H\}$ NMR (CD_2Cl_2): −1.1, 2.2 IR (Nujol): 1056 (ν(BF)), 1936 (ν(CO)) [1]
	X = O_3SCF_3	like Method Ia except for using TlO_3SCF_3 (yield: 81%) IR (Nujol): 1938 (ν(CO)) [3]
*4	[(π-$HC{\equiv}CC_3H_7$-n)Re_2(CO)((C_6H_5)$_2PCH_2P$(C_6H_5)$_2$)$_2Cl_3$]X	
	X = PF_6	Ia (74%) ^1H NMR (CD_2Cl_2): 11.1 (m, CH≡) $^{31}P\{^1H\}$ NMR (CD_2Cl_2): −2.6, 0.3 IR (Nujol): 1944 (ν(CO)) CV: $E_{p,c}$ = −1.32, $E_{1/2}$ (red.) = −0.50 [1]
	X = O_3SCF_3	like Method Ia except for using TlO_3SCF_3 (yield: 87%) IR (Nujol): 1942 (ν(CO)) [3]
5	[(π-$C_2H_5C{\equiv}CC_2H_5$)Re_2(CO)((C_6H_5)$_2PCH_2P$(C_6H_5)$_2$)$_2Cl_3$]PF_6	Ia (69%) $^{31}P\{^1H\}$ NMR ($CDCl_3$): −1.8, 2.0 IR (Nujol): 1941 (ν(CO)) CV: $E_{p,c}$ = −1.47, $E_{1/2}$ (red.) = −0.52 [1]
*6	[(π-$HC{\equiv}CC_4H_9$-n)Re_2(CO)((C_6H_5)$_2PCH_2P$(C_6H_5)$_2$)$_2Cl_3$]PF_6	Ia [3]
*7	[(π-$HC{\equiv}CC_6H_5$)Re_2(CO)((C_6H_5)$_2PCH_2P$(C_6H_5)$_2$)$_2Cl_3$]PF_6	Ia (77%) ^1H NMR (CD_2Cl_2): 11.4 (m, CH≡) $^{31}P\{^1H\}$ NMR (CD_2Cl_2): −3.5 IR (Nujol): 1950 (ν(CO)) CV: $E_{p,c}$ = −1.30, $E_{1/2}$ (red.) = −0.45 [1]

References on p. 109

Table 6 (continued)

No.	compound	method of preparation (yield) properties and remarks
8	$[(\pi\text{-}HC{\equiv}CC_6H_4CH_3\text{-}4)Re_2(CO)((C_6H_5)_2PCH_2P(C_6H_5)_2)_2Cl_3]PF_6$	Ia (66%) ^1H NMR (CD_2Cl_2): 10.4 (m, CH≡) $^{31}P\{^1H\}$ NMR (CD_2Cl_2): −3.5 IR (Nujol): 1943 (ν(CO)) CV: $E_{p,c}$ = −1.21, $E_{1/2}$ (red.) = −0.45 [1]
9	$[(\pi\text{-}CH_3C{\equiv}CC_6H_5)Re_2(CO)((C_6H_5)_2PCH_2P(C_6H_5)_2)_2Cl_3]PF_6$	Ia (61%) $^{31}P\{^1H\}$ NMR (CD_2Cl_2): −1.4, 2.9 IR (Nujol): 1936 (ν(CO)) CV: $E_{p,c}$ = −1.54, $E_{1/2}$ (red.) = −0.50 [1]
10	$[(\pi\text{-}HC{\equiv}CH)Re_2(CO)((C_6H_5)_2PCH_2P(C_6H_5)_2)_2Br_3]PF_6$	Ia (79%) $^{31}P\{^1H\}$ NMR (CD_2Cl_2): −13.5, −5.1 IR (Nujol): 1937 (ν(CO)) CV: $E_{p,c}$ = −1.25, $E_{1/2}$ (red.) = −0.45 [1]
compounds with ^1L = RNC		
*11	$[(\pi\text{-}HC{\equiv}CH)Re_2(CNC_4H_9\text{-}t)((C_6H_5)_2PCH_2P(C_6H_5)_2)_2Cl_3]O_3SCF_3$	Ia (72%) pale orange crystals; contain lattice CH_2Cl_2 ^1H NMR (acetone-d_6): 1.63 (s); 5.88, 6.12; 7.0 to 8.3 (m), 9.89 (m, CH) $^{31}P\{^1H\}$ NMR (CD_2Cl_2): −8.55, 0.12 IR (Nujol): 2114 (ν(CN)) CV: $E_{p,c}$ = −0.88; $E_{1/2}$ (ox.) = 1.32 [2]
12	$[(\pi\text{-}HC{\equiv}CH)Re_2(CNC_4H_9\text{-}t)((C_6H_5)_2PCH_2P(C_6H_5)_2)_2Br_3]O_3SCF_3$	Ia (72%) ^1H NMR (acetone-d_6): 1.97 (s); 5.90, 6.27; 7.0 to 8.3 (m), 9.71 (m, CH) $^{31}P\{^1H\}$ NMR (acetone-d_6): −16.09, 1.68 IR (Nujol): 2100 (ν(CN)) CV: $E_{p,c}$ = −0.87; $E_{1/2}$ (ox.) = 1.34 [2]
13	$[(\pi\text{-}HC{\equiv}CH)Re_2(CNC_6H_3(CH_3)_2\text{-}2,6)((C_6H_5)_2PCH_2P(C_6H_5)_2)_2Cl_3]O_3SCF_3$	Ia (42%) yellow solid ^1H NMR (acetone-d_6): ca. 2.6 (br); 5.91, 6.09; 6.9 to 8.2 (m), 10.13 (m, CH) $^{31}P\{^1H\}$ NMR (acetone-d_6): −4.34, 9.60 IR (Nujol): 2064 (ν(CN)) CV: $E_{p,c}$ = −0.79; $E_{1/2}$ (ox.) = 1.32 [2]
14	$[(\pi\text{-}HC{\equiv}CH)Re_2(CNC_6H_3(CH_3)_2\text{-}2,6)((C_6H_5)_2PCH_2P(C_6H_5)_2)_2Br_3]O_3SCF_3$	Ia (63%); compound probably not pure, because additional weak absorptions show up in the ^{13}C NMR and IR spectra

References on p. 109

Table 6 (continued)

No.	compound	method of preparation (yield) properties and remarks
14 (continued)		yellow solid ^{1}H NMR (acetone-d_6): 3.05 (br); 5.97, 6.31; 6.8 to 8.2 (m), 10.05 (m, CH) ^{31}P {^{1}H} NMR (acetone-d_6): −14.85, 5.32; additional weak peaks at −9.7, 4.2 IR (Nujol): 1968 (w), 2040 (ν(CN)) CV: $E_{p,c}$ = −0.76; $E_{1/2}$ (ox.) = 1.33 [2]
15	$[(\pi\text{-HC{\equiv}CC_3H_7\text{-}n})Re_2(CNC_6H_3(CH_3)_2\text{-}2,6)((C_6H_5)_2PCH_2P(C_6H_5)_2)_2Cl_3]O_3SCF_3$	Ia (57%) ^{1}H NMR (acetone-d_6): 0.674 (t); 0.93, 1.65 (m's); ca. 3.1 (br); 5.70, 5.98; 6.24 (m); 6.7 to 8.7 (m), 10.29 (m, CH) ^{31}P {^{1}H} NMR (acetone-d_6): −3.64, 12.21 IR (Nujol): 2070 (ν(CN)) CV: $E_{p,c}$ = −0.85; $E_{1/2}$ (ox.) = 1.23 [2]

* Further information:

$[(\pi\text{-HC{\equiv}CR'})Re_2(CO)((C_6H_5)_2PCH_2P(C_6H_5)_2)_2Cl_3]X$ (Table **6**, Nos. **1**, **4**, **6**, and **7** with R′ = H, C_3H_7-n, C_4H_9-n, C_6H_5; X = PF_6, O_3SCF_3). Treatment with phosphanes, PR''_3 (R''_3 = $(CH_3)_n$-

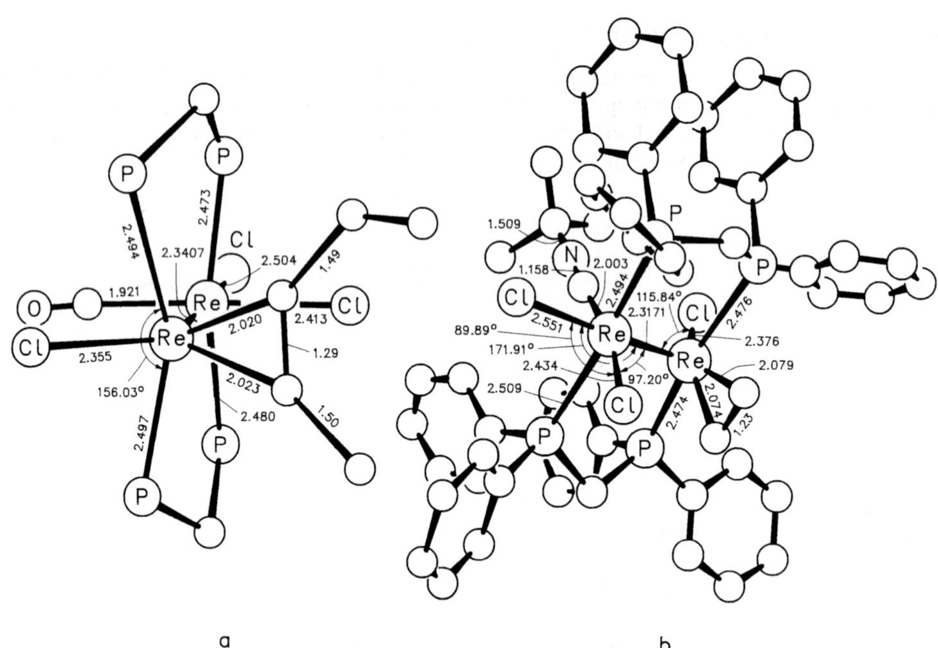

Fig. 51. Molecular structures of
a) $[(\pi\text{-CH_3C{\equiv}CC_2H_5})Re_2(CO)((C_6H_5)_2PCH_2P(C_6H_5)_2)_2Cl_3]^+$ (phenyl groups are omitted) [1],
b) $[(\pi\text{-HC{\equiv}CH})Re_2(CNC_4H_9\text{-t})((C_6H_5)_2PCH_2P(C_6H_5)_2)_2Cl_3]^+$ [2].

References on p. 109

$(C_6H_5)_{3-n}$ (n = 3, 2, 1) or $(C_2H_5)_3$), in THF gave ylides with the composition [R''_3PCH-(R')C=$Re_2(CO)((C_6H_5)_2PCH_2P(C_6H_5)_2)_2Cl_3$]X via a formal nucleophilic attack of the phosphane on the ≡CH group of the coordinated alkyne [3].

[(π-$CH_3C≡CC_2H_5$)$Re_2(CO)((C_6H_5)_2PCH_2P(C_6H_5)_2)_2Cl_3$]X (Table **6**, No. **3** with X = PF_6, O_3SCF_3). The PF_6 salt crystallizes in the monoclinic space group $P2_1/n-C^5_{2h}$ (No. 14) with a = 1.885(2), b = 19.702(3), c = 23.830(3) Å, β = 99.05(1)°; Z = 4 formula units per unit cell, D_{calc} = 1.794 g/cm³. The structure of the cation is shown in **Fig. 51a** [1].

The reaction of the O_3SCF_3 salt with $P(CH_3)_2C_6H_5$ in refluxing THF was not straightforward. The resulting oil did not contain an ylide complex as did mixtures obtained from reactions of compounds bearing terminal alkynes (see above) [3].

[(π-HC≡CH)$Re_2(CNC_4H_9$-t)$((C_6H_5)_2PCH_2P(C_6H_5)_2)_2Cl_3$]$O_3SCF_3$ · $CH_3CO_2C_2H_5$ (Table **6**, No. **11**) crystallizes in the monoclinic space group $P2_1/n-C^5_{2h}$ (No. 14) with a = 16.505(2), b = 15.172(2), c = 25.826(3) Å, β = 105.661(9)°; Z = 4 formula units per unit cell, D_{calc} = 1.700 g/cm³. The cation structure is depicted in **Fig. 51b** [2].

References:

[1] Shih, K.-Y.; Fanwick, P. E.; Walton, R. A. (Organometallics **12** [1993] 347/51).

[2] Kort, D. A.; Wu, W.; Fanwick, P. E.; Walton, R. A. (Transition Met. Chem. [London] **20** [1995] 625/9).

[3] Shih, K.-Y.; Tylicki, R. M.; Wu, W.; Fanwick, P. E.; Walton, R. A. (Inorg. Chim. Acta **229** [1995] 105/12).

2.2.3.1.2.2 Compounds of the Type [(RC≡CR)$_2$ReO]$_2$

General. The compounds described in this section are composed of two (RC≡CR)$_2$ReO fragments connected by an unsupported Re-Re single bond (see Formula I). R and R′ may be identical or different. In addition, there is also one compound with unsymmetrical acetylene ligands; their pairwise combination at the Re centers led to an isomer mixture.

O O
Re Re
R R' R R' R R' R' R' R

I

Preparation. Various reactions produced the compounds, but most of them are only briefly mentioned. The compounds could be obtained by subjecting (RC≡CR)$_2$Re(O)I to a one-electron reduction (with $(t\text{-}C_4H_9)_2Zn$, $t\text{-}C_4H_9ZnCl$, Na, $NaC_{10}H_8$, $(C_5H_5)_2Co$, or electrochemically) [5] or by subjecting the salts Na[(RC≡CR)$_2$ReO] to a one-electron oxidation (with air, $[(C_5H_5)_2Fe]^+$, C_6H_5I, $t\text{-}C_4H_9I$, or even $[(C_5H_5)_2Co]^+$) [4, 5] in C_6H_6, THF, or CH_3CN (oxidation of the salts with primary and secondary alkyl halides R′X such as C_2H_5I, $i\text{-}C_3H_7I$, CH_2=$CHCH_2I$, $C_6H_{11}Br$, or CH_2=$CH(CH_2)_4I$ led to mixtures of the title compounds and (RC≡CR)$_2$Re(O)R′ [4]). The compounds are also accessible by treating (RC≡CR)$_2$Re(O)I with Na[(R′C≡CR′)$_2$ReO] [5].

References on p. 113

Most reactions (except those producing the derivative with four $CH_3C{\equiv}CCH_3$ ligands) yielded mixtures consisting of the title compounds and compounds of the type $(RC{\equiv}CR)_2Re$-(μ-O)(μ-RC≡CR)Re(RC≡CR)(O) ("asymmetric isomers"; see Section 2.2.3.2.2.1) in a ratio somewhere between 0.75:1 and 10:1 (the portion of the title compounds was larger in unpolar solvents such as benzene) [5]. Upon thermolysis the asymmetric isomers generally transform to the title complexes; thus, the present "symmetric isomers" may be considered thermodynamically favored [1, 5].

Preparative details are given for the following procedures:

Method I: Treatment of $(RC{\equiv}CR)_2Re(O)I$

a. with $(t\text{-}C_4H_9)_2Zn$ (in situ prepared from $ZnCl_2$ and $t\text{-}C_4H_9Li$) in ether at −78 °C. After warming to room temperature, the solvent was evaporated and the residue was extracted into CH_2Cl_2 [1, 5].

b. with 1 equivalent $NaC_{10}H_8$ ($C_{10}H_8$ = naphthalenide anion radical) in THF at −78 °C. The mixture was allowed to slowly reach room temperature, and the solvent was subsequently evaporated [5].

In both cases the product was separated chromatographically on silica using ethyl acetate/hexane mixtures [5].

Except for the derivative with four $C_2H_5C{\equiv}CC_2H_5$ ligands, all other compounds are diamagnetic, air-stable solids. They are soluble in CH_2Cl_2 and benzene and at least sparingly soluble in CH_3CN and $CHCl_3$ [5].

Spectroscopy. The ^{13}C NMR spectra of the homo-ligand compounds display signals due to four different acetylenic C atoms, and the mixed-ligand derivatives display eight acetylenic resonances, suggesting that there is only one isomer and no ligand scrambling between the Re centers during synthesis. All compounds are rigid on an NMR time scale. The IR spectra exhibit strong ν(ReO) absorptions in the range 930 to 970 cm^{-1} [5].

Reaction with Halogens. All compounds react with X_2 (X = Br, I) to quantitatively give $(RC{\equiv}CR)_2Re(O)X$. Reactions with Br_2 are faster than reactions with I_2 [5].

$[(CH_3C{\equiv}CCH_3)_2ReO]_2$ was obtained by Method Ia [1, 5] or Method Ib [5] with 93% or 82% yield, respectively [1, 5]. It also formed by treating a mixture of $(CH_3C{\equiv}CCH_3)_2Re(O)I$ and $(C_2H_5C{\equiv}CC_2H_5)_2Re(O)I$ (mole ratio 1:1) according to Method Ib [5]. Otherwise, treatment of $(CH_3C{\equiv}CCH_3)_2Re(O)I$ with $(i\text{-}C_3H_7)_2Zn$ (benzene, 1 h, chromatographic workup with CH_2Cl_2) yielded only 31% (main product: $(CH_3C{\equiv}CCH_3)_2Re(O)C_3H_7\text{-}i$) [3]. The following reactions also produced $[(CH_3C{\equiv}CCH_3)_2ReO]_2$: treatment of $(CH_3C{\equiv}CCH_3)_2Re(O)OH$ with CO (C_6D_6, atmospheric pressure, 4 d; yield: 79%) [2]; decomposition of $(CH_3C{\equiv}CCH_3)_2Re(O)OC_2H_5$ [2] or $(CH_3C{\equiv}CCH_3)_2Re(O)OC(O)H$ [3] in hot benzene [3]; treatment of $Na[(CH_3C{\equiv}CCH_3)_2ReO]$ with C_6H_5I in CH_3CN (main product: $(CH_3C{\equiv}CCH_3)_2Re(O)CH_2CN$) [4]. The title complex forms a tan solid [5].

^1H NMR spectrum ($CDCl_3$): δ = 1.03, 2.80, 3.06, 3.16 (q's, CH_3; $^5J(H,H)$ = 1 Hz) ppm [1, 5]. $^{13}C\{^1H\}$ NMR spectrum ($CDCl_3$): δ = 13.17, 15.13, 15.30, 23.10 (CH_3); 140.78, 143.33, 146.50, 148.01 (C≡C) ppm [5]. IR spectrum (Nujol): 805; 941, 951 (ν(ReO)) [1, 5]; 1035, 1155; 1766 (ν(C≡C)) [5] cm^{-1}. Mass spectrum: $[M]^+$ [5].

The compound crystallizes in the monoclinic space group $P2_1/c-C_{2h}^5$ (No. 14) with a = 17.530(4), b = 7.264(2) c = 18.638(5) Å, β = 131.72(2)°; Z = 4 molecules per unit cell, D_{calc} = 2.327 g/cm^3 [1]. The molecular structure is depicted in **Fig. 52**. The molecule adopts an ethane-like conformation with the two roughly tetrahedral $Re(CH_3C{\equiv}CCH_3)_2O$ fragments in a staggered gauche conformation (O=Re−Re=O torsion angle: 74.2°) [1, 5].

References on p. 113

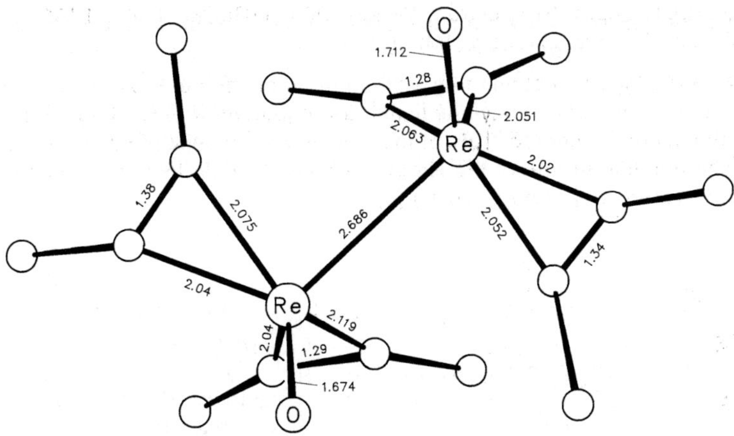

Fig. 52. Molecular structure of $[(CH_3C{\equiv}CCH_3)_2ReO]_2$ [1, 5].

Thermolysis in hot benzene did not affect the compound. Under an O_2 atmosphere, a benzene solution slowly formed hexamethylbenzene at 25 °C. The compound reacts with strong acids like $HOSO_2CF_3$ or HCl, but is unreactive towards H_2, $(n\text{-}C_4H_9)_3SnH$, CH_3CO_2H, CF_3CO_2H, CH_3I, or $CH_3SO_3CF_3$ [5]. Reduction with Na in the presence of a catalytic amount of naphthalene in THF yielded $Na[(CH_3C{\equiv}CCH_3)_2ReO]$ [6].

$[(C_2H_5C{\equiv}CC_2H_5)_2ReO]_2$ was obtained by Method Ia [1, 5] with 88% yield [5]. It also formed when treating a mixture of $(CH_3C{\equiv}CCH_3)_2Re(O)I$ and $(C_2H_5C{\equiv}CC_2H_5)_2Re(O)I$ (mole ratio 1:1) according to Method Ib [5]. Otherwise, treatment of $(C_2H_5C{\equiv}CC_2H_5)_2Re(O)I$ with $(i\text{-}C_3H_7)_2Zn$ (benzene, 1 h; chromatographic workup with CH_2Cl_2) yielded only 23% (main product: $(C_2H_5C{\equiv}CC_2H_5)_2Re(O)C_3H_7\text{-}i$) [3]. As already mentioned in the introduction above, thermolysis of $(C_2H_5C{\equiv}CC_2H_5)_2Re(\mu\text{-}O)(\mu\text{-}C_2H_5C{\equiv}CC_2H_5)Re(C_2H_5C{\equiv}CC_2H_5)(O)$ at 100 °C for 1 h yielded the title complex quantitatively [1]. The complex forms a pale golden oil [5].

^1H NMR spectrum (C_6D_6): δ = 0.69 (t, CH_3; J(H,H) = 7 Hz); 1.14 (q, 2 CH_2); 1.19, 1.43, 1.45 (t's, CH_3; J(H,H) = 7); 3.02, 3.17, 3.30, 3.45 (dq's, CH_2); 3.77 (4 H) ppm. $^{13}C\{^1H\}$ NMR spectrum (C_6D_6): δ = 14.30, 14.55, 15.10 (CH_3); 24.01, 24.78, 25.15, 33.25 (CH_2) [5]; 144.0, 147.2, 150.3, 150.8 (C≡C) [1, 5] ppm. IR spectrum (neat): 770, 803, 934, 955 (ν(ReO)) [1, 5]; 1050, 1090, 1138, 1251, 1372, 1455, 1758 (ν(C≡C)); 2865, 2940, 2965 [5] cm^{-1}. Mass spectrum: $[M]^+$ [5].

Thermolysis in hot benzene did not affect the compound. Treatment with HCl gas in toluene-d_8 resulted in $(C_2H_5C{\equiv}CC_2H_5)_2Re(O)Cl$ [5]. Treatment with Na in the presence of a catalytic amount of naphthalene in THF yielded $Na[(C_2H_5C{\equiv}CC_2H_5)_2ReO]$ [6].

$(C_2H_5C{\equiv}CC_2H_5)_2Re(O)Re(O)(CH_3C{\equiv}CCH_3)_2$ was obtained with 25% yield by treating a 1:1 molar mixture of $(CH_3C{\equiv}CCH_3)_2Re(O)I$ and $(C_2H_5C{\equiv}CC_2H_5)_2Re(O)I$ according to Method Ib in benzene at 5 °C. The two preceding homo-ligand derivatives were also separated.

^1H NMR spectrum (CD_3CN): δ = 0.71 (t, $\mathbf{CH_3}CH_2$; J = 7 Hz); 0.93, 1.06 (m's, 1 H); 1.10 (q, ≡CCH_3; J = 1 Hz); 1.26, 1.33, 1.54 (t's, $\mathbf{CH_3}CH_2$; J = 7 Hz); 2.80, 2.95, 3.13 (q's, ≡$C\mathbf{CH_3}$; J = 1 Hz); 3.22 (m, 1 H); 3.45 (m, 3 H); 3.50, 3.70 (m's, 1 H) ppm. $^{13}C\{^1H\}$ NMR spectrum (C_6D_6): δ = 14.1 (≡$C\mathbf{C}H_3$); 14.2, 14.3, 14.6, 15.0 ($\mathbf{C}H_3CH_2$); 15.2, 15.3, 23.4 (≡$C\mathbf{C}H_3$); 23.9, 24.7, 24.9, 33.0 (CH_2); 140.6, 143.0 (≡$\mathbf{C}CH_3$); 143.7 (≡$\mathbf{C}CH_2$); 146.5, 147.1 (≡$\mathbf{C}CH_3$); 147.4,

References on p. 113

150.4, 150.8 (≡CCH_2) ppm. IR spectrum (Nujol): 963 (ν(ReO)); 1048, 1158, 1252, 1304; 1761 (ν(C≡C)) cm^{-1}. Mass spectrum: $[M]^+$ [5].

[(t-C_4H_9C≡CH)$_2$ReO]$_2$ was obtained by Method Ia as an isomer mixture. The main isomer is said to have an antiparallel acetylene ligand arrangement at each Re atom (see Formula II), whereas the minor isomer (ca. 10% of the main isomer) supposedly has a parallel acetylene ligand coordination at one of the Re atoms (see Formula III). The isomers did not interconvert at elevated temperature (70 °C) [5].

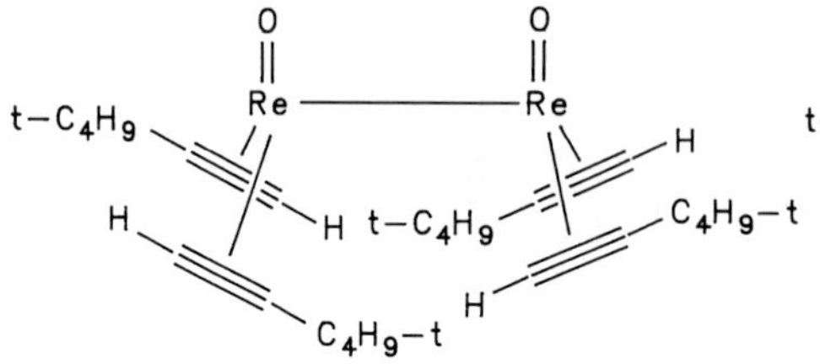

II (main isomer) 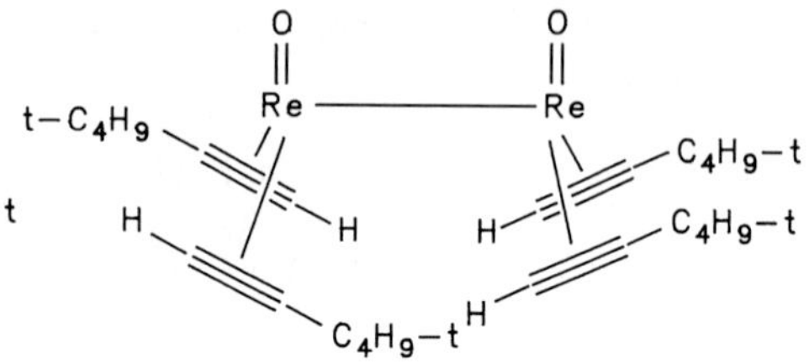 III (minor isomer)

The isomers were separately characterized as follows (abbreviations and units on p. X):

Main isomer: Yield: 19%. – Pale tan solid. – ^{1}H NMR: 1.41, 1.52 (s, t-C_4H_9); 5.20, 9.66 (s, each 2 H, ≡CH). – ^{13}C {^{1}H} NMR (CD_2Cl_2): 30.8, 31.4 (q, CH_3; J = 128); 37.0, 37.2 (s, ≡C**C**); 125.8 (d, ≡CH; J = 214); 130.4 (d,≡CH; J = 207); 156.3, 175.11 (s, ≡C**C**). – IR (neat): 782, 965 (ν(ReO)), 1038, 1215, 1235, 1360, 1465, 1462 (?), 1655 (ν(C≡C)), 2900, 2960. – MS: $[M]^+$.
Treatment with I_2 gave only the asymmetric form of (t-C_4H_9C≡CH)Re(O)I [5].

Minor isomer: ^{1}H NMR: 0.66, 1.46, 1.67, 1.69 (each s, 9 H, CH_3); 5.95, 9.88, 9.89, 10.26 (s's, ≡CH).
Treatment with I_2 yielded both the asymmetric and symmetric isomers of (t-C_4H_9C≡CH)$_2$Re(O)I in approximately equal amounts [5].

[(C_6H_5C≡CC_6H_5)$_2$ReO]$_2$ was obtained by Method Ib with ca. 47.6% yield. Off-white solid.

^{1}H NMR spectrum (C_6D_6): δ = 6.36 (4 H), 6.70 (t, 1 H), 6.77 (t, 2 H), 6.90 (t, 4 H), 6.92 (t, 2 H), 6.99 (t, 6 H), 7.12 (t, 2 H), 7.21 (d, 8 H), 7.38 (t, 4 H), 8.45 (4 H) ppm; all J(H,H) = 7 Hz. ^{13}C {^{1}H} NMR spectrum (CD_2Cl_2): δ = 127.0 to 131.7 (C_6H_5); 133.3, 134.0, 134.4, 135.0 (C_{ipso}); 142.2, 149.6, 150.6, 154.2 (C≡C) ppm. IR spectrum (neat): 695, 700, 720, 910, 925, 972, 1028, 1072, 1180, 1270, 1440, 1480, 1550, 1590, 2920, 3020, 3058, 3080 cm^{-1} [5].

Treatment with Na in the presence of a catalytic amount of naphthalene in THF yielded Na[(C_6H_5C≡CC_6H_5)$_2$ReO] as a solvated solid-state dimer (see p. 115) [6].

(C_6H_5C≡CC_6H_5)$_2$Re(O)Re(O)(CH_3C≡CCH_3)$_2$ was obtained by heating the unsymmetrical derivative (C_6H_5C≡CC_6H_5)$_2$Re(μ-O)(μ-CH_3C≡CCH_3)Re(CH_3C≡CCH_3)(O) (see p. 122) in a sealed tube at 110 °C for 3 h in the presence of CD_2Cl_2. After that, 25% of the unsymmetrical derivative were converted, and the compounds could be separated chromatographically on silica using CH_2Cl_2. Off-white solid.

^{1}H NMR spectrum (C_6D_6): δ = 1.64, 2.06, 2.28, 3.55 (q's, CH_3; J(H,H) = 1 Hz); 6.77, 6.88 (m, C_6H_5); 6.19, 7.33 (m, 5 H); 7.19, 8.15 (d's, 2 H; J = 7 Hz) ppm. ^{13}C {^{1}H} NMR spectrum (CD_2Cl_2): δ = 15.0, 15.3, 15.6, 24.53 (CH_3); 127.0 to 135.0 (C_6H_5); 141.17, 141.9,

144.9, 145.9, 146.7, 148.4, 149.9, 154.2 (C≡C) ppm. IR spectrum (neat): 688, 740, 765, 808; 958 (ν(ReO)); 1025, 1070, 1152, 1270, 1360, 1440, 1480, 1675 (?), 1592; 1752 cm^{-1} [5].

References:

[1] Valencia, E.; Santarsiero, B. D.; Geib, S. J.; Rheingold, A. L.; Mayer, J. M. (J. Am. Chem. Soc. **109** [1987] 6896/8).
[2] Erikson, T. K. G.; Bryan, J. C.; Mayer, J. M. (Organometallics **7** [1988] 1930/8).
[3] Spaltenstein, E.; Erikson, T. K. G.; Critchlow, S. C.; Mayer, J. M. (J. Am. Chem. Soc. **111** [1989] 617/23).
[4] Conry, R. R.; Mayer, J. M. (Organometallics **10** [1991] 3160/6).
[5] Spaltenstein, E.; Mayer, J. M. (J. Am. Chem. Soc. **113** [1991] 7744/53).
[6] Cundari, T. R.; Conry, R. R.; Spaltenstein, E.; Critchlow, S. C.; Hall, K. A.; Tahmassebi, S. K.; Mayer, J. M. (Organometallics **13** [1994] 322/31).

2.2.3.1.2.3 Miscellaneous Compounds

This section presents several other compounds with terminal alkynes. Some of them were prepared by the following methods:

Method I: Reduction of $(t\text{-}C_4H_9CH_2C{\equiv}CCH_2C_4H_9\text{-}t)_nRe({=}NC_6H_3(C_3H_7\text{-}i)_2\text{-}2,6)_{3-n}Cl$ (n = 1, 2) with 1 equivalent Na/Hg in THF. After evaporating the solvent, the slurry was extracted with ether (for n = 1) or pentane (for n = 2) [9].

Method II: Treatment of $(RC{\equiv}CR)Re(O)_2CH_3$ (R = CH_3, C_2H_5) with 1 equivalent of polymer-bonded $P(C_6H_5)_3$ in toluene (reflux, 2 d). The mixture was filtered and evaporated, and the residue was recrystallized from hexane. Further purification could be achieved by chromatography on silylated silica with toluene [5].

(R = $C_6H_3(CH_3)_2$-2,6)

I

[π-HC≡CHRe$_2$(CNC$_6$H$_3$(CH$_3$)$_2$-2,6)(CO)((C$_6$H$_5$)$_2$PCH$_2$P(C$_6$H$_5$)$_2$)$_2$Cl$_3$]PF$_6$ (see Formula I) was obtained, when 2,6-$(CH_3)_2C_6H_3NCRe_2(CO)((C_6H_5)_2PCH_2P(C_6H_5)_2)_2Cl_4$ was treated with acetylene in the presence of $TlPF_6$ for ca. 5 h (reaction time was critical) [8, 11]. Yield: 74% [11]. The salt forms orange microcrystals [8, 11].

^1H NMR spectrum (CD_2Cl_2): δ = 4.0, 5.1 (m's, CH_2), 6.6 to 8.0 (C_6H_5), 13.4 (CH≡) ppm. $^{13}C\{^1H\}$ NMR spectrum (CD_2Cl_2): δ = 193.3 (t, CO) ppm. $^{31}P\{^1H\}$ NMR spectrum (CD_2Cl_2): δ = −29.5, −13.0 (AA′BB′ pattern) ppm [11]. IR spectrum (CH_2Cl_2): 1880 (ν(CO)), 2138 (ν(CN)) cm^{-1} [8, 11]. Cyclic voltammetry (CH_2Cl_2/0.1 M $[N(C_4H_9\text{-}n)_4]PF_6$, Pt bead): $E_{p,c}$ =

References on pp. 116/7

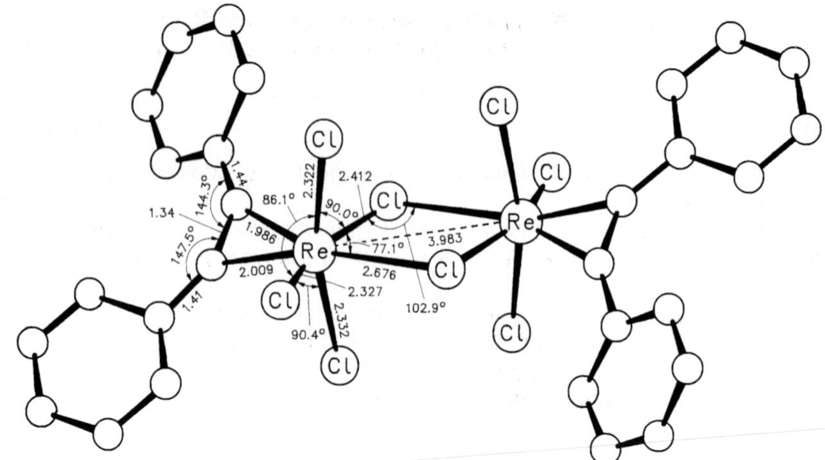

Fig. 53. Molecular structure of $[(C_6H_5C{\equiv}CC_6H_5)ReCl_3]_2(\mu\text{-}Cl)_2$ [6].

−0.96 V, $E_{1/2}$ (red.) = −0.25 V, $E_{1/2}$ (ox.) = 1.03 V (vs. Ag/AgCl). Conductivity (10^{-3} M in acetone): $\Lambda_M = 125\ cm^2 \cdot \Omega^{-1} \cdot mol^{-1}$ [11].

The salt is sparely soluble in CH_2Cl_2 or CH_3CN. When stored in CH_2Cl_2, it converts into $[(\mu\text{-}COCH{=}CH)Re_2(CNC_6H_3(CH_3)_2\text{-}2,6)((C_6H_5)_2PCH_2P(C_6H_5)_2)_2Cl_3]PF_6$ (see p. 6) within 2 d [11].

$[(C_6H_5C{\equiv}CC_6H_5)ReCl_3]_2(\mu\text{-}Cl)_2 \cdot 2\ CH_2Cl_2$ was produced by treating a 1:1 molar mixture of $ReCl_5$ and $POCl_3$ with excess $C_6H_5C{\equiv}CC_6H_5$ at 0°C. The brown precipitate, which formed after 24 h, was filtered off, washed with CCl_4, and dried [1, 6]. Yield: 63% [1]. The solid appears black. Single crystals readily lose solvate molecules [6].

IR spectrum (KBr): 215, 229, 255, 279 (ν(ReClRe)); 300, 318, 337, 353, 365, 397 (ν(ReCl)); 1175 (ν_{sym}(CC)), 1271 (ν_{asym}(CC)), 1690 (ν(C≡C)) cm^{-1} [1].

Single crystals are monoclinic with a = 9.872(2), b = 15.339(1), c = 11.938(2) Å, β = 90.17(1)°; space group $P2_1/c-C^5_{2h}$ (No. 14); Z = 2 formula units per unit cell, D_{calc} = 2.17 g/cm^3. The molecular structure is depicted in **Fig. 53**. There are different Re-Cl bond lengths in the molecule; the longer ones are trans to the side-on-bonded alkyne ligands [6].

Molar susceptibility measurements in the range 4.2 K to 293 K revealed a minimum for $1/\chi_M$ at T ca. 13 K. Above 50 K the molar susceptibility is linearly dependent on the temperature. The magnetic moment μ_{eff} amounts to 1.3 μ_B at 50 K and 1.6 μ_B at 25°C. The results suggest a d^1 electron configuration at the Re atoms [7].

A treatment with $[P(C_6H_5)_4]Cl$ yielded $[(C_6H_5C{\equiv}CC_6H_5)ReCl_5]^-$ [1].

$[(t\text{-}C_4H_9CH_2C{\equiv}CCH_2C_4H_9\text{-}t)Re({=}NC_6H_3(C_3H_7\text{-}i)_2\text{-}2,6)_2]_2Hg$ was obtained via Method I. The ethereal extract was reevaporated, and the residue extracted with pentane, leaving red cubes with 76% yield.

^1H NMR spectrum (C_6D_6): δ = 1.03, 1.06 (s's, CH_3 of C_4H_9-t); 1.16, 1.21 (d's, CH_3 of C_3H_7-i); 3.35, 3.65 (s's, CH_2); 3.78 (sept, CH), 7.01 (m, C_6H_3) ppm. $^{13}C\{^1H\}$ NMR spectrum (C_6D_6): δ = 23.7, 23.8 (CH_3 of C_3H_7-i); 28.1 (CH); 30.2, 30.7 (CH_3 of C_4H_9-t); 34.5, 34.9 (**C**$(CH_3)_3$); 46.5, 54.2 (CH_2); 122.7, 126.4 (C_6H_3); 143.5 (C≡C), 144.5 (C_6H_3), 154.2 (C≡C) ppm. IR spectrum (Nujol): 1769 (ν(C≡C)) cm^{-1} [9].

References on pp. 116/7

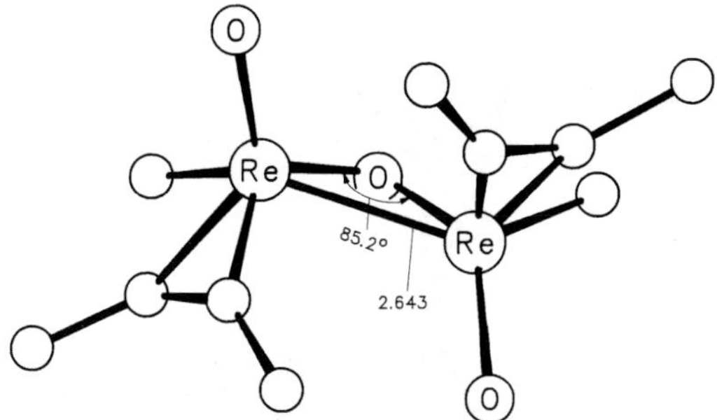

Fig. 54. Molecular structure of $[(CH_3C{\equiv}CCH_3)Re(O)CH_3]_2O$ [5].

$[(t\text{-}C_4H_9CH_2C{\equiv}CCH_2C_4H_9\text{-}t)_2Re{=}NC_6H_3(C_3H_7\text{-}i)_2\text{-}2,6]_2Hg$ was obtained via Method I. Evaporation of the pentane extract gave the compound with 85% yield as a cherry red solid. It could not be obtained in analytically pure form.

^1H NMR spectrum (C_6D_6): δ = 1.03 (br s, C_4H_9-t), 1.32 (d, C_3H_7-i), 3.63 (sept, CH); 3.72, 3.88, 3.98, 4.15 (all: d's, CH_2); 7.01 (m, C_6H_3) ppm. ^{13}C {^1H} NMR spectrum (C_6D_6): δ = 24.0 (CH_3 of C_3H_7-i), 27.9 (CH of C_3H_7-i); 30.7, 30.9 (CH_3 of C_4H_9-t); 33.9, 34.8 (**C**C_4H_9-t); 45.9, 52.5 (CH_2); 122.4, 124.8, 144.5, 152.0 (C_6H_3); 160.1, 167.9 (C≡C) ppm. IR spectrum (Nujol): 1725, 1743 (ν(C≡C)) cm^{-1}. EI mass spectrum: $[M]^+$ [9].

$[(CH_3C{\equiv}CCH_3)Re(O)CH_3]_2O$ was obtained via Method II with 90% yield. Orange solid; dec. p. 185 °C.

^1H NMR spectrum ($CDCl_3$): δ = 2.76 (s, $ReCH_3$); 2.88, 3.25 (q's, CCH_3; ^5J(H,H) = 1.0 Hz) ppm. ^{13}C NMR spectrum ($CDCl_3$): δ = 10.8, 15.9, 23.6 (q's, ≡CCH_3, $ReCH_3$, and ≡CCH_3 with J(C,H) = 130.4, 132.9, and 130.4 Hz, respectively); 143.3, 148.5 (s's, C≡C) ppm. IR spectrum (KBr): 720 (ν(ReORe)), 964 (ν(ReO)) [3], 1720 (ν(C≡C)) cm^{-1} [5]. EI mass spectrum: $[M]^+$ [5].

X-ray crystallography revealed an orthorhombic lattice with a = 32.481(7), b = 11.807(3), c = 7.071(1) Å; space group Fdd2$-C_{2v}^{19}$ (No. 43). Due to the poor quality of the crystal, only the molecular geometry could be determined; see **Fig. 54** [3, 5].

$[(C_2H_5C{\equiv}CC_2H_5)Re(O)CH_3]_2O$ was obtained via Method II with 90% yield. Orange solid.

^1H NMR spectrum ($CDCl_3$): δ = 1.41, 1.44 (t's; J = 7.5 Hz); 2.74 ($ReCH_3$); 3.32, 3.49 (m's, $\mathbf{CH_2}CH_3$) ppm. ^{13}C {^1H} NMR spectrum ($CDCl_3$): δ = 13.70, 14.76 (CH_3); 15.12, 19.87 (CH_2); 33.19 ($ReCH_3$); 147.0, 152.7 (C≡C) ppm. IR spectrum (KBr): 971, 977 (v(ReO)) cm^{-1}. EI mass spectrum: $[M]^+$ [5].

$[(CH_3C{\equiv}CCH_3)_2Re(O)]_2O$ formed after concentrating or simply leaving a solution containing in situ-prepared $(CH_3C{\equiv}CCH_3)_2Re(O)OH$ (from $(CH_3C{\equiv}CCH_3)_2Re(O)X$ (X = OC_2H_5, $NHCH_3$) and H_2O) stand. The reaction took place even at −80 °C.

^1H NMR spectrum (C_6D_6): δ = 1.99, 2.53 (q's; J = 1 Hz) ppm. ^{13}C {^1H} NMR spectrum ($CDCl_3$): δ = 7.2, 17.5 (q's, CH_3C≡; J(C,H) = 129 Hz); 144.9, 157.4 (C≡C) ppm. IR spectrum (Nujol): 621, 799 (756), 946 (896), 958 (907), 1040, 1156, 1779, 1789 cm^{-1} (in parentheses: the respective bands of the ^{18}O-enriched derivative). Mass spectrum: $[M - n\,(CH_3C{\equiv}CCH_3)]^+$ (n = 1, 2). The compound slowly decomposed in solution [2].

$\{Na[(C_6H_5C{\equiv}CC_6H_5)_2ReO]\cdot 2\,CH_3CN\}_2$. A series of salts of the type $Na[(RC{\equiv}CR)_2ReO]$ was prepared by reducing compounds of the type $(RC{\equiv}CR)_2Re(O)I$ or $[(RC{\equiv}CR)_2ReO]_2$ (R = CH_3,

 References on pp. 116/7

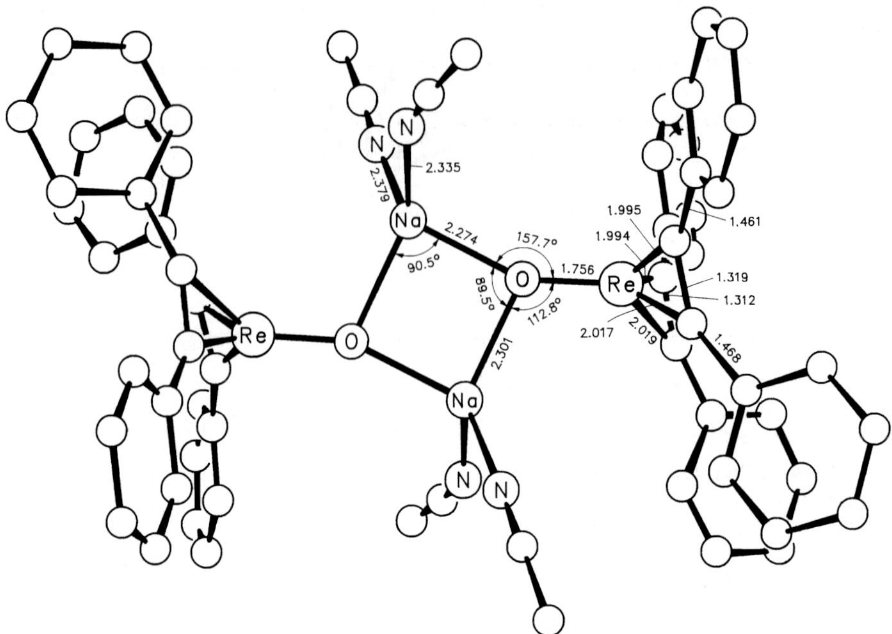

Fig. 55. Molecular structure of $\{Na[(C_6H_5C{\equiv}CC_6H_5)_2ReO] \cdot 2\ CH_3CN\}_2$ [4, 10].

C_2H_5, C_6H_5) with 2 equivalents Na or $NaC_{10}H_8$ ($C_{10}H_8$ = naphthalenide) in THF at −78 °C. The red-purple, C_6H_5-substituted derivative was crystallized from CH_3CN and thereby obtained as a CH_3CN solvate. An X-ray crystallographic study revealed a dimeric structure in the solid state.

^1H NMR spectrum (CD_3CN): δ = 7.14, 7.29 (t's); 7.67 (d) ppm; all J(H,H) = 8 Hz. $^{13}C\{^1H\}$ NMR spectrum (CD_3CN): δ = 172.2 (C≡C) ppm. IR spectrum (Nujol): 824 (ν(ReO)) cm^{-1} [4].

Single crystals are monoclinic with a = 11.640(2), b = 12.648(2), c = 19.668(4) Å, β = 99.19(2)° [4] (71.61(1)° [10]); Z = 4 formula units per unit cell, D_{calc} = 1.54 g/cm³. The centrosymmetric dimer is depicted in **Fig. 55**. The Re atoms are surrounded by the acetylene and oxo ligands in a roughly trigonal-planar arrangement. The C≡C bonds are roughly perpendicular to the Re-O bond, but they are not parallel [4, 10].

References:

[1] Hey, E.; Weller, F.; Dehnicke, K. (Z. Anorg. Allg. Chem. **514** [1984] 25/38).

[2] Erikson, T. K. G.; Mayer, J. M. (Angew. Chem. **100** [1988] 1585/6; Angew. Chem. Int. Ed. Engl. **27** [1988] 1527).

[3] Felixberger, J. K.; Kuchler, J. G.; Herdtweck, E.; Paciello, R. A.; Herrmann, W. A. (Angew. Chem. **100** [1988] 975/8; Angew. Chem. Int. Ed. Engl. **27** [1988] 946).

[4] Spaltenstein, E.; Conry, R. R.; Critchlow, S. C.; Mayer, J. M. (J. Am. Chem. Soc. **111** [1989] 8741/2).

[5] Herrmann, W. A.; Felixberger, J. K.; Kuchler, J. G.; Herdtweck, E. (Z. Naturforsch. **45b** [1990] 876/86).

[6] Swidersky, H.-W.; Kindel, O.; Weller, F.; Dehnicke, K. (Z. Anorg. Allg. Chem. **580** [1990] 18/26).
[7] Swidersky, H.-W.; Pebler, J.; Dehnicke, K.; Fenske, D. (Z. Naturforsch. **45b** [1990] 1227/34).
[8] Shih, K.-Y.; Fanwick, P. E.; Walton, R. A. (J. Am. Chem. Soc. **115** [1993] 9319/20).
[9] Williams, D. S.; Schrock, R. R. (Organometallics **12** [1993] 1148/60).
[10] Cundari, T. R.; Conry, R. R.; Spaltenstein, E.; Critchlow, S. C.; Hall, K. A.; Tahmassebi, S. K.; Mayer, J. M. (Organometallics **13** [1994] 322/31).

[11] Shih, K.-Y.; Fanwick, P. E.; Walton, R. A. (Organometallics **13** [1994] 1235/42).

2.2.3.2 Compounds with Bridging ²L Ligands

2.2.3.2.1 Compounds with Re-Bonded Alkenes

This section presents miscellaneous compounds.

(μ-C_4H_6O)$Re_2(CO)_8$ (see Formula I) formed by irradiating a mixture of $(CO)_{10}Re_2$, NO, and cyclooctatetraene in THF at −30 °C for 3.5 h. Subsequent chromatographic workup (silica, hexane/ether mixtures) separated 7 fractions, which were all purified by another chromatographic separation. The title complex was thereby obtained with 4% yield [2]. It also formed when butadiene was employed in place of cyclooctatetraene. Yield: 2% [1, 2]. Pale yellow solid; m.p. 108 °C (dec.) [2].

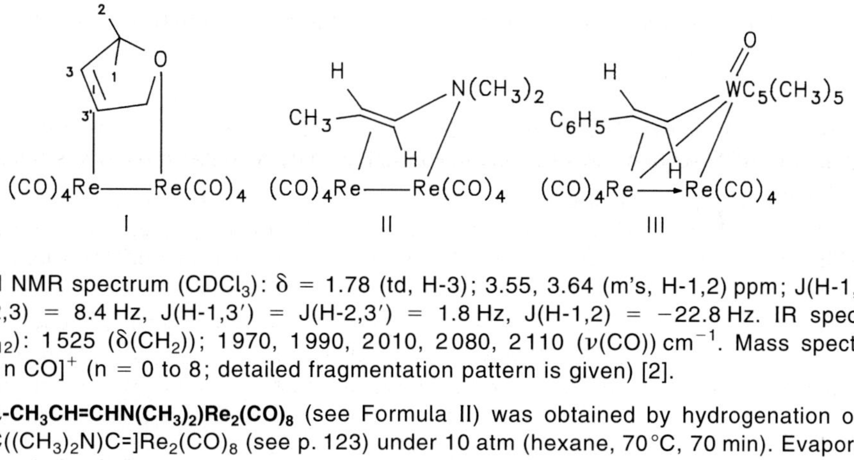

$(CO)_4Re—Re(CO)_4$ I; $(CO)_4Re—Re(CO)_4$ II; $(CO)_4Re→Re(CO)_4$ III

^{1}H NMR spectrum ($CDCl_3$): δ = 1.78 (td, H-3); 3.55, 3.64 (m's, H-1,2) ppm; J(H-1,3) = J(H-2,3) = 8.4 Hz, J(H-1,3′) = J(H-2,3′) = 1.8 Hz, J(H-1,2) = −22.8 Hz. IR spectrum (C_6H_{12}): 1525 (δ(CH_2)); 1970, 1990, 2010, 2080, 2110 (ν(CO)) cm^{-1}. Mass spectrum: $[M - n\,CO]^+$ (n = 0 to 8; detailed fragmentation pattern is given) [2].

(μ-CH_3CH=$CHN(CH_3)_2$)$Re_2(CO)_8$ (see Formula II) was obtained by hydrogenation of [μ-$CH_3C((CH_3)_2N)C$=]$Re_2(CO)_8$ (see p. 123) under 10 atm (hexane, 70 °C, 70 min). Evaporation of the solution followed by preparative TLC on silica using hexane/CH_2Cl_2 (4:1) yielded 41% of the title compound together with (μ-$CH_3C(H)CN(CH_3)_2$)$Re_2(CO)_8$(μ-H) (see p. 58).

^{1}H NMR spectrum ($CDCl_3$): δ = 2.06 (d, CCH_3; J = 5.7 Hz), 2.17 (dq, CH; J = 10.5, 5.7 Hz); 2.25, 3.26 (s's, NCH_3); 4.28 (d, CH; J = 10.5 Hz) ppm. IR spectrum (n-hexane): 1932, 1967, 1989, 2040, 2086 (ν(CO)) cm^{-1}. Mass spectrum: $[M - n\,CO]^+$ (n = 0 to 8).

(μ-CH_3CH=$CHN(CH_3)_2$)$Re_2(CO)_8$ crystallizes in the monoclinic space group $P2_1/n-C_{2h}^5$ (No. 14) with a = 7.686(2), b = 16.372(4), c = 14.080(3) Å, β = 100.43(1)°; Z = 4 molecules per unit cell. The molecular structure is depicted in **Fig. 56** [3].

[μ-($C_5(CH_3)_5$)W(O)CH=CHC_6H_5]$Re_2(CO)_8$ (see Formula III) was produced with 28% yield by treating $(CO)_8Re_2(\mu_3\text{-}CCC_6H_5)W(C_5(CH_3)_5)(O)$ (see Formulas IVa, IVb; R = C_6H_5; p. 126)

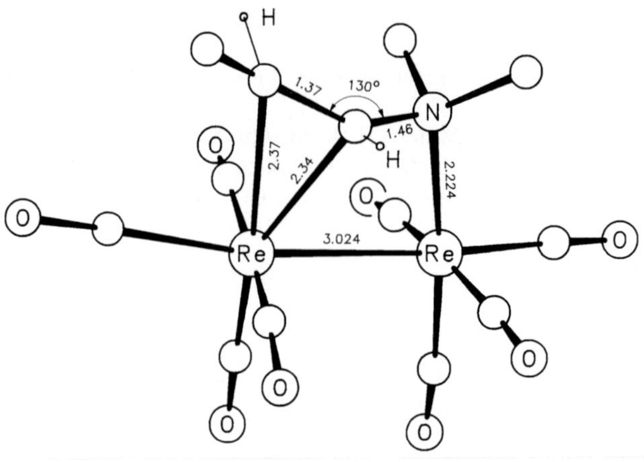

Fig. 56. Molecular structure of (μ-$CH_3CH{=}CHN(CH_3)_2$)$Re_2(CO)_8$ [3].

with H_2 in refluxing heptane for 40 min. It could also be prepared by treating [μ-($C_5(CH_3)_5$)-W(O)(C≡CC_6H_5)H]$Re_2(CO)_6$(μ-H) (see p. 99) with CO in refluxing CH_2Cl_2. Orange crystals.

^{1}H NMR spectrum ($CDCl_3$): δ = 2.22 (s, CH_3); 4.72, 5.93 (d's, HCCH; J = 13.6 Hz); 7.21 to 7.42 (m, C_6H_5) ppm. ^{13}C {^{1}H} NMR spectrum ($CDCl_3$): δ = 76.8 (ReC), 161.3 (WC; J(W,C) = 115 Hz) ppm (other peaks not given). IR spectrum (C_6H_{12}): 1919, 1942, 1961, 1981, 1992, 2008, 2043, 2088 (ν(CO)) cm^{-1}.

Hydrogenation yielded $(CO)_8Re_2$(μ-H)W(O)($C_5(CH_3)_5$)(μ-$CHCH_2C_6H_5$), whereas treatment with D_2 gave $(CO)_8Re_2$(μ-D)W(O)($C_5(CH_3)_5$)(μ-$CHCHDC_6H_5$) (see Formula IV) [6].

[μ-$(C_5H_5)_2Ni_2(C_6H_5C{=}CCH{=}CHC_6H_5)$]$Re_2(CO)_6$ (see Formula V) was obtained by reacting equimolar amounts of $(C_5H_5)_2Ni$ and (μ-$\eta^{2:1}$-$C_6H_5C{\equiv}CH$)$Re_2(CO)_8$(μ-H) in xylene (reflux, 4 h). Chromatographic separation of the mixture (alumina, hexane/benzene mixtures) gave some $(CO)_{10}Re_2$ [4, 5], the title product (yield: 30% [5], 60% [4]), $(C_5H_5)_3Ni_3(CO)_2$ (15% [4], 26% [5]), and [μ-$(C_5H_5)_2Ni_2CCHC_6H_5$]$Re_2(CO)_6$(μ-CO) (see Formula VIII, p. 95) [5]. The title compound forms air-stable, black-green crystals [4, 5].

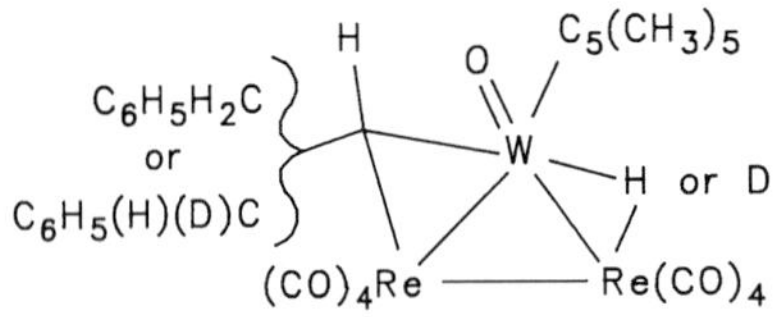

IV

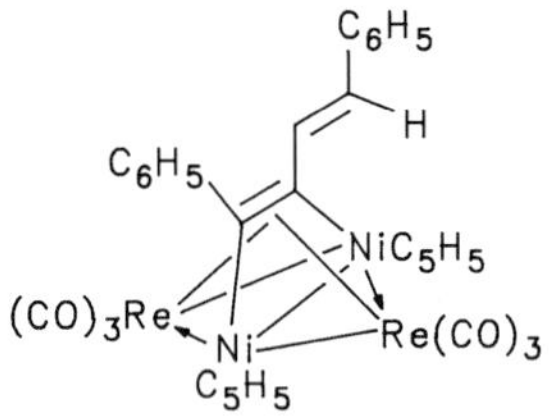

V

IR spectrum (KBr): 685, 745, 805, 820; 1868, 1885, 1900, 1980 (ν(CO)) cm^{-1} [4, 5].

Single crystals have a monoclinic lattice with a = 9.399(9), b = 16.572(28), c = 18.886(23) Å, β = 95.542(9)°, space group $P2_1/c-C_{2h}^5$ (No. 14), and Z = 4 molecules per

References on p. 119

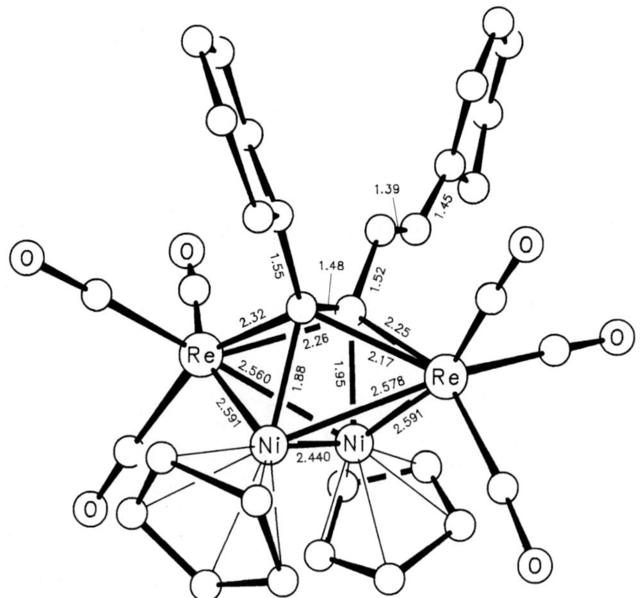

Fig. 57. Molecular structure of $[\mu\text{-}(C_5H_5)_2Ni_2(C_6H_5C{=}CCH{=}CHC_6H_5)]Re_2(CO)_6$ [4].

unit cell. The molecular structure is depicted in **Fig. 57**. It has a butterfly Re_2Ni_2 core. The two $ReNi_2$ planes form a dihedral angle of 122°. There are two short Ni-C σ bonds [4].

References:

[1] Balbach, B. K.; Helus, F.; Oberdorfer, F.; Ziegler, M. L. (Angew. Chem. **93** [1981] 479/80; Angew. Chem. Int. Ed. Engl. **20** [1981] 470/1).

[2] Oberdorfer, F.; Balbach, B.; Ziegler, M. L. (Z. Naturforsch. **37b** [1982] 157/67).

[3] Adams, R. D.; Chen, G.; Yin, J. (Organometallics **10** [1991] 2087/8).

[4] Shaposhnikova, A. D.; Kamalov, G. L.; Stadnichenko, R. A.; Pasynskii, A. A.; Eremenko, I. L.; Nefedov, S. E.; Struchkov, Yu. T.; Yanovsky, A. I. (J. Organomet. Chem. **405** [1991] 111/20).

[5] Pasynskii, A. A.; Eremenko, I. L.; Nefedov, S. E.; Yanovskii, A. I.; Struchkov, Yu. T.; Shaposhnikova, A. D.; Stadnichenko, R. A. (Zh. Neorg. Khim. **38** [1993] 455/65; Russ. J. Inorg. Chem. [Engl. Transl.] **38** [1993] 423/32).

[6] Chi, Y.; Cheng, P.-S.; Wu, H.-L.; Hwang, D.-K.; Su, P.-C.; Peng, S.-M.; Lee, G. H. (J. Chem. Soc. Chem. Commun. **1994** 1839/40).

2.2.3.2.2 Compounds with Re-Bonded Alkynes

Bridging alkynes have been observed to coordinate to Re_2 groupings in different ways. The most commonly encountered forms are the parallel mode (see Formula I; these compounds are described in the section "σ,σ-Bonded Compounds") and the perpendicular mode (see Formula III). In addition, there also exist compounds with an intermediate, "twisting" bonding mode (see Formula II). Compounds with the two latter coordination modes are described in the following sections.

References on p. 123

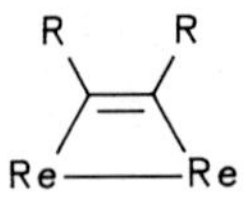

I 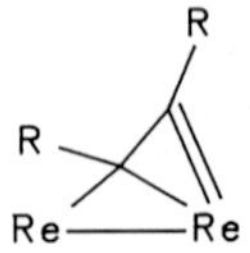 II 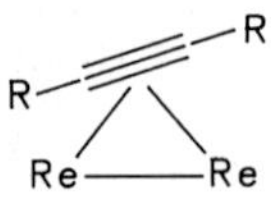 III

2.2.3.2.2.1 Compounds of the Type $(RC{\equiv}CR)_2Re(\mu\text{-}O)(\mu\text{-}RC{\equiv}CR)Re(RC{\equiv}CR)(O)$

General. Structure. The compounds presented in this section each have an $(RC{\equiv}CR)_2Re$ and $(RC{\equiv}CR)ReO$ unit, connected by one bridging oxo and one bridging acetylene ligand (see Formulas Ia to Id).

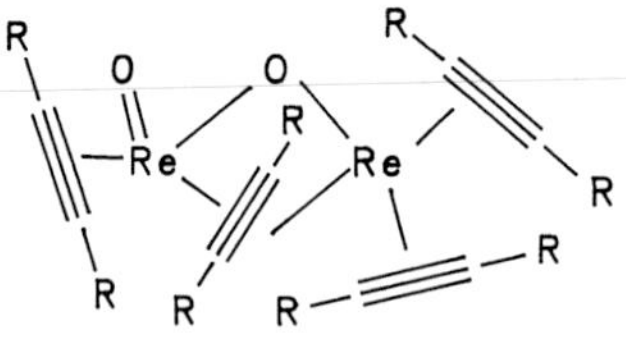

Ia 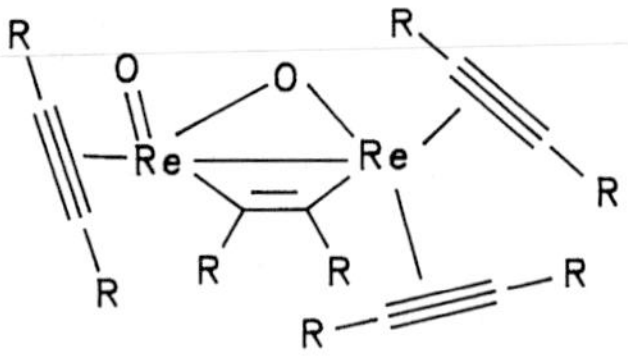 Ib

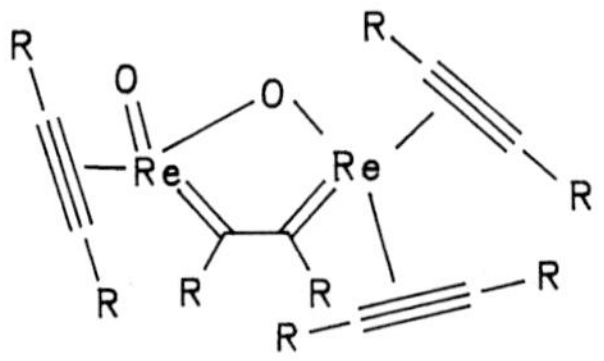

Ic 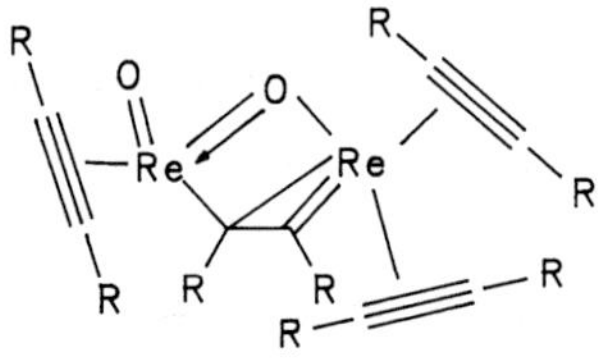 Id

A crystal-structure determination of one derivative established the following features: an unsymmetrically twisted coordination of the bridging acetylene ligand with four significantly different Re-C_μ bond lengths, an unsymmetrically bridging oxo ligand, and a short Re-Re distance. The structure cannot be described by a single valence bond picture; rather, it is best rationalized as a composite of the resonance forms Ia to Id, whereby the short Re-Re distance mainly arises from a contribution of structure Ib and the asymmetry of both bridging ligands from a contribution of structure Id. When considering the surroundings of each Re atom, the compounds can be described as mixed-valence Re^{I}-Re^{III} complexes [3].

Preparation. The compounds were produced by various reactions, but most of them are only briefly mentioned. The compounds could be obtained by subjecting $(RC{\equiv}CR)_2Re(O)I$ to a one-electron reduction (with $(t\text{-}C_4H_9)_2Zn$, $t\text{-}C_4H_9ZnCl$, Na, $NaC_{10}H_8$, $(C_5H_5)_2Co$, or electrochemically) [3] or by subjecting the salts $Na[(RC{\equiv}CR)_2Re(O)]$ to a one-electron oxidation (with air, $[(C_5H_5)_2Fe]^+$, $[(C_5H_5)_2Co]^+$, C_6H_5I, or $t\text{-}C_4H_9I$) [2, 3] in C_6H_6, ether, THF, or CH_3CN (oxidation of the salts with primary and secondary alkyl halides R′X such as C_2H_5I, $i\text{-}C_3H_7I$, $CH_2{=}CHCH_2I$, $C_6H_{11}Br$, or $CH_2{=}CH(CH_2)_4I$ led to mixtures of the title compounds and $(RC{\equiv}CR)_2Re(O)R'$ [2]). The compounds are also accessible by directly treating $(RC{\equiv}CR)_2Re(O)I$ with $Na[(R'C{\equiv}CR')_2ReO]$ [3].

References on p. 123

It should be pointed out that the title products were generally obtained together with complexes of the type $[(RC{\equiv}CR)_2ReO]_2$ ("symmetric isomers"; see Section 2.2.3.1.2.2) in a ratio somewhere between 1:0.75 and 1:10. Yields of the title compounds were usually higher in coordinating solvents [3].

Preparative details are given for the following procedure:

Method I: $(RC{\equiv}CR)_2Re(O)I$ was treated
a. with $t\text{-}C_4H_9ZnCl$ (in situ prepared from $ZnCl_2$ and $t\text{-}C_4H_9Li$) in ether at −78 °C. After warming to room temperature, the mixture was evaporated and the residue extracted into CH_2Cl_2.
b. with 1 equivalent $NaC_{10}H_8$ ($C_{10}H_8$ = naphthalenide anion radical) in THF at −78 °C. The mixture was allowed to slowly reach room temperature and was subsequently evaporated.

In any case, the product was separated by chromatography on silica using ethyl acetate/hexane mixtures [3].

All compounds are air-stable, bright yellow solids [3].

Fluxionality. All compounds are nonrigid as was shown by variable-temperature NMR methods: While the low-temperature ^{13}C NMR spectra display 8 signals due to the nonequivalent acetylenic C atoms (consistent with the solid-state structure), the room-temperature spectra display only four of these signals clearly and the other four are broad. At elevated temperature the broad signals coalesce into two. The changes are completely reversible on cooling. These observations can best be rationalized in terms of a coupled rotation of the two terminal acetylene ligands bonded to the Re atom not bearing the terminal oxo ligand, whereas the two other acetylene ligands remain rigid [3].

Thermal Behavior. Thermolysis (70 to 100 °C) of most compounds provided the symmetric isomers $[(RC{\equiv}CR)_2ReO]_2$. Details are given for each derivative.

$(t\text{-}C_4H_9C{\equiv}CH)_2Re(\mu\text{-}O)(\mu\text{-}t\text{-}C_4H_9C{\equiv}CH)Re(t\text{-}C_4H_9C{\equiv}CH)(O)$ was obtained by Method Ia with 26% yield [3].

^1H NMR spectrum (CD_2Cl_2, −40 °C): δ = 0.62, 1.25, 1.69, 1.73 (s's, CH_3); 9.85, 10.08, 10.16, 10.74 (s's, HC≡C) ppm; (25 °C): δ = 0.88 (br s); 1.30 (s); 1.58 (br s), 1.82 (s); 9.99, 10.72 (s, HC≡C) ppm. ^{13}C {^1H} NMR spectrum (CD_2Cl_2, −40 °C): δ = 134.5, 156.7 (d's, H**C**≡C; J = 209, 201 Hz); 163.9 (s, **C**C_4H_9); 166.4, 170.1 (d's; J = 204, 185 Hz); 180.0, 183.1, 183.8 (s's, **C**C_4H_9-t) ppm; (25 °C): δ = 134.5 (d; J = 214 Hz); 156.9 (br d, H**C**≡C; J = 197 Hz), 164.2 (s), 171.1 (d; J = 185 Hz); 184.3 (s) ppm. IR spectrum (neat): 905, 945, 980, 1040, 1215, 1240, 1365, 1460, 1480, 2920, 2990 cm^{-1}. Mass spectrum: $[M]^+$.

Thermolysis yielded the main isomer of $[(t\text{-}C_4H_9C{\equiv}CH)_2ReO]_2$ exclusively. Treatment with I_2 gave exclusively the asymmetric form of $(t\text{-}C_4H_9C{\equiv}CH)_2Re(O)I$ [3].

$(C_2H_5C{\equiv}CC_2H_5)_2Re(\mu\text{-}O)(\mu\text{-}C_2H_5C{\equiv}CC_2H_5)Re(C_2H_5C{\equiv}CC_2H_5)(O)$ was obtained with 37% yield [3] by Method Ia [1, 3].

^1H NMR spectrum (CD_2Cl_2, −40 °C): δ = 0.46, 0.84, 0.95, 1.05, 1.08, 1.42, 1.51, 1.63 (t's, CH_3; J = 7 Hz); 1.76, 1.96 (m's, 1 H); 2.80 (m); 2.97 (m, 1 H); 3.12 (m, CH_2); 3.25 (m, 3 H); 3.40, 3.51 (m's, 1 H); 3.60 (m, 3 H); 3.97 (m, 1 H) ppm [3]. ^{13}C {^1H} NMR spectrum (CD_2Cl_2, −40 °C): δ = 12.89, 14.17 (3 C unresolved), 14.40, 14.54, 14.69, 15.43 (CH_3); 21.78, 22.87, 23.82, 24.82, 27.67, 29.79, 30.54, 31.89 (CH_2) [3]; 150.09, 153.52, 165.06, 168.62, 169.71, 180.75, 192.90, 196.47 (C≡C) [1, 3] ppm; (25 °C): δ = 150.1, 153.4, 169.9, 193.6 (C≡C) [3] ppm. Based on the coalescence of the ethyl resonances in the ^1H NMR

References on p. 123

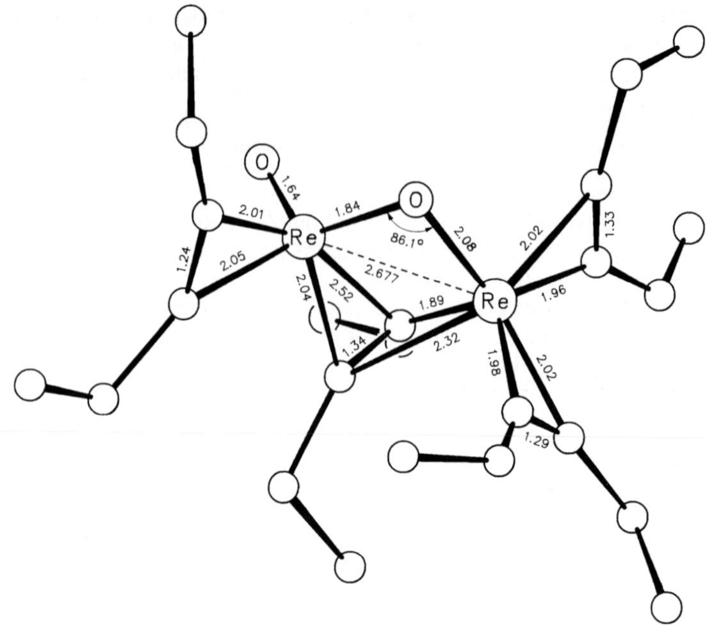

Fig. 58. Molecular structure of $(C_2H_5C{\equiv}CC_2H_5)_2Re(\mu\text{-}O)(\mu\text{-}C_2H_5C{\equiv}CC_2H_5)Re(C_2H_5C{\equiv}CC_2H_5)(O)$ [1, 3].

spectrum, the barrier for the fluxional process (see above) is estimated to be $\Delta G^{\neq} = 62.8$ kJ/mol [3]. IR spectrum (neat): 700 (ν(ReORe)), 928 (ν(ReO)) [1, 3]; 1056, 1085, 1145, 1245, 1302, 1370, 1455, 1759, 2865, 2935, 2962 cm^{-1} [3]. Mass spectrum: $[M]^+$ [3].

An X-ray structure analysis showed the compound to crystallize in the monoclinic lattice with a = 9.424(1), b = 13.833(1), c = 20.067(2) Å, β = 94.05(1)°, space group $P2_1/c-C^5_{2h}$ (No. 14), and D_{calc} = 1.866 g/cm^3 [1]. A view of the molecular structure is presented in **Fig. 58** [1, 3]. The bridging ligands are asymmetrically bonded, with the briging O atom being closer to the Re atom which bears the terminal oxo atom, indicating some π bonding. The bridging acetylene ligand is twisted ca. 34° away from an orientation perpendicular to the Re-Re vector. There are four significantly different Re-C_μ distances. The shorter bonds to the bridging acetylene originate from the Re atom that bears the two terminal acetylene ligands [3].

Treatment with $^{18}OH_2$ in CH_2Cl_2 for 3 d provided **$(C_2H_5C{\equiv}CC_2H_5)_2Re(\mu\text{-}O)(\mu\text{-}C_2H_5C{\equiv}CC_2H_5)$-$Re(C_2H_5C{\equiv}CC_2H_5)(^{18}O)$**, whose IR spectrum displays a ν(Re^{18}O) absorption band at 880 cm^{-1} (the bridging O atom was not affected) [3]. Thermolysis at 100 °C for 1 h quantitatively gave the symmetrical isomer $[(C_2H_5C{\equiv}CC_2H_5)_2ReO]_2$ [1, 3]. The reaction with I_2 yielded at least 10 products, but $(C_2H_5C{\equiv}CC_2H_5)_2Re(O)I$ clearly predominated [3]. Reduction with Na metal in the presence of a catalytic amount of naphthalene in THF gave $Na[(C_2H_5C{\equiv}CC_2H_5)_2ReO]$ [4].

$(C_6H_5C{\equiv}CC_6H_5)_2Re(\mu\text{-}O)(\mu\text{-}CH_3C{\equiv}CCH_3)Re(CH_3C{\equiv}CCH_3)(O)$ was obtained by successively adding at −78 °C 2.1 equivalents $NaC_{10}H_8$ and 1 equivalent $(CH_3C{\equiv}CCH_3)_2Re(O)I$ to a THF solution containing $(C_6H_5C{\equiv}CC_6H_5)_2Re(O)I$. After the mixture had slowly reached room temperature, it was worked up by column chromatography on silica with CH_2Cl_2. Yield: 67%.

References on p. 123

The mixed-ligand compound is also accessible by treating $(CH_3C{\equiv}CCH_3)_2Re(O)I$ with Na-$[(C_6H_5C{\equiv}CC_6H_5)_2ReO]$.

^1H NMR spectrum (CD_2Cl_2, −40 °C): δ = 2.51, 2.59, 2.84, 3.49 (q's, CH_3; J = 1 Hz); 6.10 to 7.23 (m, C_6H_5); 6.92, 6.94 (d's, 2 H; J = 7 Hz); 7.40 (br m, 3 H); 7.54 (t, 2 H); 7.84, 7.98 (d's, 2 H) ppm; (25 °C): δ = 2.54, 2.61, 2.90, 3.47 (q's, CH_3; J = 1 Hz), 7.01 (br, 4 H); 6.10 to 7.30 and 7.36 to 7.50 (each br, 6 H); 7.87 (br, 4 H) ppm. ^{13}C NMR spectrum (CD_2Cl_2, −60 °C): δ = 14.3, 16.1, 18.6, 21.1 (q's, CH_3); 125.7 to 139.8 (C_6H_5); 145.6, 152.8 ($CH_3\mathbf{C}{\equiv}$); 163.9 ($C_6H_5\mathbf{C}$); 164.7 ($CH_3\mathbf{C}{\equiv}$); 166.9, 176.4, 190.0 ($C_6H_5\mathbf{C}{\equiv}$); 194.1 ($CH_3\mathbf{C}{\equiv}$) ppm; (25 °C): δ = 15.1, 15.6, 18.9, 21.5 (CH_3); 127.6 to 131.6 (C_6H_5); 145.3, 152.5, 165.1, 195.6 ($CH_3\mathbf{C}{\equiv}$) ppm; estimated barrier for the fluxional process (see above): $\Delta G^{\neq}$ = 62.8 kJ/mol [3]. IR spectrum (neat): 692, 770, 932, 968, 1030, 1050, 1072, 1160, 1178, 1275, 1440, 1480, 1590, 2910, 3060 cm^{-1}.

Thermolysis induced partial rearrangement to the symmetrical isomer $(C_6H_5C{\equiv}CC_6H_5)_2$-$Re(O)Re(O)(CH_3C{\equiv}CCH_3)_2$ [3].

$(C_6H_5C{\equiv}CC_6H_5)_2Re(\mu\text{-}O)(\mu\text{-}C_6H_5C{\equiv}CC_6H_5)Re(C_6H_5C{\equiv}CC_6H_5)(O)$ was obtained by Method Ib.

^1H NMR spectrum (CD_2Cl_2, −80 °C): δ = 6.18 (d, 2 H), 6.38 (t, 2 H), 6.59 (t, 1 H), 6.97 (t, 2 H); 6.91 to 7.41 (21 H); 7.47 (t, 1 H), 7.60 (m, 1 H), 7.77 (m, 3 H), 7.86 (t, 1 H); 8.22, 8.91 (d's, 1 H) ppm; all J(H,H) = 7 Hz; (25 °C): δ = 6.44 (d, 2 H), 6.61 (t, 2 H), 6.74 (t, 1 H); 6.90 to 7.45 (26 H); 7.59 (m, 2 H), 7.77 (t, 1 H), 7.86 (d, 2 H), 7.89 (t, 4 H), 8.90 (d, 2 H) ppm; all J(H,H) = 7 Hz. $^{13}C\{^1H\}$ NMR spectrum (CD_2Cl_2, −80 °C): δ = 149.2, 157.9, 159.7, 161.4, 166.5, 182.6, 194.4, 195.2 ppm; (25 °C): δ = 149.9, 157.6, 162.6, 195.4 ppm. IR spectrum (neat): 682, 760, 835, 925, 995, 1020, 1065, 1155, 1175, 1280, 1440, 1470, 1570, 1590, 1660, 1745, 1804, 1895, 1958, 2920, 3010, 3070 cm^{-1}.

The compound was recovered unchanged after 4 d at 100 °C. Treatment with I_2 quantitatively gave $(C_6H_5C{\equiv}CC_6H_5)_2Re(O)I$ [3]. Reduction with Na metal in the presence of a catalytic amount of naphthalene in THF gave $Na[(C_6H_5C{\equiv}CC_6H_5)_2ReO]$ which could be crystallized from CH_3CN as a solid-state dimer (see pp. 115/6) [4].

References:

[1] Valencia, E.; Santarsiero, B. D.; Geib, S. J.; Rheingold, A. L.; Mayer, J. M. (J. Am. Chem. Soc. **109** [1987] 6896/8).
[2] Conry, R. R.; Mayer, J. M. (Organometallics **10** [1991] 3160/6).
[3] Spaltenstein, E.; Mayer, J. M. (J. Am. Chem. Soc. **113** [1991] 7744/53).
[4] Cundari, T. R.; Conry, R. R.; Spaltenstein, E.; Critchlow, S. C.; Hall, K. A.; Tahmassebi, S. K.; Mayer, J. M. (Organometallics **13** [1994] 322/31).

2.2.3.2.2.2 Miscellaneous Compounds

This section presents several other compounds whose structures are illustrated by Formulas I to III. Most of them are heterometal-containing clusters.

$[\mu\text{-}CH_3C((CH_3)_2N)C{=}]Re_2(CO)_8$ (see Formula I) was prepared by treating (μ-$\eta^{2:1}$-n-C_4H_9-$CH{=}CH)Re_2(CO)_8(\mu\text{-}H)$ with a 5-fold molar excess of $CH_3C{\equiv}CN(CH_3)_2$ (hexane, room temperature, 24 h). Workup of the mixture by evaporation and chromatography (hexane/CH_2Cl_2 (5:1)) yielded 25%. The reaction was also performed at 50 °C; under these conditions, not only the title compound formed (yield: 32%), but also 3 isomers of $[C_4(CH_3)_2$-$(N(CH_3)_2)_2]Re_2(CO)_7$ (see Table 7, Nos. 1 to 3, p. 150). Orange solid.

References on pp. 127/8

$N(CH_3)_2$
CH_3
$(CO)_4Re$ — $Re(CO)_4$
I

R
$(CH_3)_5C_5W$ — $Re(CO)_3$
$(CO)_2$
CO
Re
$(CO)_3$
II

R
$(CH_3)_5C_5W$ — $Re(CO)_3$
$(CO)_2$
H
H
Re
$(CO)_3$
III

^1H NMR spectrum ($CDCl_3$): $\delta = 3.30, 3.48, 3.53$ (s's, CH_3) ppm [2]. ^{13}C {^1H} NMR spectrum ($CDCl_3$): $\delta = 32.7$ ($\mathbf{C}H_3CC$); 46.6, 47.5 (NCH_3); 121.6 ($CH_3\mathbf{C}C$); 191.3, 191.4, 193.4, 194.4, 197.2 (CO); 229.5 ($C\mathbf{C}NCH_3$) ppm [4]. IR spectrum (n-hexane): 1933, 1950, 1961, 1988, 1994, 2046, 2086 (ν(CO)) cm^{-1} [2]. Mass spectrum: $[M - n\ CO]^+$ (n = 0 to 8) [2].

X-ray crystallography showed the compound to crystallize in the monoclinic space group $P2_1/c-C^5_{2h}$ (No. 14) with a = 9.776(2), b = 11.315(4), c = 15.695(4) Å, $\beta = 97.56(2)°$; Z = 4 molecules per unit cell, $D_{calc} = 2.62\ g/cm^3$. The molecular structure is shown in **Fig. 59**. It can be seen that the bridging ligand coordinates asymmetrically (twisted). The N atom has planar geometry, and the C-N bond is very short [2].

The cause for the twist of the ynamine ligand away from the perpendicular and parallel orientation (compare with Formulas I to III, p. 120) was investigated by extended Hückel MO calculations, which were performed with the alkyne in each of the perpendicular, twisted, and parallel orientations. It appears that the twisted coordination permits a strong stabilizing interaction between the one of the π^* orbitals of the ligand and the metallic orbital that is principally responsible for the formation of the Re-Re bond. The HOMO-LUMO gap is widened [1].

Treatment with excess $CH_3C{\equiv}CN(CH_3)_2$ in refluxing hexane yielded three isomers of the type $[C_4(CH_3)_2(N(CH_3)_2)_2]Re_2(CO)_7$ [2]. Hydrogenation in refluxing hexane (10 atm) yielded (μ-$CH_3C(H)CN(CH_3)_2)Re_2(CO)_8(\mu$-H) and ($\mu$-$CH_3CH{=}CHN(CH_3)_2)Re_2(CO)_8$ (see

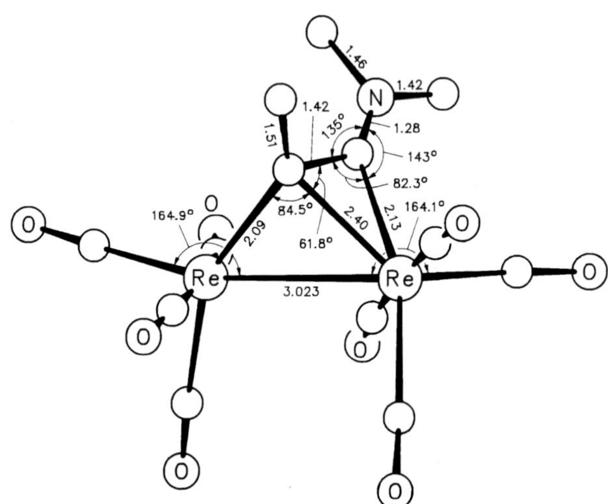

Fig. 59. Molecular structure of [μ-$CH_3C((CH_3)_2N)C{=}]Re_2(CO)_8$ [2].

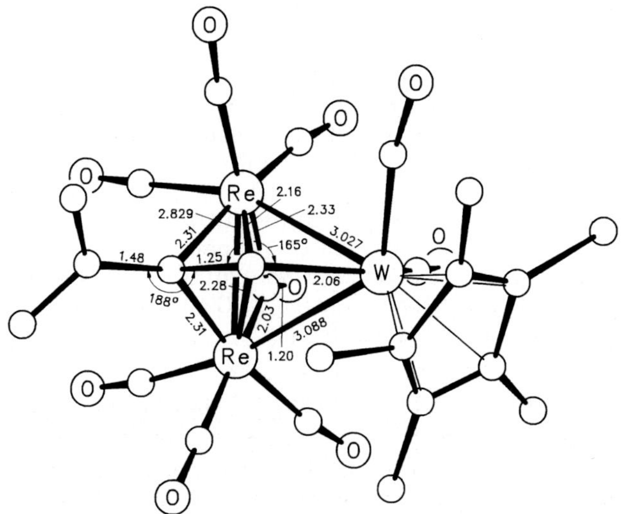

Fig. 60. Molecular structure of [μ-$(C_5(CH_3)_5)W(CO)_2C{\equiv}CC(CH_3){=}CH_2]Re_2(CO)_6$(μ-CO) [7].

pp. 58, 117), while hydrogenation in refluxing heptane under atmospheric pressure gave (μ-$CH_3CH{=}CHN(CH_3)_2)Re_2(CO)_8$ and a compound suggested to be $(CO)_8Re_2$(μ-H)(μ-C(H)-$(C_2H_5){-}N(CH_3)_2$) (see "Organorhenium Compounds" 5, 1994, pp. 384/5) [3]. Thermolysis in refluxing cyclohexane led to the ^{3}L complexes (μ-$CH_2C((CH_3)_2N)C{=})Re_2(CO)_7$(μ-H), (μ-$CH_2{=}CH((CH_3)_2N)C{=})Re_2(CO)_8$ (see p. 134), and to (μ-$CH_2{=}CCN(CH_3)_2)Re_2(CO)_8$(μ-H) (see p. 59) [4].

[μ-$(C_5(CH_3)_5)W(CO)_2C{\equiv}CR]Re_2(CO)_6$(μ-CO) (see Formula II). Compounds of this type were prepared by treating $(CO)_8Re_2(NCCH_3)_2$ with $(C_5(CH_3)_5)W(CO)_3C{\equiv}CR$ (R = $C(CH_3){=}CH_2$, C_6H_9-cyclo, $CH{=}CHOCH_3$, C_6H_5) in toluene (reflux, 1 h) [5, 7]. The resulting mixture was separated by preparative TLC (silica, CH_2Cl_2/hexane (1:2)), and the products were purified by recrystallization from CH_2Cl_2/CH_3OH [7].

The compounds have the following properties:

R	yield, physical and chemical properties, remarks (for abbreviations and units see p. X)
$C(CH_3){=}CH_2$	Yield: 31% [5], 54% [7]. – Red-orange crystals [5, 7]. ^{1}H NMR ($CDCl_3$): 2.22 (C_5CH_3), 2.31 (s, CH_3); 5.19, 5.35 (s's, CH_2=). – ^{13}C {^{1}H} NMR ($CDCl_3$): 12.1 (C_5**C**H_3), 27.3 (CH_3); 94.7 (WC≡**C**C=C; J(W,C) = 24), 103.0 (C_5), 118.2 (=CH_2), 140.6 (WC≡C**C**=C), 160.8 (W**C**≡CC=C; J(W,C) = 1324); 185.7 (br), 187.2, 189.2, 195.4 (ReCO; 2:1:2:2 C); 209.3 (WCO; J(W,C) = 171). – IR (CH_2Cl_2): 1891, 1935, 1950, 1974, 1987, 2025, 2064 (ν(CO)). – FAB MS: $[M]^+$ [5, 7]. Single-crystal data: monoclinic, a = 36.091(9), b = 14.277(4), c = 17.239(3) Å, β = 115.26(3)°; space group $Cc{-}C_s^4$ (No. 9); Z = 12 molecules per unit cell, D_{calc} = 2.503 g/cm^3 [5]. The structure is depicted in **Fig. 60** [7]. The bridging W-C≡C-R unit coordinates perpendicular to the Re-Re bond [5].

References on pp. 127/8

R	yield, physical and chemical properties, remarks (for abbreviations and units see p. X)
$C(CH_3)=CH_2$ (continued)	Treatment with H_2 generated an inseparable mixture of two isomers of $(CO)_6Re_2(\mu\text{-}H)(\mu\text{-}CO)W(C_5(CH_3)_5)(\mu\text{-}C_4(CH_3)H_3)$ (see pp. 250/1) in the ratio 48:52. Treatment with ROH (R = CH_3, C_2H_5, C_6H_5) yielded $[\mu\text{-}(CH_3)_2CCCW(C_5(CH_3)_5)(CO)_2]Re_2(CO)_6(\mu\text{-}OR)$ (see pp. 136/8) [5, 7]. Treatment with O_2 or N_2O (80 °C, 3 h) gave $(CO)_8Re_2(\mu_3\text{-}CCC(CH_3)=CH_2)\text{-}W(C_5(CH_3)_5)(O)$ (see Formulas IVa, IVb; R = $C(CH_3)=CH_2$) [6]. Short-term heating with C_6H_5SH provided $[\mu\text{-}(C_5(CH_3)_5)W(=CCH=C(CH_3)_2)(SC_6H_5)CO]\text{-}Re_2(CO)_6(\mu\text{-}SC_6H_5)$ (see p. 54) [8].
C_6H_9-cyclo	^1H NMR ($CDCl_3$): 1.64 to 1.68 and 1.75 to 1.80 (m's, CH_2); 2.20 (s, C_5CH_3); 2.27 to 2.31 and 2.46 to 2.49 (m's, CH_2); 6.11 (m, C=CH). – IR (C_6H_{12}): 1896, 1951, 1978, 2026, 2066 (ν(CO)). – FAB MS: $[M]^+$ [7]. Hydrogenation yielded $[\mu\text{-}(C_5(CH_3)_5)W(CO)_2(C{\equiv}CC_6H_9)H]Re_2(CO)_6(\mu\text{-}H)$ [7]. Short treatment with excess C_6H_5SH provided $[\mu\text{-}(C_5(CH_3)_5)W(=CCH=C(CH_2)_5)(SC_6H_5)CO]Re_2(CO)_6(\mu\text{-}SC_6H_5)$ (see pp. 54/5) [8].
$CH=CHOCH_3$	^1H NMR ($CDCl_3$): 2.09 (s, C_5CH_3), 3.76 (s, OCH_3); 5.92, 6.12 (d's, CH=; J = 5.6). – $^{13}C\{^1H\}$ NMR ($CDCl_3$): 10.3 ($(CH_3)_5$), 59.3 (OCH_3), 77.6 (WC≡**C**CH=C), 99.1 (WC≡C**C**H=C), 101.0 (C_5), 151.3 (WC≡CCH=**C**HO), 156.7 (W**C**≡CCH=CHO; J(W,C) = 134); 184.2, 185.6, 187.4, 194.6 (ReCO, 2:1:2:2 C); 210.2 (WCO; J(W,C) = 172). – IR (C_6H_{12}): 1894, 1949, 1960, 1976, 2025, 2059 (ν(CO)). – FAB MS: $[M]^+$ [7]. Short-term exposure to an H_2 atmosphere yielded $[\mu\text{-}(C_5(CH_3)_5)W(CO)_2(C{\equiv}CCH=CHOCH_3)H]Re_2(CO)_6(\mu\text{-}H)$, while long-term exposure provided $(CO)_6Re_2(\mu\text{-}H)(\mu\text{-}CO)W(C_5(CH_3)_5)(\mu\text{-}C_4H_4)$ (see p. 250) [7]. Treatment with excess C_6H_5SH gave $[\mu\text{-}(C_5(CH_3)_5)W(SC_6H_5)\text{-}(HC=CCH=CHOCH_3)CO]Re_2(CO)_6$ (see p. 136) [8].
C_6H_5	Preparation as above (no details) [6, 8]. Treatment with excess C_6H_5SH yielded both $[\mu\text{-}(C_5(CH_3)_5)W(CO)_2CH=CC_6H_5]Re_2(CO)_6(\mu_3\text{-}SC_6H_5)$ (see p. 80) and $[\mu\text{-}(C_5(CH_3)_5)W(SC_6H_5)(HC=CC_6H_5)CO]Re_2(CO)_6$ (see p. 136) [8]. Treatment with O_2 or N_2O (80 °C, 3 h) gave $(CO)_8Re_2(\mu_3\text{-}CCC_6H_5)W(C_5(CH_3)_5)(O)$ (see Formulas IVa, IVb; R = C_6H_5) [6].

(R = $C(CH_3)=CH_2$; C_6H_5)

IVa IVb

References on pp. 127/8

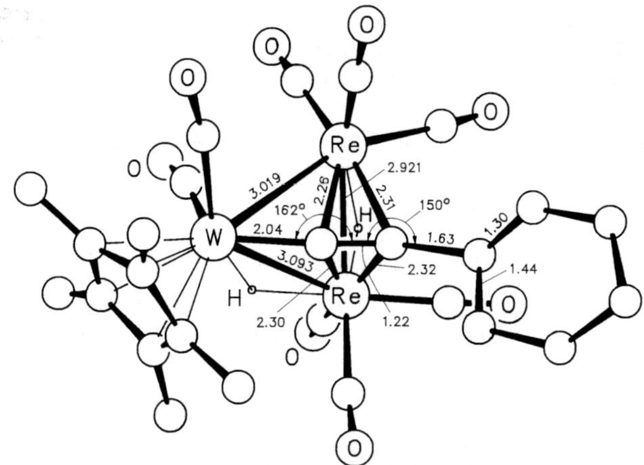

Fig. 61. Molecular structure of [μ-($C_5(CH_3)_5$)W$(CO)_2$(C≡CC_6H_9-cyclo)H]$Re_2(CO)_6$(μ-H) [7].

[μ-($C_5(CH_3)_5$)W$(CO)_2$(C≡CR)H]$Re_2(CO)_6$(μ-H) (see Formula III). Compounds with this composition were obtained by exposing [μ-($C_5(CH_3)_5$)W$(CO)_2$C≡CR]$Re_2(CO)_6$(μ-CO) (R = C_6H_9-cyclo, CH=CHOCH_3) to an H_2 atmosphere (1 atm) in toluene (reflux, 10 to 30 min). Separation was accomplished by preparative TLC on silica with CH_2Cl_2/hexane (1:1) [7].

R = C_6H_9-cyclo: Yield: 75% (after 30 min). – Pale yellow solid.
^{1}H NMR ($CDCl_3$): −17.89 (ReHRe), −15.73 (ReHW; J(W,H) = 42); 1.63 to 1.70 and 1.77 to 1.85 (m's, CH_2); 2.20 ($(CH_3)_5$); 2.26 to 2.31 and 2.46 to 2.49 (m's, CH_2); 6.44 (br m, =CH). – ^{13}C {^{1}H} NMR ($CDCl_3$): 11.3 (C_5**C**H_3); 22.0, 23.7, 26.5, 34.6 (CH_2); 100.8 (WC≡**C**), 103.7 (C_5), 125.3 (W**C**≡C; J(W,C) = 111), 129.4 (**C**=CH), 135.3 (C=**C**H); 183.5, 185.5, 185.9, 191.8, 192.4, 197.5 (ReCO); 207.3, 214.9 (WCO; J(W,C) = 148 and 135). – IR (CH_2Cl_2): 1915, 1929, 1952, 1984, 2014, 2043 (ν(CO)). – FAB MS: $[M]^+$.
Single-crystal data: orthorhombic; a = 20.170(6), b = 16.854(3), c = 9.320(3) Å; space group $Pna2_1-C_{2v}^9$ (No. 33); Z = 4 molecules per unit cell, D_{calc} = 2.322 g/cm^3. The structure is shown in **Fig. 61**.
Long exposure to an H_2 atmosphere in hot toluene provided two isomers of $(CO)_6Re_2$(μ-H)(μ-CO)W($C_5(CH_3)_5$)(μ-CHCH(C_6H_8)) (see p. 251) [7].

R = CH=CHOCH_3: Yield: ca. 20% (after 10 min). The compound was contaminated with several unidentified impurities which could not be removed.
^{1}H NMR: −17.60 (ReHRe), −15.64 (ReHW; J(W,H) = 42).
Exposure to an H_2 atmosphere in refluxing toluene yielded $(CO)_6Re_2$(μ-H)(μ-CO)W($C_5(CH_3)_5$)(μ-C_4H_4) (see p. 250) [7].

References:

[1] Adams, R. D.; Chen, G.; Chen, L.; Yin, J.; Halet, J.-F. (J. Cluster Sci. **2** [1991] 83/103).
[2] Adams, R. D.; Chen, G.; Yin, J. (Organometallics **10** [1991] 1278/82).
[3] Adams, R. D.; Chen, G.; Yin, J. (Organometallics **10** [1991] 2087/8).
[4] Adams, R. D.; Chen, G.; Chi, Y.; Wu, W.; Yin, J. (Organometallics **11** [1992] 1480/6).
[5] Cheng, P.-S.; Chi, Y.; Peng, S.-M.; Lee, G.-H. (Organometallics **12** [1993] 250/2).

[6] Chi, Y.; Cheng, P.-S.; Wu, H.-L.; Hwang, D.-K.; Su, P.-C.; Peng, S.-M.; Lee, G. H. (J. Chem. Soc. Chem. Commun. **1994** 1839/40).
[7] Peng, J.-J.; Horng, K.-M.; Cheng, P.-S.; Chi, Y.; Peng, S.-M.; Lee, G.-H. (Organometallics **13** [1994] 2365/74).
[8] Peng, J.-J.; Peng, S.-M.; Lee, G.-H.; Chi, Y. (Organometallics **14** [1995] 626/33).

2.3 Compounds with Ligands Bonded to Rhenium by Three C Atoms (^{3}L Compounds)

There are only a few compounds with ^{3}L ligands. With one exception, all of them also bear CO groups. The ^{3}L ligand can coordinate to a single Re atom ("terminal") or can coordinate to both Re atoms ("bridging"), thereby showing various bonding modes.

2.3.1 Compounds with Terminal ^{3}L Ligands

(π-C_3H_5)$_4$$Re_2$ was prepared by reacting $ReCl_5$ with a 5-fold excess of C_3H_5MgCl in ether. Combination of the reactants at −78°C followed by slowly warming the mixture to room temperature produced a deep orange solution which was filtered and evaporated. Extraction with petroleum ether, concentration, and cooling gave the complex with ca. 46% yield.

The orange crystals decompose at 120°C, but sublime at 80°C/10^{-3} Torr [1].

The ^{1}H NMR spectrum (CS_2) shows two independent A_2B_2X systems with signals at δ = −0.53 ("q", CH_2), 4.12 ("t", CH_2), and 5.77 (m, CH) ppm; J(A,B) and J(B,X) ca. 5 Hz, J(A,X) ca. 9 Hz [2] (see also [1]). The spectrum is unchanged in the range −85 to +80°C [1]. IR spectrum (Nujol): 540, 680, 750, 885, 912, 875 (misprint?), 1198, 1375, 1455, 2865, 2920 cm^{-1}; (CS_2): 335, 540, 675, 750, 885, 910, 918, 975, 1255, 1362, 2940, 3300 cm^{-1} [1]. Photoelectron spectrum (He I): E = 6.73 (narrow), 7.77 (br), 7.98 (narrow), 8.94 (sh), ca. 9.50 (v br) eV [2, 7]. Mass spectrum: $[M - n\ C_3H_5]^+$ with n = 0 to 3 [1].

The compound crystallizes in the monoclinic space group C2−C_2^3 (No. 5) with a = 12.218(3), b = 8.692(2), c = 18.763(4) Å, and β = 97.28(2)°. (π-C_3H_5)$_4$$Re_2$ possesses a structure with two π-bonded allyl ligands connected to each Re atom. Within the unit cell, there are four molecules on general positions and two on special positions having C_2 symmetry. The structures of these two types of molecule are essentially identical. **Fig. 62** illustrates two views of one of those molecules with a crystallographically imposed C_2 axis [2].

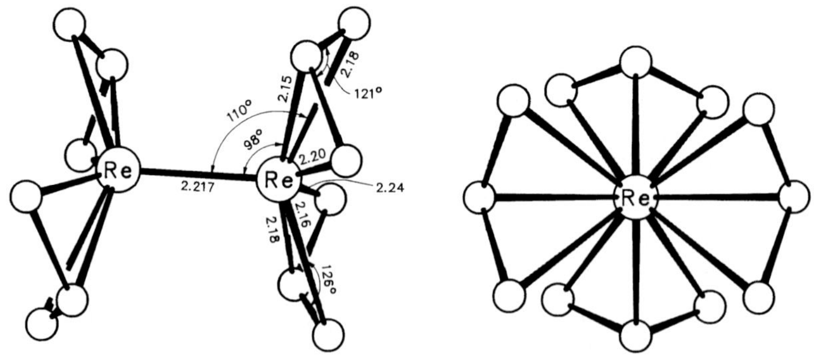

Fig. 62. Two views of the structure of (π-C_3H_5)$_4$$Re_2$ [2].

References on p. 139

MO calculations were made using 1) atomic coefficients for Tc instead of Re and 2) geometric parameters obtained from the X-ray structure analysis of $(C_3H_5)_4Re_2$. The results suggest that the Re-Re bond can be regarded as a triple bond with strong σ ($2a_1$) and π (2e) components. The highest occupied levels are of the δ ($2b_2$) and δ^* ($3a_1$) type, cancelling each other [2]. Another study, taking also relativistic effects into account, similarly revealed the ($\sigma^2\pi^4\delta^2\delta^{*2}$) Re≡Re triple bond character and, furthermore, made clear that a substantial amount of π back-bonding is occurring, mostly via interaction of the δ and δ^* orbitals with the allyl π^* levels [7].

The crystals slowly decompose in air, but are pyrophoric when well dried. The compound is soluble in most organic solvents. Solutions are very sensitive [1].

I

CH_2=$CHC_2H_4(C_2H_5O)C$=$Re(CO)_3-Re(CO)_5$ (see Formula I). Stirring a THF solution containing CH_2=$CHC_2H_4(C_2H_5O)C$=$Re(CO)_4-Re(CO)_5$ with a large excess of $(CH_3)_3NO \cdot 2\,H_2O$ for 18 h followed by chromatographic workup (silica, petroleum ether, CH_2Cl_2) gave the compound as a yellow oil with ca. 30% yield.

^{1}H NMR spectrum (C_6D_6): δ = 0.94 (t, CH_3); 0.94, 1.72, 2.10, 2.35 (all 1 H, C_2H_4); 2.74, 2.90 (=CH_2); 3.77 (m, OCH_2), 4.44 (m, =CH) ppm. ^{13}C {^{1}H} NMR spectrum (C_6D_6): δ = 14 (CH_3); 30, 50, 53 (C-5,4,7, respectively); 76 (CH_2), 81 (C-6); 196 to 199 (CO); 321 (Re=C) ppm. The mass spectrum shows $[M]^+$ [4, 5].

The complex induces the polymerization of hept-1-yne (hexane, 20 °C, 24 h). The polymer obtained under these conditions has an average molecular mass of 60000 Daltons and ca. 70% of the double bonds are trans-configurated [4].

2.3.2 Compounds with Bridging ^{3}L Ligands

Most compounds were prepared by the following methods:

Method I: Photochemical reaction of $(CO)_{10}Re_2$ with the respective allene CH_2=C=CRR′ (R = H, R′ = H, CH_3 [15], C_6H_5 [8]; R = CH_3, R′ = CH_3 [16]), CH_3CH=C=CRR′ (R = H, R′ = CH_3 [15]; R = CH_3, R′ = CH_3 [16]) in hexane (temperature and irradiation time are given for each product). The resulting complex mixture (for product pattern, see "Chemical Behavior of $(CO)_{10}Re_2$" in "Organorhenium Compounds" 6) was separated chromatographically on alumina using CH_2Cl_2/hexane mixtures [8, 15, 16].

Method II: Thermolysis of the respective complexes (μ-$CH_2CCHR)Re_2(CO)_9$ or (μ-$CH_3CHCCHCH_3)Re_2(CO)_9$ in isooctane at 80 °C for 1 h initiated quantitative decarbonylation. Side products were not observed [8, 15].

Method III: Thermolysis of [μ-$CH_3C((CH_3)_2N)C$=]$Re_2(CO)_8$ (see p. 123) in refluxing cyclohexane for 2 h. The mixture was separated chromatographically with hexane/CH_2Cl_2 (4:1) on silica TLC plates [11].

References on p. 139

Method IV: Treatment of $[\mu\text{-}(C_5(CH_3)_5)W(CO)_2C{\equiv}CR]Re_2(CO)_6(\mu\text{-}CO)$ (R = $CH{=}CHOCH_3$, C_6H_5) (see p. 126) with excess thiophenol in refluxing toluene for 10 min. The resulting mixture was separated by preparative TLC using CH_2Cl_2/hexane [17].

Method V: Refluxing a mixture of $[\mu\text{-}(C_5(CH_3)_5)W(CO)_2C{\equiv}CC(CH_3){=}CH_2]Re_2(CO)_6(\mu\text{-}CO)$ (see p. 125), toluene, and a large excess of anhydrous ROH (R = CH_3, C_2H_5, C_6H_5) for ca. 35 min, during which period the color changed from orange to brown. The resulting mixture was separated by preparative TLC on silica using CH_2Cl_2/hexane (1:3). Yields ranged from ca. 30 to ca. 60% [14].

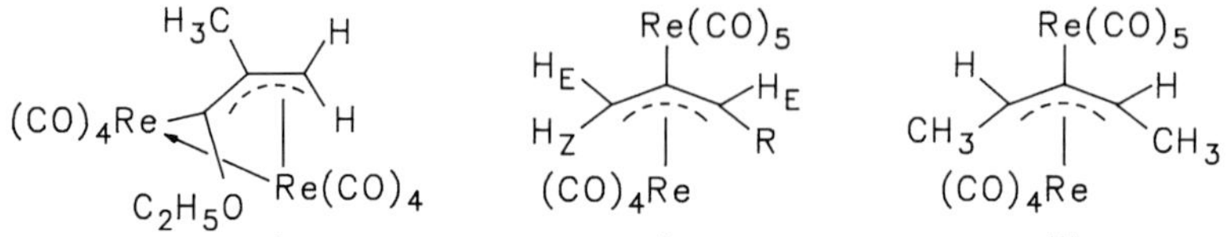

I II III

$(\mu\text{-}\eta^{3:1}\text{-}C_2H_5OCC(CH_3)CH_2)Re_2(CO)_8$ (see Formula I) was obtained along with $[C_2H_5OCC(CH_3)CH_2W(CO)_3(\mu\text{-}CO)]Re_2(CO)_9$ (see p. 53), $(CO)_8W_2(C_2H_5OCC(CH_3){=}CH_2)$, and a considerable amount of $(CO)_6W$ by photochemically reacting $(CO)_{10}Re_2$ with $CH_2{=}(CH_3)C(C_2H_5O)C{=}W(CO)_5$ in pentane at −50 °C for 70 min. The resulting mixture was at first roughly separated by column chromatography on silica using CH_2Cl_2/pentane mixtures. A pure product could only be obtained after carefully repeating the chromatographic separation with pentane/CH_2Cl_2 (9:1) at −10 °C. This eventually yielded 0.2% of light red crystals.

^1H NMR spectrum (C_6D_6): δ = 1.01 (t, CH_3; J = 7.0 Hz), 1.96 (s, $CH_3C{\doteq}$); 2.02, 2.54 (d's, ${\doteq}CH_2$; J = 1.5 Hz); 3.82, 3.90 (dq's, OCH_2; J = 7.0 and 9.3 Hz) ppm. IR spectrum (n-hexane): 1947, 1959, 1990, 2000, 2020, 2056, 2100 (ν(CO)) cm^{-1} [12].

$(\mu\text{-}C_3H_4)Re_2(CO)_9$ (see Formula II, R = H_Z) was prepared by Method I (−50 °C, 35 min) with 6% yield. Yellow-brown crystals.

^1H NMR spectrum (CD_2Cl_2): δ = 2.65 (H_E) and 2.87 (H_Z) ppm (AA′BB′ pattern); $^2J(H,H)$ = 2 Hz, $^4J(H\text{-}E,Z)$ = 1 Hz, $^4J(H\text{-}E,E)$ = 2 Hz, and $^4J(H\text{-}Z,Z)$ = 1 Hz. IR spectrum (n-hexane): 1959, 1991, 2029, 2080, 2151 (ν(CO)) cm^{-1}.

Thermolysis according to Method II quantitatively gave $(\mu\text{-}C_3H_4)Re_2(CO)_8$ [15].

$(\mu\text{-}CH_2CCHCH_3)Re_2(CO)_9$ (see Formula II, R = CH_3) was obtained by Method I (−40 °C, 35 min) with 5% yield. Pale yellow crystals.

^1H NMR spectrum (CD_2Cl_2): δ = 1.92 (d, CH_3; J = 6.0 Hz), 2.26 (dd, H_{1E}; J = 2.5, 1.7 Hz), 2.55 (dd, H_{1Z}; J = 2.5, 0.8 Hz), 2.94 (ddq, H-3; J = 6.0, 1.7, 0.8 Hz) ppm. IR spectrum (n-hexane): 1956, 1991, 2024, 2082, 2155 (ν(CO)) cm^{-1}.

Thermolysis according to Method II gave $(\mu\text{-}CH_2{=}C{=}CHCH_3)Re_2(CO)_8$ [15].

$(\mu\text{-}CH_2CCHC_6H_5)Re_2(CO)_9$ (see Formula II, R = C_6H_5) was obtained according to Method I (−25 °C, 75 min) with 7% yield [8] or also by photochemically treating α-diimine complexes of the type $(CO)_5Re{-}Re(CO)_3(^4D)$ (^4D = 1,10-phenanthroline or $4\text{-}CH_3C_6H_4N{=}CHCH{=}N\text{-}C_6H_4CH_3\text{-}4$) with phenylallene at −33 °C [9]. Yellow needles [8].

^1H NMR spectrum (CD_2Cl_2): δ = 2.43 (dd; J = 2.0, 1.5 Hz), 2.53 (dd; J = 2.0, 0.6 Hz), 3.81 (dd; J = 1.5, 0.6 Hz), 7.23 (m, C_6H_5) ppm [8]. IR spectrum (n-hexane): 1949, 1990, 1996, 2038, 2076, 2138 (ν(CO)) cm^{-1} [8]; also given: 2063, 2133 (highest wave numbers of the $Re(CO)_4$ and the $Re(CO)_5$ unit, respectively) [9].

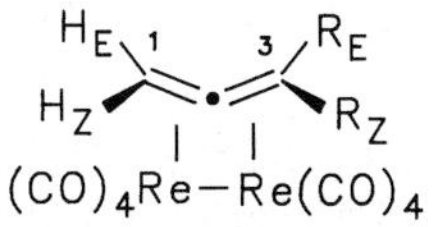

IV 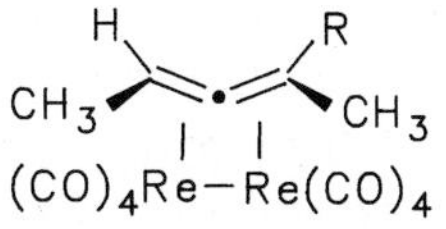 V 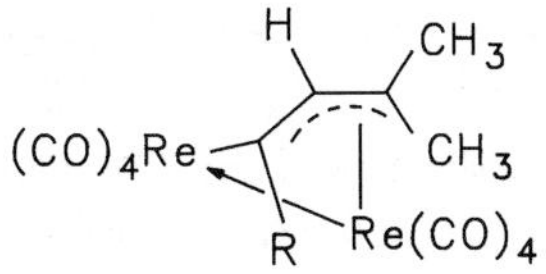 VI

Thermolysis in isooctane quantitatively gave (μ-CH_2=C=CHC_6H_5)$Re_2(CO)_8$ [8].

(μ-$CH_3CHCCHCH_3$)$Re_2(CO)_9$ (see Formula III) was obtained by Method I (−30 °C, 45 min) with only 2% yield. Pale yellow needles.

^1H NMR spectrum (CD_2Cl_2): δ = 1.95 (CH_3), 2.46 (CH) ppm; J(CH_3,H) = 6.4, ^4J(CH,CH) = 1.4 Hz. IR spectrum (n-hexane): 1945, 1986, 1998, 2024, 2076, 2118 (ν(CO)) cm^{-1}.

Thermolysis according to Method II gave (μ-CH_3CH=C=$CHCH_3$)$Re_2(CO)_8$ [15].

(μ-C_3H_4)$Re_2(CO)_8$ (see Formula IV, R_E, R_Z = H) was obtained by Method I (−50 °C, 35 min) with 6% yield or from (μ-C_3H_4)$Re_2(CO)_9$ by employing Method II. Yellow crystals.

^1H NMR spectrum (CD_2Cl_2): δ = 3.96 and 4.14 ppm (AA′BB′ pattern); ^2J(H,H) = 4.1 Hz, ^4J(H-E,E) = 0.1 Hz, ^4J(H-Z,Z) = 3.2 Hz, and ^4J(H-E,Z) = 1.9 Hz [15].

(μ-CH_2=C=$CHCH_3$)$Re_2(CO)_8$ (see Formula IV, R_E = CH_3, R_Z = H and vice versa) was obtained by Method I (−40 °C, 35 min, yield: 8%) or Method II as pale yellow rhombes. In both cases ^1H NMR spectroscopy confirmed the presence of two isomers in the ratio 4:1 (for abbreviations and units see p. X):

Main isomer: ^1H NMR (CD_2Cl_2): 1.74 (d, CH_3); 2.05, 3.68 (both dd, CH_2); 4.48 (ddq, H-3); coupling constants: ^3J(CH_3,H) = 6.8, ^2J(H-E,Z) = 3.8, ^4J(H-E,Z) = 2.8, ^4J(H-E,E) = 0.2. A structure with the CH_3 group in Z position was suggested [15].

Minor isomer: ^1H NMR (CD_2Cl_2): 1.98 (dd, CH_3); 3.61, 3.97 (both dd, CH_2); 4.64 (ddq, H-3); coupling constants: ^3J(CH_3,H) = 6.0, ^5J(CH_3,H_E) = 0.7, ^2J(H-E,Z) = 4.2, ^4J(H-Z,Z) = 1.6, ^4J(H-E,Z) = 2.8. A structure with the CH_3 group in E position was suggested [15].

IR spectrum (n-hexane, mixture): 1969, 1972, 1987, 2001, 2061, 2091 (ν(CO)) cm^{-1} [15].

(μ-CH_2=C=CHC_6H_5)$Re_2(CO)_8$ (see Formula IV, R_E = H, R_Z = C_6H_5) was obtained by Method I (−25 °C, 75 min) with 7% yield or by Method II. Furthermore, (μ-$\eta^{2:1}$-C_6H_5CH=C=CH)Re_2-$(CO)_8$(μ-H) (see p. 71), when kept in solution, rearranged into the title compound either at room temperature during 4 d or at 80 °C during 1 h. The complex forms yellow crystals. Like the foregoing complex, two isomers are possible, but only one of them appears to form. Spectroscopic data account for a Z position of the phenyl group.

^1H NMR spectrum: δ = 4.17 (dd; J = 4.3, 1.5 Hz), 4.26 (dd; J = 4.3, 3.0 Hz), 5.51 (dd; J = 3.0, 1.5 Hz); 7.38 (m, C_6H_5) ppm. IR spectrum (n-hexane): 1961, 1986, 1988, 2017, 2067, 2105 (ν(CO)) cm^{-1} [8].

(μ-CH_2=C=$C(CH_3)_2$)$Re_2(CO)_8$ (see Formula IV, R_E, R_Z = CH_3) was obtained by Method I (−30 °C, 60 min) with 4% yield. Lemon yellow crystals.

^1H NMR spectrum (CD_2Cl_2): δ = 2.25, 2.70 (br s's, CH_3); 3.35, 3.65 (dqq, CH_2; J = 4.4, ca. 1.0, ca. 0.8 Hz) ppm. IR spectrum (n-hexane): 1967, 1984, 1989, 2000, 2059, 2106 (ν(CO)) cm^{-1} [16].

References on p. 139

(μ-CH_3CH=C=$CHCH_3$)$Re_2(CO)_8$ (see Formula V, R = H) was obtained by Method I (−30 °C, 45 min) with 18% yield or by Method II. Yellow-orange crystals.

^{1}H NMR spectrum (CD_2Cl_2): δ = 1.94 (CH_3), 4.34 (CH) ppm; J(CH_3,H) = 7.0 Hz, J(CH,CH) = 1.8 Hz. IR spectrum (n-hexane): 1967, 1985, 2000, 2058, 2093 (ν(CO)) cm^{-1} [15].

(μ-CH_3CH=C=C$(CH_3)_2$)$Re_2(CO)_8$ (see Formula V, R = CH_3) was obtained by Method I (−25 °C, 45 min) with 15% yield. Pale yellow crystals.

^{1}H NMR spectrum (CD_2Cl_2): δ = 1.78, 2.18, 2.67 (d's, CH_3; J = 6.2, 0.7, 1.0 Hz, respectively); 4.07 (qqq, CH) ppm. IR spectrum (n-hexane): 1966, 1983, 2000, 2058, 2109 (ν(CO)) cm^{-1} [16].

(μ-$(CH_3)_2$CCHCH)$Re_2(CO)_8$ (see Formula VI, R = H) formed by stirring a hexane solution containing (μ-$\eta^{2:1}$-C_2H_5CH=CH)$Re_2(CO)_8$(μ-H) and excess 3,3-dimethyl-cyclopropene at room temperature for 2 d. Subsequent filtration at −20 °C separated the yellow solid, which was recrystallized from ether/hexane (1 : 1). Yield: 78%.

^{1}H NMR spectrum ($CDCl_3$): δ = 1.94, 2.09 (s's, CH_3); 4.53, 7.65 (d's, CHRe and CCHC; J = 12.45 Hz) ppm. ^{13}C {^{1}H} NMR spectrum ($CDCl_3$): δ = 23.3, 29.2 (CH_3); 90.2 (**C**$(CH_3)_2$); 105.3 (CCHC; ^{1}J(C,H) = 156.3 Hz), 139.0 (CHRe; ^{1}J(C,H) = 138 Hz, ^{2}J(C,H) = 9.2 Hz); 185.8, 186.9, 187.3, 188.9, 190.0, 191.9, 193.4, 195.1 (CO) ppm. IR spectrum (n-hexane): 1941, 1961, 1973, 1985, 2000, 2059, 2097 (ν(CO)) cm^{-1} [3, 6].

The complex crystallizes in the monoclinic space group $P2_1/c-C^5_{2h}$ (No. 14) with a = 8.576(3), b = 9.682(7), c = 19.623(10) Å, β = 99.54(3)°; Z = 4 molecules per unit cell, D_{calc} = 2.75 g/cm^3 [3, 6]. The molecular structure is illustrated in **Fig. 63** [6].

Uptake of one CO group was achieved under 100 atm of CO yielding (μ-$(CH_3)_2$-CCHCH)$Re_2(CO)_9$ [3, 6].

(μ-$(CH_3)_2$CCHCCH_3)$Re_2(CO)_8$ (see Formula VI, R = CH_3) was obtained by Method I (−25 °C, 45 min). Yield: 4%. Brown-orange crystals.

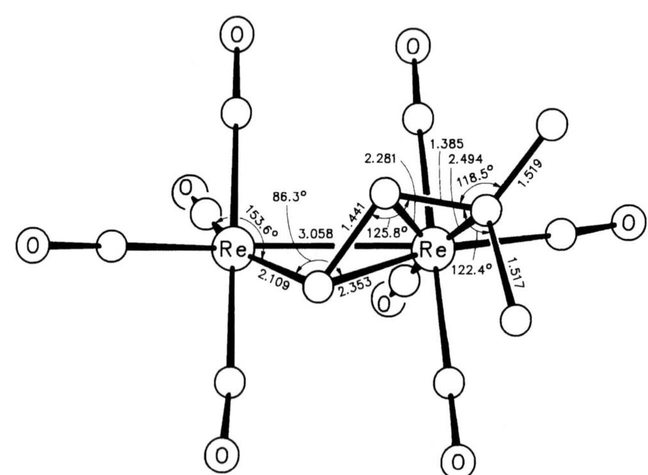

Fig. 63. Molecular structure of (μ-$(CH_3)_2$CCHCH)$Re_2(CO)_8$ [6].

References on p. 139

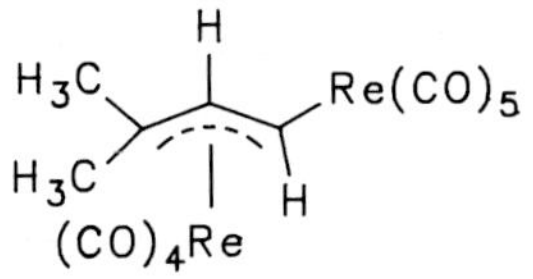

VII 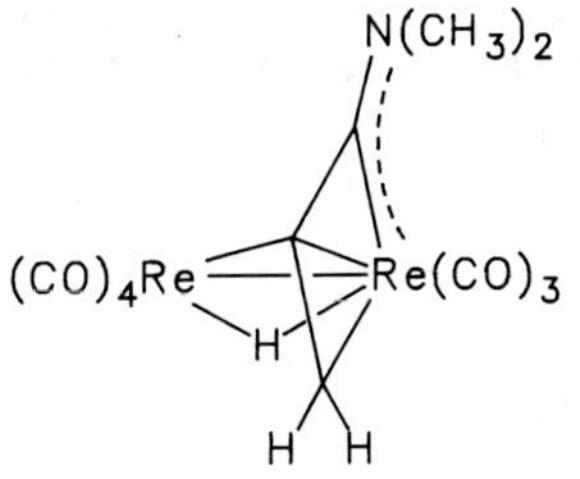 VIII 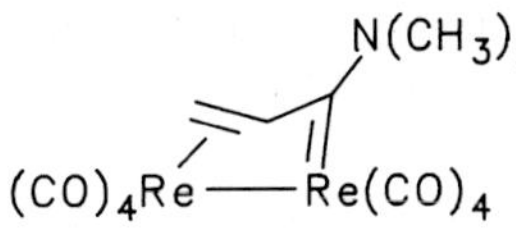 IX

^{1}H NMR spectrum (CD_2Cl_2): δ = 2.14, 2.26 (s's); 2.98 (d; J = 0.6 Hz), 4.23 (br s, CH) ppm. IR spectrum (n-hexane): 1950, 1966, 1980, 1990, 1998, 2063, 2103 (ν(CO)) cm^{-1} [16].

(μ-$(CH_3)_2$CCHCH)$Re_2(CO)_9$ (see Formula VII) was obtained by pressurizing a hexane solution containing (μ-$(CH_3)_2$CCHCH)$Re_2(CO)_8$ with 100 atm CO at room temperature for 16 h [3, 6]. The resulting mixture was evaporated, and the residue was recrystallized from hexane/ether (4:1). Yield: 90% [6].

^{1}H NMR spectrum ($CDCl_3$): δ = 1.74, 2.08 (s's, CH_3); 2.52, 4.83 (d's, ReCH and CCHC; J = 14.65 Hz) ppm. ^{13}C NMR spectrum ($CDCl_3$): δ = 21.7 (C**C**HC; ^{1}J(C,H) = 138 Hz, ^{2}J(C,H) = 7.4 Hz); 25.2, 32.3 (CH_3); 67.8 (**C**$(CH_3)_2$); 111.7 (ReCH; J(C,H) = 147 Hz); 179.7, 183.1, 191.7, 192.8, 193.6, 195.1 (CO) ppm. IR spectrum (n-hexane): 1941, 1963, 1977, 1985, 2017, 2021, 2069 (ν(CO)) cm^{-1} [3, 6].

The compound crystallizes in the triclinic space group $P\bar{1}-C_i^1$ (No. 2) with a = 9.754(5), b = 9.034(3), c = 11.168(5) Å, α = 90.30(3)°, β = 112.95(4)°, γ = 93.40(4)°; Z = 2 molecules per unit cell, D_{calc} = 2.55 g/cm^3. The molecular structure with selected intramolecular bond parameters is illustrated in **Fig. 64**. The Re···Re distance amounts to 4.190 Å [6].

Even though the compound is thermally stable, room-temperature irradiation provided (μ-$\eta^{2:1}$-CH_2=C(CH_3)CH=CH)$Re_2(CO)_8$(μ-H) which in solution exists in form of two rotameric species (see p. 70) [3, 6].

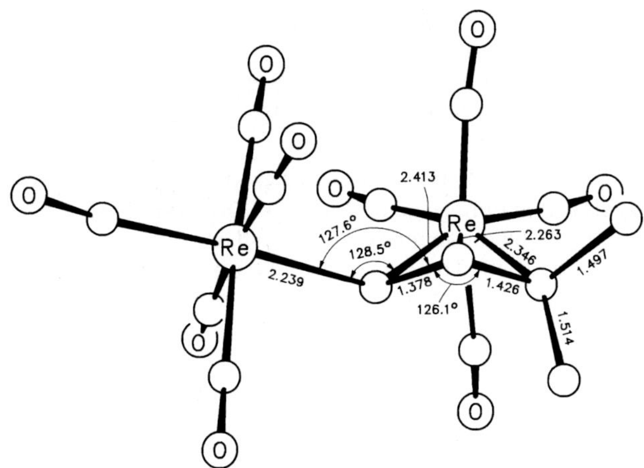

Fig. 64. Molecular structure of (μ-$(CH_3)_2$CCHCH)$Re_2(CO)_9$ [6].

References on p. 139

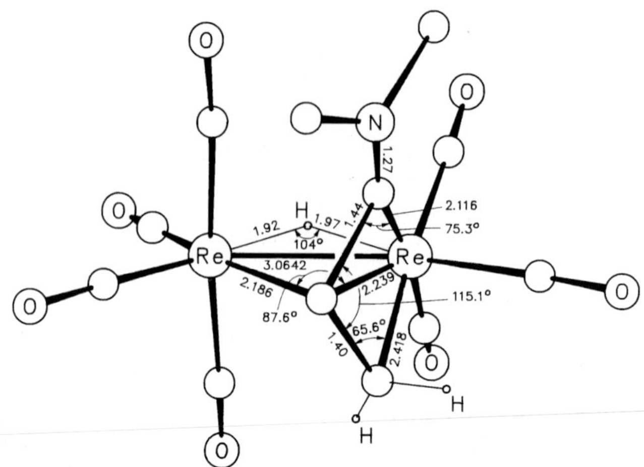

Fig. 65. Molecular structure of (μ-$CH_2C((CH_3)_2N)C$=)$Re_2(CO)_7$(μ-H) [11].

(μ-$CH_2C((CH_3)_2N)C$=)$Re_2(CO)_7$(μ-H) (see Formula VIII) was obtained by Method III with 17% yield. Likewise, thermolysis of (μ-CH_2=$CCN(CH_3)_2$)$Re_2(CO)_8$(μ-H) (see p. 59) for 6 h yielded 9.5% of the compound. Colorless solid.

^{1}H NMR spectrum ($CDCl_3$): δ = −15.66 (s, μ-H); 3.43, 3.66 (s's, CH_3); 4.12, 5.01 (d's; J = 2.9 Hz) ppm. ^{13}C {^{1}H} NMR spectrum ($CDCl_3$): δ = 46.7, 48.7 (CH_3); 62.1 (H_2**C**CC), 109.7 (H_2C**C**C); 182.8, 183.4, 185.2, 186.2, 188.8, 191.9, 192.4 (CO); 225.3 (H_2CC**C**) ppm. IR spectrum (C_6H_{12}): 1654 (ν(CN)); 1940, 1953, 1961, 1976, 1998, 2031, 2092 (ν(CO)) cm^{-1}.

The complex crystallizes in the triclinic space group $P\bar{1}-C_i^1$ (No. 2) with a = 9.572(2), b = 9.922(2), c = 9.193(2) Å, α = 107.91(2)°, β = 101.25(1)°, γ = 92.63(2)°; Z = 2 molecules per unit cell, D_{calc} = 2.67 g/cm^3. The molecular structure is depicted in **Fig. 65**.

Thermolysis (100°C, 10 min) in the absence of CO provided a mixture containing still substantial amounts of the title product, (μ-CH_2=$CH((CH_3)_2N)C$=)$Re_2(CO)_8$ (following compound), (μ-$\eta^{2:1}$-CHCHCH=$N(CH_3)_2$)$Re_2(CO)_8$ (see p. 79), and (μ-$CHCHCHN(CH_3)_2$)-$Re_2(CO)_6$(μ-CO) in the ratio 0.54 : 0.05 : 0.14 : 0.27. Stirring a benzene/C_6H_{12} solution under CO (1 atm) at room temperature achieved nearly complete conversion into (μ-CH_2=CC-$N(CH_3)_2$)$Re_2(CO)_8$(μ-H) (main product, see p. 59) and (μ-CH_2=$CH((CH_3)_2N)C$=)$Re_2(CO)_8$ [11].

(μ-CH_2=$CH((CH_3)_2N)C$=)$Re_2(CO)_8$ (see Formula IX) was obtained according to Method III with 25% yield. A similar thermolysis of (μ-CH_2=$CCN(CH_3)_2$)$Re_2(CO)_8$(μ-H) (see p. 59) yielded 4.7% after 6 h, but time-dependent NMR studies revealed an increasing yield with time of heating (54% after 18 h). The compound was also observed with low yield (ca. 1%) when (μ-$CH_2C((CH_3)_2N)C$=)$Re_2(CO)_7$(μ-H) (foregoing complex) was treated with CO under atmospheric pressure. Pale yellow crystals.

^{1}H NMR spectrum ($CDCl_3$): δ = 1.40 (dd; J = 2.2, 13.2 Hz), 1.98 (dd; J = 2.2, 8.6 Hz), 3.33 (dd; J = 8.6, 13.2 Hz); 3.42, 3.43 (CH_3) ppm. IR spectrum (C_6H_{12}): 1522 (ν(CN)); 1931, 1949, 1964, 1974, 1983, 1991, 2044, 2085 (ν(CO)) cm^{-1}.

The complex crystallizes in the monoclinic space group $P2_1/a-C_{2h}^5$ (No. 14) with a = 15.428(2), b = 16.047(6), c = 15.341(4) Å, β = 115.03(2)°; Z = 8 molecules per unit cell,

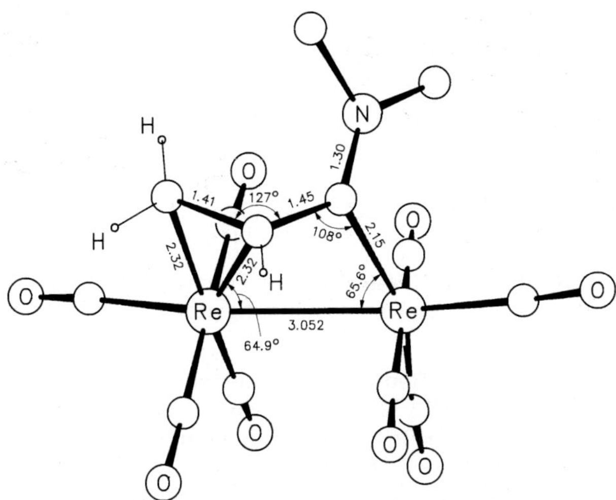

Fig. 66. Molecular structure of (μ-CH_2=CH((CH_3)$_2$N)C=)$Re_2(CO)_8$ [11].

D_{calc} = 2.62 g/cm^3. There are two symmetry-independent molecules in the asymmetric unit. Both are structurally similar. A view of the structure of one of these molecules is given in **Fig. 66**.

Thermolysis (100 °C, 35 h) provided a mixture of (μ-$\eta^{2:1}$-CHCHCH=N(CH_3)$_2$)$Re_2(CO)_8$ (see p. 79) and (μ-CHCHCHN(CH_3)$_2$)$Re_2(CO)_6$(μ-CO) (following complex) in the ratio 0.58 : 1. When the compound was similarly heated under CO, (μ-$\eta^{2:1}$-CHCHCH=N(CH_3)$_2$)-$Re_2(CO)_8$ almost exclusively formed [11].

X XI XII

(μ-CHCHCHN(CH_3)$_2$)$Re_2(CO)_6$(μ-CO) (see Formula X) formed as the main product when either (μ-CH_2C((CH_3)$_2$N)C=)$Re_2(CO)_7$(μ-H) or (μ-CH_2=CH((CH_3)$_2$N)C=)$Re_2(CO)_8$ (the two preceding compounds) was thermolyzed in C_6D_6 at 100 °C. In the latter case, chromatographic workup on silica with hexane/CH_2Cl_2 (3 : 1) yielded 53%. Light yellow solid.

^{1}H NMR spectrum (C_6D_6): δ = 1.48, 1.95 (CH_3); 2.90 (dd; J = 2.6, 3.2 Hz), 6.20 (dd; J = 3.2, 6.3 Hz), 7.66 (dd; J = 2.6, 6.3 Hz) ppm. IR spectrum (C_6H_{12}): 1896, 1944, 1947, 1972, 2004, 2026, 2078 (ν(CO)) cm^{-1}. The mass spectrum shows the complete series [M − n CO]$^+$ (n = 0 to 7).

The complex did not react with CO (1 atm) at 100 °C [11].

$(\mu\text{-}\eta^{2:2}\text{-}C_3H_4)(\mu\text{-}\eta^{3:1}\text{-}C_3H_4)Re_2(CO)_6$ (see Formula XI) was one product obtained by photochemically reacting $(CO)_{10}Re_2$ with allene according to Method I (−50 °C, 35 min). It was separated with 8% yield as yellow rhombes.

^{1}H NMR spectrum (C_6D_6): ($\eta^{2:2}$-C_3H_4): δ = 1.20 (d; J = 4.0 Hz), 1.62 (d; J = 2.8 Hz), 2.38 (dd; J = 4.0, 3.8 Hz), 2.50 (dd; J = 2.8, 3.8 Hz) ppm; ($\eta^{3:1}$-C_3H_4): δ = 1.56 (dd, H_{3Z}), 2.18 (dd, H_{3E}), 2.66 (d, H-1), 4.23 (ddd, H-2) ppm; coupling constants: J(H-1,2) = 11.0 Hz, J(H-2,3E) = 7.0 Hz, J(H-2,3Z) = 12.0 Hz, J(H-3E,3Z) = 1.7 Hz. IR spectrum (n-hexane): 1967, 1973, 1980, 1991, 1998, 2091 (ν(CO)) cm^{-1} [15].

$[\mu\text{-}(C_5(CH_3)_5)W(SC_6H_5)(HC{=}CR)CO]Re_2(CO)_6$ (see Formula XII). Compounds of this type were obtained by Method IV. They exhibit the following properties [17] (for abbreviations and units see p. X):

R = $CH{=}CHOCH_3$: Yield: 17%. − Red-brown crystals from CH_2Cl_2/heptane.
^{1}H NMR ($CDCl_3$): 2.04 (CH_3), 3.80 (s, OCH_3), 4.95 (d, **CH**=$CHOCH_3$), 5.37 (d, CH=**CH**OCH_3; J = 6); 6.88 (d) and 7.10 to 7.26 (m, C_6H_5); 12.32 (s, **H**C=CCH=CHO). − IR (C_6H_{12}): 1769 ($\nu(CO)_\mu$); 1907, 1924, 1949, 1960, 2011, 2042 (ν(CO)). − FAB MS: $[M]^+$ [17].

R = C_6H_5: Yield: 31%. − Black crystals. Another compound, $[\mu\text{-}(C_5(CH_3)_5)W(CO)_2CH{=}CC_6H_5]Re_2(CO)_6(\mu_3\text{-}SC_6H_5)$ (see p. 80) was also separated, but thermolysis of this product (toluene, 1 h) followed by chromatographic workup gave the title compound with 47% yield. This reaction was partially reversible by heating the title complex under CO.
^{1}H NMR ($CDCl_3$): 2.07 (CH_3), 6.81 to 7.15 (m, C_6H_5), 12.03 (s, CH=). − ^{13}C {^{1}H} NMR ($CDCl_3$): 10.7 (CH_3), 106.2 (C_5); 127.5, 128.0, 128.1, 128.3, 128.4, 131.1, 132.5, 138.8 (C_6H_5); 143.1 (=**C**C_6H_5), 184.3 (CH=; J(W,C) = 56), 186.9, 193.6, 194.5, 197.1 (br), 219.7 (J(W,C) = 185). − IR (C_6H_{12}): 1772 (($\nu(CO)_\mu$); 1907, 1926, 1953, 1965, 2015, 2045 (ν(CO)). − FAB MS: $[M]^+$.
Single-crystal data: monoclinic, space group $P2_1/n-C^5_{2h}$ (No. 14), a = 10.839(5), b = 17.112(4), c = 16.581(7) Å, β = 94.16(3)°; Z = 4 molecules per unit cell, D_{calc} = 2.380 g/cm^3.
The structure is shown in **Fig. 67** [17].

XIIIa ⇌ XIIIb XIV

$[\mu\text{-}(CH_3)_2CCCW(C_5(CH_3)_5)(CO)_2]Re_2(CO)_6(\mu\text{-}OR)$ (see Formulas XIIIa, XIIIb). Compounds of this type (R = CH_3, C_2H_5, C_6H_5) were prepared by Method V [14]. All derivatives are brown solids [13, 14].

Variable-temperature ^{1}H NMR spectra indicate a fluxional behavior: At room temperature the two methyl groups of the allenyllidene ligand give rise to two broad signals, which coalesce to a broad peak upon increasing the temperature and turn into two sharp singlets

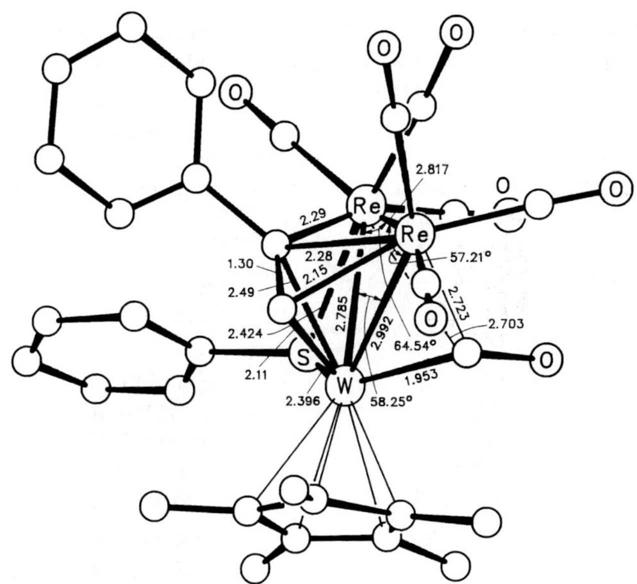

Fig. 67. Molecular structure of $[\mu\text{-}(C_5(CH_3)_5)W(SC_6H_5)(HC{=}CC_6H_5)CO]Re_2(CO)_6$ [17].

when lowering the temperature. Thus, a dynamic motion of the hydrocarbon ligand can be suggested, which is not coupled with a rotational motion of the $W(CO)_2$ unit, since the ^{13}C NMR spectrum (recorded on only 1 derivative) shows only one broad W-CO signal instead of two signals, even at −40 °C [13, 14].

The compounds exhibit the following physical and chemical properties (for abbreviations and units see p. X):

R = CH_3: Yield: 47% [14].
^1H NMR ($CDCl_3$): 2.25 ($(CH_3)_5$); 2.42, 2.63 (br s's, CCH_3); 2.87 (s, OCH_3). – $^{13}C\{^1H\}$ NMR ($CDCl_3$): 11.9 ($(CH_3)_5$); 29.5, 36.2 (CCC**C**H_3); 70.9 (OCH_3); 106.1 (C_5); 116.1 (CC**C**CH_3), 140.3 (C**C**CCH_3; J(W,C) = 32), 186.7 (**C**CCCH_3; J(W,C) = 163); 197.5, 200.4 (ReCO); 215.9 (WCO; J(W,C) = 173). – IR (C_6H_{12}): 1901, 1919, 1937, 1974, 2010, 2034 (ν(CO)). – FAB MS: $[M]^+$ [13, 14].
Single crystals: monoclinic, space group $P2_1/c-C^5_{2h}$ (No. 14), a = 16.496(1), b = 15.863(2), c = 10.409(2) Å, β = 97.05(1)°; Z = 4 molecules per unit cell, D_{calc} = 2.489 g/cm³ [13]. The structure is illustrated in **Fig. 68** [13, 14].

R = C_2H_5: Yield: 57% [14].
^1H NMR ($CDCl_3$): 0.46 (t; J = 6.9), 2.25 ($(CH_3)_5$); 2.43, 2.62 (br s's, CCH_3); 2.89 (q, OCH_2; becomes broad at −23 °C). Variable-temperature spectroscopy revealed $\Delta G^{\neq}$ = 59 kJ/mol for the dynamic motion of the hydrocarbon ligand. – IR (CH_2Cl_2): 1906, 1929, 1971, 2007, 2031 (ν(CO)). – FAB MS: $[M]^+$ [13, 14].
Heating in the presence of thiophenol yielded $(\mu\text{-}C{=}CC_3H_7\text{-i})Re(CO)_3\text{-}(\mu\text{-}SC_6H_5)W(C_5(CH_3)_5)(CO)_2$ as the only isolable compound [17].

References on p. 139

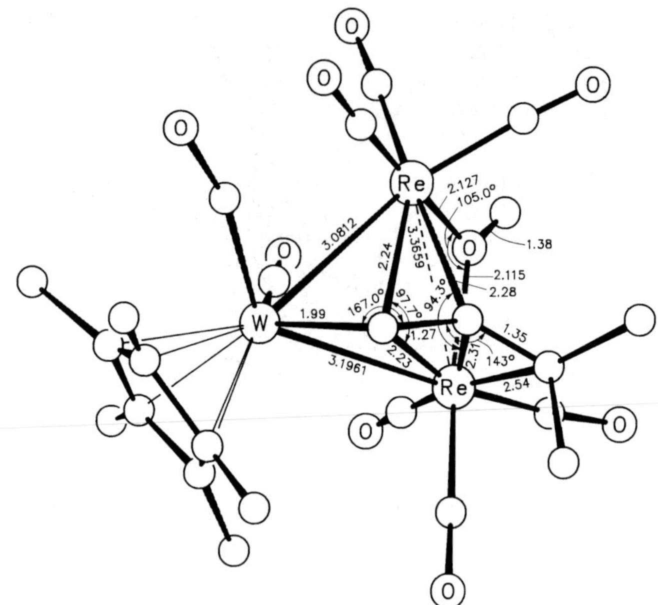

Fig. 68. Molecular structure of $[\mu\text{-}(CH_3)_2CCCW(C_5(CH_3)_5)(CO)_2]Re_2(CO)_6(\mu\text{-}OCH_3)$ [13, 14].

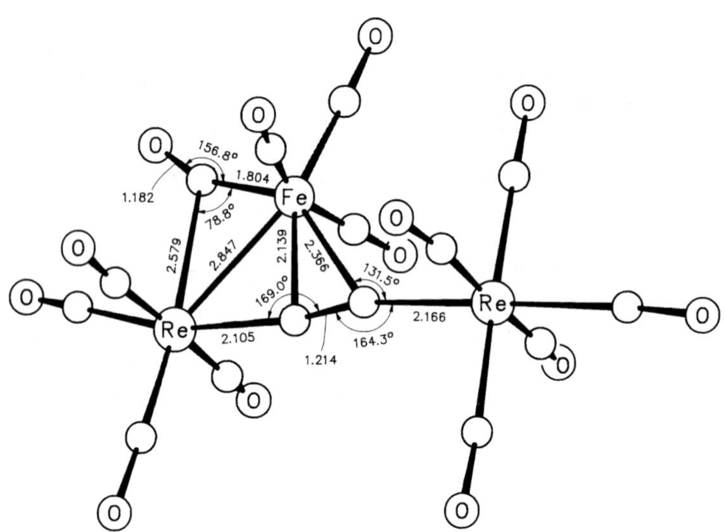

Fig. 69. Molecular structure of $[\mu\text{-}(CO)_4Fe(C{\equiv}C)]Re_2(CO)_9$ [10].

R = C_6H_5: Yield: 27% [14].
^{1}H NMR ($CDCl_3$): 2.27 ($(CH_3)_5$); 2.48, 2.68 (br s's, CH_3); 6.34 (d, 2 H; J = 7.2), 6.60 (t, 1 H; J = 6.8), 7.00 (t, 2 H; J = 6.8, 7.2). –
IR (CH_2Cl_2): 1904, 1920, 1943, 1952, 1976, 2017, 2040 (ν(CO)). –
FAB MS: $[M]^+$ [13, 14].

References on p. 139

[μ-(CO)$_4$Fe(C≡C)]Re$_2$(CO)$_9$ (see Formula XIV). $(CO)_9Fe_2$ and $(CO)_5ReC{\equiv}CRe(CO)_5$ (mole ratio 1.7:1) were combined in toluene at −78 °C. The mixture was allowed to warm to room temperature and stirred for another 30 h. The product was separated chromatographically on silica using CH_2Cl_2. This gave a yellow-brown fraction, from which orange-red crystals with m.p. 90 °C (dec.) could be isolated. Yield: 58%.

^{13}C NMR spectrum (CD_2Cl_2): δ = 69.0, 126.3 (C≡C); 179.0, 179.3 ($Re(CO)_5$); 188.4, 190.8, 192.3 ($Re(CO)_4$); 210 (FeCO) ppm. IR spectrum (CH_2Cl_2): 1931, 1970, 1998, 2019, 2039, 2063, 2102, 2148 (ν(CO)) cm^{-1}. FD mass spectrum: $[M]^+$.

A single-crystal X-ray study revealed the compound to crystallize in the monoclinic space group $P2_1/n-C^5_{2h}$ (No. 14) with a = 7.2388(9), b = 24.703(3), c = 11.5846(12) Å, β = 105.025(9)°; Z = 4 molecules per unit cell, D_{calc} = 2.710 g/cm^3. The structure of the molecule is depicted in **Fig. 69**. The Re-Fe bond is supported by a semi-bridging CO group [10].

References:

[1] Masters, A. F.; Mertis, K.; Gibson, J. F.; Wilkinson, G. (Nouv. J. Chim. **1** [1977] 389/95).

[2] Cotton, F. A.; Extine, M. W. (J. Am. Chem. Soc. **100** [1978] 3788/92).

[3] Green, M.; Orpen, A. G.; Schaverien, C. J.; Williams, I. D. (J. Chem. Soc. Chem. Commun. **1983** 1399/401).

[4] Rudler, H.; Fontanille, M.; Soum, A.; Harzallah, F.; Siove, A. (Fr. Demande 2566778 [1984/86] 1/16; C.A. **105** [1986] No. 61049).

[5] Alvarez-Toledano, C.; Parlier, A.; Rudler, H.; Rudler, M.; Daran, J. C.; Knobler, C.; Jeannin, Y. (J. Organomet. Chem. **328** [1987] 357/73).

[6] Green, M. L.; Orpen, A. G.; Schaverien, C. J.; Williams, I. D. (J. Chem. Soc. Dalton Trans. **1987** 1313/8).

[7] Cotton, F. A.; Stanley, G. G.; Cowley, A. H.; Lattman, M. (Organometallics **7** [1988] 835/40).

[8] Kreiter, C. G.; Michels, W.; Exner, R. (Z. Naturforsch. **45b** [1990] 793/802).

[9] van Houwelingen, T.; Stufkens, D. J.; Oskam, A. (Coord. Chem. Rev. **111** [1991] 325/30).

[10] Weidmann, T.; Weinrich, V.; Wagner, B.; Robl, C.; Beck, W. (Chem. Ber. **124** [1991] 1363/8).

[11] Adams, R. D.; Chen, G.; Chi, Y.; Wu, W.; Yin, J. (Organometallics **11** [1992] 1480/6).

[12] Kreiter, C. G.; Schufft, S.; Heckmann, G. (J. Organomet. Chem. **424** [1992] 163/72).

[13] Cheng, P.-S.; Chi, Y.; Peng, S.-M.; Lee, G.-H. (Organometallics **12** [1993] 250/2).

[14] Peng, J.-J.; Horng, K.-M.; Cheng, P.-S.; Chi, Y.; Peng, S.-M.; Lee, G.-H. (Organometallics **13** [1994] 2365/74).

[15] Kreiter, C. G.; Michels, W.; Heeb, G. (Z. Naturforsch. **50b** [1995] 649/60).

[16] Kreiter, C. G.; Michels, W.; Heeb, G. (J. Organomet. Chem. **502** [1995] 9/18).

[17] Peng, J.-J.; Peng, S.-M.; Lee, G.-H.; Chi, Y. (Organometallics **14** [1995] 626/33).

2.4 Compounds with Ligands Bonded to Rhenium by Four C Atoms (^{4}L Compounds)

In most compounds the ^{4}L ligand is bonded to both Re atoms ("bridging"). Only two compounds bear ^{4}L ligands bonded to only one Re atom ("terminal").

References on p. 140

2.4.1 Compounds with Terminal ^{4}L Ligands

π-$C_4H_6Re(CO)_3$–$Re(CO)_5$ (see Formula I, R = H) was one of the products obtained by irradiating a hexane solution containing $(CO)_{10}Re_2$ and butadiene at −85 or −40 °C for 20 min. The resulting mixture was separated by HPLC using hexane. Only traces were obtained at −40 °C; however, a yield of 21% was achieved at −85 °C. The compound forms pale yellow crystals.

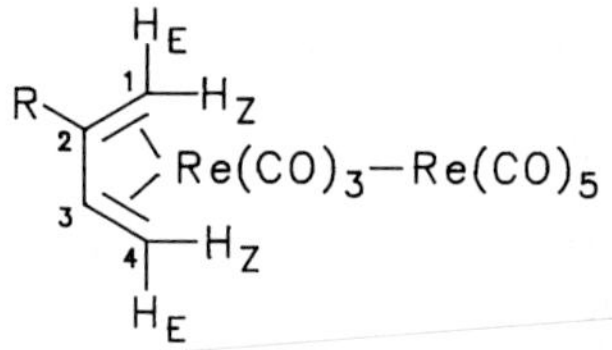

R, H_E, H_Z, 1, 2, 3, 4, $Re(CO)_3$–$Re(CO)_5$, H_Z, H_E

I

The ^{1}H NMR spectrum ($CDCl_3$) displays three multiplets at δ = 0.50 (H-1Z,4Z), 2.20 (H-1E,4E), and 5.16 (H-2,3) ppm. Coupling constants (by simulation): J(H-2,3) = 4.6 Hz, J(H-1E,1Z) = 2.2 Hz, J(H-1Z,2) = 9.8 Hz, J(H-1E,2) = 7.3 Hz, J(H-1Z,3) = −0.9 Hz, J(H-1Z,4Z) = 0.04 Hz, J(H-1Z,4E) = 0.01 Hz, J(H-4E,2) = 1.3 Hz, J(H-1E,4E) = −0.08 Hz (the values suggest an s-cis conformation). ^{13}C NMR spectrum (CD_2Cl_2, −70 °C): δ = 39.54 (CH_2; J(C,H) = 159 Hz), 79.32 (CH; J(C,H) = 171 Hz); 184.09 ($Re(CO)_5$), 193.59 ($Re(CO)_3$), 194.02 ($Re(CO)_5$), 197.97 ($Re(CO)_3$; ratio 1:1:4:2) ppm. At room temperature the former peaks are unchanged, but only 3 peaks due to δ(CO) show up at 183.43, 193.36, and ca. 195.85 ppm (ratio 1:4:3), suggesting a hindered rotation at the $Re(CO)_3$ moiety (estimated barrier: $\Delta G^{\neq}$ = 56 ± 4 kJ/mol). IR spectrum (n-hexane): 1940, 1950, 1977, 1992, 2005, 2100 (ν(CO)) cm^{-1} [1].

π-CH_2=$CHC(CH_3)$=$CH_2Re(CO)_3$–$Re(CO)_5$ (see Formula I, R = CH_3). Irradiation of a hexane solution containing $(CO)_{10}Re_2$ and a large excess of 2-methyl-butadiene at −80 °C followed by HPLC separation using hexane/CH_2Cl_2 mixtures separated four compounds. One of them was the title compound in form of golden yellow crystals. Yield: 3.8%.

^{1}H NMR spectrum (CD_2Cl_2): δ = 0.41 (H-4Z), 0.53 (H-1Z), 2.13 (H-4E), 2.32 (H-1E), 2.70 (CH_3), 5.33 (H-3) ppm. Coupling constants: J(H-1E,1Z) = 2.2 Hz, J(CH_3,1Z) = 0.7 Hz, J(CH_3,1E) = 1.4 Hz, J(H-4E,4Z) = 2.5 Hz, J(H-3,4E) = 7.4 Hz, J(H-3,4Z) = 9.4 Hz. IR spectrum (n-hexane): 1940, 1947, 1979, 1992, 2010, 2100 (ν(CO)) cm^{-1} [2].

References:

[1] Franzreb, K.-H.; Kreiter, C. G. (Z. Naturforsch. **37b** [1982] 1058/69).
[2] Kreiter, C. G.; Franzreb, K. H.; Sheldrick, W. S. (Z. Naturforsch. **41b** [1986] 904/14).

2.4.2 Compounds with Bridging ^{4}L Ligands

2.4.2.1 Compounds with Bridging Hydrocarbons

The structure of the compounds presented in this section is illustrated in Formulas I to VIII. Most compounds were prepared by one of the following methods:

Method I: Irradiation of a pentane solution containing $(CO)_{10}Re_2$ and butadiene at −30 °C for 90 min, while additional butadiene was bubbled through the solution. The

References on p. 145

mixture was evaporated and the residue sublimed (35 °C/10^{-3} Torr, 24 h). The yellow sublimate was separated by chromatography (silica, hexane) [3].

Method II: Irradiation of a mixture containing $(CO)_{10}Re_2$ and the respective allene CH_3-CH=C=CHCH$_3$ [12] or CRR′=C=C$(CH_3)_2$ (R = H, R′ = H, CH_3; R = R′ = CH_3) [13] in hexane at −30 to −25 °C for ca. 45 min. The resulting mixture was roughly separated by column chromatography on alumina using CH_2Cl_2/hexane mixtures and the portions eluted were purified by preparative HPLC [12, 13].

Method III: Irradiation of a mixture of $(CO)_{10}Re_2$ and cyclooctatetraene in pentane at −30 °C for 90 min. The deep red mixture was separated by column chromatography on silica with hexane/ether mixtures [3] (see also [11]).

Method IV: Treatment of $(CO)_5ReFBF_3$ with the respective diene in CH_2Cl_2 at room temperature. The product precipitated along with compounds of the type [(π-diene)Re$(CO)_5$]BF_4. These mononuclear derivatives could be removed by extracting the mixture with acetone [8].

Methods I to III provided multi-component mixtures. Surveys of all products are given in the chapter “Chemical Behavior of $(CO)_{10}Re_2$” in “Organorhenium Compounds” 6.

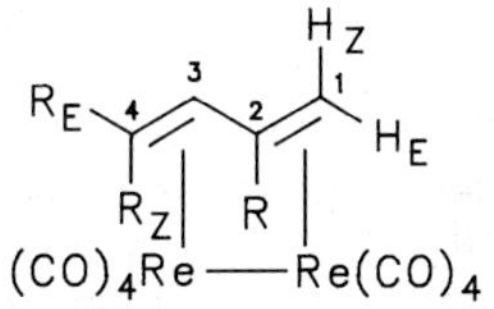

I 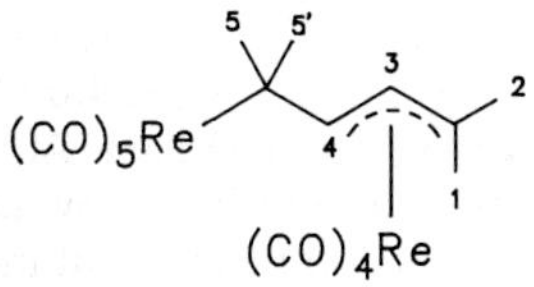 II 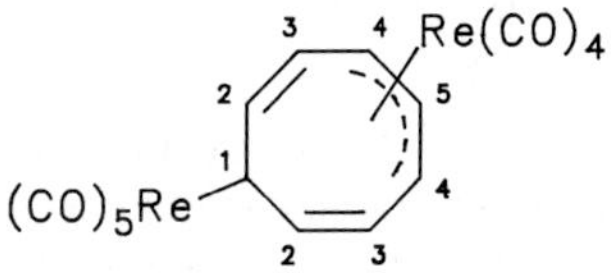 III

(μ-η$^{2:2}$-C_4H_6)$Re_2(CO)_8$ (see Formula I, all R's = H) was obtained with 5% yield according to Method I [3]. The compound also formed by irradiating $(CO)_{10}Re_2$ in the presence of NO and cyclooctatetraene (THF, −30 °C, ca. 4 h; yield: 5%) [2, 5] or, alternatively, in the presence of ethylene (hexane, −50 °C, 80 min; yield: 10%) [6, 10]. In each case the resulting mixture was chromatographically separated. Larger yields could be obtained when starting from more complex compounds: Thus, treatment of (μ-η$^{2:1}$-CH_2=CH)$Re_2(CO)_8$(μ-H) with butadiene for 30 to 40 h yielded the title complex quantitatively [7]. Otherwise, leaving (μ-η$^{2:1}$-CH_2=CHCH=CH)$Re_2(CO)_8$(μ-H) stand in polar solvents produced (μ-η$^{2:2}$-C_4H_6)-$Re_2(CO)_8$ as a result of an isomerization process [4]. The compound forms pale yellow [3, 5], colorless [6, 10] crystals; m.p. 160 °C [3, 5].

^{1}H NMR spectrum ($CDCl_3$): δ = 2.21 [3, 5], 2.27 (?) [5] or 2.77 [3] (m's, 4 and 2 H) ppm. A more accurate study elucidated the following signals and coupling constants (acetone-d_6): δ = 2.54 (H-1Z,4Z), 2.73 (H-2,3), 3.06 (H-1E,4E) ppm; J(H-1Z,2) = 12.7 Hz, J(H-1E,2) = 8.1 Hz, J(H-1Z,1E) = −0.87 Hz, J(H-2,3) = 11.3 Hz, J(H-1Z,4Z) = −0.01 Hz, J(H-1Z,4E) = −0.07 Hz, J(H-1E,4E) = 0.5 Hz, J(H-1E,3) = −0.52 Hz, J(H-1Z,3) = −0.45 Hz. ^{13}C NMR spectrum ($CDCl_3$): δ = 26.52 (C-1,4; J(C,H) = 164 Hz), 67.14 (C-2,3; J(C,H) = 154 Hz); 187.49, 188.36, 188.97, 192.68 (CO, all of equal intensity) ppm [4]. IR spectrum (n-hexane): 1958, 1978, 1997, 2055, 2098 (ν(CO)) cm^{-1} [3, 5]. UV spectrum: λ_{max} (log ε) = 262 (4.1), 299 (3.98), 320 (3.95) nm [3].

The complex crystallizes in the monoclinic space group $P2_1/n-C_{2h}^5$ (No. 14) with a = 7.455(3), b = 12.986(6), c = 15.554(5) Å, β = 92.78(6)°; Z = 4 molecules per unit cell, D_{calc} = 2.87, D_{meas} = 3.1 g/cm^3. The molecular structure is illustrated in **Fig. 70**. It can be

References on p. 145

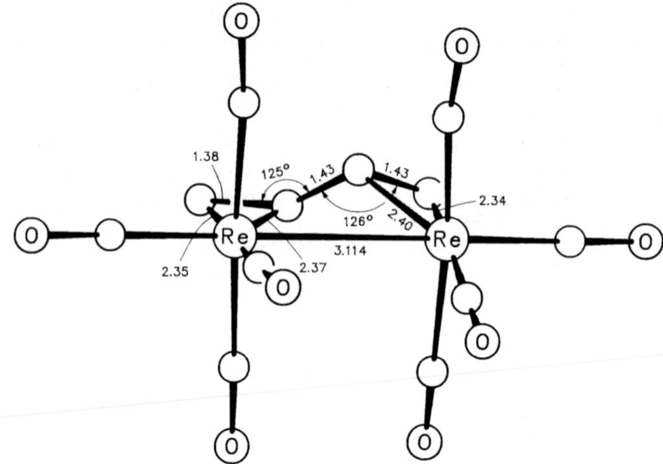

Fig. 70. Molecular structure of $(\mu\text{-}\eta^{2:2}\text{-}C_4H_6)Re_2(CO)_8$ [3].

seen that the butadiene ligand is fixed in an s-trans conformation. The CO groups are in a nearly eclipsed arrangement (dihedral angle OC-Re-Re-CO only 4.9°) [3].

$(\mu\text{-}\eta^{2:2}\text{-}CH_2{=}CHC(CH_3){=}CH_2)Re_2(CO)_8$ (see Formula I; R-2 = CH_3, R_E = R_Z = H) was produced with 4% yield via Method II when starting from $CH_2{=}C{=}C(CH_3)_2$ [13]. It could also be prepared by irradiating $(CO)_{10}Re_2$ and 50 equivalents 2-methylbuta-1,3-diene (hexane, −80 °C) followed by separation by HPLC with hexane/CH_2Cl_2 mixtures. Yield: 4.1% [10]. It forms a yellow oil [10]; otherwise, it was also described as gray-yellow crystals [13].

^{1}H NMR spectrum (CD_2Cl_2): δ = 1.61 (CH_3); 1.95 (H-4Z), 2.37 (H-3), 2.45 (H-1Z), 2.70 (H-4E), 3.38 (H-1E) ppm; J(H-4E,4Z) = 1.0 Hz, J(H-4E,3) = 8.3 Hz, J(H-4Z,3) = 13.3 Hz, J(H-1Z,1E) = 0.8 Hz (compare with [13]). ^{13}C NMR spectrum (?): δ = 18.9 (q, CH_3; J(C,H) = 127 Hz), 20.4 (t, C-4; J(C,H) = 158 Hz), 40.0 (t, C-1; J(C,H) = 159 Hz), 68.1 (d, C-3; J(C,H) = 159 Hz), 82.7 (C-2); 187.8, 188.1, 188.7, 189.3, 191.4, 192.0, 193.5 (CO) ppm [10]. IR spectrum (n-hexane): 1957, 1977, 1995, 2050, 2092 (ν(CO)) cm^{-1} [10] (similar in [13]).

$(\mu\text{-}\eta^{2:2}\text{-}CH_2{=}CHCH{=}C(CH_3)_2)Re_2(CO)_8$ (see Formula I; R-2 = H, R_E = R_Z = CH_3) was produced with 6% yield via Method II when employing $CH_3CH{=}C{=}C(CH_3)_2$. Yellow crystals.

^{1}H NMR spectrum (CD_2Cl_2): δ = 2.08 (dd); 2.35, 2.38 (s's, CH_3); 2.40 (d), 2.68 (dd), 2.91 (ddd) ppm. IR spectrum (n-hexane): 1962, 1976, 1980, 1996, 2048, 2099 (ν(CO)) cm^{-1} [13].

$(\mu\text{-}\eta^{2:1}\text{-}CH_2{=}C(CH_3)CH{=}C(CH_3)_2)Re_2(CO)_8$ (see Formula I; R-2, R_E, and R_Z = CH_3) was obtained with 27% yield via Method II when employing $(CH_3)_2C{=}C{=}C(CH_3)_2$. Yellow-green crystals.

^{1}H NMR spectrum (CD_2Cl_2): δ = 2.07 (CH_3-2), 2.28 (H-1Z), 2.31 (CH_3-4Z), 2.42 (H-3), 2.55 (CH_3-4E), 3.15 (H-1E) ppm (all signals are singlets). IR spectrum (n-hexane): 1959, 1976, 1989, 1998, 2040, 2090 (ν(CO)) cm^{-1} [13].

$(\mu\text{-}\eta^{3:1}\text{-}C_4H_6)Re_2(CO)_9$ (see Formula II) was observed to form with 2% yield via Method I [3]. Other authors who carried out the same reaction did not observe the compound (probably due to a different workup procedure) [4]. The complex forms pale yellow crystals [3].

References on p. 145

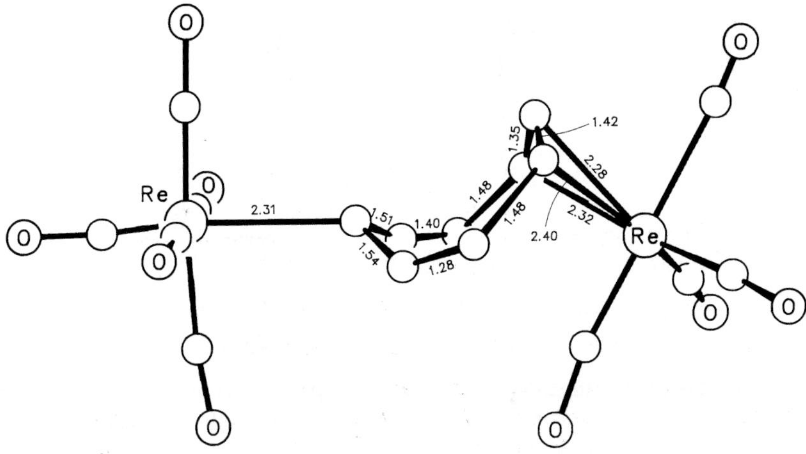

Fig. 71. Molecular structure of $(\mu\text{-}\eta^{3:1}\text{-}C_8H_8)Re_2(CO)_9$ [3].

^{1}H NMR spectrum ($CDCl_3$): δ = 1.9 ("d", H-5,5′), 2.9 (H-1,4), 3.2 (H-2), 4.4 (H-3) ppm. IR spectrum (n-hexane): 1942, 1968, 1993, 2008, 2050, 2073, 2098, 2115 (ν(CO)) cm^{-1}. Mass spectrum: $[M - n\ CO]^+$ (n = 0 to 9); $[Re_2]^+$ not observed (fragmentation pattern is given).

In solution the complex slowly deteriorated into $(CO)_{10}Re_2$, even under an Ar atmosphere [3].

$(\mu\text{-}\eta^{3:1}\text{-}C_8H_8)Re_2(CO)_9$ (see Formula III) was one of the products obtained by Method III. Yield: 4% [3]. However, other researchers, when carrying out an analogous reaction, did not report this compound [6]. The compound forms light yellow crystals; m.p. 125 °C [3].

^{1}H NMR spectrum ($CDCl_3$): δ = 4.44 (s, H-4), 5.80 (s, H-1), 5.87 (d, H-2), 5.95 (t, H-5), 6.70 (d, H-3) ppm; J(H-2,3) = 11 Hz. IR spectrum (n-hexane): 1608 (ν(C=C)); 1943, 1975, 1993, 2005, 2040, 2065, 2093, 2105 (ν(CO)) cm^{-1}. UV spectrum (C_6H_{12}): λ_{max} (log ε) = 242 (3.52), 249 (3.47), 290 (3.05), 322 (3.98) nm. Mass spectrum: $[M - n\ CO]^+$ (n = 1 to 9), $[M - Re(CO)_5 - n\ CO]^+$ (n = 0 to 4); intensity pattern is given [3].

The complex crystallizes in the monoclinic space group $P2_1/n-C^5_{2h}$ (No. 14) with a = 14.794(5), b = 6.787(2), c = 20.012(8) Å, β = 94.41(3)°; Z = 4 molecules per unit cell, D_{calc} = 2.41, D_{meas} = 2.50 g/cm^3. **Fig. 71** illustrates the molecular structure.

The compound is only slightly air-sensitive but nevertheless unstable in solution, where it decomposes into $(CO)_{10}Re_2$ [3].

$[(\mu\text{-}\eta^{2:2}\text{-}C_6H_{10})Re_2(CO)_{10}][BF_4]_2$ (see Formula IV) was produced by Method IV.

$[Re(CO)_5 \ldots Re(CO)_5]^{2+}$ — IV

$[(CO)_5Re — C_8H_8 — Re(CO)_5]^{2+}$ — V

References on p. 145

VI VII VIII

IR spectrum (Nujol): 1535 (ν(C=C)); 2055, 2165 (ν(CO)) cm^{-1} [8].

[(μ-$\eta^{2:2}$-C_8H_{12})$Re_2(CO)_{10}$][BF_4]$_2$ (see Formula V) was obtained by Method IV.

IR spectrum (Nujol): 1529 (ν(C=C)); 2040, 2165 (ν(CO)) cm^{-1} [8].

(μ-$\eta^{2:1:2:1}$-$C_6(CH_3)_4R_2H_2$)$Re_2(CO)_6$(μ-CO) (see Formula VI; R = H or CH_3). Compounds with this composition were obtained via Method II when starting from CH_3CH=C=$CHCH_3$ [12] or CH_3CH=C=$C(CH_3)_2$ [13]. They are characterized as follows (for abbreviations and units see p. X):

R = H: Yield: 1.6%. – Red cubes.
^{1}H NMR (CD_2Cl_2): 0.90 (m, CH_3), 1.65 (d, CH_3), 3.13 (m, C**H**CH_3), 4.82 (dq, CH=); $^3J(CH_3$,H-2) = 6, $^3J(CH_3$,H-4) = 6.7, J(H-2,4) = 0.9, J(H-4,5) = 3.0 Hz. – IR (n-hexane): 1976, 2024 (ν(CO)).
Single-crystal data: triclinic, a = 9.527(3), b = 13.948(2), c = 7.621(3) Å, α = 90.37(1)°, β = 108.97(2)°, γ = 83.39(1)°; space group $P\bar{1}-C_i^1$ (No. 2); Z = 2 molecules per unit cell, D_{calc} = 2.47 g/cm^3. The molecular structure is illustrated in **Fig. 72**. Due to the poor crystal quality, the bonding distances and angles could not be resolved [12].

R = CH_3: Yield: 1%. – Red rhomboids.
^{1}H NMR (CD_2Cl_2): 1.08 (m, CH_3-4,5; J = 6.6); 2.34, 2.36 (s's, CH_3-2,7); 3.16 (m, CH; J = 6.6, 4.5). – IR (n-hexane): 1969, 2023 (ν(CO)) [13].

(μ-cyclo-C_7H_8)$Re_2(CO)_8$ (see Formula VII) is the structure suggested for one of the products obtained by irradiating $(CO)_{10}Re_2$ with excess cycloheptatriene in petroleum ether for

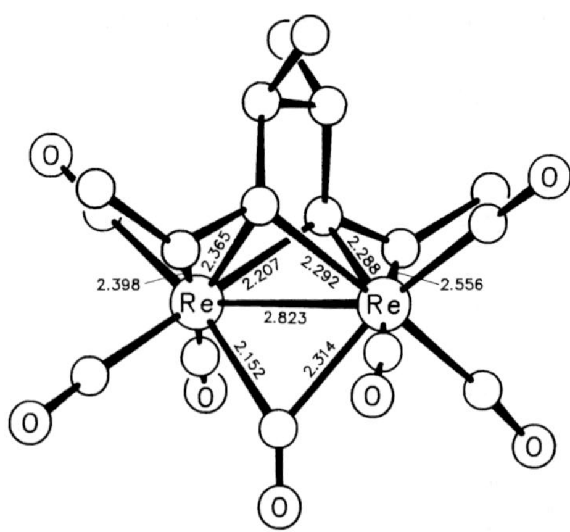

Fig. 72. Molecular structure of (μ-$\eta^{2:1:2:1}$-$C_6(CH_3)_4H_4$)$Re_2(CO)_6$(μ-CO) [12].

9 h. The compound was contaminated with $C_7H_9Re(CO)_3$. The mixture showed IR absorption bands at 1969, 1977, 1980, 1996, 2014, 2080, and 2108 (ν(CO)) cm^{-1}. Low-voltage mass spectroscopy (12 eV) clearly confirmed the molecular ions of each component in the mixture [1].

In contrast, other researchers reported the two isomers of (μ-$\eta^{2:1}$-C_7H_7)$Re_2(CO)_8$(μ-H) (see Table 4, pp. 72/3) to be the products of the photolysis of $(CO)_{10}Re_2$ with cycloheptatriene. A compound with the title composition was not mentioned [6, 9].

(μ-$\eta^{3:1}$-C_8H_8)Re_2(η^2-C_8H_8)$(CO)_8$ (see Formula VIII) was reported to be one of the products formed by Method III. Purification by preparative TLC was required. Yield: 2% [3] (see also [11]). The compound forms pale yellow crystals; m.p. 225 °C (dec.) [3]. However, other researchers did not observe the complex, when they investigated the reaction again [6].

IR spectrum (C_6H_{12}): 1610, 1627, 1925, 1965, 1985, 2010, 2040, 2100 (ν(CO)) cm^{-1}. The UV spectrum does not exhibit a band assignable to a $\sigma \rightarrow \sigma^*$ transition. The mass spectum shows the series $[M - n\ CO]^+$ (n = 0 to 8) and $[C_8H_8Re(CO)_n]^+$ (n = 0 to 3); intensities are given [3].

References:

[1] Davis, R.; Ojo, I. A. O. (J. Organomet. Chem. **110** [1976] C 39/C 41).
[2] Balbach, B. K.; Helus, F.; Oberdorfer, F.; Ziegler, M. L. (Angew. Chem. **93** [1981] 479/80; Angew. Chem. Int. Ed. Engl. **20** [1981] 470/1).
[3] Guggolz, E.; Oberdorfer, F.; Ziegler, M. L. (Z. Naturforsch. **36b** [1981] 1060/8).
[4] Franzreb, K.-H.; Kreiter, C. G. (Z. Naturforsch. **37b** [1982] 1058/69).
[5] Oberdorfer, F.; Balbach, B.; Ziegler, M. L. (Z. Naturforsch. **37b** [1982] 157/67).
[6] Franzreb, K.-H.; Kreiter, C. G. (J. Organomet. Chem. **246** [1983] 189/95).
[7] Nubel, P. O.; Brown, T. L. (J. Am. Chem. Soc. **106** [1984] 644/52).
[8] Beck, W.; Raab, K.; Nagel, U.; Sacher, W. (Angew. Chem. **97** [1985] 498/9; Angew. Chem. Int. Ed. Engl. **24** [1985] 505).
[9] Kreiter, C. G.; Franzreb, K.-H.; Michels, W.; Schubert, U.; Ackermann, K. (Z. Naturforsch. **40b** [1985] 1188/98).
[10] Kreiter, C. G.; Franzreb, K. H.; Sheldrick, W. S. (Z. Naturforsch. **41b** [1986] 904/14).
[11] Caballero, C.; Oberdorfer, F.; Guggolz, E.; Nuber, B.; Korswagen, R. P.; Ziegler, M. L. (Bol. Soc. Qim. Peru **53** [1987] 135/49; C.A. **110** [1989] No. 213022).
[12] Kreiter, C. G.; Michels, W.; Heeb, G. (Z. Naturforsch. **50b** [1995] 649/60).
[13] Kreiter, C. G.; Michels, W.; Heeb, G. (J. Organomet. Chem. **502** [1995] 9/18).

2.4.2.2 Compounds with Rhena-β-diketone and Rhena-β-ketoimine Groups

Compounds with rhena-β-diketone groups have the structures depicted in Formulas I to III, while compounds with rhena-β-ketoimine groups have the structures given in Formulas IV and V.

Most of the compounds with rhena-β-diketone groups were prepared as follows:

Method I: Reaction of $(CO)_4Re(C(R)O)(C(CH_3)O)\cdots H$ (R = CH_3, C_3H_7-i; see "Organorhenium Compounds" 1, 1989, p. 388) with 0.5 equivalent of $(C_9H_{14}N)_3Cr$, $Cu(OCH_3)_2$, $(C_2H_5)_2Zn$, or $(C_6H_5)_2Mg$ in benzene, diethyl ketone, hexane, or ether (solvent specified with each compound) at room temperature. Workup by evaporation and recrystallization of the residue [1, 3, 4].

References on pp. 148/9

Compounds represented by Formulas II and III are fairly stable in air [1, 4]. They are well-soluble in organic solvents, even in aliphatic hydrocarbons, with the exception of the Mg derivative which only dissolves in aromatic hydrocarbons and polar solvents. Thus, some ionic or polymeric character can be suggested for this derivative [4].

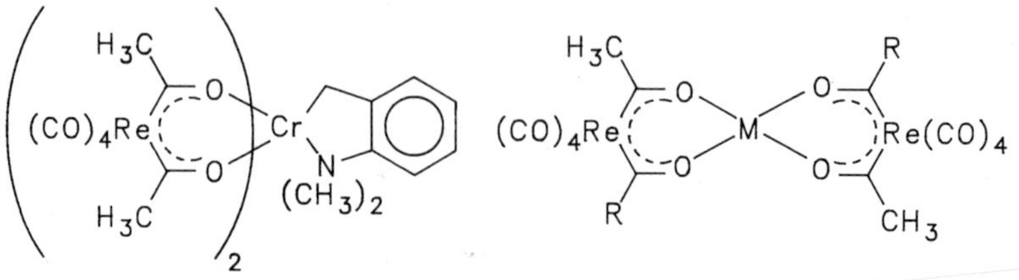

I II

$[(CO)_4Re(C(CH_3)O)_2]_2CrC_9H_{12}N$ (see Formula I, $C_9H_{14}N$ = N,N-dimethyl-o-toluidyl) was obtained with 36% yield via Method I (solvent: benzene). The compound also formed from $(CO)_4Re(C(CH_3)O)_2Cr(C_9H_{14}N)_2$ upon standing in solution. It forms air-stable (for at least 4 h), brown crystals; m.p. 114 to 115°C.

The central Cr^{3+} ion has a high-spin electron configuration: μ_{eff} = 3.48 μ_B. IR spectrum (n-hexane): 1510 (ν(C⩦O)); 1959, 1978, 1987, 1995, 2100 (ν(CO)) cm^{-1} [3].

$[(CO)_4Re(C(CH_3)O)_2]_2Cu$ (see Formula II, M = Cu, R = CH_3). Application of Method I using diethyl ketone as solvent and recrystallizing from toluene gave the compound with 18% yield [1, 4]. $[(CO)_4Re(C(CH_3)O)_2]_2Cu$ could also be synthesized with 92% yield by reacting $[N(CH_3)_4][(CO)_4Re(C(CH_3)O)_2]$ with $(CO)CuOSO_2CF_3$ in THF. The resulting mixture was evaporated, and the residue was recrystallized from pentane [7]. The compound forms deep red needles [1, 4, 7]; m.p. 120 to 121 [1, 4], 135°C [7].

Magnetic behavior (25°C): μ = 1.76 μ_B [1, 4]. ESR spectrum (powder): $g_\parallel$ = 2.21, $g_\perp$ = 2.09 [1, 4]; ($CHCl_3$/toluene): g_{iso} = 2.118, a_{iso} (Cu) = 80.7×10^{-4} cm^{-1}; (−166°C, $CHCl_3$/toluene glass): $g_\parallel$ = 2.252, $a_\parallel$ (Cu) = 194.5×10^{-4}, $a_\parallel$ (Re) = 22.1×10^{-4} cm^{-1}; $g_\perp$ = 2.044, $a_\perp$ (Cu) = 23.8×10^{-4}, $a_\perp$ (Re) = 21.0×10^{-4} cm^{-1} [4]. IR spectrum (n-hexane): 1506 (ν(C⩦O)); 1965, 1975, 2087, 1991 (ν(CO)) cm^{-1} [4]. UV spectrum ($CHCl_3$): λ_{max} (log ε) = 288 (4.31, Re → CO), 313 (4.13), 415 (3.94), 440 (3.98) nm [4] (similar in [1]).

The compound crystallizes in the monoclinic space group $P2_1/n-C_{2h}^5$ (No. 14) with a = 11.2231(9), b = 7.3648(5), c = 13.9179(10) Å, β = 90.791(6)°; Z = 2 molecules per unit cell, D_{calc} = 2.403 g/cm^3. The molecular structure is depicted in **Fig. 73**. The Cu atom resides at a crystallographic inversion center. Thus, the CuO_4 plane is essentially planar, but the rhena-acetylacetonato ligand is significantly distorted from planarity [5].

Cyclic voltammetry (CH_2Cl_2/0.1 M $[N(C_4H_9\text{-}n)_4]ClO_4$): $E_{p,c}$ = −0.54 V, $E_{p,a}$ = 0.35 (reoxidation of the reduced species), 0.54, 1.09 V vs. Ag/AgCl [6].

Addition of excess pyridine gave the adduct $[(CO)_4Re(C(CH_3)O)_2]_2Cu \cdot C_5H_5N$. Treatment with $NaOCH_3$ in hexane yielded the dimer $[(CO)_4Re(C(CH_3)O)_2Cu(\mu\text{-}OCH_3)]_2$ [4].

$[(CO)_4Re(C(CH_3)O)_2]_2Cu \cdot C_5H_5N$ formed by treating the foregoing complex with pyridine, causing the color to turn from red-brown into lime yellow.

ESR spectrum ($CHCl_3$/toluene (2:3)): g_{iso} = 2.154, a_{iso} (Cu) = 55.3×10^{-4}; (−166°C): $g_\parallel$ = 2.263, $g_\perp$ = 2.048; $a_\perp$ (N) = 13.38×10^{-4} cm^{-1}. The ESR data suggest a square-pyramidal arrangement around Cu.

References on pp. 148/9

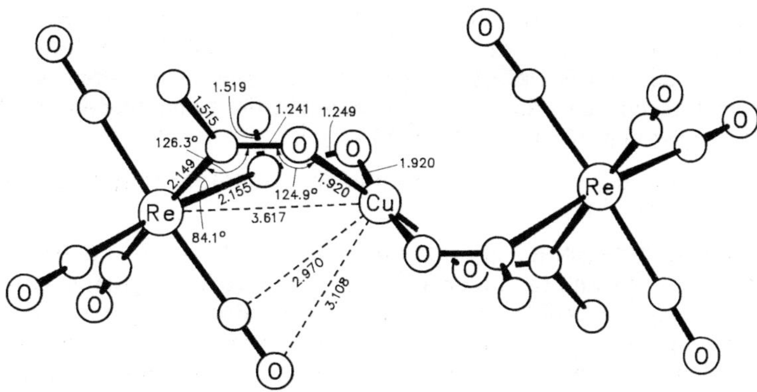

Fig. 73. Molecular structure of $[(CO)_4Re(C(CH_3)O)_2]_2Cu$ [5].

Exposure to air released ca. 50% of the ligand after 15 min [4].

$[(CO)_4Re(C(CH_3)O)C(C_3H_7\text{-i})O]_2Cu$ (see Formula II, M = Cu, R = C_3H_7-i) was synthesized by Method I (solvent: diethyl ketone, recrystallization from toluene) with 19% yield. The compound forms air-stable, deep red needles; m.p. 93.5 to 94 °C.

Magnetic behavior: μ_{eff} = 2.2 μ_B. IR spectrum (n-hexane): 1506 (ν(C⩦O)); 1965, 1975, 1991, 2087 (ν(CO)) cm^{-1} [4].

$[(CO)_4Re(C(CH_3)O)_2]_2Zn$ (see Formula II, M = Zn, R = CH_3) was prepared by Method I (solvent: hexane). Evaporation of the mixture after 4 h of stirring gave a semisolid residue which was extracted with hexane. Reevaporation followed by recrystallization from CH_2Cl_2/hexane (1:4) eventually yielded 20%. When the reaction was conducted in ether, an ionic compound with the suggested composition $[((CO)_4Re(C(CH_3)O)_2)_3Zn]^-$ instead formed. The title complex is a white solid; m.p. 132 to 135 °C.

^1H NMR spectrum ($CDCl_3$): δ = 2.78 (s) ppm. IR spectrum (n-hexane): 1492 (ν(C⩦O)); 1963, 1985, 1995, 2095 (ν(CO)) cm^{-1} [4].

$[(CO)_4Re(C(CH_3)O)_2]_2Mg$ (see Formula II, M = Mg, R = CH_3) was prepared according to Method I (solvent: ether). Recrystallization of the yellow residue from hexane/toluene (1:1) separated white crystals; m.p. 120 to 169 °C; yield: 14%.

^1H NMR spectrum ($CDCl_3$): δ = 2.63 ppm. IR spectrum (ether): 1563 (ν(C⩦O)); 1930, 1953, 1968, 2078 (ν(CO)) cm^{-1}.

The complex is soluble in aromatic hydrocarbons or polar organic solvents [4].

$[(CO)_4Re(C(CH_3)O)_2Cu(\mu\text{-}OCH_3)]_2$ (see Formula III) formed by modifying Method I in that only 1.2 equivalents $(CO)_4Re(C(CH_3)O)_2\cdots H$ were reacted with $Cu(OCH_3)_2$ in ethereal solu-

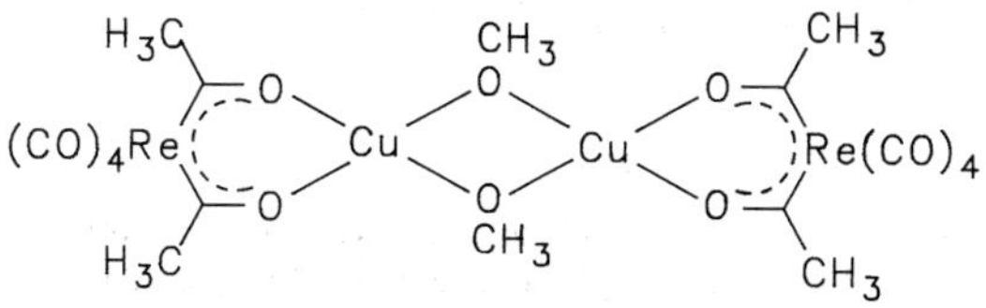

III

References on pp. 148/9

tion. The pale blue-green material, solidified after 30 min, was recrystallized from ether at −20 °C. Yield: 5%. The complex was also obtained when treating $[(CO)_4Re(C(CH_3)O)_2]_2Cu$ with a methanolic solution of $NaOCH_3$ in hexane. The powder left after evaporation had identical IR-spectroscopic data as a sample prepared by the former procedure. The complex forms green crystals; m.p. 105 to 190 °C (dec.).

Magnetic behavior: $\mu_{eff} = 1.2\ \mu_B$. IR spectrum (n-hexane): 1510 (ν(C⋯O)); 1972, 1980, 1997, 2090 (ν(CO)) cm^{-1} [4].

$[(CO)_4Re(C(CH_3)O)C(CH_3){=}NHCH_2]_2(CH_2)_n$ (see Formulas IV, V; n = 0, 1). Treatment of $(CO)_4Re(C(CH_3)O)_2\cdots H$ with ca. 0.5 equivalents $H_2NCH_2(CH_2)_nCH_2NH_2$ in CH_2Cl_2 at room temperature resulted in a Schiff base condensation reaction, giving a mixture of the geometrical isomers of the bis(rhena-β-ketoimine) complexes. The isomerism results from a 180° rotation around the C=N bond. Only in the case n = 0, one isomer was obtained in pure form after two crystallizations from ether. The spectroscopic properties below are for mixtures of the isomers (for abbreviations and units see p. X).

The isomers can be readily assigned based on their ^1H NMR spectra: Peaks at ca. 13.4 and ca. 10.2 ppm result from the enolic NH proton of the Formula IV and Formula V isomer, respectively. The formation reaction yielded them in the ratio 3:2. The IR spectra exhibit the ν(CO) bands ca. 25 cm^{-1} lower than the respective bands in the rhena-β-diketone complexes, consistent with the more negative charge at the Re atoms than that in the rhena-β-diketone complexes, confirming the description of the intramolecular bonding by a zwitterionic electron structure.

The compounds are air-stable for several hours. They are well soluble in common organic solvents [2].

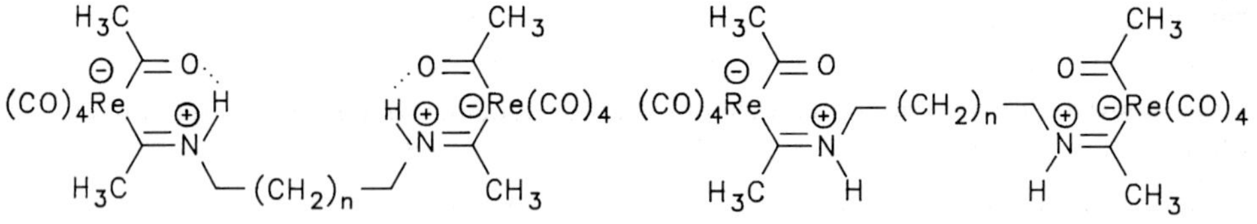

IV V

n = 0: Workup: Evaporation, recrystallization from ether. Yield: 38%. – Yellow solid; m.p. 132 to 133 °C.
^1H NMR ($CDCl_3$): 2.54, 2.65 (s's, CH_3CO); 2.80, 2.97 (s's, CH_3CN); 3.03, 4.30 (complex m's, CH_2); 10.05, 13.16 (br s's, =NH). – IR (CH_2Cl_2): 1550 (ν(C=N) + ν(C=O)); 1935, 1970, 2075 (ν(CO)) [2].

n = 1: Workup: Addition of hexane, filtration, and cooling. Yield: 34%. – Pale yellow solid; m.p. 118 to 119 °C.
^1H NMR ($CDCl_3$): 2.32 (complex m, CH_2); 2.58, 2.68 (s's, CH_3CO); 2.78, 2.89 (s's, CH_3CN); 3.76, 4.06 (complex m's, NCH_2); 10.22, 13.33 (br s's, =NH). – IR (CH_2Cl_2): 1555 (ν(C=N) + ν(C=O)); 1940, 1975, 1985, 2080 (ν(CO)) [2].

References:

[1] Lukehart, C. M.; Torrence, G. P. (J. Chem. Soc. Chem. Commun. **1978** 183/4).
[2] Lukehart, C. M.; Zeile, J. V. (Inorg. Chem. **17** [1978] 2369/74).
[3] Darst, K. P.; Hobbs, D. T.; Lukehart, C. M. (J. Organomet. Chem. **179** [1979] C 9/C 12).

[4] Lukehart, C. M.; Torrence, G. P. (Inorg. Chem. **18** [1979] 3150/5).
[5] Lenhart, P. G.; Lukehart, C. M.; Warfield, L. T. (Inorg. Chem. **19** [1980] 311/5).
[6] Beaver, B. D.; Hall, L. C.; Lukehart, C. M.; Preston, L. D. (Inorg. Chim. Acta **47** [1981] 25/30).
[7] Lippmann, E.; Robl, C.; Berke, H.; Kaesz, H. D.; Beck, W. (Chem. Ber. **126** [1993] 933/40).

2.4.2.3 Compounds Containing the Rhenacyclopentadiene Fragment

General. Structure. The compounds dealt with in this section consist of a rhenacyclopentadiene ring which is attached to a second Re atom in the η^4 mode (see Formula I). There is an Re-Re bond. In one case (No. 9) two CO groups of the $Re(CO)_4$ moiety are replaced by a side-on-coordinated acetylene ligand.

R R $Re(CO)_4$ R R $Re(CO)_3$

I

Five compounds were structurally characterized. All have a similar $C_4R_4Re_2(CO)_7$ core geometry. The Re-Re bond length is shorter than in $(CO)_{10}Re_2$. The rhenacyclopentadiene ring has an envelope conformation and shows alternating C-C bond lengths. One of the CO ligands of the $(CO)_4Re$ fragment is nonlinear and oriented in the direction of the other Re atom, suggesting a slightly semi-bridging character [4, 5, 9, 11].

Preparation. Some compounds were obtained by special procedures given in the table or in "Further information". In addition, the following general methods were applied:

Method I: Reaction between excess $CH_3C{\equiv}CN(CH_3)_2$ and
a. $[\mu\text{-}CH_3C((CH_3)_2N)C{=}]Re_2(CO)_8$ (see p. 123) in refluxing hexane
b. $(\mu\text{-}\eta^{2:1}\text{-n-}C_4H_9CH{=}CH)Re_2(CO)_8(\mu\text{-}H)$ in hexane at 50°C
for ca. 1.5 h. In either case, the resulting mixture contained Nos. 1 to 3. Separation was achieved by chromatography (silica, hexane/CH_2Cl_2 (5:1)). In the case of Method Ib, $[\mu\text{-}CH_3C((CH_3)_2N)C{=}]Re_2(CO)_8$ was the main product [4].

Method II: Reaction of diphenylacetylene
a. photochemically (λ = 366 nm) with $(CO)_{12}Re_3(\mu\text{-}H)_3$ under any of the following conditions: 1) pentane, −25°C, 12 to 18 h [3]; 2) isooctane, room temperature [1]. The products of the room-temperature reaction were initially assigned formulas with hydride ligands (see below) [1].
b. thermally with $(CO)_8Re_2(NCCH_3)_2$ (CH_2Cl_2, reflux, 24 h) [11].
Either reaction gave a mixture containing Nos. 6 and 9 together with $(C_6(C_6H_5)_6)Re_2(C_6H_5C{\equiv}CC_6H_5)(CO)_4$ (see p. 264). Separation was accomplished by chromatography using CH_2Cl_2/hexane mixtures [3, 11]. Method IIa also produced $(CO)_{10}Re_2$ (>25%), $(CO)_5ReH$ [1, 3], and $(CO)_{14}Re_3H$ [1].

Initially, the actual composition of two of the compounds was not recognized. Based on mass-spectroscopic results, formulas with hydride ligands, viz. $\mathbf{(C_2(C_6H_5)_2)_2Re_2(CO)_7H_2}$ and $\mathbf{(C_2(C_6H_5)_2)_3Re_2(CO)_5H_2}$, were assigned to Nos. 6 and 9 [1].

References on p. 155

Table 7
Compounds with the Rhenacyclopentadiene Fragment.
An asterisk indicates further information at the end of the table.
For explanations, abbreviations, and units see p. X.

No.	compound	method of preparation (yield) properties and remarks
*1	$[C(CH_3)=C(N(CH_3)_2)C(N(CH_3)_2)=C(CH_3)]Re_2(CO)_7$ 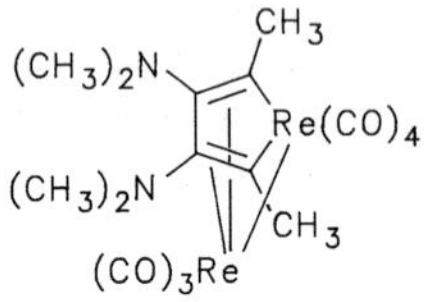	Ia (54%); Ib (6%) yellow solid ^1H NMR ($CDCl_3$): 2.73, 2.89 (CH_3; ratio 6:12) IR (n-hexane): 1925, 1996, 2004, 2028, 2082 (ν(CO)) MS: $[M - n\ CO]^+$ (n = 0 to 7) [4]
2	$[C(N(CH_3)_2)=C(CH_3)C(CH_3)=C(N(CH_3)_2)]Re_2(CO)_7$ 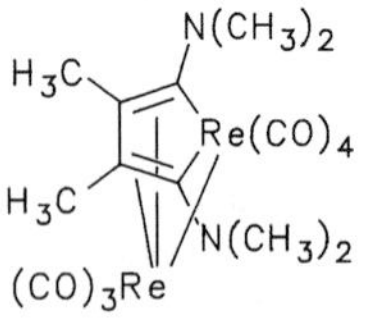	Ia (17%); Ib (3%) orange solid ^1H NMR ($CDCl_3$): 2.40, 2.47 (CH_3; ratio 6:12) IR (n-hexane): 1930, 1955, 1964, 1994, 2022, 2073 (ν(CO)) MS: $[M - n\ CO]^+$ (n = 0 to 7) [4]
3	$[C(N(CH_3)_2)=C(CH_3)C(N(CH_3)_2)=C(CH_3)]Re_2(CO)_7$ 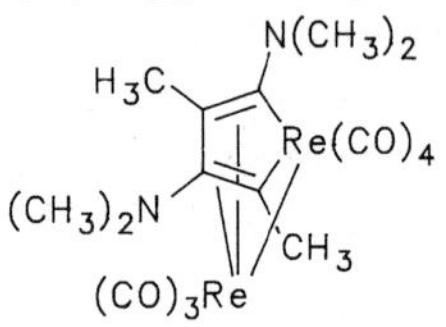	Ia (9.5%); Ib (1%) yellow solid ^1H NMR ($CDCl_3$): 2.38, 2.57, 2.79, 2.90 (s's, CH_3; ratio 3:6:3:6) IR (n-hexane): 1930, 1959, 1966, 2000, 2024, 2078 (ν(CO)) MS: $[M - n\ CO]^+$ (n = 0 to 7) [4]
*4	$[CH=C(CO_2CH_3)CH=C(CO_2CH_3)]Re_2(CO)_7$ 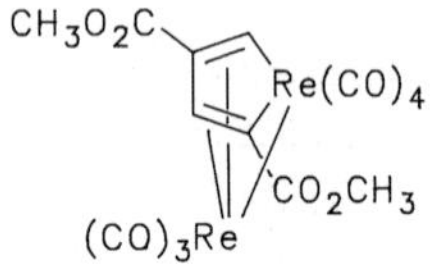	for preparation see "Further information" ^1H NMR ($CDCl_3$): 3.75, 3.87 (s's, OCH_3); 7.71, 8.20 (d's; $^4J(H,H)$ = 1.9) IR (n-hexane): 1719, 1740, 1747, 1947, 1960, 1977, 1991, 2030, 2051, 2103 (ν(CO)) MS: $[M - n\ CO]^+$ (n = 0 to 7) [8]
*5	$(C_4(CO_2C_2H_5)_4)Re_2(CO)_7$ 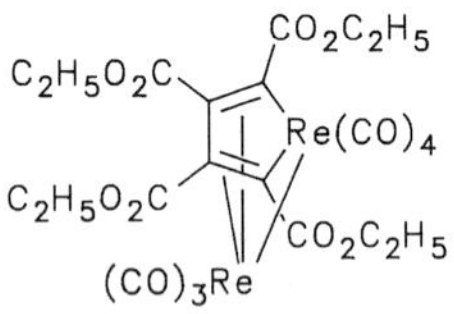	for preparation, see "Further information" ^1H NMR ($CDCl_3$): 1.29, 1.31 (t's, CH_3); 4.16, 4.17 (q's, CH_2; J = 7.2) IR (n-hexane): 1719, 1739, 1746, 1960, 1977, 1992, 2030, 2051, 2103 (ν(CO)) MS: $[M]^+$ observed [7]
*6	$(C_4(C_6H_5)_4)Re_2(CO)_7$ 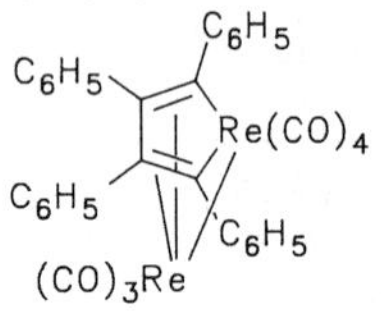	IIa [1]; IIa (0.4%) [3]; IIb (8%) [11] ^1H NMR (CD_2Cl_2): 7.2 (m) [2] IR (n-hexane): 1948, 1958, 1979, 1992, 2015, 2035, 2085 (ν(CO)) [3] (almost similar in [1]); 1935, 1966, 1978, 2016, 2034, 2086 (ν(CO)) [2] (almost identical in [11])

References on p. 155

Table 7 (continued)

No.	compound	method of preparation (yield) properties and remarks
		MS: $[M - n\ CO]^+$ [2, 3, 11], $[M - n\ CO - C_2(C_6H_5)_2]^+$ (n = 0 to 7) [2]
*7	$(C_{32}H_{20})Re_2(CO)_7$ 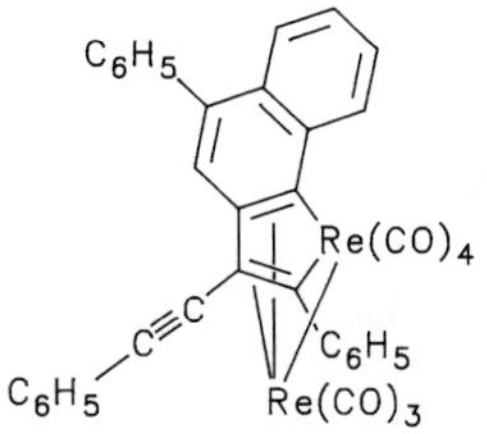	by thermolyzing $(CO)_5Re{-}C{\equiv}CC_6H_5$ in refluxing toluene followed by column-chromatographic workup (compound was one of the main products) [5]; also isolated during workup of the reaction leading to No. 8 [9] light yellow-green solid [5] ^1H NMR (acetone-d_6): 7.25 to 8.29 IR (n-hexane): 1937, 1945, 1961, 1970, 1974, 2017, 2036, 2091 (ν(CO)) MS: $[M - n\ CO]^+$ (n = 0 to 7) [5]
*8	$(C_{40}H_{30}Fe_2)Re_2(CO)_7$ 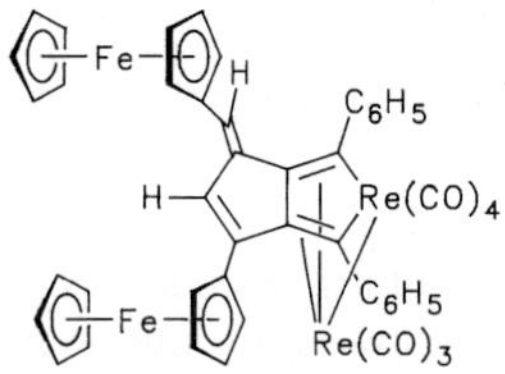	from $(CO)_5Re{-}C{\equiv}CC_6H_5$ and 2 equivalents $HC{\equiv}C{-}C_5H_4FeC_5H_5$ (refluxing toluene, 3 h); chromatography (silica, petroleum ether/benzene mixtures) gave $(CO)_{10}Re_2$, No. 7, and the title complex (yield of crude product: 25%; further purification by preparative TLC was required) deep red, almost black crystals ^1H NMR (acetone-d_6): 3.00, 3.87 (m's, 1 H); 4.09 (C_5H_5), 4.10 (m, 1 H), 4.14 (C_5H_5), 4.31 (m, 1 H), 4.52 (m, 3 H), 4.65 (m, 1 H), 6.12 (s, 1 H), 6.9 to 7.7 (m, 10 H), 7.90 (s, 1 H) [9] IR (n-hexane): 1937, 1945, 1961, 1970, 1984, 2017, 2036, 2091 (ν(CO)) [9] (similar in [6])
another compound with a rhenacyclopentadiene ligand		
*9	$(C_4(C_6H_5)_4)Re_2(C_6H_5C{\equiv}CC_6H_5)(CO)_5$ 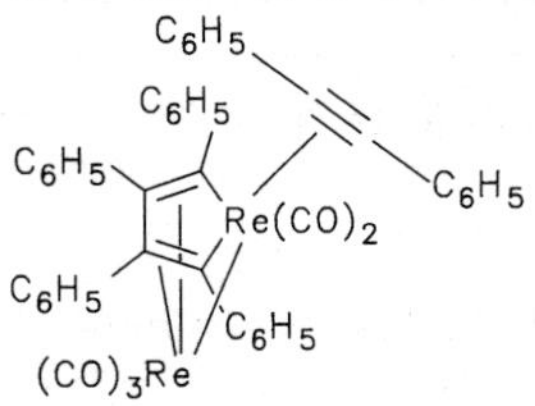	IIa [1]; IIa (6.5%) [3]; IIb (7%) [11] yellow solid [1], orange plates [3] IR (n-hexane): 1928, 1947, 1967, 2019, 2046 (ν(CO)) [3] (nearly identical in [1] and in [11]) MS: $[M - n\ CO]^+$ (n = 0 to 5), $[M - 5\ CO - C_2(C_6H_5)_2]^+$ [3] only slightly soluble in hexane [1]

*Further information:

$[C(CH_3){=}C(N(CH_3)_2)C(N(CH_3)_2){=}C(CH_3)]Re_2(CO)_7$ (Table **7**, No. **1**) crystallizes in the orthorhombic space group $Pca2_1{-}C_{2v}^5$ (No. 29) with a = 17.668(4), b = 8.510(2), c = 14.058(4) Å; Z = 4 molecules per unit cell, D_{calc} = 2.31 g/cm^3. The molecular structure is depicted in **Fig. 74** [4].

$[CH{=}C(CO_2CH_3)CH{=}C(CO_2CH_3)]Re_2(CO)_7$ (Table **7**, No. **4**) was obtained when treating $(CO)_4ReO{=}(CH_3O)CC[CH{=}C(Re(CO)_5)CO_2CH_3]{=}CH$-cyclo (see p. 42) with 1 equivalent

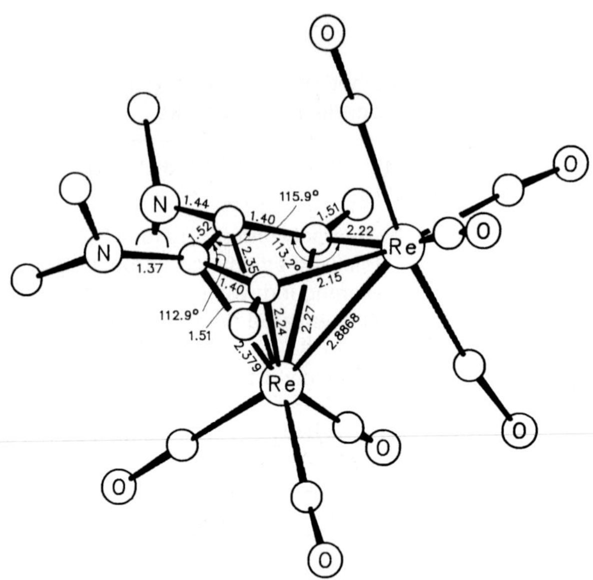

Fig. 74. Molecular structure of $[C(CH_3)=C(N(CH_3)_2)C(N(CH_3)_2)=C(CH_3)]Re_2(CO)_7$ [4].

$(CH_3)_3NO$ in CH_3CN for 30 min. The mixture was evaporated, and the residue redissolved in CH_2Cl_2. Subsequent heating at reflux temperature for 3 h followed by chromatographic workup yielded 50%. The yield dropped to 30% upon carrying out the heating in the presence of $HC{\equiv}CCO_2CH_3$. Small amounts of No. 4 (9%) were obtained by treating $(CO)_4ReO{=}(CH_3O)$-$CC[Re(CO)_4NCCH_3]{=}CH$-cyclo (see p. 41) with a ca. 10-fold excess of $HC{\equiv}CCO_2CH_3$ in CH_2Cl_2 under an inert gas atmosphere at 40 °C. In contrast, the compound did not form at 25 °C or when the reaction was conducted under CO (for all products of the reaction, see pp. 47/8) [8].

$(C_4(CO_2C_2H_5)_4)Re_2(CO)_7$ (Table **7**, No. **5**) formed as a by-product when treating $(CO)_9Re_2NCCH_3$ with $C_2H_5O_2CC{\equiv}CCO_2C_2H_5$ (mole ratio 2:3) in refluxing hexane for 1.5 h [7, 10]. The main product was $(CO)_4ReO{=}C(OC_2H_5)C[{=}C(CO_2C_2H_5)Re(CO)_5]$-cyclo (see p. 41) [10] (caution: wrong composition in [7]). Evaporation followed by redissolving in CH_2Cl_2/hexane (1:1) and cooling removed the main product, while No. 5 could be isolated by chromatographic workup of the mother liquor with hexane/CH_2Cl_2 (1:1). Yield: 8% [7].

$(C_4(C_6H_5)_4)Re_2(CO)_7$ (Table **7**, No. **6**) was also produced by heating $(CO)_{10}Re_2$ with diphenylacetylene (bomb tube, hexane, 190 °C, 16 h). Separation by preparative TLC using hexane/CH_2Cl_2 (3:2) not only gave No. 6 but also $(C_6(C_6H_5)_6)Re_2(C_6H_5C{\equiv}CC_6H_5)(CO)_4$ and $(C_6(C_6H_5)_6)Re_2(CO)_6$ (see p. 264) with a combined yield of ca. 80%, but individual yields varied. No. 6 converted into the two last-named products upon heating with excess diphenylacetylene [2].

The compound crystallizes in the monoclinic space group $P2_1/c-C_{2h}^5$ (No. 14) with a = 10.597(2), b = 19.80(2), c = 16.141(3) Å, β = 114.34(2)°; Z = 4 molecules per unit cell, and D_{calc} = 1.99 g/cm^3. The molecular structure is illustrated in **Fig. 75** [11].

$(C_{32}H_{20})Re_2(CO)_7$ (Table **7**, No. **7**). An X-ray structure analysis showed the compound to crystallize in the triclinic space group $P\bar{1}-C_i^1$ (No. 2) with a = 8.753(2), b = 11.676(2), c =

References on p. 155

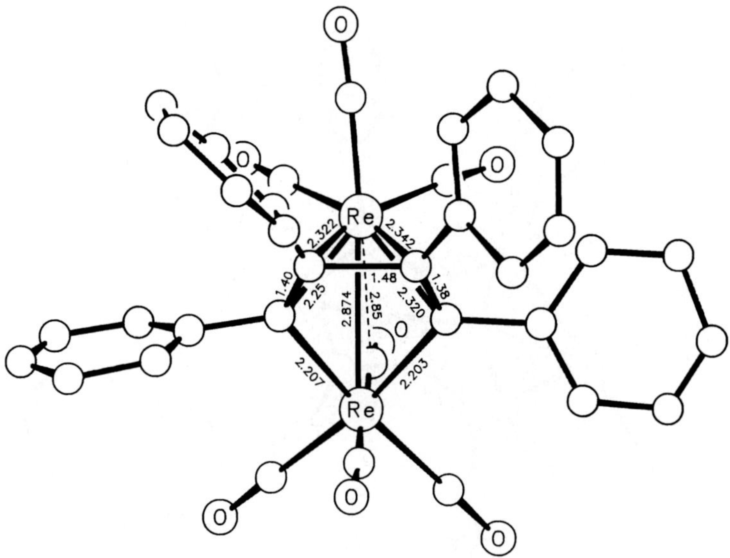

Fig. 75. Molecular structure of $(C_4(C_6H_5)_4)Re_2(CO)_7$ [11].

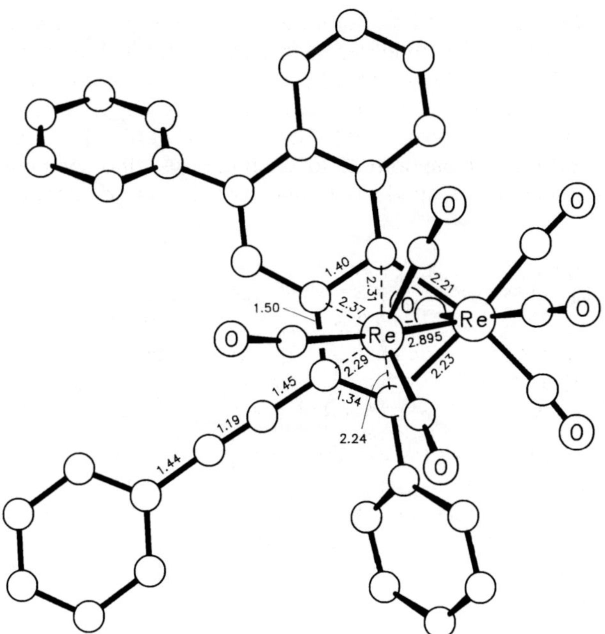

Fig. 76. Molecular structure of $(C_{32}H_{20})Re_2(CO)_7$ [5].

17.142(4) Å, α = 74.73(1)°, β = 80.50(2)°, γ = 81.81(1)°; Z = 2 molecules per unit cell, D_{calc} = 1.89 g/cm^3. A view of the molecular structure is shown in **Fig. 76** [5].

References on p. 155

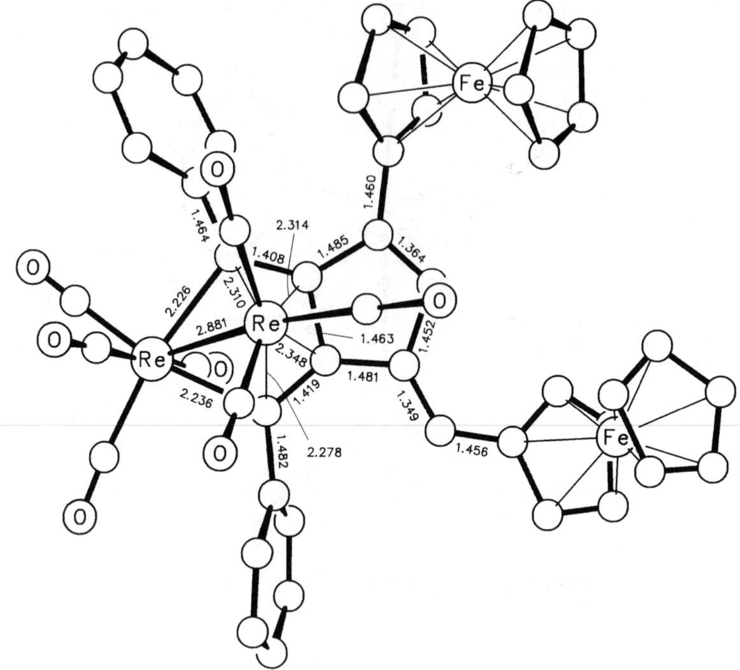

Fig. 77. Molecular structure of $(C_{40}H_{30}Fe_2)Re_2(CO)_7$ [9].

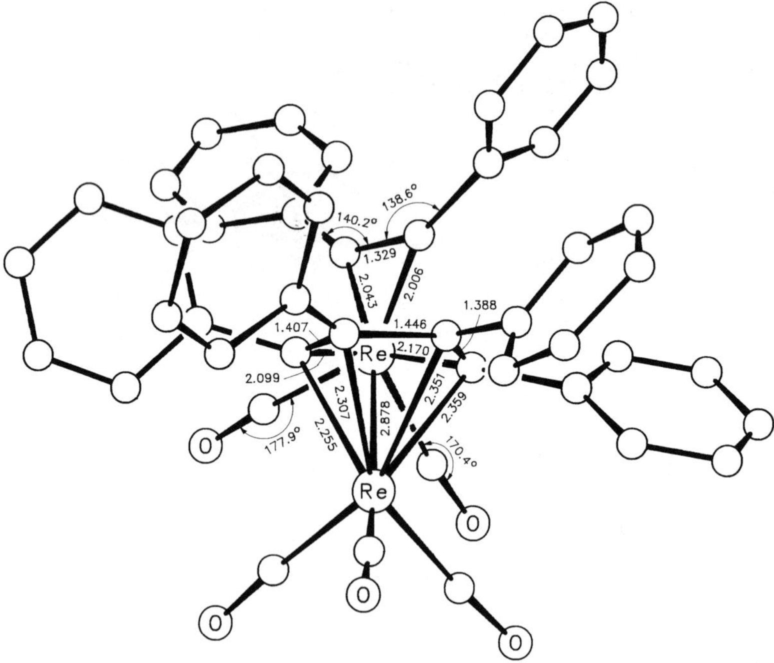

Fig. 78. Molecular structure of $(C_4(C_6H_5)_4)Re_2(C_6H_5C{\equiv}CC_6H_5)(CO)_5$ [3].

References on p. 155

$(C_{40}H_{30}Fe_2)Re_2(CO)_7 \cdot O{=}C(CH_3)_2$ (Table **7**, No. **8**). Single crystals containing 1 mole acetone per complex are monoclinic with a = 20.363(5), b = 10.121(2), c = 22.952(7) Å, β = 112.49(2)°; space group $P2_1/n-C_{2h}^5$ (No. 14), Z = 4 formula units per unit cell, and D_{calc} = 1.928 g/cm^3 [6, 9]. The structure of the molecule is illustrated in **Fig. 77**. The folding angle of the ReC_4 envelope along the C···C vector is 16.4°. The C_5 cycle fused at the ReC_4 ring is almost planar but forms a small dihedral angle (4.7°) with the metallacycle [9].

$(C_4(C_6H_5)_4)Re_2(C_6H_5C{\equiv}CC_6H_5)(CO)_5$ (Table 7, No. **9**) crystallizes in the triclinic space group $P\bar{1}-C_i^1$ (No. 2) with a = 11.277(3), b = 12.301(3), c = 14.851(3) Å, α = 97.37(2)°, β = 104.68(2)°, γ = 104.49(2)°; Z = 2 molecules per unit cell, D_{calc} = 1.845 g/cm^3. The molecular structure is illustrated in **Fig. 78**. Intramolecular bond parameters are not very different from those found in No. 6. The side-on-coordinated alkyne can be considered a 4-electron donor ligand [3].

References:

[1] Epstein, R. A.; Gaffney, T. R.; Geoffroy, G. L.; Gladfelter, W. L.; Henderson, R. S. (J. Am. Chem. Soc. **101** [1979] 3847/52).
[2] Mays, M. J.; Prest, D. W.; Raithby, P. R. (J. Chem. Soc. Dalton Trans. **1981** 771/6).
[3] Pourreau, D. B.; Whittle, R. R.; Geoffroy, G. L. (J. Organomet. Chem. **273** [1984] 333/46).
[4] Adams, R. D.; Chen, G.; Yin, J. (Organometallics **10** [1991] 1278/82).
[5] Koridze, A. A.; Zdanovich, V. I.; Batsanov, A. S.; Struchkov, Yu. T. (Mendeleev Commun. **1991** 126/7).
[6] Koridze, A. A.; Zdanovich, V. I.; Polyakova, N. V.; Yanovskii, A. I.; Struchkov, Yu. T. (Izv. Akad. Nauk SSSR Ser. Khim. **1992** 1689/91; Bull. Russ. Acad. Sci. Div. Chem. Sci. [Engl. Transl.] **41** [1992] 1314).
[7] Adams, R. D.; Chen, L.; Wu, W. (Organometallics **12** [1993] 1257/65).
[8] Adams, R. D.; Chen, L.; Wu, W. (Organometallics **12** [1993] 1623/8).
[9] Koridze, A. A.; Zdanovich, V. I.; Kizas, O. A.; Yanowsky, A. I.; Struchkov, Yu. T. (J. Organomet. Chem. **464** [1993] 197/201).
[10] Adams, R. D.; Chen, L. (Organometallics **13** [1994] 1264/71).
[11] Bruce, M. I.; Low, P. J.; Skelton, B. W.; White, A. H. (J. Organomet. Chem. **464** [1994] 191/5).

2.4.2.4 Miscellaneous Compounds

This section comprises several other compounds with ^4L ligands. Their structures are illustrated in Formulas I to VIII. Most of them were determined by crystal structure analysis.

The compounds of the type $(\mu\text{-}(C_5H_5)_2M_2(CO)_3(CR)_2)Re_2(CO)_6$ (see Formula I) can be regarded as possessing either two face-bridging alkylidyne groups and two semi-bridging

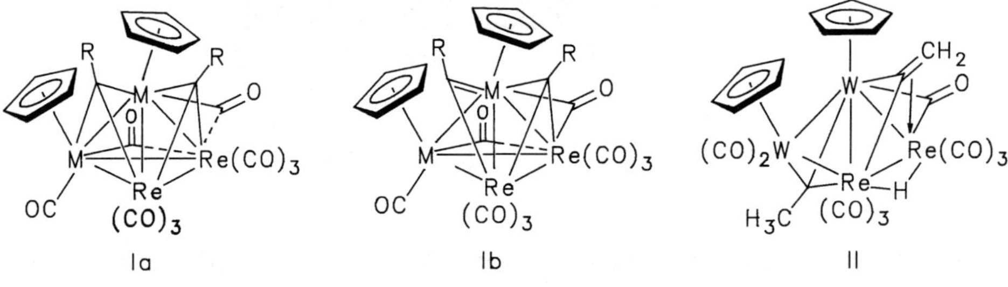

References on p. 163

CO groups (Formula Ia) or one edge- and one face-bridging alkylidyne group together with one semi- and one ordinarily bridging CO group (Formula Ib). The solid-state structure agrees with that of the former, but the structure in solution appears more to be in accord with the latter way of looking at it [3].

Compounds represented by Formulas I and II were prepared as follows:

Method I: Treatment of (μ-$\eta^{2:1}$-n-$C_4H_9CH{=}CH)Re_2(CO)_8(\mu$-H) with 2 equivalents C_5H_5-$M(CO)_2(\equiv CR)$ (M = Mo, R = $C_6H_4CH_3$-4; M = W, R = CH_3, $C_6H_4CH_3$-4) in refluxing THF for 12 to 18 h. The resulting mixture was separated by column chromatography on alumina using CH_2Cl_2/petroleum ether mixtures [3].

(μ-$(C_5H_5)_2Mo_2(CO)_3(CC_6H_4CH_3-4)_2)Re_2(CO)_6$ · 1.5 CH_2Cl_2 (see Formula I, M = Mo, R = $C_6H_4CH_3$-4) was obtained along with $(C_5H_5)_2Mo_2(CO)_4(\mu$-$C_2(C_6H_4CH_3$-4$)_2)$ and other undefined material via Method I. Yield: 15%. Green microcrystals.

^{1}H NMR spectrum ($CDCl_3$): δ = 2.36, 2.37 (s's, CH_3); 5.28, 5.42 (s's, C_5H_5); 7.00 (m, C_6H_4) ppm. ^{13}C {^{1}H} NMR spectrum (CD_2Cl_2/CH_2Cl_2, −40 °C): δ = 21.2, 21.3 (CH_3); 95.8, 101.2 (C_5H_5); 127.5, 128.4, 128.7, 128.9, 129.5, 136.1, 137.9 (C_6H_4); 163.2, 165.9 (C_{ipso}); 193.9, 197.0 (ReCO); 223.5, 232.9 (MoCO); 264.9 (μ-CO), 293.0 (μ_3-C), 355.3 (μ-C) ppm. IR spectrum (CH_2Cl_2): 1775, 1832, 1951, 2007, 2039 (ν(CO)) cm^{-1}.

The compound crystallizes in the monoclinic space group $P2_1/c-C^5_{2h}$ (No. 14) with a = 18.841(5), b = 19.317(10), c = 21.148(7) Å, β = 91.96(2)°, Z = 8 molecules per unit cell, D_{calc} = 2.21 g/cm^3. The asymmetric unit contains two very similar cluster molecules and three CH_2Cl_2 molecules. The structure of one of the cluster molecules is illustrated in **Fig. 79**. The metal atoms form a tetrahedron wherein two edges are supported by semi-bridging CO groups [3].

Extraction with petroleum ether yielded traces of a brown solid having an IR spectrum almost identical with that of the title compound, except that a band at 1735 cm^{-1} was also present. Its ^{1}H NMR spectrum shows signals at 2.37, 2.38 (s's, CH_3) and 5.51, 5.65 (s's,

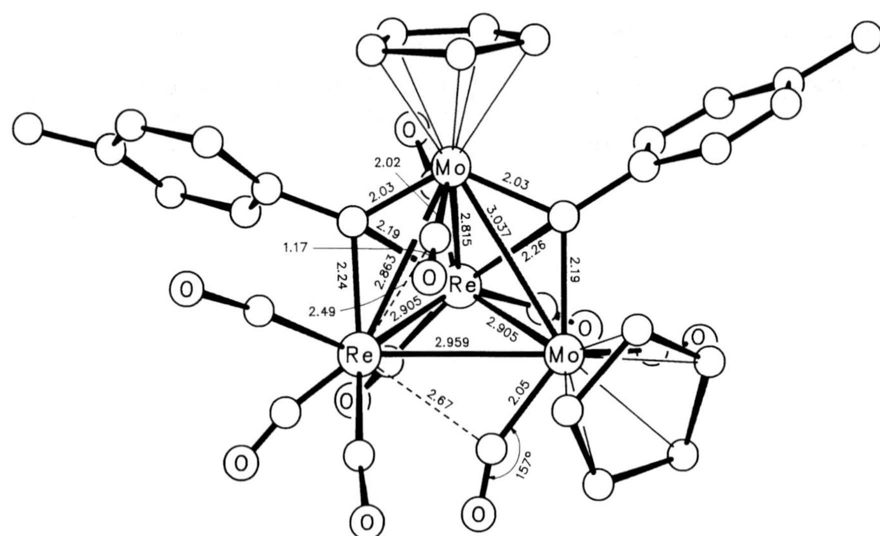

Fig. 79. Molecular structure of (μ-$(C_5H_5)_2Mo_2(CO)_3(CC_6H_4CH_3-4)_2)Re_2(CO)_6$ [3].

C_5H_5) ppm. This compound was suggested to be an isomer with a yet unknown structure. The two compounds could not be interconverted [3].

(μ-(C_5H_5)$_2$$W_2(CO)_3$($CCH_3$)$_2$)$Re_2(CO)_6$ · CH_2Cl_2 (see Formula I, M = W, R = CH_3) was obtained by Method I with 41% yield. The reaction gave also (CO)$_8$$Re_2$W($C_5H_5$)(CO)$_2$($\mu_3$-$CCH_3$) (see "Organorhenium Compounds" 5, 1994, p. 394) and [μ-(C_5H_5)$_2$$W_2(CO)_3$($CCH_3$)(C=$CH_2$)]-$Re_2(CO)_6$(μ-H) (see Formula II). Nevertheless, the latter could be partially converted into the title compound by heating it in THF. Brown crystals.

^{1}H NMR spectrum ($CDCl_3$): δ = 3.77, 3.92 (s's, CH_3); 5.23, 5.64 (s's, C_5H_5) ppm; also δ = 5.30 (s) ppm (from CH_2Cl_2). The compound is too insoluble for ^{13}C {^{1}H} NMR spectroscopy. IR spectrum (CH_2Cl_2): 1735, 1840, 1918, 1949, 1998, 2035 (ν(CO)) cm^{-1}.

The compound decomposed into unidentified products in refluxing THF [3].

(μ-(C_5H_5)$_2$$W_2(CO)_3$($CC_6H_4CH_3$-4)$_2$)$Re_2(CO)_6$ · 1.5 CH_2Cl_2 (see Formula I, M = W, R = $C_6H_4CH_3$-4) was prepared by varying the procedure described in the Method I (reaction in a closed vessel; solvent: toluene). This gave green crystals with 48% yield.

^{1}H NMR spectrum ($CDCl_3$): δ = 2.33, 2.37 (s's, CH_3); 5.39, 5.48 (s's, C_5H_5); 7.14 (m, C_6H_4) ppm. ^{13}C {^{1}H} NMR spectrum (CH_2Cl_2/CD_2Cl_2; −40°C): δ = 21.1, 21.3 (CH_3); 93.6, 99.4 (C_5H_5); 128.5, 128.7, 128.8, 129.0, 129.1, 135.7, 137.4 (C_6H_4); 165.2, 168.4 (C_{ipso}); 193.0, 198.3 (ReCO); 206.5, 222.1 (WCO); 253.7 (μ-CO), 262.1 (μ_3-C), 337.0 (μ-C) ppm. IR spectrum (CH_2Cl_2): 1775, 1830, 1920, 1946, 2003, 2035 (ν(CO)) cm^{-1} [3].

[μ-(C_5H_5)$_2$$W_2(CO)_3$($CCH_3$)(C=$CH_2$)]$Re_2(CO)_6$(μ-H) (see Formula II) was obtained with 35% yield by reacting (μ-$\eta^{2:1}$-n-C_4H_9CH=CH)Re_2(CO)$_8$(μ-H) with C_5H_5W(≡CCH_3)(CO)$_2$ as described in Method I. Red crystals.

^{1}H NMR spectrum ($CDCl_3$): δ = −19.57 (μ-H), 4.65 (s, CH_3), 5.29 (d, CH_2); 5.34, 5.68 (s's, C_5H_5); 6.19 (d, CH_2; J = 1 Hz) ppm. ^{13}C {^{1}H} NMR spectrum (CD_2Cl_2, −40°C): δ = 52.7 (μ_3-C**C**H_3), 78.0 (C=**C**H_2); 93.4, 98.4 (C_5H_5); 186.5, 187.0, 188.2, 188.5, 190.1, 194.2 (ReCO); 207.7, 224.3, 224.7, 229.6, 238.7 (3 WCO + μ_3-**C**CH_3 + **C**=CH_2) ppm. IR spectrum (CH_2Cl_2): 1836, 1915, 1940, 1955, 1972, 2012, 2037 (ν(CO)) cm^{-1}.

The compound crystallizes in the monoclinic space group $P2_1/n-C_{2h}^5$ (No. 14) with a = 15.774(4), b = 8.690(3), c = 18.370(12) Å, β = 104.22(4)°; Z = 4 molecules per unit cell, D_{calc} = 3.20 g/cm^3. The structure of the molecule is shown in **Fig. 80**. The Re_2W_2 core can be best described as two triangles sharing a common edge (dihedral angle between the W_2Re and Re_2W planes: 19.4°). One Re-W bond is supported by a semi-bridging CO group.

Partial conversion into (μ-(C_5H_5)$_2$$W_2(CO)_3$($CCH_3$)$_2$)$Re_2(CO)_6$ occurred in boiling THF [3].

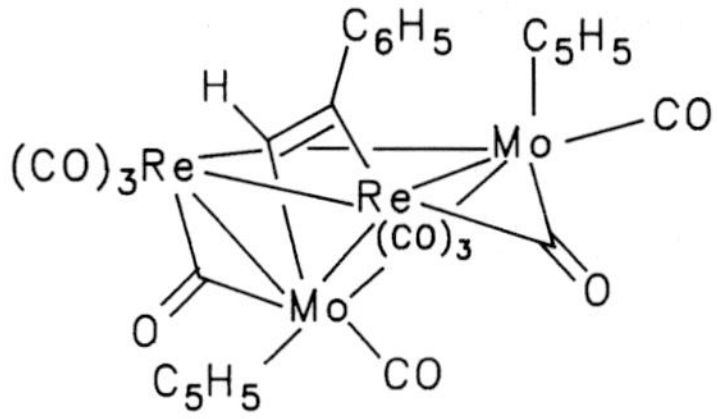

III 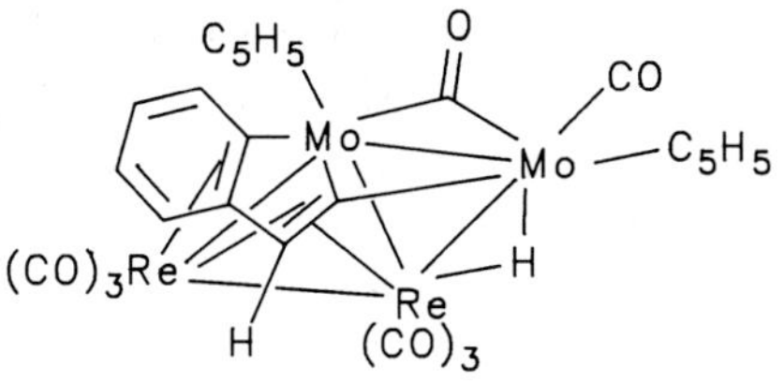 IV 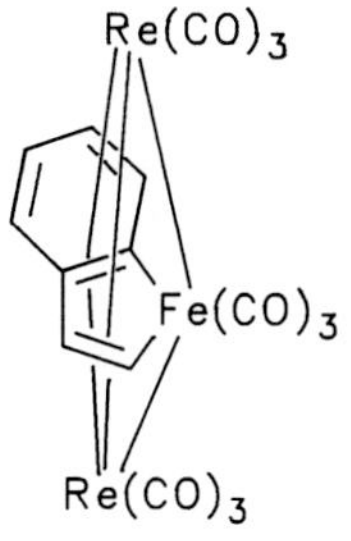 V

References on p. 163

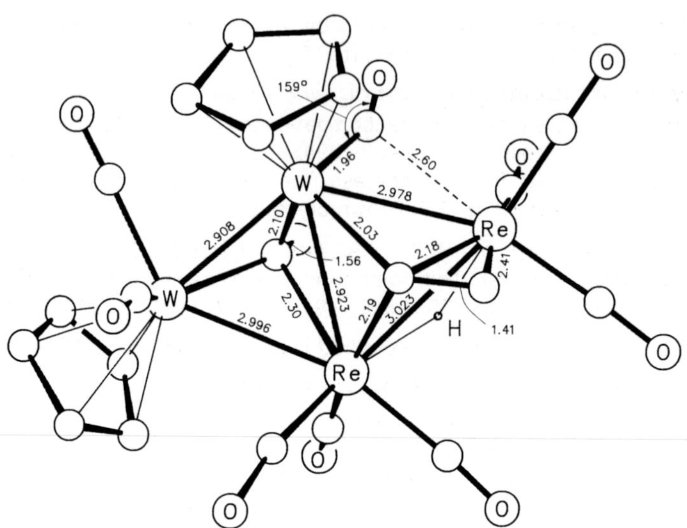

Fig. 80. Molecular structure of [μ-$(C_5H_5)_2W_2(CO)_3(CCH_3)(C{=}CH_2)$]$Re_2(CO)_6$($\mu$-H) [3].

[$(C_5H_5)_2Mo_2(CO)_2(CH{=}CC_6H_5)(\mu\text{-}CO)_2$]$Re_2(CO)_6$ (see Formula III) was produced by reacting (μ-$\eta^{2:1}$-$C_6H_5C{\equiv}C$)$Re_2(CO)_8$(μ-H) [2] or (μ-$\eta^{2:1}$-$C_6H_5C{\equiv}C$)$Re_2(CO)_6(NCCH_3)_2$(μ-H) [4] with $(C_5H_5)_2Mo_2(CO)_6$ in refluxing xylene. The product separated as black cubic crystals from the solution upon cooling [2, 4]. Yield: 28% [4].

^{1}H NMR spectrum: δ = 11.83 (CH=) ppm [2]. IR spectrum (KBr): 695, 760, 830, 1010; 1760, 1825, 1860, 1915, 1935, 1995, 2035 (ν(CO)) cm^{-1}. Mass spectrum: $[Re_2(CO)_n]^+$ (n = 10, 7 to 3), $[C_5H_5Re(CO)_3]^+$ [4].

The compound crystallizes in the orthorhombic space group Pcab (non-standard setting of Pbca$-D_{2h}^{15}$ (No. 61)) with a = 17.395(5), b = 16.053(3), c = 19.677(5) Å; Z = 8 molecules per unit cell. The molecular structure, which is illustrated in **Fig. 81**, consists of an Re_2Mo_2 butterfly unit with two triangular Re_2Mo and Mo_2Re wings having an interfacial angle of 110.7°. The $CH{=}CC_6H_5$ ligand is σ-bonded to an Re and an Mo atom forming an $ReMoC_2$ 4-membered cycle which is π-bonded to the other Re and Mo atoms [2, 4].

Rearrangement of the bridging vinylidene ligand accompanied with the loss of two CO groups occurred upon further heating the complex in xylene. This produced [μ-$(C_5H_5)_2$-$Mo_2H(CO)_2(C{=}CHC_6H_4)$]$Re_2(CO)_6$ (following compound) [5, 6].

[μ-$(C_5H_5)_2Mo_2H(CO)_2(C{=}CHC_6H_4)$]$Re_2(CO)_6$ (see Formula IV) was obtained by thermolyzing [$(C_5H_5)_2Mo_2(CO)_2(CH{=}CC_6H_5)(\mu\text{-}CO)_2$]$Re_2(CO)_6$ (preceding complex) in refluxing xylene for 5 to 10 h followed by column-chromatographic workup on alumina using hexane/benzene (1:4). Yield: 41%. Black-green, air-stable crystals [5, 6].

^{1}H NMR spectrum (CD_2Cl_2): δ = −13.31 (μ-H), 7.84 (=CHC_6H_4) ppm [5]. IR spectrum (?): 675, 750, 810, 830, 1010; 1805, 1880, 1900, 1950, 1995, 2035 cm^{-1} [6]; (KBr): 1805, 1910, 1925, 1950, 1995, 2035 (ν(CO)) cm^{-1} [5]. Mass spectrum: $[M - 2\,CO]^+$, $[M - H - 8\,CO]^+$, $[M - H - 8\,CO - C_2H_4]^+$ (most intense peaks) [5].

The complex crystallizes in the monoclinic space group $P2_1/c-C_{2h}^5$ (No. 14) with a = 10.621(3), b = 10.979(4), c = 26.106(6) Å, β = 96.91(1)°; Z = 4 molecules per unit cell. The

References on p. 163

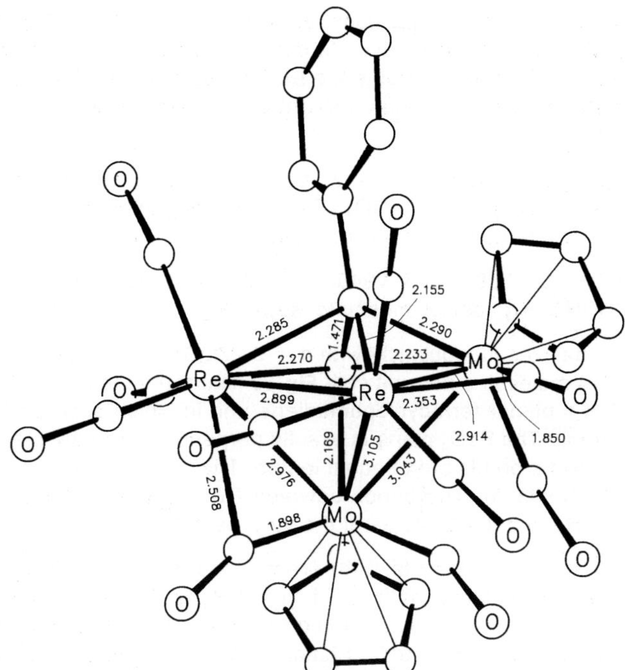

Fig. 81. Molecular structure of $[(C_5H_5)_2Mo_2(CO)_2(CH{=}CC_6H_5)(\mu\text{-}CO)_2]Re_2(CO)_6$ [2].

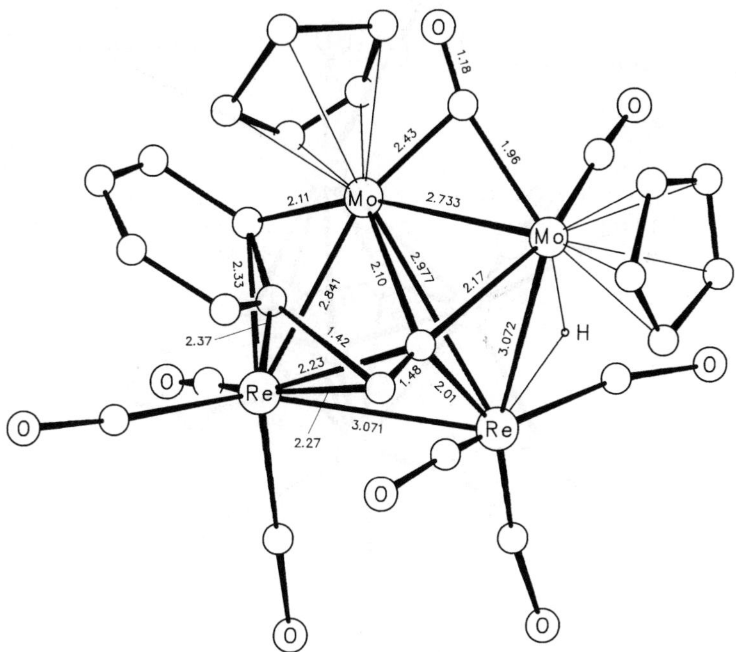

Fig. 82. Molecular structure of $[\mu\text{-}(C_5H_5)_2Mo_2H(CO)_2(C{=}CHC_6H_4)]Re_2(CO)_6$ [6].

References on p. 163

molecular structure along with selected bond lengths [5] is depicted in **Fig. 82**. The molecule consists of an Re_2Mo_2 butterfly core having a dihedral angle of 119.5° between the Re_2Mo and Mo_2Re planes. The Re and Mo atoms are bridged by a phenylvinylidene ligand coordinated to all four metal atoms. In addition, the ortho-metalated phenyl ring is π-coordinated to an Re atom [5, 6].

(μ-C_8H_6Fe(CO)$_3$)Re_2(CO)$_6$ (see Formula V) formed by reacting equimolar quantities of (μ-$\eta^{2:1}$-$C_6H_5C{\equiv}C$)$Re_2(CO)_8$(μ-H) and $(CO)_{12}Fe_3$ in refluxing toluene for 8 h. Yellow-orange, air-stable crystals precipitated upon cooling. Yield: 70%.

IR spectrum (KBr): 700, 750; 1915, 1945, 2010, 2065 (ν(CO)) cm^{-1}. The mass spectrum contains $[Re_2FeCHCHC_6H_4]^+$ as the most intense peak.

The complex crystallizes monoclinicly with a = 9.276(9), b = 9.961(9), c = 10.667(1) Å, β = 97.97(8)°; Z = 4 molecules per unit cell. **Fig. 83** illustrates the molecular structure. The molecule consists of a planar ferracyclopentadiene unit in which each C=C bond is coordinated to both Re atoms. The FeC_4 plane is exactly perpendicular to the plane formed by the Re_2Fe atoms. Both Re-Fe bonds are of equal length. The Re atoms are equidistant from the FeC_4 ring (Re···Re 3.893 Å). The bonds between Re and the C atoms of the arene are larger than between Re and the C atoms of the vinyl part [8].

(C_5H_6O)Re_2(CO)$_9$ (see Formulas VIa, VIb) precipitated when combining cooled (−40°C) solutions containing equimolar quantities of [η^2-$C_4H_6Re(CO)_5$]BF_4 in CH_3CN and Na[$(CO)_5Re$] in THF. The yellow solid was filtered off after 3 h of stirring. It could be purified by chromatography on silica at −10°C. Elution with CH_2Cl_2 yielded trans-$(CO)_5ReCH_2$-

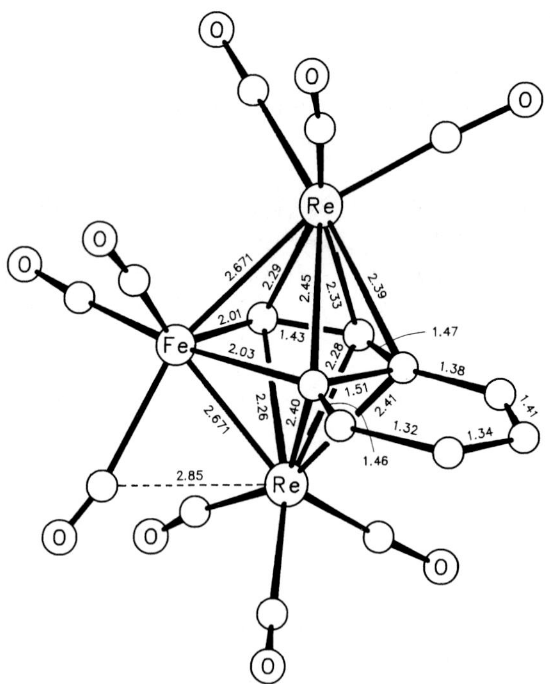

Fig. 83. Molecular structure of (μ-C_8H_6Fe(CO)$_3$)Re_2(CO)$_6$ [8].

References on p. 163

$CH=CHCH_2Re(CO)_5$, and subsequent elution with acetone gave the title compound with 42 to 63% yield. $(C_5H_6O)Re_2(CO)_9$ converts into trans-$(CO)_5ReCH_2CH=CHCH_2Re(CO)_5$ upon increasing the temperature without evolution of CO.

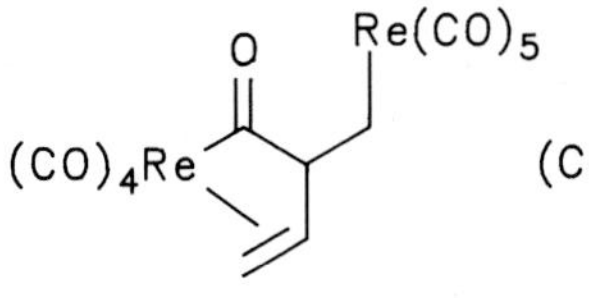

VIa 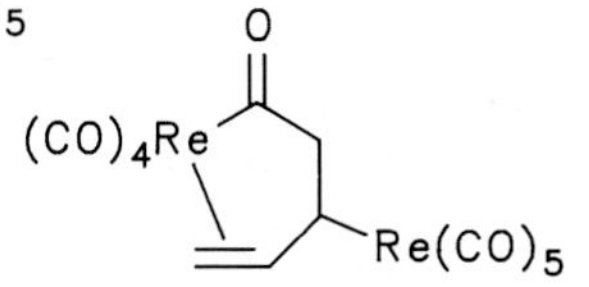 VIb 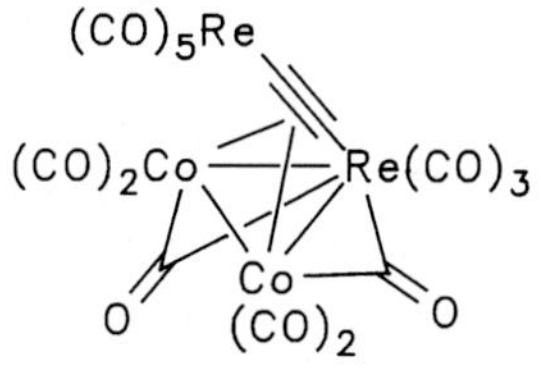 VII

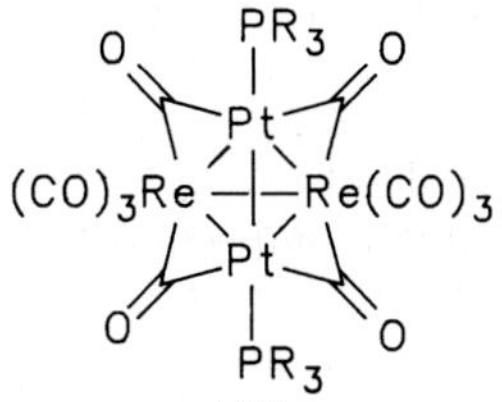

VIII

^{1}H NMR spectrum: allegedly recorded but not given. IR spectrum (KBr): 1514 (ν(C=C)), 1595 (ν(C=O)); 1920, 1932, 1981, 2007, 2017, 2037, 2067, 2082, 2132 (ν(CO)) cm^{-1}. Each of the structures given in Formulas VIa and VIb agrees with the spectroscopic data. The homogeneity of the compound was established [1].

[μ-$(CO)_6Co_2(C{\equiv}C)$]$Re_2(CO)_8$ (see Formula VII) was produced by reacting equimolar amounts of $(CO)_5ReC{\equiv}CRe(CO)_5$ and $(CO)_8Co_2$ in CH_2Cl_2 at room temperature for 3 h. The mixture was subsequently evaporated, and the residue was extracted with pentane. The remainder was recrystallized from CH_2Cl_2. Yield: 60%. Black-green plates; m.p. 150 °C (dec.).

IR spectrum (CH_2Cl_2): 1974, 2014, 2031, 2047, 2054, 2079, 2149 (ν(CO)) cm^{-1}. FD mass spectrum: $[M]^+$ observed.

Single crystals are triclinic with a = 9.474(2), b = 10.283(2), c = 13.102(2) Å, α = 68.46(1)°, β = 85.61(2)°, γ = 71.24(2)°; space group $P\bar{1}-C_i^1$ (No. 2), Z = 2 molecules per unit cell, D_{calc} = 2.681 g/cm^3. The molecular structure is depicted in **Fig. 84** [9].

[μ-$(CO)_4Pt_2(P(C_6H_{11})_3)_2$]$Re_2(CO)_6 \cdot CH_2Cl_2$ (see Formula VIII, R = C_6H_{11}) was obtained by treating $(CO)_{10}Re_2$ with 2 equivalents $(\pi\text{-}C_2H_4)_2PtP(C_6H_{11})_3$ in CH_2Cl_2. After adding hexane the cluster precipitated as orange needles. Yield: 76%. Detailed studies showed that the reaction proceeds via [μ-$(C_6H_{11})_3PPt(CO)_2$]$Re_2(CO)_8$ (see p. 55).

IR spectrum (CH_2Cl_2): 1802, 1840, 1853, 1956, 1979, 2011, 2056 (ν(CO)) cm^{-1}.

The solvate crystallizes in the triclinic space group $P\bar{1}-C_i^1$ (No. 2) with a = 11.567(4), b = 15.308(6), c = 16.178(6) Å, α = 89.87(3)°, β = 108.13(3)°, γ = 90.28(3)°; Z = 2 formula units per unit cell and D_{calc} = 2.060 g/cm^3. The molecule consists of a tetrahedrally arranged Re_2Pt_2 unit in which each Re-Pt bond is bridged by a CO group. The structure is very similar with that of the following cluster (compare with Fig. 85). Selected bond lengths are as follows: Re-Re 3.0560, Pt-Pt 3.0417, Re-Pt 2.7448 (av.), Pt-P 2.316 (av.), Re-(μ-C) 2.195 (av.), Pt-(μ-C) 2.005 (av.) Å [10].

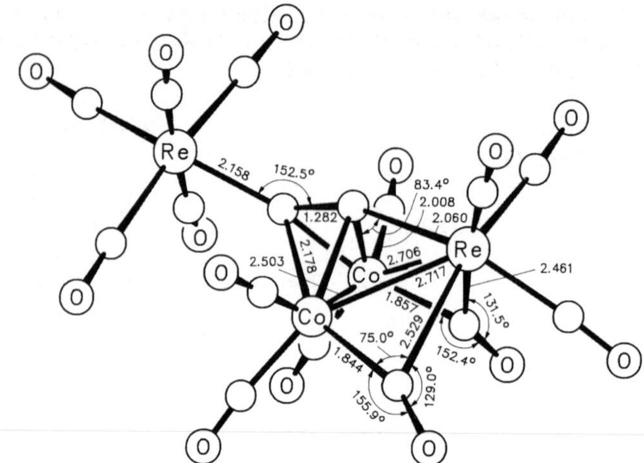

Fig. 84. Molecular structure of [μ-$(CO)_6Co_2(C{\equiv}C)$]$Re_2(CO)_8$ [9].

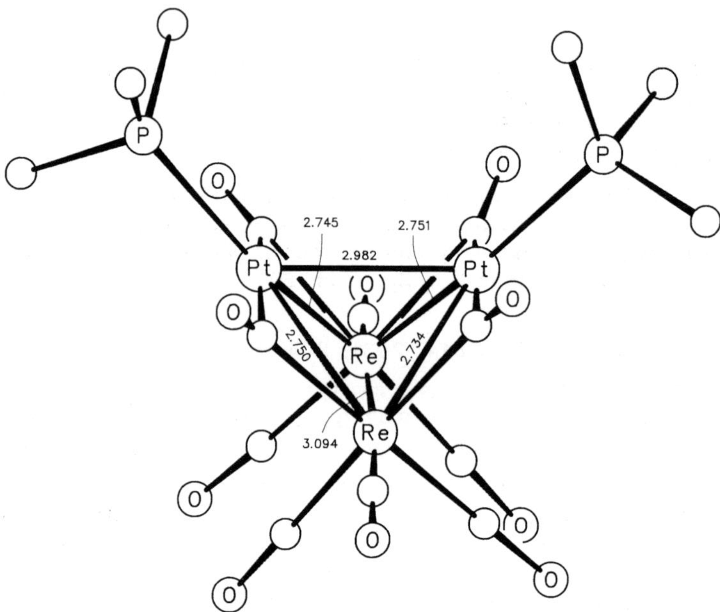

Fig. 85. Molecular structure of [μ-$(CO)_4Pt_2(P(C_6H_5)_3)_2$]$Re_2(CO)_6$ (phenyl rings represented by C_{ipso} only) [7].

[μ-$(CO)_4Pt_2(P(C_6H_5)_3)_2$]$Re_2(CO)_6$ · 0.5 CH_2Cl_2 (see Formula VIII, R = C_6H_5) was produced by reacting $(CO)_{10}Re_2$ with π-$C_2H_4Pt(P(C_6H_5)_3)_2$ in CH_2Cl_2 solution while bubbling nitrogen through it. When using an excess of the Pt compound, the Re_2Pt_3 cluster [μ-$(CO)_6$-$Pt_3(P(C_6H_5)_3)_3$]$Re_2(CO)_4$ (see p. 271) simultaneously formed, and flash-chromatographic separation was required for isolating the title compound (yield: 50%). However, when using a deficient amount of the Pt starting material, the title cluster formed exclusively. Yellow solid.

IR spectrum (CH_2Cl_2): 1817, 1850, 1864, 1963, 1991, 2020, 2064 (ν(CO)) cm^{-1}.

The cluster crystallizes in the orthorhombic space group Pcab (non-standard setting of Pbca$-D_{2h}^{15}$ (No. 61)) with a = 18.004(3), b = 11.212(2), c = 46.573(7) Å; Z = 8 formula units per unit cell, D_{calc} = 2.274 g/cm^3. Analogous to the foregoing cluster, the molecular structure consists of an approximately tetrahedral Re_2Pt_2 core (dihedral angle between the $PtRe_2$ planes: 82.2°) with all the Re-Pt edges being asymmetrically bridged by CO (Re-C (av. 2.18 Å) is longer than Pt-C (av. 2.03 Å)). The structure is illustrated in **Fig. 85** [7].

References:

[1] Beck, W.; Raab, K.; Nagel, U.; Sacher, W. (Angew. Chem. **97** [1985] 498/9; Angew. Chem. Int. Ed. Engl. **24** [1985] 505).

[2] Shaposhnikova, A. D.; Stadnichenko, R. A.; Bel'skii, V. K.; Pasynskii, A. A. (Izv. Akad. Nauk SSSR Ser. Khim. **1987** 1913/4; Bull. Acad. Sci. USSR Div. Chem. Sci. [Engl. Transl.] **36** [1987] 1776/7).

[3] Jeffery, J. C.; Parrott, M. J.; Pyell, U.; Stone, F. G. A. (J. Chem. Soc. Dalton Trans. **1988** 1121/9).

[4] Shaposhnikova, A. D.; Stadnichenko, R. A.; Bel'skii, V. K.; Pasynskii, A. A. (Metalloorg. Khim. **1** [1988] 945/51; Organomet. Chem. USSR [Engl. Transl.] **1** [1988] 522/6).

[5] Shaposhnikova, A. D.; Kamalov, G. L.; Stadnichenko, R. A.; Pasynskii, A. A.; Eremenko, I. L.; Struchkov, Yu. T.; Yanovsky, A. I.; Petrovskii, P. V. (J. Organomet. Chem. **378** [1989] 67/72).

[6] Shaposhnikova, A. D.; Stadnichenko, R. A.; Pasynskii, A. A.; Eremenko, I. L.; Yanovskii, A. I.; Struchkov, Yu. T. (Metalloorg. Khim. **2** [1989] 932/3; Organomet. Chem. USSR [Engl. Transl.] **2** [1989] 489/90).

[7] Ciani, G.; Moret, M.; Sironi, A.; Beringhelli, T.; d'Alfonso, G.; Della Pergola, R. (J. Chem. Soc. Chem. Commun. **1990** 1668/70).

[8] Shaposhnikova, A. D.; Kamalov, G. L.; Stadnichenko, R. A.; Pasynskii, A. A.; Eremenko, I. L.; Nefedov, S. E.; Struchkov, Yu. T.; Yanovsky, A. I. (J. Organomet. Chem. **405** [1991] 111/20).

[9] Weidmann, T.; Weinrich, V.; Wagner, B.; Robl, C.; Beck, W. (Chem. Ber. **124** [1991] 1363/8).

[10] Powell, J.; Brewer, J. C.; Gulia, G.; Sawyer, J. F. (J. Chem. Soc. Dalton Trans. **1992** 2503/16).

2.5 Compounds with Ligands Bonded to Rhenium by Five C Atoms (^{5}L Compounds)

2.5.1 Compounds with Terminal ^{5}L Ligands

2.5.1.1 Compounds with One ^{5}LRe Fragment

There are not so many binuclear compounds which bear only one terminal ^{5}L ligand. In all these compounds, there is an additional ^{n}L ligand (n = 1 to 3) connecting the ^{5}LRe fragment with the other Re center. There are no Re-Re bonds.

However, the nonexisting low/high-valent complex **$C_5H_5Re(CO)_2$=Re(≡CH)$(OH)_2$** was theoretically investigated using Fenske-Hall MO calculations. An eclipsed conformation of the

References on p. 164

C_5H_5-Re and HC≡Re bonds was thereby suggested. The computations concluded in that the Re-Re σ bond should be composed of nearly equal contributions from the two fragments while the Re-Re π bonding should be strongly polarized toward the low-valent fragment $C_5H_5Re(CO)_2$. Approximately 0.2 e is transferred from the high-valent to the low-valent fragment [1].

Reference:

[1] Barckholtz, T. A.; Bursten, B. E.; Niccolai, G. P.; Casey, C. P. (J. Organomet. Chem. **478** [1994] 153/60).

2.5.1.1.1 Compounds with the $^5LRe(\mu\text{-}^1L)Re$ Skeleton

$(C_5(CH_3)_5)Re(CO)(NO)(\mu\text{-}\eta^{2:1}\text{-}CO_2)Re(CO)_n(P(C_6H_5)_3)_{4-n}$ (see Formula I, n = 2, 3). Compounds of this type were obtained by treating $(C_5(CH_3)_5)Re(CO)(NO)CO_2H$ with either $(CO)_4Re(P(C_6H_5)_3)FBF_3$ [1] or $(CO)_3Re(P(C_6H_5)_3)_2OC(O)CF_3$ [3] in the presence of Na_2CO_3. Reaction conditions and workup procedures are given with the compounds.

CO, O, Re, Re(CO)$_n$(P(C$_6$H$_5$)$_3$)$_{4-n}$, O, NO

I

CO, O—Re(CO)$_5$, Re, O, NO

II

The compounds were characterized as follows (abbreviations and units on p. X):

n = 2: Synthesis conditions: toluene, 85 °C, 16 h; purification by recrystallization from CH_2Cl_2/hexane. – Orange blocks [3]. – IR (KCl): 1278, 1435 ($\nu(CO_2)$) [2].
Single-crystal data: monoclinic; a = 13.236(3), b = 17.146(5), c = 20.781(6) Å, β = 103.04(2)°; space group $P2_1/n\text{-}C^5_{2h}$ (No. 14); Z = 4 molecules per unit cell, D_{calc} = 1.721 g/cm^3. The structure is shown in **Fig. 86a**. As a remarkable feature, the CO_2 plane is almost perpendicular to the OC-Re-NO plane [3].

n = 3: Synthesis conditions: CH_2Cl_2, warming from 0 °C to room temperature. Workup by filtration, evaporation of the filtrate, and recrystallization from ether. Yield: 71%. Orange crystals; m.p. 208 to 210 °C (dec.).
^{1}H NMR (CD_2Cl_2): 2.08 (s, CH_3), 7.46 (m, C_6H_5). – ^{13}C {^{1}H} NMR (CD_2Cl_2): 10.44 (CH_3), 104.8 (C_5), 128.82 (d; J(P,C) = 9.8), 130.86 (s); 131.55, 134.68, 193.19, 197.23, 197.67 (d's; J(P,C) = 43.8, 11.2, 76.3, 7.2, 7.3); 206.37 (s), 219.17 (d; J(P,C) = 2.1). – ^{31}P {^{1}H} NMR (CD_2Cl_2): 24.43 (s). – IR (KCl): 1282, 1437 ($\nu(CO_2)$); (CH_2Cl_2): 1710, 1900, 1920, 1980, 2020 (ν(CO)).
Single-crystal data: triclinic; a = 11.117(3), b = 16.892(4), c = 10.121(3) Å, α = 104.67(2)°, β = 107.16(2)°, γ = 97.86(2)°; space group $P\bar{1}\text{-}C^1_i$ (No. 2); Z = 2 molecules per unit cell, D_{calc} = 1.86 g/cm^3. The molecular structure is depicted in **Fig. 86b**. Contrasting to the foregoing compound, the CO_2 plane is almost coincident with the Re-N-O plane [1].

$(C_5(CH_3)_5)Re(CO)(NO)(\mu\text{-}\eta^{1:1}\text{-}CO_2)Re(CO)_5$ (see Formula II) formed by combining equimolar amounts of $(C_5(CH_3)_5)Re(CO)(NO)CO_2H$ and $[\pi\text{-}C_2H_4Re(CO)_5]BF_4$ in the presence of

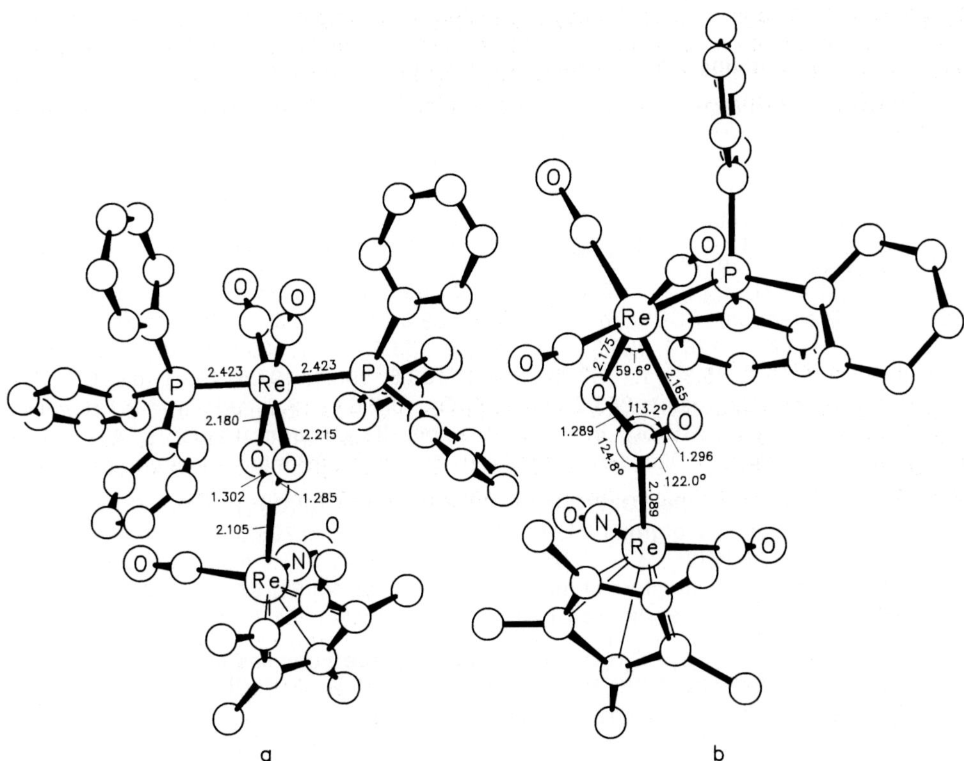

Fig. 86. Molecular structures of $(C_5(CH_3)_5)Re(CO)(NO)(\mu\text{-}\eta^{2:1}\text{-}CO_2)Re(CO)_n(P(C_6H_5)_3)_{4-n}$ (a. n = 2 [3]; b. n = 3 [1]).

excess Na_2CO_3 (acetone, 0 °C, 1 h). After filtration, the filtrate was evaporated and the residue extracted into ether. Evaporation of the extract gave the product with 82% yield.

Orange solid; m.p. 132 °C (dec.). ^1H NMR spectrum (CD_2Cl_2, −10 °C): δ = 2.09 (s) ppm. ^{13}C {^1H} NMR spectrum (CD_2Cl_2, −10 °C): δ = 10.48 (CH_3), 104.46 (C_5); 182.84, 183.49 (CO); 192.72 (CO_2), 210.61 (CO) ppm. IR spectrum (KCl): 1182, 1514 (ν(OCO)) cm^{-1}; (CH_2Cl_2): 1690 (ν(NO)); 1970, 2028, 2138 (ν(CO)) cm^{-1}.

Neither solution- (57 °C, 45 min) nor solid-state thermolysis (105 °C, 30 min) produced a derivative with a bridging $\eta^{2:1}$-coordinated CO_2 ligand; rather, degradation occurred [2].

References:

[1] Gibson, D. H.; Mehta, J. M.; Ye, M.; Richardson, J. F.; Mashuta, M. S. (Organometallics **13** [1994] 1070/2).

[2] Gibson, D. H.; Franco, J. O.; Mehta, J. M.; Mashuta, M. S.; Richardson, J. F. (Organometallics **14** [1995] 5068/72).

[3] Gibson, D. H.; Bardon, R. F.; Mehta, J. M.; Mashuta, M. S.; Richardson, J. F. (Acta Crystallogr. C **52** [1996] 852/4).

2.5.1.1.2 Compounds with the $^5LRe(\mu\text{-}^2L)Re$ Skeleton

2.5.1.1.2.1 The Bridging Ligand is Only C-Bonded

$(C_5(CH_3)_5)Re(NO)(P(C_6H_5)_3)-C(O)CH_2-Re(CO)_4P(C_6H_5)_3$ was obtained by successively treating $(C_5(CH_3)_5)Re(NO)(P(C_6H_5)_3)C(O)CH_3$ with n-C_4H_9Li (THF, −78°C) and $(CO)_4Re(P(C_6H_5)_3)OSO_2CF_3$. After warming the mixture to room temperature, the solvent was removed. Chromatographic workup on silica with hexane/THF (4:3) separated the compound, which was twice recrystallized from $CHCl_3$/hexane. Yield: 22%.

Orange crystals; m.p. 147 to 151 °C (dec.). ^1H NMR spectrum ($CDCl_3$): δ = 1.68 (s, CH_3); 1.86, 2.32 (dd's, CH_2; J_{gem} = 11.7, J(P,H) = 8.2 and 5.8 Hz); 7.25 to 7.39 (m, C_6H_5) ppm. $^{13}C\{^1H\}$ NMR spectrum ($CDCl_3$): δ = 9.8 (CH_3), 39.8 ("t", CH_2; J = 3.0, 2.6 Hz), 101.8 (d, C_5; J = 2.0 Hz); 127.7, 128.4, 129.5, 130.1, 133.6, 133.6, 134.1, 135.6 (d's, C_6H_5; J = 10.1, 10.1, none, 2.0, 10.1, 45.3, 11.1, 53 Hz, resp.); 188.5, 190.1, 192.8, 193.1 (d's, CO; J = 7.0, 53.4, 9.8, 10.1 Hz, resp.); 263.9 (dd; ReC(O); $^2J(P,C)$ = 8.2, $^3J(P,C)$ = 3.1 Hz) ppm. $^{31}P\{^1H\}$ NMR spectrum ($CDCl_3$): δ = 8.5, 23.2 ppm. IR spectrum (CH_2Cl_2): 1431, 1481, 1625 (ν(C=O) + ν(NO)); 1920, 1975, 2075 (ν(CO)) cm^{-1}. FAB mass spectrum: $[MH]^+$ [4].

The solvate **$(C_5(CH_3)_5)Re(NO)(P(C_6H_5)_3)-C(O)CH_2-Re(CO)_4P(C_6H_5)_3 \cdot 0.7\ CH_2Cl_2$** crystallizes in the triclinic space group $P\bar{1}-C_i^1$ (No. 2) with a = 10.708(3), b = 10.846(3), c = 22.163(12) Å, α = 98.82(3)°, β = 93.35(4)°, γ = 90.71(2)°; Z = 2 molecules per unit cell, D_{calc} = 1.702 g/cm³. The structure of the molecule is shown in **Fig. 87**. The ON-Re-C=O torsion angle is almost 180° and the Re-CH_2-C=O torsion angle is 54.7° [4].

$(C_5(CH_3)_5)Re(NO)(P(C_6H_5)_3)-C_3H_2O_2-Re(CO)_4P(CH_3)_3$ (see Formulas Ia, Ib; keto and enol tautomer). $[(C_5(CH_3)_5)Re(NO)(P(C_6H_5)_3)-C(O)CH_2C(O)-Re(CO)_4P(CH_3)_3 \cdot LiO_3SCF_3]_2$ was treated with 4,7,13,18-tetraoxa-1,10-diazabiccylo[8.8.8]eicosane (mole ratio 1:1) in THF for 30 min. The mixture was filtered and evaporated, and the residue was extracted with dibutyl ether. After filtration through Celite and cooling, the crude product precipitated. Recrystallization from hexane/ether gave red and orange crystals which could be separated by hand.

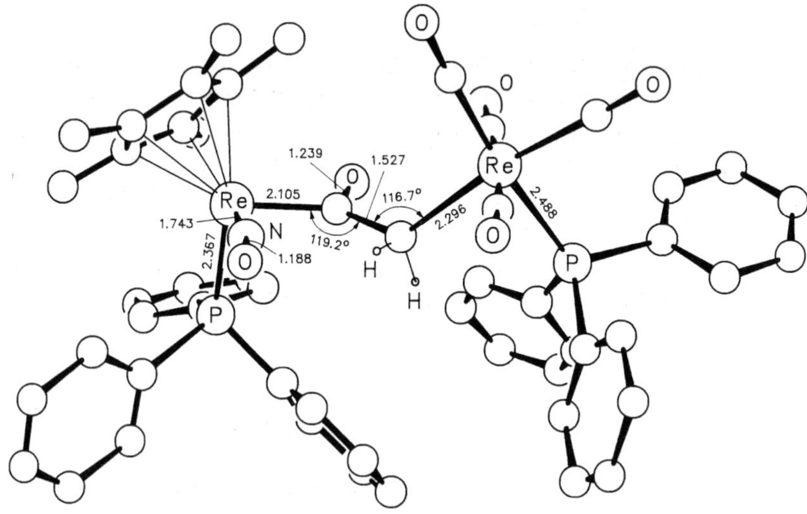

Fig. 87. Molecular structure of $(C_5(CH_3)_5)Re(NO)(P(C_6H_5)_3)-C(O)CH_2-Re(CO)_4P(C_6H_5)_3$ [4].

References on p. 170

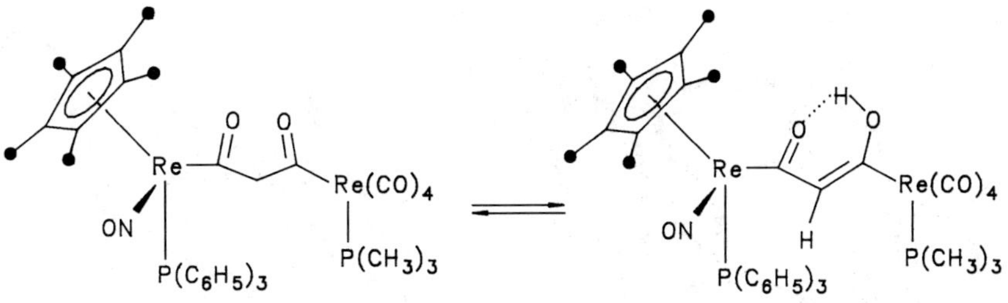

Ia (keto form) ⇌ Ib (enol form)

THF-d_8 solutions of the red or the orange crystals were kept at room temperature. After one week each solution contained an identical mixture of both tautomers. Therefrom the equilibrium constant K_{eq} = [enol tautomer]/[keto tautomer] = 0.65 could be determined [3] (abbreviations and units on p. X).

keto form: Orange crystals. – ^1H NMR ($CDCl_3$): 1.61 (d, PCH_3; J = 9.3), 1.70 ($(CH_3)_5$); 2.73, 3.94 (d's, CH_2; J = 15.3). – $^{13}C\{^1H\}$ NMR ($CDCl_3$, 15 °C): 9.7, 18.1 (d, PCH_3; J(P,H) = 33.0), 102.3 (C_5), 102.5 (CH_2); 128.0. 128.1, 129.9, 133.8 (C_6H_5); 188.3, 188.9, 189.0, 189.1, 190.0, 190.1, 190.2, 190.3 (4 d's, CO); 250.4, 260.4 (d's, **C**(O)CH_2**C**(O); J(P,C) = 10.9 and 9.1).
Single-crystal data: triclinic; a = 8.912(5), b = 11.277(5), c = 23.282(13) Å, α = 78.54(4)°, β = 89.63(4)°, γ = 67.99(4)°; space group $P\bar{1}-C_i^1$ (No. 2); Z = 2 molecules per unit cell, D_{calc} = 1.657 g/cm^3. The structure of the molecule is depicted in **Fig. 88a**. Two torsion angles are noteworthy: C-C-C=O: 91°, P-Re···Re-P: 133.7°. The twist of the C=O bond out of the Re-C-O plane places the two malonyl O atoms at a 3.38 Å nonbonding distance [3].

enol form: Red crystals. – ^1H NMR (THF-d_8): 1.62 (d, PCH_3; J = 9.2), 1.69 ($(CH_3)_5$), 6.10 (CH), 16.44 (O···H···O). – $^{13}C\{^1H\}$ NMR (THF-d_8): 101.5 (C_5); 127.8, 127.9, 129.6, 140.2 (m, CH=), 195.1 (dd, ReCO; J = 12.6, 5.0), 249.3 (d, PRe**C**(O)CH=; J(P,C) = 9.2).
Single-crystal data: monoclinic; a = 25.163(4), b = 9.631(10), c = 16.605(4) Å, β = 96.76(2)°; space group $P2_1/c-C_{2h}^5$ (No. 14); Z = 4 molecules per unit cell, D_{calc} = 1.759 g/cm^3. The structure of the molecule is depicted in **Fig. 88b**. Compared with the keto form, the torsion angles P-Re···Re-P (15.2°) and C-C-C=O (9.6°) are remarkably different. Furthermore, the enol ligand exists as a localized structure because both C=O bond lengths are unequal [3].

Li[($C_5(CH_3)_5$)Re(NO)($P(C_6H_5)_3$)–C(O)CH_2C(O)–Re$(CO)_4$Br] · 2 C_4H_8O (see Formula II) was obtained by treating ($C_5(CH_3)_5$)Re(NO)($P(C_6H_5)_3$)C(O)CH_2–Li with $(CO)_5$ReBr in THF at –78 °C. The solution was warmed to room temperature and concentrated. Addition of hexane precipitated the compound with 79% yield. Yellow, air-stable crystals; m.p. 185 to 200 °C (dec.). The solvent-free compound could be obtained by twice precipitating the THF adduct from CH_2Cl_2/hexane.

^1H NMR spectrum (THF-d_8): δ = 1.72 (s); 2.29, 5.90 (d's, CH_2; J_{gem} = 15.6 Hz); 7.41 (C_6H_5) ppm. IR spectrum (KBr): 1433, 1480, 1518, 1649; 1914, 1976, 2085 (ν(CO)) cm^{-1} [2].

Exposure to HCl gas gave a mixture of [($C_5(CH_3)_5$)Re(NO)($P(C_6H_5)_3$)=C(OH)CH_3]Cl, $(CO)_5$ReBr, CH_3(HO)C=Re$(CO)_4$Br, and [($C_5(CH_3)_5$)Re(NO)($P(C_6H_5)_3$)CO]Cl [5]. Treatment with $MgBr_2$ · $O(C_2H_5)_2$ in CH_2Cl_2 yielded [MgBr][($C_5(CH_3)_5$)Re(NO)($P(C_6H_5)_3$){C(O)CH_2-

References on p. 170

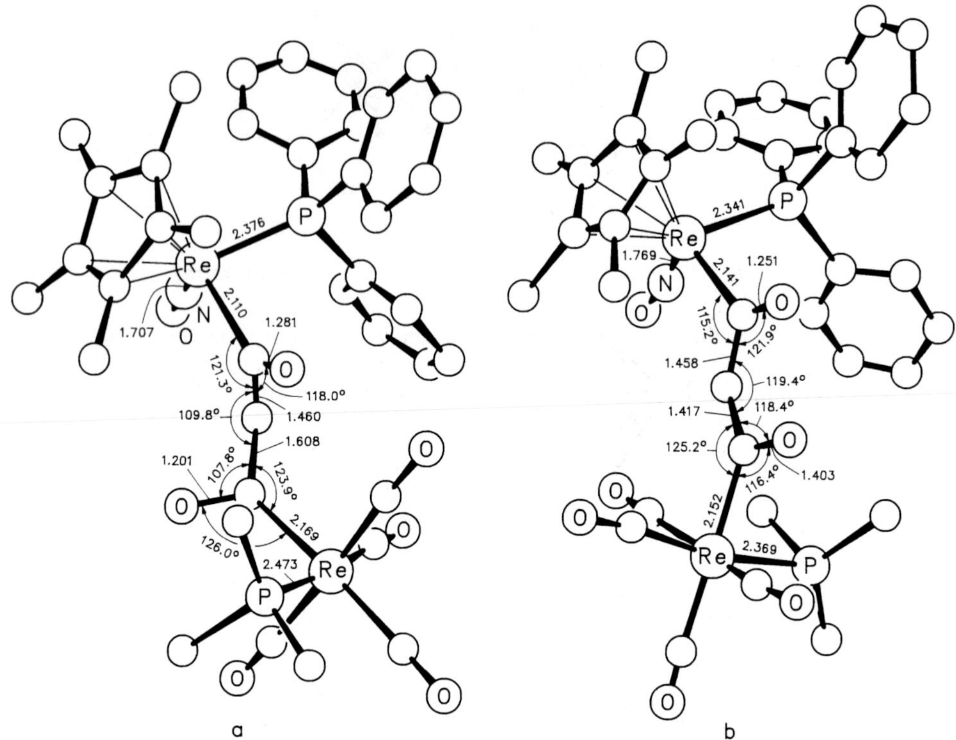

Fig. 88. Molecular structure of
$(C_5(CH_3)_5)Re(NO)(P(C_6H_5)_3)-C_3H_2O_2-Re(CO)_4P(CH_3)_3$ (a. keto form; b. enol form [3]).

C(O)}Re(CO)$_4$Br]; otherwise, the compound slowly decomposed in the presence of 12-crown-4, giving a mixture of $(C_5(CH_3)_5)Re(NO)(P(C_6H_5)_3)\{C(O)CH_2C(O)\}Re(CO)_4$, $(CO)_5$-ReBr, and $(C_5(CH_3)_5)Re(NO)(P(C_6H_5)_3)C(O)CH_3$ [2]. Addition of $(CH_3)_2NC_2H_4N(CH_3)_2$ formed the following diamine adduct [2]. Treatment with CH_3I for at least 1 d gave the two compounds $(C_5(CH_3)_5)Re(NO)(P(C_6H_5)_3)\{C(O)C(R)(CH_3)C(O)\}Re(CO)_4$ (R = H, CH_3) and other unidentified products [5].

^{2}D ^{2}D Li O O H H Re Re(CO)$_4$ ON P(C$_6$H$_5$)$_3$ Br

II

MgBr O O H H Re Re(CO)$_4$ ON P(C$_6$H$_5$)$_3$ Br

III

$Li[(C_5(CH_3)_5)Re(NO)(P(C_6H_5)_3)-C(O)CH_2C(O)-Re(CO)_4Br]\cdot N_2C_6H_{16}$ (see Formula II) formed by adding 2 equivalents $(CH_3)_2NC_2H_4N(CH_3)_2$ to a solution of the preceding Li salt.

References on p. 170

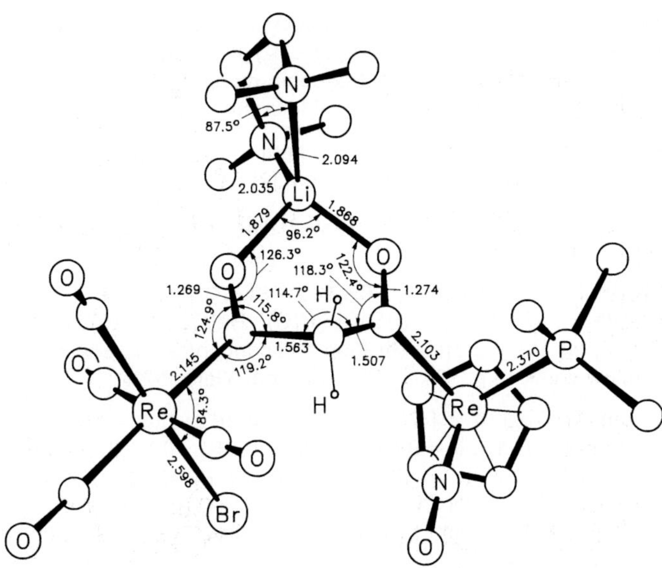

Fig. 89. Molecular structure of $Li[(C_5(CH_3)_5)Re(NO)(P(C_6H_5)_3)-C(O)CH_2C(O)-Re(CO)_4Br] \cdot N_2C_6H_{16}$ (C_6H_5 groups at P represented by C_{ipso}; CH_3 groups at C_5 not shown) [2].

^1H NMR spectrum (CD_2Cl_2): δ = 1.70 (s); 1.93, 5.82 (d's; J_{gem} = 15.1 Hz); 7.40 (m) ppm. ^{13}C NMR spectrum (CD_2Cl_2, 14 °C): δ = 9.9 (q, $(CH_3)_5$); 45.8 (qm) and 54.4 (tm, CH_2NCH_3); 97.0 (dd, CH_2; J = 138.1, 120.8 Hz), 102.8 (C_5); 128.4, 128.6, 130.6, 133.9 (C_6H_5); 189.4, 189.5, 190.2, 190.6 (CO); 270.7 (C=O), 270.8 (d, PReC(O); J(P,C) = 9.5 Hz) ppm. IR spectrum (CH_2Cl_2): 1434, 1460, 1467; 1482, 1560 (ν(C=O)); 1648 (ν(NO)); 1910, 1985, 2085 (ν(CO)) cm^{-1} [2].

At −100 °C, the adduct crystallizes in the monoclinic space group $P2_1/n-C^5_{2h}$ (No. 14) with a = 10.977(5), b = 28.094(15), c = 17.204(8) Å, β = 95.73(4)°; Z = 4 formula units per unit cell; D_{calc} = 1.74 g/cm³. The structure is depicted in **Fig. 89**. The 6-membered LiO_2C_3 ring adopts a slightly twisted boat conformation [2].

$Li[(C_5(CH_3)_5)Re(NO)(P(C_6H_5)_3)-{}^{13}C(O)CH_2C(O)-Re(CO)_4Br] \cdot N_2C_6H_{16}$ was obtained by an analogous reaction starting from $(C_5(CH_3)_5)Re(NO)(P(C_6H_5)_3){}^{13}C(O)CH_2-Li$ [2].

M.p. 180 to 192 °C. ^1H NMR spectrum (CD_2Cl_2): δ = 1.70 ($(CH_3)_5$), 1.93 (dd; J(H,H) = 15.2, J(C,H) = 5.3 Hz); 2.20, 2.39 (br s's, CH_2NCH_3); 5.81 (d; J(H,H) = 15.2; J(C,H) < 1 Hz), 7.40 (C_6H_5) ppm. $^{13}C\{^1H\}$ NMR spectrum (CD_2Cl_2): δ = 9.9, 45.8, 57.4, 97.0 (d; J(C,C) = 18.5 Hz), 102.8; 128.5, 128.6, 130.6, 134.0; 189.4, 189.5, 190.2, 190.6; 270.8 (d; J(P,C) = 9.6 Hz) ppm; ^1H-coupled: 270.8 (dd; J(P,C) = 9.6, J(C,H) = 5.1 Hz). IR spectrum (CH_2Cl_2): 1435, 1460, 1467, 1561, 1649, 1910, 1985, 2085 cm^{-1} [2].

$Li[(C_5(CH_3)_5)Re(NO)(P(C_6H_5)_3)-C(O)CH_2\ {}^{13}C(O)-Re({}^{13}CO)_4Br] \cdot N_2C_6H_{16}$ was similarly obtained when starting from $({}^{13}CO)_5ReBr$ [2].

M.p. 182 to 192 °C (dec.). ^1H NMR spectrum (CD_2Cl_2): δ = 1.70 (s), 1.93 (dd; J(H,H) = 15.2, J(C,H) = 4.8 Hz), 5.81 (d; J(H,H) = 15.2 Hz, J(C,H) < 1 Hz), 7.40 (m) ppm. $^{13}C\{^1H\}$ NMR spectrum (CD_2Cl_2): δ = 270.7 (s) ppm. IR spectrum (CH_2Cl_2): 1433, 1460, 1467, 1480, 1525, 1560, 1649, 1866, 1900, 1940, 1960, 2040, 2065 cm^{-1} [2].

References on p. 170

$Li[(C_5(CH_3)_5)Re(NO)(P(C_6H_5)_3)\{C(O)CH_2C(O)\}Re(CO)_4OSO_2CF_3]$ was suggested to be the intermediate species in the reaction of $(C_5(CH_3)_5)Re(NO)(P(C_6H_5)_3)C(O)CH_2-Li$ with $(CO)_5ReOSO_2CF_3$ in THF at −80°C. Spectroscopic monitoring during the reaction in THF recorded the following values attributed to the title compound.

^1H NMR spectrum: δ = 5.58 (d; J = 16 Hz) ppm. $^{13}C\{^1H\}$ NMR spectrum: δ = 267.5, 271.6 ppm. $^{31}P\{^1H\}$ NMR spectrum: δ = 17.9 ppm [2].

Addition of $P(CH_3)_3$ at −80°C yielded $[(C_5(CH_3)_5)Re(NO)(P(C_6H_5)_3)\{C(O)CH_2C(O)\}Re(CO)_4P(CH_3)_3 \cdot LiO_3SCF_3]_2$ [1, 2].

$[MgBr][(C_5(CH_3)_5)Re(NO)(P(C_6H_5)_3)-C(O)CH_2C(O)-Re(CO)_4Br]$ (see Formula III) formed when $Li[(C_5(CH_3)_5)Re(NO)(P(C_6H_5)_3)-C(O)CH_2C(O)-Re(CO)_4Br]$ was treated with 1 equivalent $MgBr_2$ in CH_2Cl_2 for 10 min. After filtration and concentration, addition of hexane separated the bright yellow solid; m.p. 184 to 191°C (dec.). Yield: 57%.

^1H NMR spectrum (THF-d_8): δ = 1.73 (s); 2.58, 6.42 (d's; J_{gem} = 15.9 Hz); 7.42 (m) ppm. $^{13}C\{^1H\}$ NMR spectrum (THF-d_8): δ = 9.5 (CH_3), 97.2 (CH_2), 103.3 (C_5); 129.0, 130.8, 132.7, 134.6 (C_6H_5); 188.7, 189.5, 189.6, 189.7 (CO); 273.1 (d, PReC(O); J(P,C) = 7.3 Hz), 283.1 (BrReC(O)) ppm. IR spectrum (CH_2Cl_2): 1377, 1440, 1479; 1661 (ν(NO)); 1941, 2005, 2102 (ν(CO)) cm^{-1}. Molecular weight by osmometry in CH_2Cl_2: 1233 (calc. 1186) g/mol [2].

$[(C_5(CH_3)_5)Re(NO)(P(C_6H_5)_3)-C(O)CH_2C(O)-Re(CO)_4P(CH_3)_3 \cdot LiO_3SCF_3]_2$ was prepared from $(C_5(CH_3)_5)Re(NO)(P(C_6H_5)_3)C(O)CH_2-Li$ and $(CO)_5ReOSO_2CF_3$ in the presence of a large excess of $P(CH_3)_3$ in THF or also independently from the Li compound and $[(CO)_5Re-P(CH_3)_3]OSO_2CF_3$ (yield: 42%) [1]. Because the compound is a solid-state dimer, it will be described in full in the chapter "Tetranuclear Compounds".

References:

[1] O'Connor, J. M.; Uhrhammer, R.; Rheingold, A. L.; Staley, D. L. (J. Am. Chem. Soc. **111** [1989] 7633/4).

[2] O'Connor, J. M.; Uhrhammer, R.; Rheingold, A. L.; Staley, D. L.; Chadha, R. K. (J. Am. Chem. Soc. **112** [1990] 7585/98).

[3] O'Connor, J. M.; Uhrhammer, R.; Rheingold, A. L.; Roddick, D. M. (J. Am. Chem. Soc. **113** [1991] 4530/44).

[4] O'Connor, J. M.; Uhrhammer, R.; Chadha, R. K. (Polyhedron **12** [1993] 527/32).

[5] O'Connor, J. M.; Uhrhammer, R.; Chadha, R. K.; Tsuie, B. (J. Organomet. Chem. **455** [1993] 143/56).

2.5.1.1.2.2 The Bridging Ligand is C- and O-Bonded

General. Each of the compounds described here has one $(C_5(CH_3)_5)Re(NO)P(C_6H_5)_3$ and one $Re(CO)_4$ fragment, which are connected by a substituted malonyl ligand which in addition is coordinatively bonded to that Re atom bearing the CO ligands. The structures and interconversions of the compounds are shown in Scheme 3. The complete reaction sequence I → II → III, IV exists for the cases R = H and R = CH_3.

$(C_5(CH_3)_5)Re(NO)(P(C_6H_5)_3)\{C(O)CH_2C(O)\}Re(CO)_4$ (see Formula Ia, R = H) was prepared by combining $(C_5(CH_3)_5)Re(NO)(P(C_6H_5)_3)C(O)CH_3$ with n-C_4H_9Li and $(CO)_5ReOSO_2CF_3$ in THF at −78°C followed by slowly warming the solution to room temperature. After concentration, addition of hexane precipitated an initial portion. Another quantity was obtained by evaporating and reprecipitating the residue from CH_2Cl_2/hexane. The combined yield was

References on p. 179

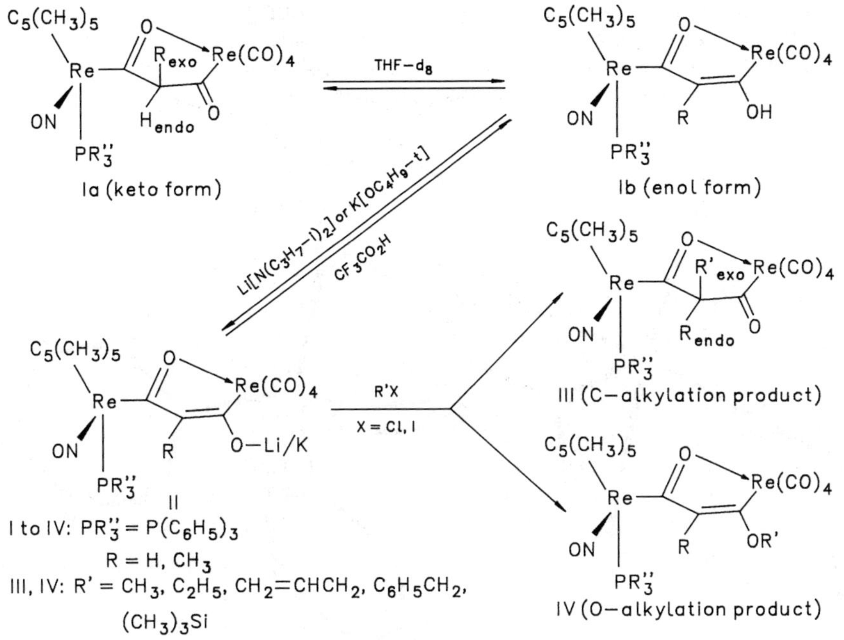

Ia (keto form)

Ib (enol form)

II

III (C–alkylation product)

IV (O–alkylation product)

I to IV: $PR''_3 = P(C_6H_5)_3$
$R = H, CH_3$
III, IV: $R' = CH_3, C_2H_5, CH_2{=}CHCH_2, C_6H_5CH_2, (CH_3)_3Si$

Scheme 3

71% [1, 5]. An analogous treatment of $(C_5(CH_3)_5)Re(NO)(P(C_6H_5)_3)^{13}C(O)CH_3$ led to the specifically labeled derivative **$(C_5(CH_3)_5)Re(NO)(P(C_6H_5)_3)\{^{13}C(O)CH_2C(O)\}Re(CO)_4$** [4, 5]. The complex was also formed with 77% yield along with $(C_5(CH_3)_5)Re(NO)(P(C_6H_5)_3)C(O)CH_3$ and $(CO)_5ReBr$ by treating $Li[(C_5(CH_3)_5)Re(NO)(P(C_6H_5)_3)-C(O)CH_2C(O)-Re(CO)_4Br]$ with 12-crown-4 (CH_2Cl_2, 119 h) [5]. It is an air-stable, yellow solid [1, 5]; m.p. 195 to 200°C [5].

^1H NMR spectrum (nonlabeled, THF-d_8): δ = 1.76 (CH_3); 2.47, 3.13 (d's, CH_2); 7.4 to 7.9 ppm [1, 5]; (^{13}C-labeled, $CDCl_3$): δ = 1.75 (CH_3); 2.49, 3.13 (dd's, $^{13}CH_2$; J_{gem} = 20.6 Hz, $^2J(C,H)$ = 2.2 and 1.6 Hz, resp.); 7.30, 7.43 (br s's, C_6H_5) ppm [5] (see also [4]). $^{13}C\{^1H\}$ NMR spectrum (nonlabeled, $CDCl_3$): δ = 9.9 (CH_3), 95.7 (CH_2), 103.6 (C_5); 128.7, 128.8, 130.9, 133.3 (C_6H_5); 190.9, 191.4, 191.6, 194.9 (CO); 275.6 ($(CO)_4Re$**C**(O)), 289.8 (d, PReC(O); J(P,C) = 7.4 Hz) ppm [1, 3, 5]. $^{31}P\{^1H\}$ NMR spectrum (nonlabeled, THF-d_8): δ = 19.4 (s) ppm [3, 5, 6]. IR spectrum (CH_2Cl_2, nonlabeled): 1344, 1374, 1394, 1432, 1480, 1615 (ν(C=O)), 1664 (ν(NO)); 1923, 1972, 2080 (ν(CO)) cm^{-1}; (^{13}C-labeled): 1336 (shifted from 1374), 1365 (shifted from 1394), 1390, 1418, 1434, 1480, 1617 (ν(C=O)); 1665 (ν(NO)); 1925, 1972, 2080 (ν(CO)) cm^{-1} [5]. FAB mass spectrum: $[M]^+$ [5].

The spectroscopic data are consistent with a carbenoid character of the Re-C bonds (see Formulas Va to Vc; [Re] = $(C_5(CH_3)_5)Re(NO)P(C_6H_5)_3$, [Re′] = $Re(CO)_4$) [5].

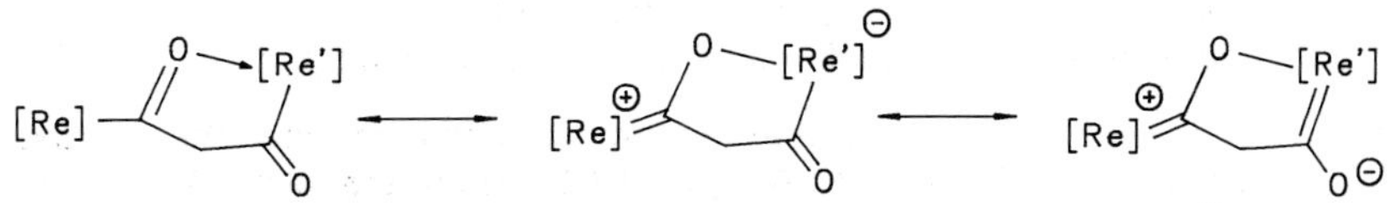

Va Vb Vc

References on p. 179

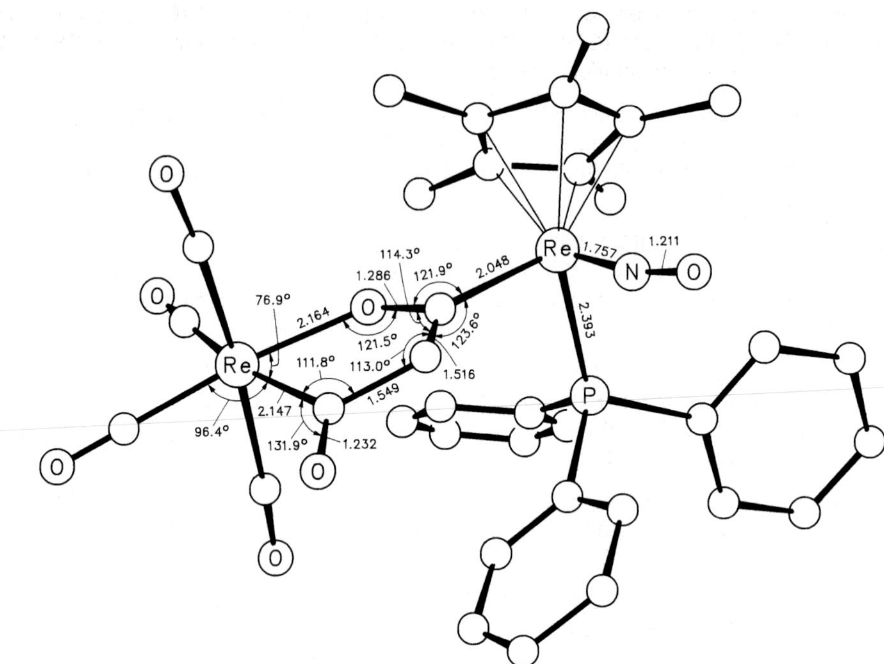

Fig. 90. Molecular structure of $(C_5(CH_3)_5)Re(NO)(P(C_6H_5)_3)\{C(O)CH_2C(O)\}Re(CO)_4$ [5].

The compound crystallizes in the triclinic space group $P\bar{1}-C_i^1$ (No. 2) with a = 10.5112(11), b = 11.0281(17), c = 16.7597(22) Å, α = 74.860(9)°, β = 83.512(8)°, γ = 68.508(8)°; Z = 2 molecules per unit cell, D_{calc} = 1.87 g/cm^3. The structure of the molecule is illustrated in **Fig. 90**. The ON-Re-C=O torsion angle is 178.6° which places the acyl C carbonyl anti to the NO ligand [5].

The compound is soluble in CH_2Cl_2, $CHCl_3$, and THF but insoluble in ether, benzene, CH_3CN, or acetone [5]. In $CHCl_3$ or CH_2Cl_2, the compound exclusively exists in the keto form (even in the presence of added H_2O) [6]; in contrast, in THF-d_8 one observes partial enolization to $(C_5(CH_3)_5)Re(NO)(P(C_6H_5)_3)\{C(O)CH{=}C(OH)\}Re(CO)_4$ (by NMR) [3, 5, 6]. The equilibrium constant K_{eq} = [enol form]/[keto form] is 0.66 at 23 °C [3, 5, 6] and 2.3 at −80 °C [6]. When the THF-d_8 is replaced by $CDCl_3$, the ^{1}H NMR spectrum again indicated only the presence of the keto form [3, 5].

Treatment with D_2O in THF incorporated deuterium in both positions of the CH_2 group (the enol form is also affected; see below), but the H atoms are exchanged at significantly different rates [3, 6]. H_{exo} undergoes H/D exchange ca. 30 times faster than H_{endo} ($t_{1/2}$ ca. 4 min vs. 120 min; 95% formation of the monodeuterated product is observed before detectable conversion into the bideuterated product occurred) [6]. Conversely, addition of H_2O to the deuterated derivative in CD_2Cl_2 reformed the protio compound [3, 6].

When treated with HCl gas, NMR-monitoring demonstrated the formation of $[(C_5(CH_3)_5)Re(NO)(P(C_6H_5)_3)\{C(O)CH_2C(OH)\}Re(CO)_4]Cl$, but the title compound could be largely recovered after evaporation of the mixture. However, long-term exposure to a large

References on p. 179

excess of HCl gas initiated degradation to $[(C_5(CH_3)_5)Re(NO)(P(C_6H_5)_3)CO]Cl$ and $[(C_5(CH_3)_5)Re(NO)(P(C_6H_5)_3)=C(OH)CH_3]Cl$ (ratio 9:91); a small quantity of cis-$CH_3(HO)$-C=$Re(CO)_4Cl$ (tentative composition) was also observed [7].

The compound did not react with $P(CH_3)_3$ [5]. Treatment with $Li[N(C_3H_7\text{-}i)_2]$ [1], $Li[OC_4H_9\text{-}t]$ [6], or $K[OC_4H_9\text{-}t]$ [2, 7] generated the respective enolate $(C_5(CH_3)_5)Re(NO)(P(C_6H_5)_3)\{C(O)CH=C(OM)\}Re(CO)_4$ (see below). The reaction could be reversed with CF_3CO_2H at −80 °C [6].

$(C_5(CH_3)_5)Re(NO)(P(C_6H_5)_3)\{C(O)CH=C(OH)\}Re(CO)_4$ (see Formula Ib, R = H) is the enol form of $(C_5(CH_3)_5)Re(NO)(P(C_6H_5)_3)\{C(O)CH_2C(O)\}Re(CO)_4$. Both tautomers simultaneously exist in THF solution; for the equilibrium constant see the description of the keto form. The enol tautomer was spectroscopically characterized as follows (solvent: THF-d_8):

^1H NMR spectrum: δ = 1.73 (CH_3), 6.28 (CH=), 8.91 (=COH) ppm [3, 5, 6]. $^{13}C\{^1H\}$ NMR spectrum (−80 °C): δ = 10.1 (CH_3), 103.2 (C_5); 128.7 to 139 (C_6H_5); 189.6, 191.3, 193.3, 194.7 (CO); 224.0 (=COH), 270.3 (d, PRe**C**(O); J(P,C) = 8.5 Hz) ppm [6]. $^{31}P\{^1H\}$ NMR spectrum: δ = 22.7 ppm [3, 5, 6].

Treatment with D_2O in THF-d_8 initiated complete H/D exchange of the hydroxyl H atom within 3 min; whereas H/D exchange in the vinyl site of the enol form occurred at a significantly slower rate ($t_{1/2}$ ca. 2 h) [3, 6].

$[(C_5(CH_3)_5)Re(NO)(P(C_6H_5)_3)\{C(O)CH_2C(OH)\}Re(CO)_4]Cl$ formed with ca. 87% yield by treating $(C_5(CH_3)_5)Re(NO)(P(C_6H_5)_3)\{C(O)CH_2C(O)\}Re(CO)_4$ with excess gaseous HCl in $CDCl_3$. The compound was not isolated. Evaporation largely formed the parent complex again.

^1H NMR spectrum: δ = 1.77 (s), 2.96 (br s, H_{endo}), 3.57 (br s, H_{exo}) ppm. $^{13}C\{^1H\}$ NMR spectrum ($CDCl_3$, −40 °C): δ = 9.9 ($(CH_3)_5$), 88.7 (CH_2), 104.3 (C_5); 128.6 to 135.5 (C_6H_5); 185.4, 185.9, 186.4, 191.3 (CO); 286.6 (d, PReC(O); J(P,C) = 8.8 Hz), 321.7 (COH) ppm. IR spectrum (CH_2Cl_2): 1343, 1372, 1391, 1481, 1615, 1667, 1923, 1951, 2000, 2045, 2085, 2100 cm^{-1} [7].

$(C_5(CH_3)_5)Re(NO)(P(C_6H_5)_3)\{C(O)CH=C(OM)\}Re(CO)_4$ (see Formula II; R = H, M = Li, K) formed by treating $(C_5(CH_3)_5)Re(NO)(P(C_6H_5)_3)\{C(O)CH_2C(O)\}Re(CO)_4$ with $Li[N(C_3H_7\text{-}i)_2]$ [1] or $M[OC_4H_9\text{-}t]$ [2, 6, 7] in THF-d_8. The Li salt was characterized as follows:

^1H NMR spectrum (THF-d_8): δ = 1.72 (s, CH_3), 5.98 (br s, CH); 7.32, 7.54 (br s's, C_6H_5) ppm. $^{13}C\{^1H\}$ NMR spectrum (THF-d_8): δ = 250 (COLi) ppm. IR spectrum (KBr): 1405 (C-OLi), 1631 (ν(NO)); 1895, 1942, 2060 (ν(CO)) cm^{-1} [6].

Treatment with CF_3CO_2H (THF-d_8, −80 °C) gave the parent compound again. Similarly, treatment with methyl acetoacetate only caused protonation of the anion. Treatment with 1 equivalent dimethyl malonate only caused the protonation of 59% of the anion [6].

Alkylation with R′X (R′ = CH_3, CD_3, C_2H_5, CH_2=$CHCH_2$, $C_6H_5CH_2$; X = Cl, I) gave the two principal products $(C_5(CH_3)_5)Re(NO)(P(C_6H_5)_3)\{C(O)CH(R')C(O)\}Re(CO)_4$ and $(C_5(CH_3)_5)Re(NO)(P(C_6H_5)_3)\{C(O)CH=C(OR')\}Re(CO)_4$ via a C- and O-alkylation, respectively [2, 6, 7]. Methylation could also be carried out with $[O(CH_3)_3]BF_4$, with which the O-alkylation product preferably formed [2]. Treatment with $(CH_3)_3SiCl$ gave $(C_5(CH_3)_5)Re(NO)(P(C_6H_5)_3)\{C(O)CH=C(OSi(CH_3)_3)\}Re(CO)_4$ exclusively [1, 6].

$(C_5(CH_3)_5)Re(NO)(P(C_6H_5)_3)\{C(O)CH(R')C(O)\}Re(CO)_4$ (see Formula III, R_{endo} = H). These are the C-alkylation products obtained by treating in situ-prepared $(C_5(CH_3)_5)Re(NO)(P(C_6H_5)_3)\{C(O)CH=C(OM)\}Re(CO)_4$ (M = Li, K) with R′X (R′ = CH_3, CD_3, C_2H_5, CH_2=$CHCH_2$; X = Cl, I) in THF [7].

References on p. 179

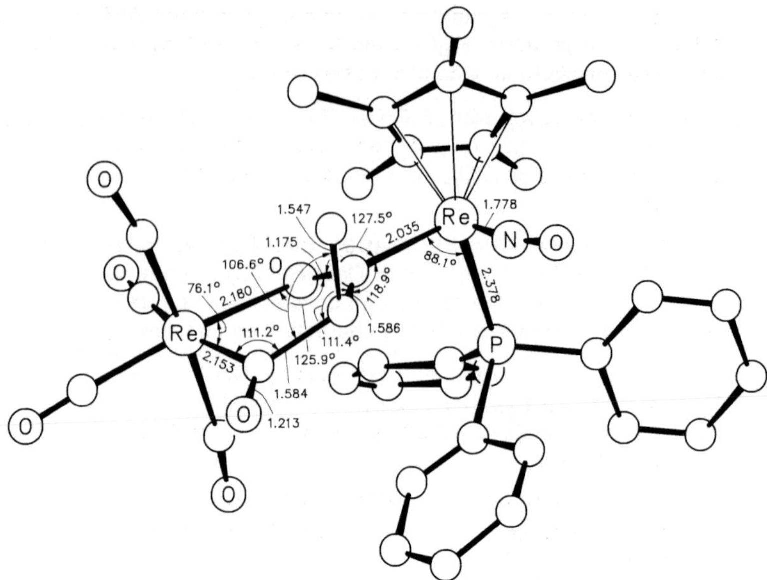

Fig. 91. Molecular structure of $(C_5(CH_3)_5)Re(NO)(P(C_6H_5)_3)\{C(O)CH(CH_3)C(O)\}Re(CO)_4$ [2, 7].

Several compounds are in an equilibrium with the respective enol tautomer, but an equilibrium constant is only given for the derivative with R′ = CH_3.

The following compounds were prepared and characterized:

R′	yield; other preparations; spectroscopic and structural properties; remarks (for abbreviations and units see p. X)
CH_3	Yield: 69% (with $Li[N(C_3H_7\text{-}i)_2]$ [1]), 91% (with $K[OC_4H_9\text{-}t]$ [2, 6, 7]); also from $Li[(C_5(CH_3)_5)Re(NO)(P(C_6H_5)_3)\{C(O)CH_2C(O)\}Re(CO)_4Br]$ (see p. 167) and CH_3I (THF, ca. 1 d) with less than 10% yield [7]. – Yellow [7] to orange [1] crystals; m.p. 196 to 200°C (dec.) [2, 7]. ^1H NMR (THF-d_8): 1.01 (d, CCH_3), 1.80 ($(CH_3)_5$), 2.07 (q; CH), 7.41 (br m, C_6H_5); J(CH,CH_3) = 7.9 [1, 6] (similar in CD_2Cl_2 [7]). – $^{13}C\{^1H\}$ NMR (CD_2Cl_2): 10.1, 15.4, 94.7, 104.2; 128.9, 129.1, 131.3 (C_6H_5); 134.5 (br m); 191.8, 192.7, 192.9, 195.0 (CO); 276.5 ($(CO)_4Re\mathbf{C}(O)$), 298.7 (d, PReC(O); J(P,C) = 7) [2, 7] (slightly different in THF-d_8 [6]). – IR (CH_2Cl_2): 1602, 1654, 1917, 1962, 2075 (ν(CO)) [2, 7]. – FAB MS: $[M]^+$ [2, 7]. Single-crystal data: triclinic; a = 10.295(3), b = 11.096(3), c = 17.367(6) Å, α = 73.30(6)°, β = 87.39(2)°, γ = 70.62(2)°; space group P1–C_1^1 (No. 1); Z = 2 molecules per unit cell, D_{calc} = 1.848 g/cm^3. The ON-Re-C=O torsion angle is 179°. The CH_3 group sits on the less congested exo-face of the nearly planar chelate ring (compare with **Fig. 91**) [2, 7]. In $CDCl_3$ the complex exclusively exists in the keto form [2], whereas in THF-d_8 the compound is in an equilibrium with the enol tautomer

R′	yield; other preparations; spectroscopic and structural properties; remarks (for abbreviations and units see p. X)
	$(C_5(CH_3)_5)Re(NO)(P(C_6H_5)_3)\{C(O)C(CH_3)=C(OH)\}Re(CO)_4$ (see below). The equilibrium constant K_{eq} = [enol form]/[keto form] = 0.1 was established at 23 °C [2, 6], but at −80 °C the enol form almost exclusively exists [6]. Slow H/D exchange in the acidic H site occurred ($t_{1/2}$ >2 weeks) upon treatment with excess D_2O in THF-d_8 [2, 6]. Addition of $K[OC_4H_9\text{-}t]$ to the solution accelerated the rate of D incorporation ($t_{1/2}$ ca. 4.5 h) [2]. The enolate $(C_5(CH_3)_5)Re(NO)(P(C_6H_5)_3)\{C(O)C(CH_3)=C(OK)\}Re(CO)_4$ formed upon treatment with $K[OC_4H_9\text{-}t]$. Reprotonation of the enolate with CF_3CO_2H at −76 °C initially gave the enol tautomer, which slowly converted into the keto form upon warming [2, 6]. Alkylation of the enolate with CH_3I yielded $(C_5(CH_3)_5)Re(NO)(P(C_6H_5)_3)\{C(O)C(CH_3)_2C(O)\}Re(CO)_4$ and $(C_5(CH_3)_5)Re(NO)(P(C_6H_5)_3)\{C(O)C(CH_3)=C(OCH_3)\}Re(CO)_4$ (by-product) [2, 7], but using $[O(CH_3)_3]BF_4$ gave a larger quantity of the latter product [7]. Alkylation with CD_3I yielded two diastereomers of $(C_5(CH_3)_5)Re(NO)(P(C_6H_5)_3)\{C(O)C(CH_3)(CD_3)C(O)\}Re(CO)_4$ in a 38:1 ratio (see p. 177) [2, 7]. Alkylation with C_2H_5I only provided $(C_5(CH_3)_5)Re(NO)(P(C_6H_5)_3)\{C(O)C(CH_3)C(OC_2H_5)\}Re(CO)_4$ [7].
CD_3	^1H NMR ($CDCl_3$): 1.76, 2.17 (br s), 7.41. − ^2H {^1H} NMR: 1.00. − IR (CH_2Cl_2): 1610, 1660, 1930, 1975, 2080. Successive treatment with $K[OC_4H_9\text{-}t]$ and CH_3I gave mainly $(C_5(CH_3)_5)Re(NO)(P(C_6H_5)_3)\{C(O)C(CD_3\text{-endo})(CH_3\text{-exo})C(O)\}Re(CO)_4$ [7].
C_2H_5	Yield: 10%. − Orange solid; m.p. 182 to 191 °C (dec.). ^1H NMR ($CDCl_3$): 1.28 (t; J = 7.1), 1.76 ($(CH_3)_5$); 3.74, 3.89 (8 line m's; CCH_2); 7.37 (br s, C_6H_5). − ^{13}C {^1H} NMR ($CDCl_3$): 9.8, 9.9, 67.8, 102.8; 128.0, 128.1, 130.1, 133.8, 135.6 (C_6H_5); 188.7, 189.5, 193.1, 193.3 (CO); 224.4, 270.4 (d; J(P,C) = 9.3). − IR (CH_2Cl_2): 1380, 1384, 1645, 1914, 1979, 2070. − FAB MS: $[M]^+$ [7].
$CH_2CH=CH_2$	Formed when using $CH_2=CHCH_2Br$. The compound was not isolated, NMR integration indicated a yield of 27% [7].

$(C_5(CH_3)_5)Re(NO)(P(C_6H_5)_3)\{C(O)C(CH_3)=C(OH)\}Re(CO)_4$ (see Formula Ib, R = CH_3) is the enol tautomer of $(C_5(CH_3)_5)Re(NO)(P(C_6H_5)_3)\{C(O)CH(CH_3)C(O)\}Re(CO)_4$. Both forms simultaneously exist in THF at room temperature [2, 6], but upon lowering the temperature to −80°C the enol form clearly predominated [6]. It was exclusively formed by treating the enolate $(C_5(CH_3)_5)Re(NO)(P(C_6H_5)_3)\{C(O)C(CH_3)=C(OK)\}Re(CO)_4$ with CF_3CO_2H at −80 °C, but then slowly converted into the keto tautomer upon warming the mixture [2].

^1H NMR spectrum (THF-d_8): δ = 1.97 (CCH_3), 8.92 (COH) ppm. ^{13}C {^1H} NMR spectrum (toluene-d_8): δ = 219.3 (=COH), 279.1 (d, PRe**C**(O)CH; J(P,C) = 9.1 Hz) ppm [6].

$(C_5(CH_3)_5)Re(NO)(P(C_6H_5)_3)\{C(O)CH=C(OR')\}Re(CO)_4$ (see Formula IV, R = H). These O-alkylation products were obtained by treating in situ-prepared $(C_5(CH_3)_5)Re(NO)(P(C_6H_5)_3)$-$\{C(O)CH=C(OM)\}Re(CO)_4$ (M = Li, K) with R′X (R′ = CH_3, C_2H_5, $CH_2=CHCH_2$; X = Cl, I) [7] or $(CH_3)_3SiCl$ [1, 6] in THF.

References on p. 179

The following compounds were obtained and characterized:

R′	yield; spectroscopic and structural properties; remarks (for abbreviations and units see p. X)
CH_3	Yield: 4% [2, 7]; improved yield (18%) when methylating with $[O(CH_3)_3]BF_4$ [2]; moreover, when using $CF_3SO_2OCH_3$, the yield could be forced up to 42% [7]. – Orange solid from hexane [7]; m.p. 192 to 197 °C (dec.) [2, 7]. ^1H NMR (THF-d_8): 1.75, 1.97, 7.41 (m), 8.92 (s) [2]; ($CDCl_3$): 1.76 ($(CH_3)_5$), 3.51 (CH_3), 6.13 (s, CH), 7.4 (C_6H_5) [7]. – ^{13}C {^1H} NMR ($CDCl_3$): 9.9 ($(CH_3)_5$), 58.4 (OCH_3), 103.0 (C_5); 128.1, 128.2, 130.2, 133.9 (C_6H_5); 134.4 (d, =CH; J = 10.6); 188.8, 189.3, 193.0, 193.4 (CO); 224.4 (Re**C**OCH_3), 271.4 (d, PReC(O); J(P,C) = 8.8). – IR (CH_2Cl_2): 1467, 1647 (ν(NO)); 1913, 1975, 2080 (ν(CO)). – FAB MS: $[M]^+$ [2, 7].
C_2H_5	Yield: 77%; could not completely be separated from the C-alkylation product. A yellow solid containing the products in the ratio 12:1 was thus characterized. M.p. 200 to 210 °C (dec.) [7]. ^1H NMR ($CDCl_3$): 0.89 (t; J = 7.3), 1.60 (m, OCH_2), 1.76 ($(CH_3)_5$), 1.87 (dd; J = 9.2, 3.7), 7.40 (C_6H_5). – ^{13}C {^1H} NMR ($CDCl_3$): 9.4, 11.7, 22.1, 100.1, 103.3, 128.1, 128.2, 130.2, 130.5, 130.7, 133.7; 190.9, 191.4, 192.4, 194.2; 274.3, 297.9 (d; J(P,C) = 8.2). – IR (CH_2Cl_2): 1371, 1393, 1610, 1661, 1922, 1970, 2078 [7].
$CH_2CH{=}CH_2$	Yield: 57%. – Orange solid; m.p. 162 to 164 °C (dec.). ^1H NMR ($CDCl_3$): 1.74 ($(CH_3)_5$); 4.18, 4.34 (dd's; J = 13.4, 4.9); 5.17 (d; J = 10.4), 5.28 (dd; J = 17.6, 1.0), 5.95 (m), 6.23 (s), 7.37 (C_6H_5). – ^{13}C {^1H} NMR ($CDCl_3$): 9.9, 72.6, 102.9, 116.5, 128.1, 128.2, 130.2, 133.7 (d; J = 10.2), 134.1, 135.9; 188.6, 189.4, 193.0, 193.3; 224.1, 270.9, 271.1 (d; J = 9.5). – IR (CH_2Cl_2): 1311, 1448, 1469, 1649, 1918, 1979, 2081. – FAB MS: $[M]^+$ [7].
$CH_2C_6H_5$	Yield: 76% (using $C_6H_5CH_2Br$; sole product). – Orange solid; m.p. 200 to 206 °C (dec.). ^1H NMR ($CDCl_3$): 1.74 ($(CH_3)_5$); 4.69, 4.86 (d's, CH_2; J = 12.4); 6.30 (s, CH), 7.40 (C_6H_5). – ^{13}C {^1H} NMR ($CDCl_3$): 9.9, 73.3; 103.0 (C_5); 127.0, 127.2, 128.0, 128.2, 130.2, 138.8, 136.0, 138.0 (C_6H_5); 188.6, 189.3, 192.9, 193.3 (CO); 223.9, 271.3 (d; J(P,C) = 9.5). – IR (CH_2Cl_2): 1444, 1470, 1650, 1915, 1975, 2082 [7].
$Si(CH_3)_3$	Yield: 74% [1, 6]. – Yellow [6] to orange [1] solid. ^1H NMR (CD_2Cl_2): 0.27 ($SiCH_3$), 6.47 (=CH) [6]. – ^{13}C {^1H} NMR (CD_2Cl_2): 0.8 ($SiCH_3$), 10.0 ($(CH_3)_5$), 103.2 (C_5); 128.5 (d; J(P,C) = 10.1), 130.6, 134.1, 145.2 (C_6H_5); 189.3, 190.8, 193.87, 193.9 (CO) [3]; 223.6 (=COSi), 272.6 (d, PReC(O); J(P,C) = 11.0) [3, 6]. Single-crystal data: monoclinic; a = 17.741(5), b = 11.745(3), c = 20.199(7) Å, β = 108.14(2)°; space group $P2_1/n-C^5_{2h}$ (No. 14); Z = 4 molecules per unit cell, D_{calc} = 1.755 g/cm^3. A view of the molecular structure is depicted in **Fig. 92**. In the solid state, the s-trans silyloxy conformation is preferred. The ON-Re-C=O torsion angle is 161.8° [6]. Hydrolysis gave $(C_5(CH_3)_5)Re(NO)(P(C_6H_5)_3)\{C(O)CH_2C(O)\}Re(CO)_4$; the rate constant $k_{obs} = 1.7 \times 10^5\ s^{-1}$ was determined [6].

References on p. 179

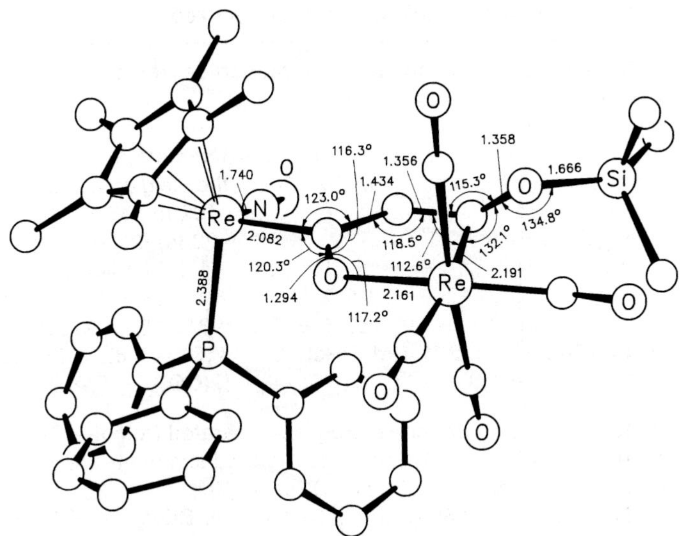

Fig. 92. Molecular structure of $(C_5(CH_3)_5)Re(NO)(P(C_6H_5)_3)\{C(O)CH{=}C(OSi(CH_3)_3)\}Re(CO)_4$ [6].

$(C_5(CH_3)_5)Re(NO)(P(C_6H_5)_3)\{C(O)C(CH_3)_2C(O)\}Re(CO)_4$ (see Formula III, R = R′ = CH_3) was obtained from $(C_5(CH_3)_5)Re(NO)(P(C_6H_5)_3)\{C(O)CH(CH_3)C(O)\}Re(CO)_4$ when the complex was successively treated with K[OC_4H_9-t] and CH_3I in THF. Yield: 85% [2, 7]. The labeled analog **$(C_5(CH_3)_5)Re(NO)(P(C_6H_5)_3)\{C(O)C(CH_3)(CD_3)C(O)\}Re(CO)_4$** was also prepared by employing CD_3I. This reaction even yielded two isomers, dependent on wether the CD_3 group entered in the exo or endo position (compare with Formula III). It appeared that the product with the CD_3 group in the exo position was significantly favored (ratio 38:1). Conversely, by treating $(C_5(CH_3)_5)Re(NO)(P(C_6H_5)_3)\{C(O)CH(CD_3)C(O)\}Re(CO)_4$ with K[OC_4H_9-t] and CH_3I, the opposite CD_3-labeled isomer could mainly be prepared.

The three compounds were separately characterized as follows:

R_{exo}	R_{endo}	preparation; spectroscopic and structural poperties; remarks (for abbreviations and units see p. X)
CH_3	CH_3	Also from Li[$(C_5(CH_3)_5)Re(NO)(P(C_6H_5)_3)\{C(O)CH_2C(O)\}Re(CO)_4Br$] (see p. 167) and CH_3I (THF, ca. 1 d; yield: 23%). – Yellow crystals; m.p. 192 to 197°C (dec.) [7]. ^{1}H NMR ($CDCl_3$): 0.03 (s, CH_3-endo), 0.95 (s, CH_3-exo), 1.68 ($(CH_3)_5$), 7.41 (C_6H_5) [7]. – ^{13}C {^{1}H} NMR ($CDCl_3$): 9.9 ($(CH_3)_5$), 17.7 (CH_3-exo), 23.3 (CH_3-endo), 93.1 (**C**$(CH_3)_2$), 103.6 (C_5); 128.5, 128.6, 130.7, 133.8 (C_6H_5); 191.5, 191.7, 192.4, 195.3 (CO); 277.8 ($(CO)_4$Re**C**(O)), 304.8 (d, PRe**C**(O); J(P,C) = 8). – IR (CH_2Cl_2): 1600, 1655, 1925, 1971, 2078 cm^{-1}. – FAB MS: $[M]^+$ [2, 7]. Single-crystal data: monoclinic; a = 17.220(5), b = 12.984(2), c = 32.584(3) Å, β = 94.45(2)°; space group C2/c–C^6_{2h} (No. 15); Z = 8 molecules per unit cell, D_{calc} = 1.847 g/cm^3. The molecular structure is illustrated in **Fig. 93** [7].

References on p. 179

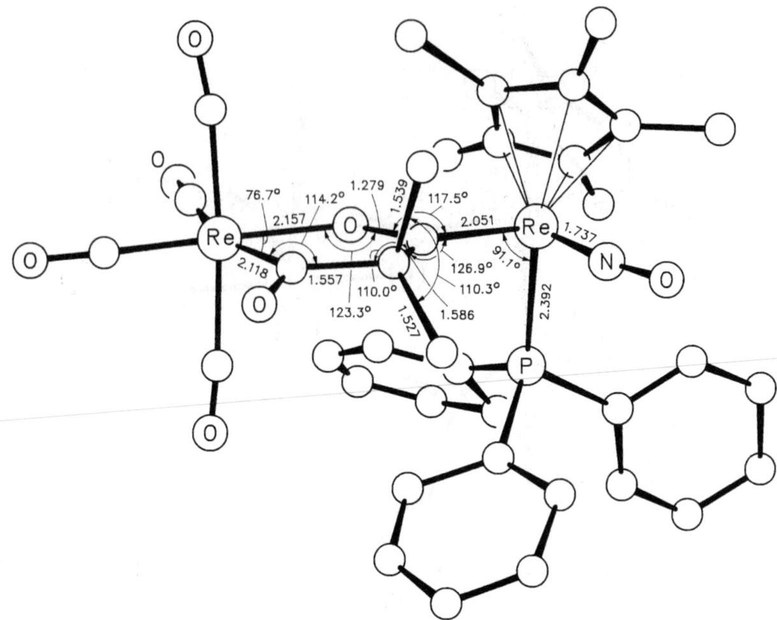

Fig. 93. Molecular structure of $(C_5(CH_3)_5)Re(NO)(P(C_6H_5)_3)\{C(O)C(CH_3)_2C(O)\}Re(CO)_4$ [7].

R_{exo}	R_{endo}	preparation; spectroscopic and structural poperties; remarks (for abbreviations and units see p. X)
CD_3	CH_3	M.p. 205 to 208 °C (dec.). – ^1H NMR ($CDCl_3$): 0.03, 1.68, 7.41. – ^2H {^1H} NMR ($CHCl_3$): 0.93 (s) [7].
CH_3	CD_3	M.p. 207 to 209 °C (dec.). – ^1H NMR ($CDCl_3$): 0.94 (s), 1.68, 7.41. – ^2H {^1H} NMR ($CHCl_3$): 0.01 (s) [7].

$(C_5(CH_3)_5)Re(NO)(P(C_6H_5)_3)\{C(O)C(CH_3)=C(OR')\}Re(CO)_4$ (see Formula IV, R = CH_3). The compounds were obtained by successively treating $(C_5(CH_3)_5)Re(NO)(P(C_6H_5)_3)\{C(O)$-$CH(CH_3)C(O)\}Re(CO)_4$ with $K[OC_4H_9$-t] and R′I in THF. The resulting mixture was worked up by chromatography using CH_2Cl_2/hexane (1 : 1) [2, 7].

R′ = CH_3: Yield: 4% [2, 7]; improved yield (43%) when the methylation was performed with $[O(CH_3)_3]BF_4$ [7]. – M.p. 69 to 80 °C [2].
^1H NMR ($CDCl_3$): 1.72 (s, $(CH_3)_5$), 1.98 (CCH_3), 3.89 (s, OCH_3), 7.40 (C_6H_5). – ^{13}C {^1H} NMR ($CDCl_3$): 9.8, 12.9, 65.9, 102.2; 127.9, 128.0, 128.2, 130.2, 134.0, 144.8; 186.6, 189.1, 193.1, 193.9, 223.2, 277.5 (d; J(P,C) = 9.0). – IR (CH_2Cl_2): 1408, 1477, 1640, 1912, 1984, 2085. – FAB MS: $[M]^+$ [2, 7].

R′ = C_2H_5: Yield: 84%. – Orange solid; m.p. 157 to 163 °C (dec.).
^1H NMR ($CDCl_3$): 1.33 (t; J = 7.0), 1.74 (s, $(CH_3)_5$), 2.01 (s, CH_3), 4.13 (m, CH_2), 7.40 (C_6H_5). – ^{13}C {^1H} NMR ($CDCl_3$): 9.8, 13.2, 15.9, 74.1, 102.1, 128.0, 128.1, 130.1, 133.6, 144.3, 186.9, 193.1, 193.9, 198.3, 222.0, 276.7 (d; J(P,C) = 9.5) [7].

References on p. 179

References:

[1] O'Connor, J. M.; Uhrhammer, R.; Rheingold, A. L. (Organometallics **6** [1987] 1987/9).
[2] O'Connor, J. M.; Uhrhammer, R.; Rheingold, A. L. (Organometallics **7** [1988] 2422/4 and Supplementary Material).
[3] O'Connor, J. M.; Uhrhammer, R. (J. Am. Chem. Soc. **110** [1988] 4448/50).
[4] O'Connor, J. M.; Uhrhammer, R.; Rheingold, A. L.; Staley, D. L. (J. Am. Chem. Soc. **111** [1989] 7633/4).
[5] O'Connor, J. M.; Uhrhammer, R.; Rheingold, A. L.; Staley, D. L.; Chadha, R. K. (J. Am. Chem. Soc. **112** [1990] 7585/98).
[6] O'Connor, J. M.; Uhrhammer, R.; Rheingold, A. L.; Roddick, D. M. (J. Am. Chem. Soc. **113** [1991] 4530/44).
[7] O'Connor, J. M.; Uhrhammer, R.; Chadha, R. K.; Tsuie, B. (J. Organomet. Chem. **455** [1993] 143/56).

2.5.1.1.3 Compound with the $^5LRe(\mu\text{-}^3L)Re$ Skeleton

$C_5H_5Re(CO)_2(\mu\text{-}\eta^{2:1}\text{-}CH_2{=}CHCH_2)Re(CO)_5$ (see Formula I) was prepared by treating $[C_5H_5\text{-}Re(CO)_2C_3H_5]BF_4$ with excess $Na[(CO)_5Re]$ in THF at −70 °C. The solution was warmed to 0 °C and subsequently evaporated. The residue was successively extracted with pentane and CH_2Cl_2. Addition of pentane to the CH_2Cl_2 fraction gave the product with 75% yield. Pale yellow solid; m.p. 150 °C (dec.).

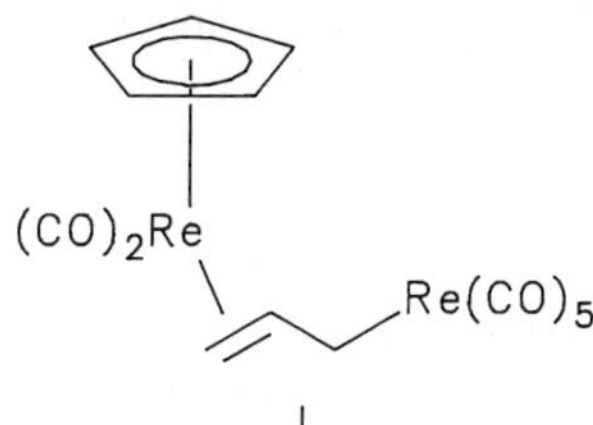

I

^1H NMR spectrum ($CDCl_3$): δ = 1.08, 2.26 (dd's, $ReCH_2$; J_{gem} = −11.7, $^3J(CH,CH_2)$ = 11.2 and 1.5 Hz); 2.33 (m, $=CH_2$), 4.07 (m, CH), 5.17 (s, C_5H_5) ppm. $^{13}C\{^1H\}$ NMR spectrum ($CDCl_3$): δ = 2.80 ($ReCH_2$), 18.78 ($=CH_2$), 60.90 (CH), 85.88 (C_5H_5), 185.89 (CO) ppm. IR spectrum (Nujol): 1850, 1933, 1970, 1989, 2043, 2125 (ν(CO)) cm^{-1} [1].

Reference:

[1] Hüffer, S.; Wieser, M.; Polborn, K.; Beck, W. (J. Organomet. Chem. **481** [1994] 45/55).

2.5.1.2 Compounds with Two ^5LRe Fragments

2.5.1.2.1 Compounds with the $^5LRe(\mu\text{-}E)Re^5L$ Skeleton

General. Structure. With one exception, the compounds described in this section have the structure shown in Formula I (R = H or CH_3; E = O, S, Se, Te, SO, or SO_2).

The solid-state structures of four compounds of the type $[C_5R_5Re(CO)_2]_2(\mu\text{-}E)$ were determined. They all proved to be very similar. The C_5R_5 rings are in a trans arrangement with respect to the central Re_2E ring, and they are nearly coplanar (angles between the normals

 References on p. 184

CO, C_5R_5, OC—Re——Re—CO, C_5R_5, E, CO

I

through the C_5 planes for E = O: 2.0° [5], Se: 8.6° [6]). The Re-Re and Re-E distances are regularly enlarged in the sequence E = O to Te, whereas the Re-E-Re angle becomes smaller. The structural principle is illustrated in **Fig. 94b** (see p. 183); selected bond lengths and angles are listed in the table at the end of this section (p. 184).

Extended Hückel calculations on the model compound series **$[C_5H_5Re(CO)_2]_2E$** (E = S, Se, Te) make plain why the molecules consist of an Re_2E triangle (other imaginable geometries are a linear or bent open-chain Re-E-Re arangement). The energy of the frontier molecular orbitals was determined for each of these molecular geometries. It appeared that the changes are not so significant when going from the linear to the bent structure, but they are very large when going from a bent to a triangular geometry. The sum of one-electron energies as a function of the Re-E-Re angle shows minima for the triangular case only [12].

Preparation. Several special reactions described in the "Further information" section produced the compounds, but the following general methods were also applied:

Method I: Treatment of $(C_5(CH_3)_5)Re(CO)_2OC_4H_8$ with S_8 or Se in THF [6, 10]. Workup of the mixture is described with each compound.

Method II: Oxidation of $[C_5R_5Re(CO)_2]_2(\mu\text{-S})$ with $3\text{-ClC}_6H_4CO_3H$ in cooled THF [10].

Table 8
^{5}L Compounds with the ^{5}LRe(μ-E)Re^{5}L Skeleton.
An asterisk indicates further information at the end of the table.
For explanations, abbreviations, and units see p. X.

No.	compound	method of preparation (yield) properties and remarks
*1	$[\{C_5H_5Re(P(C_6H_5)_3)Br\}_2(\mu\text{-O})][BF_4]_2$ [C_5H_5, C_5H_5, Br—Re——Re, O, $P(C_6H_5)_3$, Br, $P(C_6H_5)_3$]$^{2+}$	for preparation see "Further information" deep purple prisms ^{1}H NMR (CD_2Cl_2): 6.27 (C_5H_5), 6.30 to 8.30 (m, C_6H_5) $^{13}C\{^1H\}$ NMR (CD_2Cl_2): 102.2 (C_5); 123.8 to 137.3 (C_6H_5) $^{31}P\{^1H\}$ NMR (CD_2Cl_2): −0.03 [14]
*2	$[(C_5(CH_3)_5)Re(CO)_2]_2(\mu\text{-O})$	for preparation see "Further information" red crystals from CH_2Cl_2/ether; m.p. 211 °C ^{1}H NMR ($CDCl_3$): 1.98 (s, CH_3) IR (CH_2Cl_2): 1859, 1880, 1926, 1967; (KBr): 1865, 1922, 1965 (ν(CO)) MS: $[M]^+$ photolysis on air yielded $(C_5(CH_3)_5)ReO_3$

References on p. 184

Table 8 (continued)

No.	compound	method of preparation (yield) properties and remarks
		thermolysis (solid-state or in toluene) gave a mixture of $(C_5(CH_3)_5)Re(CO)_3$ and $(C_5(CH_3)_5)_2Re_2(CO)_3$ [5]
*3	$[C_5H_5Re(CO)_2]_2(\mu\text{-}S)$	for preparation see "Further information" dark green crystals from CH_2Cl_2/pentane [2] ^1H NMR ($CDCl_3$): 5.44 [2, 10] $^{13}C\{^1H\}$ NMR ($CDCl_3$): 89.3 (C_5); δ(CO) not observed [2, 10] IR (ether): 1909, 1959 [2]; (THF): 1890 (vw), 1904, 1956, 1980 (vw) [10] MS: $[M - n\ CO]^+$ (n = 0 to 4 [10]; 0, 4 [2]) well soluble in THF, CH_2Cl_2; soluble in toluene, ether [2]
4	$[(C_5(CH_3)_5)Re(CO)_2]_2(\mu\text{-}S)$	I (with a deficient amount (!) of S_8 (20°C, 4 h); separation by column chromatography (alumina, CH_2Cl_2); yield: 10%) [6], see also [10]; also by desulfurizing $(C_5(CH_3)_5)Re(CO)_2S_2$ with $P(C_6H_5)_3$ [10] brown-orange hexagonal plates from CH_2Cl_2/pentane [6, 10] ^1H NMR ($CDCl_3$): 1.97 (s) [6, 10] $^{13}C\{^1H\}$ NMR ($CDCl_3$): 10.3 (CH_3), 100.3 (C_5) [6, 10] IR (CH_2Cl_2): 1859, 1885, 1925, 1960 (ν(CO)) [6]; similar in THF [10] FD MS: $[M]^+$ [6] EI MS: $[M - n\ CO]^+$ (n = 0, 2, 3), $[M - 116]^+$, $[M - n\ CO]^{2+}$ (n = 2, 3), $[M - 116]^{2+}$ [6, 10] oxidation with $3\text{-}ClC_6H_4CO_3H$ gave No. 7 [10]
5	$[C_5H_5Re(CO)_2]_2(\mu\text{-}SO)$	II (when using 1 equivalent of the acid; not isolated) IR (THF): 1923, 1960 (ν(CO)) with excess peracid No. 6 formed [10]
6	$[C_5H_5Re(CO)_2]_2(\mu\text{-}SO_2)$	II; also from No. 5 and excess $3\text{-}ClC_6H_4CO_3H$ yellow-orange crystals ^1H NMR ($CDCl_3$): 5.59 $^{13}C\{^1H\}$ NMR ($CDCl_3$): 92.2 (C_5) IR (KBr): 1033, 1175 (ν(SO_2)); (THF): 1917, 1937, 1978, 2003 (ν(CO)) EI MS: $[M - n\ O]^+$ (n = 0 to 2), $[M - SO_2]^+$, $[M - SO_2 - n\ CO]^+$ (n = 0 to 4) [10]
7	$[(C_5(CH_3)_5)Re(CO)_2]_2(\mu\text{-}SO_2)$	II (48%) red powder ^1H NMR ($CDCl_3$): 2.01 (CH_3) $^{13}C\{^1H\}$ NMR ($CDCl_3$): 9.8 (CH_3), 102.0 (C_5)

References on p. 184

Table 8 (continued)

No.	compound	method of preparation (yield) properties and remarks
7 (continued)		IR (KBr): 1093, 1259 ($\nu(SO_2)$); (THF): 1895, 1921, 1956, 1986 (ν(CO)) EI MS: $[M]^+$, $[M - O_2]^+$, $[M - SO_2]^+$ [10]
*8	$[(C_5(CH_3)_5)Re(CO)_2]_2(\mu\text{-Se})$	I (workup by evaporation of the mixture, extraction with pentane, and cooling; yield: 30%) black-brown microcrystals ^1H NMR ($CDCl_3$): 2.00 (s) $^{13}C\{^1H\}$ NMR ($CDCl_3$): 10.6 (CH_3), 99.6 (C_5) IR (CH_2Cl_2): 1864, 1883, 1922, 1955 (ν(CO)) EI MS: $[M]^+$, $[M - n\,CO]^+$ (n = 2, 3), $[M - 116]^+$, $[M - n\,CO]^{2+}$ (n = 2, 3), $[M - 116]^{2+}$ [6]
*9	$[(C_5(CH_3)_5)Re(CO)_2]_2(\mu\text{-Te})$	for preparation see "Further information" orange needles from ether; m.p. 168 °C ^1H NMR (C_6D_6): 1.89 [9]; ($CDCl_3$): 2.06 $^{13}C\{^1H\}$ NMR ($CDCl_3$): 11.3 (CH_3), 98.9 (C_5) [6] ^{125}Te NMR (toluene-d_8): 613.7 (vs. $(CH_3)_2Te$) [9, 11] IR (THF): 1863, 1883, 1921, 1951(ν(CO)) [6] EI MS: $[M - n\,CO]^+$ (n = 0, 2, 3), $[M - 116]^+$, $[M - 116]^{2+}$ [6]

*Further information:

$[\{C_5H_5Re(P(C_6H_5)_3)Br\}_2(\mu\text{-O})][BF_4]_2$ (Table **8**, No. **1**) was obtained by treating $C_5H_5Re(=C(C_6H_5)CH(C_6H_5)\text{-cyclo})(P(C_6H_5)_3)Br$ with $HBF_4 \cdot O(C_2H_5)_2$ in ether at −78 °C followed by warming the solution to room temperature. The salt precipitated with 68% yield. Crystals were obtained from CH_2Cl_2/ether. However, when the same reaction was performed in the presence of diphenylacetylene in CH_2Cl_2, the main product became $[C_5H_5Re(C_6H_5C{\equiv}C\text{-}C_6H_5)(P(C_6H_5)_3)Br]BF_4$ and only minor quantities of No. 1 (product ratio 11:2) were formed [14] (briefly in [13]).

At −103 °C, the salt crystallizes in the monoclinic space group $C2/c-C_{2h}^6$ (No. 15) with a = 19.528(5), b = 17.211(3), c = 15.323(3) Å, β = 119.52(2)°; D_{calc} = 2.04 g/cm³. The bridging O atom is located on a two-fold axis. A view of the dication is presented in **Fig. 94a** [14].

$[(C_5(CH_3)_5)Re(CO)_2]_2(\mu\text{-O})$ (Table **8**, No. **2**) formed by photolyzing $(C_5(CH_3)_5)Re(CO)_3$ in THF (30 to 40 °C, 2 h). Workup by column chromatography separated the compound with 20 to 25% yield [5]. No. 2 was also one of the products obtained by pressurizing $(C_5(CH_3)_5)ReO_3$ with CO at 15 atm [4]. The formation of No. 2 was also observed when $(C_5(CH_3)_5)ReO_3$ was treated with $(C_5H_5)_2Ti(CO)_2$ [3, 7, 8]; yield: 80% [7].

Single-crystal data are the following: triclinic, a = 9.074(2), b = 10.402(3), c = 14.143(4) Å, α = 79.99(2)°, β = 88.42(2)°, γ = 66.18(2)°; space group $P\bar{1}-C_i^1$ (No. 2); Z = 2 molecules per unit cell, D_{calc} = 2.2 g/cm³. For selected bond lengths and angles, see the table on p. 184 [5].

References on p. 184

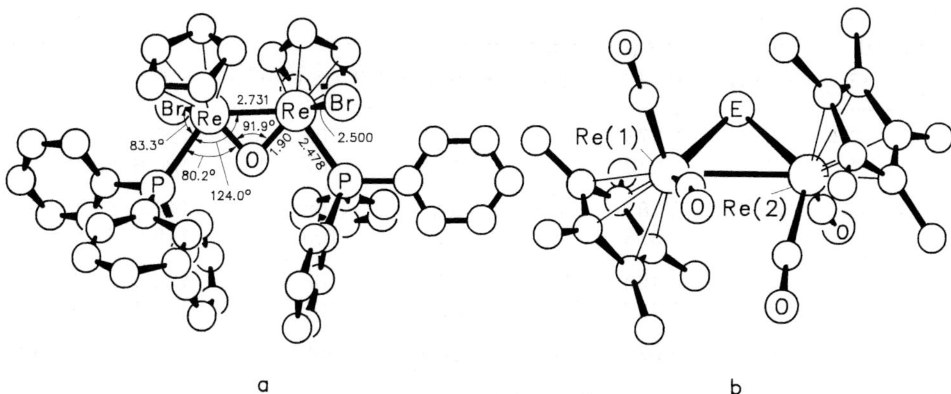

Fig. 94. Molecular structure of $[\{C_5H_5Re(P(C_6H_5)_3)Br\}_2(\mu\text{-}O)]^{2+}$ (a) [14] and the structural principle of $[C_5R_5Re(CO)_2]_2(\mu\text{-}E)$ (b; figure from [9]).

$[C_5H_5Re(CO)_2]_2(\mu\text{-}S)$ (Table **8**, No. **3**) was produced when an ethereal solution containing $C_5H_5Re(CO)_3$ was photolyzed (−30 °C, 40 min) and subsequently treated with S_8 or COS. Evaporation of the solvent followed by column-chromatographic workup (C_6H_{12}/CH_2Cl_2) [2] gave No. 3 with 11.3% [1] or 38% [2] yield. An analogous reaction was performed in THF (0 °C, irradiation time: 30 min, use of S_8), but under these conditions No. 3 was separated with only 2.6% yield (main product was $C_5H_5Re(CO)_2S_2$) [10].

No. 3 crystallizes in the monoclinic space group $P2_1/n-C_{2h}^5$ (No. 14) with a = 13.787(4), b = 12.010(4), c = 9.382(2) Å, β = 105.90(2)°; Z = 4 molecules per unit cell, D_{calc} = 2.870 g/cm³. Selected intramolecular parameters are listed in the table on p. 184 [2].

Oxidation with $3\text{-}ClC_6H_4CO_3H$ gave Nos. 5 and 6 [10].

$[(C_5(CH_3)_5)Re(CO)_2]_2(\mu\text{-}Se)$ (Table **8**, No. **8**) crystallizes in the monoclinic space group $P2_1/c-C_{2h}^5$ (No. 14) with a = 18.078(6), b = 9.374(3), c = 16.658(6) Å, β = 115.92(4)°; Z = 4 molecules per unit cell, D_{calc} = 2.181 g/cm³. The relevant bond lengths and angles are listed in the table on p. 184 [6].

$[(C_5(CH_3)_5)Re(CO)_2]_2(\mu\text{-}Te)$ (Table **8**, No. **9**) was obtained with 2% [9] or 18% [6] yield by treating $(C_5(CH_3)_5)Re(CO)_2OC_4H_8$ with TeH_2 (from Al_2Te_3 and HCl) in the dark for 4 h followed by chromatographic workup on Florisil [6, 9]. No. 9 was also produced by treating $(C_5(CH_3)_5)Re(CO)_2OC_4H_8$ with $(C_5(CH_3)_5)Re(CO)_2(TeH)H$; yield: 70% [9].

Single crystals have the following lattice parameters: monoclinic, a = 13.192(2), b = 14.021(1), c = 13.416(2) Å, β = 101.11(1)°, space group $C2/c-C_{2h}^6$ (No. 15); Z = 4 molecules per unit cell, D_{calc} = 2.407 g/cm³. Selected bond parameters are listed in the table on p. 184 [9].

The compound is well soluble in all polar organic solvents [6]. Treatment with HBF_4 or $CH_3OSO_2CF_3$ gave the salts $[\{(C_5(CH_3)_5)Re(CO)_2\}_2TeR]^+$ (R = H, CH_3) (see p. 186) [9].

The following table summarizes selected bond lengths and angles of the compounds of the type $[C_5R_5Re(CO)_2]_2(\mu\text{-}E)$ (compare with **Fig. 94b**):

References on p. 184

compound			bond lengths (Å)		bond angles (°)				Ref.
No.	E	R	Re-Re	Re-E [a]	Re-E-Re	Re-Re-E [a]	Re-Re-CO [a]	E-Re-CO [a]	
2	O	CH_3	2.817	1.973 1.972	91.1	44.4 44.4	106.3, 82.0 107.2, 81.6	120.1, 89.9 120.8, 89.1	[5]
3	S	H	2.946	2.387 2.375	76.4	51.6 52.0	108.6, 78.2 109.7, 78.0	117.6, 81.1 118.8, 82.5	[2]
8	Se	CH_3	3.033	2.498 2.487	74.9	52.3 52.7	108.5, 82.2 102.4, 70.4	120.4, 86.3 120.6, 80.0	[6]
9	Te	CH_3	3.140	2.679	71.8	54.1	not given	not given	[9]

[a] first line: at Re(1), second line: at Re(2).

References:

[1] Herberhold, M.; Reiner, D.; Thewalt, U. (Angew. Chem. **95** [1983] 1028/9; Angew. Chem. Int. Ed. Engl. **22** [1983] 1000; Angew. Chem. Suppl. **1983** 1343/52).

[2] Herberhold, M.; Reiner, D.; Ackermann, K.; Thewalt, U.; Debaerdemaeker, T. (Z. Naturforsch. **39b** [1984] 1199/205).

[3] Herrmann, W. A.; Serrano, R.; Bock, H. (Angew. Chem. **96** [1984] 364/5; Angew. Chem. Int. Ed. Engl. **23** [1984] 383).

[4] Herrmann, W. A.; Serrano, R.; Küsthardt, U.; Ziegler, M. L.; Guggolz, E.; Zahn, T. (Angew. Chem. **96** [1984] 498/500; Angew. Chem. Int. Ed. Engl. **23** [984] 515/7).

[5] Herrmann, W. A.; Serrano, R.; Schäfer, A.; Küsthardt, U.; Ziegler, M. L.; Guggolz, E. (J. Organomet. Chem. **272** [1984] 55/71).

[6] Herberhold, M.; Schmidkonz, B.; Thewalt, U.; Razari, A.; Schöllhorn, H.; Herrmann, W. A.; Hecht, C. (J. Organomet. Chem. **299** [1986] 213/21).

[7] Herrmann, W. A. (J. Organomet. Chem. **300** [1986] 111/37).

[8] Herrmann, W. A.; Küsthardt, U.; Schäfer, A.; Herdtweck, E. (Angew. Chem. **98** [1986] 818/9; Angew. Chem. Int. Ed. Engl. **25** [1986] 817).

[9] Herrmann, W. A.; Hecht, C.; Herdtweck, E.; Kneuper, H.-J. (Angew. Chem.**99** [1987] 158/60; Angew. Chem. Intern. Ed. Engl. **26** [1987] 132).

[10] Herberhold, M.; Schmidkonz, B. (J. Organomet. Chem. **358** [1988] 301/20).

[11] Herrmann, W. A.; Kneuper, H.-J. (J. Organomet. Chem. **348** [1988] 193/7).

[12] Jemmis, E. D.; Kumar, P. N. V. P.; Sastry, G. N. (J. Organomet. Chem. **478** [1994] 29/36).

[13] Carr, N.; Green, M.; Mahon, M. F.; Jones, C.; Nixon, J. F. (J. Chem. Soc. Chem. Commun. **1995** 2191/2).

[14] Carfagna, C.; Carr, N.; Deeth, R. J.; Dossett, S. J.; Green, M.; Mahon, M. F.; Vaughan, C. (J. Chem. Soc. Dalton Trans. **1996** 415/30).

2.5.1.2.2 Compounds with the $^5LRe(\mu\text{-}X)Re^5L$ Skeleton

General. The structural principles of the compounds described in this section are shown in Formulas I to III. The bridging ligand X in the headline stands for a variety of three-electron donating moieties such as halogen, TeR, or O_2CH.

References on p. 187

$$\left[[Re] \leftarrow X - [Re]\right]^+$$

$[Re] = C_5H_5Re(NO)P(C_6H_5)_3$

I

$$\left[[Re] - Te(R) - [Re]\right]^+$$

$[Re] = C_5(CH_3)_5Re(CO)_2$

II

$$\left[[Re]-O-C(H)=O-[Re]\right]^+$$

$[Re] = C_5H_5Re(CO)NO$

III

Preparation. Salts of the type $[\{C_5H_5Re(NO)P(C_6H_5)_3\}_2(\mu\text{-}X)]^+$ were prepared by the following methods:

Method I: Treatment of $[C_5H_5Re(NO)(P(C_6H_5)_3)IR]BF_4$ (R = $CH_2CH_2C_4H_9$-t [9], $C_{10}H_{15}$ (adamantyl), C_6H_{11} [10], CH_3, C_2H_5, C_3H_7-n [3, 7]) with CD_2Cl_2 at room temperature. The formation of (RR,SS)-$[\{C_5H_5Re(NO)P(C_6H_5)_3\}_2(\mu\text{-}X)]^+$ (X = Cl, I) was established after ca. 2 d by NMR spectroscopy [3, 7, 9, 10].

Method II: Treatment of rac-$C_5H_5Re(NO)(P(C_6H_5)_3)X$ (X = Br, I) with excess $AgBF_4$ (benzene, reflux, 30 min). The solvent was removed and the residue extracted into CH_2Cl_2. After filtration through Celite, addition of ether precipitated the product [3, 6].

Method III: Treatment of S-$[C_5H_5Re(NO)(P(C_6H_5)_3)ClCD_2Cl]^+$ with R-$C_5H_5Re(NO)$-$(P(C_6H_5)_3)I$ yielded SS-$[\{C_5H_5Re(NO)P(C_6H_5)_3\}_2(\mu\text{-}Cl)]^+$ and RR-$[\{C_5H_5Re(NO)$-$P(C_6H_5)_3\}_2(\mu\text{-}I)]^+$. Both products were not separated [6].

(RR,SS)-$[\{C_5H_5Re(NO)P(C_6H_5)_3\}_2(\mu\text{-}Cl)]BF_4 \cdot 0.45\ CH_2Cl_2$ was formed via Method I with the following yields: 31% (R = $CH_2CH_2C_4H_9$-t) [9], 40% (R = $C_{10}H_{15}$), 36% (R = C_6H_{11}) [10], 28% (R = C_2H_5) [7]. It was also obtained when drying $[C_5H_5Re(NO)(P(C_6H_5)_3)ClCH_2Cl]BF_4 \cdot CH_2Cl_2$ under vacuum. During the process, the solvate lost crystallinity and thereby mostly transformed to the title dimer. Yield: 65% (the condensation was not observed when similarly treating the respective PF_6 salt) [5]. Combination of $[C_5H_5Re(NO)(P(C_6H_5)_3)ClCH_2Cl]BF_4$ with $(CO)_5ReCH_2F$ also provided the title dimer only. Yield: 80% [11]. It was also obtained by adding at −60°C $HBF_4 \cdot O(C_2H_5)_2$ to a CH_2Cl_2 solution of $C_5H_5Re(NO)(P(C_6H_5)_3)CH_3$ and $C_5H_5Re(NO)(P(C_6H_5)_3)Cl$. The product precipitated upon warming the mixture to room temperature and was recrystallized from CH_2Cl_2/ether. Yield: 81% [5]. Moreover, simply dissolving $[C_5H_5Re(NO)P(C_6H_5)_3]BF_4$ in CH_2Cl_2 at −20°C initiated product formation. Yield: 64% [1]. The title compound was also formed along with the respective (μ-I) complex (ratio 56:44) by reacting $[C_5H_5Re(NO)(P(C_6H_5)_3)ClCD_2Cl]^+$ with $C_5H_5Re(NO)(P(C_6H_5)_3)I$ [6].

The salt forms fine orange needles; m.p. 244 to 248°C (dec.) [1, 5]. ^1H NMR spectrum (CD_2Cl_2): δ = 5.40 (C_5H_5); 7.10 to 7.63 (C_6H_5) ppm [5]. $^{13}C\{^1H\}$ NMR spectrum (CD_2Cl_2): δ = 93.1 (C_5); 129.1, 131.5, 133.0, 133.6 (d's, C_6H_5; J(P,C) = 11.3, 2.1, 56.0, 9.9 Hz, resp.) ppm. $^{31}P\{^1H\}$ NMR spectrum (CD_2Cl_2, −85°C): δ = 15.12 (s) ppm. IR spectrum (KBr): 1034, 1046, 1068 (BF_4); 1662 (ν(NO)) cm^{-1} [1, 5]. FAB mass spectrum: $[M]^+$, $[M - C_5H_5$-$Re(NO)P(C_6H_5)_3 - n\ Cl]^+$ (n = 0, 1) [5].

Thermolysis in CH_3CN yielded $[C_5H_5Re(NO)(P(C_6H_5)_3)NCCH_3]BF_4$ and $C_5H_5Re(NO)$-$(P(C_6H_5)_3)Cl$ [1, 5]; an investigation of the kinetics (CD_3CN, 40°C) established the rate constant $k_{obs} = (1.91 \pm 0.07) \times 10^{-4}\ s^{-1}$ [5]. Thermolysis in acetone gave $[C_5H_5Re(NO)(P(C_6H_5)_3)$-$O{=}C(CH_3)_2]BF_4$ and $C_5H_5Re(NO)(P(C_6H_5)_3)Cl$ [5].

(SS)-$[\{C_5H_5Re(NO)P(C_6H_5)_3\}_2(\mu\text{-}Cl)]BF_4$ could be obtained via Method III with 38% yield [6]. It was selectively prepared by combining S-$C_5H_5Re(NO)(P(C_6H_5)_3)CH_3$ with $HBF_4 \cdot O(C_2H_5)_2$

References on p. 187

in CH_2Cl_2 followed by warming the mixture to room temperature. After stirring for 8 h, addition of ether caused the salt to precipitate. Yield: 72%. Likewise, when adding S-C_5H_5Re(NO)(P(C_6H_5)$_3$)Cl to an HBF_4-pretreated CH_2Cl_2 solution containing S-C_5H_5Re(NO)(P(C_6H_5)$_3$)CH_3, warming to room temperature, and diluting with ether, the salt separated with 79% yield [1, 5]. Decomposition of S-[C_5H_5Re(NO)P(C_6H_5)$_3$]BF_4 in CH_2Cl_2 also yielded the optically active cation [1].

Dec.p. 145 to 150°C [5]. $[\alpha]^{23}_{589}$ = 546° or 528° [1, 5]; $[\alpha]^{22}_{589}$ = 549° [1].

(RR,SS)-[{C_5H_5Re(NO)P(C_6H_5)$_3$}$_2$(μ-Br)]BF_4 was obtained by Method II with 54% yield as fine orange feathers; m.p. 244 to 248°C (dec.) [6].

^{1}H NMR spectrum (CD_2Cl_2): δ = 5.38 (s, C_5H_5); 7.17 to 7.55 (C_6H_5) ppm. ^{13}C {^{1}H} NMR spectrum (CD_2Cl_2): δ = 92.78 (C_5); 129.24, 131.66, 133.48, 133.84 (d's, C_6H_5; J(P,C) = 10.9, 2.2, 55.9, 10.7 Hz, resp.) ppm. ^{31}P {^{1}H} NMR spectrum (CD_2Cl_2): δ = 12.8 (s) ppm. IR spectrum (KBr): 1664 (ν(NO)) cm^{-1} [6].

Thermolysis in CH_3CN yielded [C_5H_5Re(NO)(P(C_6H_5)$_3$)NCCH_3]BF_4 and C_5H_5Re(NO)(P(C_6H_5)$_3$)Br in equal amounts [6].

(RR,SS)-[{C_5H_5Re(NO)P(C_6H_5)$_3$}$_2$(μ-I)]BF_4 was obtained via Method I with the following yields: 57% (R = $CH_2CH_2C_4H_9$-t) [9], 49% (R = $C_{10}H_{15}$), 33% (R = C_6H_{11}) [10], 65% (R = C_2H_5) [3, 7], or via Method II [3, 6] with 59% yield [6]. The salt also formed with 42% yield when [C_5H_5Re(NO)(P(C_6H_5)$_3$)−ICH_2Mn(CO)$_5$]BF_4 decomposed in CH_2Cl_2 (room temperature, 13 h); other products were C_5H_5Re(NO)(P(C_6H_5)$_3$)I and $(CO)_{10}Mn_2$ [11]. It was also obtained along with the (μ-Cl) analog by reacting [C_5H_5Re(NO)(P(C_6H_5)$_3$)ClCD_2Cl]$^+$ with C_5H_5Re(NO)(P(C_6H_5)$_3$)I [6]. Light orange powder; m.p. 235 to 236°C (dec.) [6].

^{1}H NMR spectrum (CD_2Cl_2): δ = 5.40 (s, C_5H_5); 7.24 to 7.50 (C_6H_5) ppm. ^{13}C {^{1}H} NMR spectrum (CD_2Cl_2): δ = 92.09 (C_5); 129.14, 131.57, 133.77, 134.10 (d's, C_6H_5; J(P,C) = 11.0, 2.7, 10.6, 56.1 Hz, resp.) ppm. ^{31}P {^{1}H} NMR spectrum (CD_2Cl_2): δ = 12.1 (s) ppm. IR spectrum (KBr): 1675 (ν(NO)) cm^{-1} [6].

An X-ray structure determination was carried out on **[{C_5H_5Re(NO)P(C_6H_5)$_3$}$_2$(μ-I)]BF_4 · 0.7 $C_2H_2Cl_4$** (m.p. 247 to 250°C). The solvate crystallizes in the monoclinic space group $P2_1/c-C^5_{2h}$ (No. 14) with a = 14.975(4), b = 18.219(4), c = 20.855(4) Å, β = 108.20(2)°; Z = 4 formula units per unit cell, D_{calc} = 1.742 g/cm^3. A view of the cation is depicted in **Fig. 95**. It adopts a W conformation for the P-Re-I-Re-P linkage. The cyclopentadienyl rings are in an anti position with respect to each other [6].

Thermolysis in CH_3CN yielded [C_5H_5Re(NO)(P(C_6H_5)$_3$)NCCH_3]BF_4 and C_5H_5Re(NO)(P(C_6H_5)$_3$)I in equal amounts [6].

(RR)-[{C_5H_5Re(NO)P(C_6H_5)$_3$}$_2$(μ-I)]BF_4 was obtained with 62% yield via Method III [6].

[{(C_5(CH_3)$_5$)Re(CO)$_2$}$_2$TeR]X (see Formula II). Salts of this type were obtained by treating [(C_5(CH_3)$_5$)Re(CO)$_2$]$_2$(μ-Te) with HBF_4 or $CH_3OSO_2CF_3$ at −78°C [2].

R = H: (BF_4 salt). Very labile complex, protonation was reversed with weak bases [2].

R = CH_3: (O_3SCF_3 or PF_6 salt). ^{1}H NMR (CD_2Cl_2): 2.14 ((CH_3)$_5$), 2.57 (TeCH_3) [2]. − ^{125}Te NMR (acetone): 568.4 [2, 4].

[{C_5H_5Re(CO)(NO)O−}$_2$CH]PF_6 (see Formula III). Combination of C_5H_5Re(CO)(NO)H with [C(C_6H_5)$_3$]BF_4 in CH_2Cl_2 at −78°C was followed by adding 1 equivalent C_5H_5Re(CO)(NO)OC(O)H. The mixture was warmed to room temperature. Filtration, concentration of the

References on p. 187

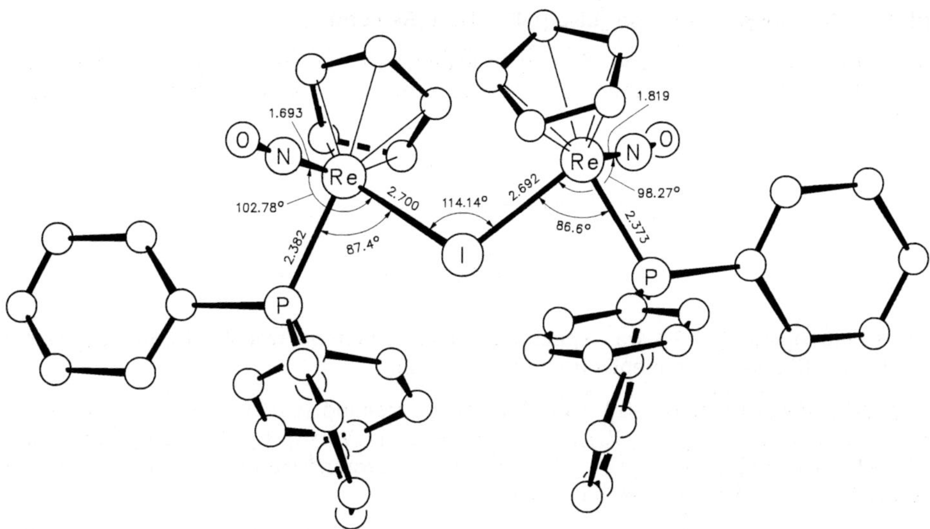

Fig. 95. Molecular structure of (RR,SS)-$[\{C_5H_5Re(NO)P(C_6H_5)_3\}_2(\mu\text{-}I)]^+$ [6].

filtrate, and addition of ether gave a yellow-brown crude precipitate, which was recrystallized from the same solvent mixture. Yield: 79%.

^{1}H NMR spectrum (CD_3NO_2): $\delta = 6.13$ (C_5H_5), 7.72 (CH) ppm. ^{13}C {^{1}H} NMR spectrum (CD_2Cl_2): $\delta = 107.4$ (C_5); 198.66, 198.96 (d's, CH; J(C,H) = 220.4 and 220.7 Hz); 209.64, 209.72 (CO) ppm. IR spectrum (KBr): 1315 ($\nu_{sym}(CO_2)$), 1560 ($\nu_{asym}(CO_2)$), 1720 (ν(NO)), 1982 (ν(CO)) cm^{-1}; (CH_2Cl_2): 1561 ($\nu(CO_2)$), 1749 (ν(NO)), 2012 (ν(CO)) cm^{-1}.

Treatment with $[N(C_4H_9\text{-}n)_4]I$, $Li[HB(C_2H_5)_3]$, or $K[HB(OC_3H_7\text{-}i)_3]$ provided a mixture of $C_5H_5Re(CO)(NO)OC(O)H$ and $C_5H_5Re(CO)(NO)X$ (X = I or H). Addition of $C_5H_5Fe(CO)_2OC(O)H$ yielded a mixture of $C_5H_5Fe(CO)(NO)OC(O)H$, $[C_5H_5Re(CO)(NO)OCHO\text{-}Fe(C_5H_5)(CO)_2]^+$, and the two reactants [8].

References:

[1] Fernández, J. M.; Gladysz, J. A. (Inorg. Chem. **25** [1986] 2672/4).
[2] Herrmann, W. A.; Hecht, C.; Herdtweck, E.; Kneuper, H.-J. (Angew. Chem. **99** [1987] 158/60; Angew. Chem. Int. Ed. Engl. **26** [1987] 132).
[3] Winter, C. H.; Arif, A. M.; Gladysz, J. A. (J. Am. Chem. Soc. **109** [1987] 7560/1).
[4] Herrmann, W. A.; Kneuper, H.-J. (J. Organomet. Chem. **348** [1988] 193/7).
[5] Fernández, J. M.; Gladysz, J. A. (Organometallics **8** [1989] 207/19).
[6] Winter, C. H.; Arif, A. M.; Gladysz, J. A. (Organometallics **8** [1989] 219/25).
[7] Winter, C. H.; Veal, W. R.; Garner, C. M.; Arif, A. M.; Gladysz, J. A. (J. Am. Chem. Soc. **111** [1989] 4766/76).
[8] Tso, C. C.; Cutler, A. R. (Inorg. Chem. **29** [1990] 471/5).
[9] Igau, A.; Gladysz, J. A. (Organometallics **10** [1991] 2327/34).
[10] Igau, A.; Gladysz, J. A. (Polyhedron **10** [1991] 1903/9).
[11] Zhou, Y.; Gladysz, J. A. (Organometallics **12** [1993] 1073/8).

2.5.1.2.3 Compounds with the ⁵LRe(μ-E-X)Re⁵L Skeleton

Formula I illustrates the type of bridging ligand arrangement present in the complexes described in this section. In one case, the structure was X-ray crystallographically established.

[Re]←X(E)[Re]

I (X, E = S, Se, Te)

$[C_5H_5Re(CO)_2]_2(\mu\text{-}\eta^{2:1}\text{-}S_2)$ was spectroscopically observed when $C_5H_5Re(CO)_3$ was irradiated in THF and subsequently treated with S_8 [2].

$[(C_5(CH_3)_5)Re(CO)_2]_2(\mu\text{-}\eta^{2:1}\text{-}S_2)$ was obtained by treating $(C_5(CH_3)_5)Re(CO)_2OC_4H_8$ with 0.86 equivalent $(C_5(CH_3)_5)Re(CO)_2S_2$ in THF for 27 h. Workup by chromatography (silica, pentane/CH_2Cl_2) separated it with 14.2% yield. The product was reprecipitated from pentane/CH_2Cl_2. It forms a blue-black powder.

^1H NMR spectrum ($CDCl_3$): δ = 1.95, 1.96 (CH_3) ppm. ^{13}C {^1H} NMR spectrum ($CDCl_3$): δ = 10.1, 10.3 $(CH_3)_5$; 96.8, 103.1 (C_5) ppm. IR spectrum (THF): 1852, 1906, 1924, 1939, 1991, 1998, 2006 cm^{-1}; (KBr): 1841, 1901, 1923, 1993 (ν(CO)) cm^{-1}. EI mass spectrum: $[M]^+$, $[M - S - n\,CO]^+$ (n = 0, 2, 3).

The complex is well soluble in CH_2Cl_2, THF, toluene; slightly soluble in ether and pentane. Toluene and pentane solutions are green; ether and THF solutions are red-violet [2].

$[(C_5(CH_3)_5)Re(CO)_2]_2(\mu\text{-}\eta^{2:1}\text{-}Te_2)$ was one of the products obtained by treating $(C_5(CH_3)_5)$-$Re(CO)_2OC_4H_8$ with TeH_2. The compound was chromatographically separated at −20°C (Florisil, hexane/toluene/ether mixtures) with 9% yield. It could also be obtained by irradiating $(C_5(CH_3)_5)Re(CO)_2(TeH)H$ for 30 min (quantitative yield) or by irradiating or thermolyzing $[(C_5(CH_3)_5)Re(CO)_2]_2(\mu\text{-}Te)_2$ in THF. Black plates [1].

^1H NMR spectrum (C_6D_6): δ = 1.60, 1.85 ppm. [1]. ^{125}Te NMR spectrum (toluene-d_8, external standard $(CH_3)_2Te$): δ = −200.3, 434.4 ppm [1, 3].

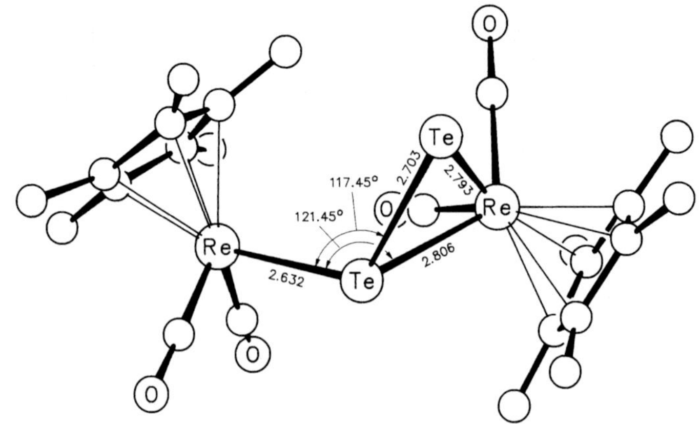

Fig. 96. Molecular structure of $[(C_5(CH_3)_5)Re(CO)_2]_2(\mu\text{-}\eta^{2:1}\text{-}Te_2)$ [1].

References on p. 189

The solvate **$[(C_5(CH_3)_5)Re(CO)_2]_2(\mu\text{-}\eta^{2:1}\text{-}Te_2) \cdot 0.5\ C_4H_8O$** crystallizes in the triclinic space group $P\bar{1}-C_i^1$ (No. 2) with a = 8.589(1), b = 12.894(2), c = 14.467(5) Å, α = 77.41(2)°, β = 74.01(2)°, γ = 89.67(2)°; Z = 2 formula units per unit cell, D_{calc} = 2.315 g/cm^3. The molecular structure is depicted in **Fig. 96**. The Re-Te_μ distances are clearly different [1].

References:

[1] Herrmann, W. A.; Hecht, C.; Herdtweck, E.; Kneuper, H.-J. (Angew. Chem. **99** [1987] 158/60; Angew. Chem. Int. Ed. Engl. **26** [1987] 132).
[2] Herberhold, M.; Schmidkonz, B. (J. Organomet. Chem. **358** [1988] 301/20).
[3] Herrmann, W. A.; Kneuper, H.-J. (J. Organomet. Chem. **348** [1988] 193/7).

2.5.1.2.4 Compounds with the $^5LRe(\mu\text{-}E)_2Re^5L$ Skeleton

General. Structure. Most of the compounds dealt with in this section have the structure illustrated in Formulas I and II, where E, E′ represent covalently bonded atoms such as O, S, Se, or Te or atom groups such as S_2, Se_2, or S_3. The terminal ligands X stand for one-electron donors such as halogen or $OReO_3$ or symbolize chelating ligands such as OC_2H_4O or $O_2S{=}O$. In several instances, two X ligands are replaced by one oxo or one CO group.

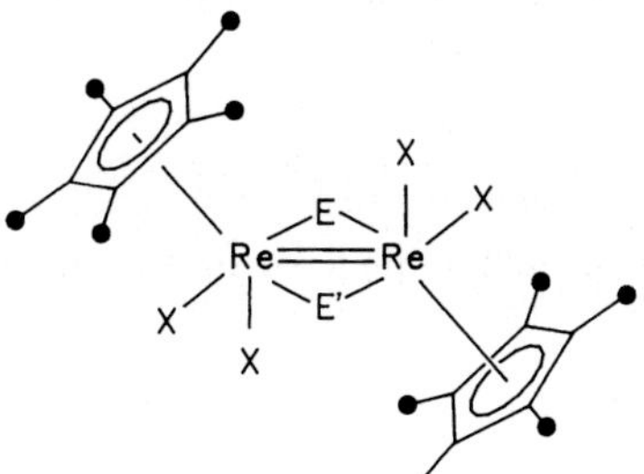

I

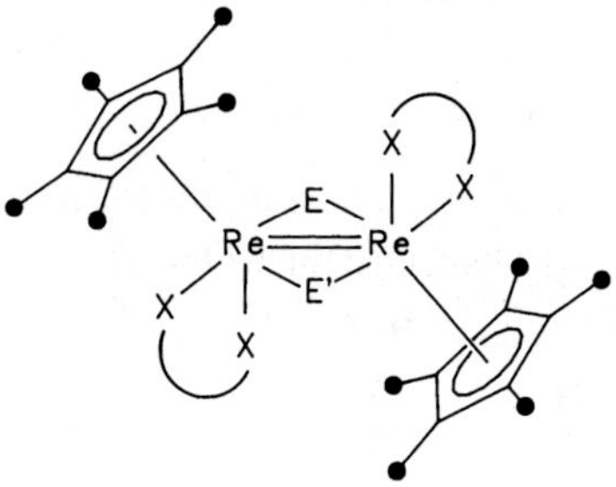

II

From electron counting, one expects an Re=Re double bond to fulfil the 18-electron rule (and all formulas in the table exhibit this double bond); however, neither solid-state structure determinations nor molecular orbital calculations in each case studied unambiguously confirm this conception. On the contrary, the compounds of the type $[C_5(CH_3)_5ReO]_2(\mu\text{-}E)_2$ (E = O, S, Se) exist as 16-electron complexes without a direct Re-Re connection (see below).

Preparation. Apart from special syntheses, the compounds were prepared by the following general methods:

Method I: Treatment of $(C_5(CH_3)_5)Re(O)Cl_2$ with Ag_2SO_3 (CH_2Cl_2, 12 h, exclusion of light). Afterwards, the mixture was evaporated and the residue separated by chromatography (silylated silica, −30°C, ether/THF (4:3)) [21].

Method II: Treatment of $(C_5(CH_3)_5)ReO_3$ with stoichiometric amounts of $C_6H_5N{=}C{=}E$ (E = O, S) and $P(C_6H_5)_3$ (toluene, room temperature, 16 h). The products were separated by chromatography. The reactions proceeded via in situ formation of

References on p. 202

$[(C_5(CH_3)_5)ReO]_2(\mu\text{-}O)_2$ (No. 9). For the case of E = O, the main product was $(C_5(CH_3)_5)Re(O)-OC(O)N(C_6H_5)$-cyclo [25].

Method III: Dropwise addition of 2 equivalents $3\text{-}RC_6H_4N{=}C{=}S$ (R = H, CF_3) to a toluene solution containing $[(C_5(CH_3)_5)ReO]_2(\mu\text{-}O)_2$ (No. 9). Stirring was continued for 15 h. The solvent was removed and the residue was chromatographically separated on silylated silica with toluene/CH_2Cl_2 mixtures. Recrystallization was from hexane/CH_2Cl_2 (1:1) [25].

Method IV: Treatment of $(C_5(CH_3)_5)Re(O)Cl_2$ with a 4-fold excess of AgX (X = CN, OCN) under exclusion of light (CH_2Cl_2, 12 h). The mixture was filtered, evaporated, and the residue was reprecipitated from CH_2Cl_2/toluene. When starting from $(C_5(CH_3)_5)Re(O^*)Cl_2$ (O^* = ^{17}O, ^{18}O), the respective labeled products could be obtained [21].

Method V: Treatment of $[(C_5(CH_3)_5)ReO]_2(\mu\text{-}O)_2$ (No. 9) with excess SO_2 (toluene, 15 min) or 4 equivalents SeO_2 (CH_2Cl_2). The solvent was removed and the residue chromatographically separated (conditions are given in the table) [21].

Method VI: Treatment of $(C_5(CH_3)_5)_2Re_2(O)(OC({=}E)N(C_6H_5)\text{-cyclo})(\mu\text{-}O)_2$ (Nos. 3, 4) with excess $C_6H_5N{=}C{=}E$ (E = O, S) in a toluene/CH_2Cl_2 mixture for 1 d. The solvent was removed and the residue separated by chromatography (silylated silica, toluene/CH_2Cl_2 mixtures). The products were recrystallized from toluene/pentane at −30 °C [25].

Method VII: Treatment of $(C_5(CH_3)_5)Re(O)E_4$ (E = S, Se) with excess $P(C_4H_9\text{-}n)_3$ in $CHCl_3$ for a short time (0.5 to 1 h). The product was separated by chromatography using CH_2Cl_2/THF (5:1) [22].

Compounds No. 1 to 8 in the following table are unsymmetrical compounds where the E ligands connect unequal molecular halves. Compounds No. 9 to 25 are symmetrical, where the two halves of the molecule are identical. In the latter subgroup, compounds with E = O precede those with E = S, Se, and Te.

Table 9
^5L Compounds with the $^5LRe(\mu\text{-}E)_2Re^5L$ Skeleton.
An asterisk indicates further information at the end of the table.
For explanations, abbreviations, and units see p. X.

No.	compound	method of preparation (yield) properties and remarks
unsymmetrical compounds		
*1	$(C_5(CH_3)_5)_2Re_2(O)Cl_2(\mu\text{-}O)_2$	I (by-product) [21]; see also "Further information" air-stable solid [16] ^1H NMR ($CDCl_3$): 1.85, 2.03 $^{13}C\{^1H\}$ NMR ($CDCl_3$): 10.52, 10.56; 107.09, 118.06 ^{17}O NMR (CD_2Cl_2): 469 IR (KBr): 290, 320, 390 (ν(ReCl)); 690, 732 (ν(ReORe)); 908 (ν(ReO)) [16]

Table 9 (continued)

No.	compound	method of preparation (yield) properties and remarks
2	$(C_5(CH_3)_5)_2Re_2(O)(O_2SO)(\mu\text{-}O)_2$	I (57%) blood-red crystals from CH_2Cl_2/hexane; dec.p. 166 °C ^1H NMR (CD_2Cl_2): 1.88, 2.06 $^{13}C\{^1H\}$ NMR ($CDCl_3$): 10.69, 10.71; 110.71, 116.95 ^{17}O NMR (CH_2Cl_2): 468.0 IR (KBr): 548, 650, 716, 731, 813, 833; 908 (ν(ReO)); 1026; 1107 (ν(SO)); 1118, 1141, 1172, 1374, 2858, 2922, 2963 thermolysis at 220 °C gave No. 9 and SO_2 treatment with excess SO_2 gave No. 13 [21]
3	$(C_5(CH_3)_5)_2Re_2(O)(OC(=O)N(C_6H_5)\text{-cyclo})(\mu\text{-}O)_2$	II (17%) ^1H NMR ($CDCl_3$): 1.77, 1.92 (CH_3); 6.65 to 7.46 (C_6H_5) IR (KBr): 899 (ν(ReO)); 1659, 1687 (ν(C=O)) FD MS: $[M]^+$ treatment with excess $C_6H_5N{=}C{=}O$ yielded No. 15 [25]
4	$(C_5(CH_3)_5)_2Re_2(O)(OC(=S)N(C_6H_5)\text{-cyclo})(\mu\text{-}O)_2$	II (76%); III (76%) dark violet crystals ^1H NMR ($CDCl_3$): 1.73, 1.99 (CH_3); 6.86 to 7.66 (C_6H_5) $^{13}C\{^1H\}$ NMR ($CDCl_3$): 10.4 (CH_3), 107.7 (C_5); 116.8, 123.5, 127.3, 145.7 (C_6H_5); 209.0 (CS) IR (KBr): 901 (ν(ReO)); 1230, 1357 (ν(C=S)) FD MS: $[M]^+$ treatment with excess $C_6H_5N{=}C{=}S$ yielded No. 16 [25]
5	$(C_5(CH_3)_5)_2Re_2(O)(OC(=S)N(C_6H_4CF_3\text{-}3)\text{-cyclo})(\mu\text{-}O)_2$	III (64%) ^1H NMR ($CDCl_3$): 1.74, 1.97 (CH_3); 7.09 to 7.71 (C_6H_4) IR (KBr): 899 (ν(ReO)); 1230, 1342 (ν(C=S)) FD MS: $[M]^+$ [25]
6	$(C_5(CH_3)_5)_2Re_2(OReO_3)Cl_3(\mu\text{-}O)_2$	by combining $AgReO_4$ (frozen CH_2Cl_2 suspension) and 0.25 equivalent $(C_5(CH_3)_5)Re(O)Cl_2$ in CH_2Cl_2 followed by warming the mixture to room temperature, filtration, and addition of ether; recrystallization from CH_2Cl_2/ether yielded 69% (no further substitution of Cl upon using a large excess of $AgReO_4$)

References on p. 202

Table 9 (continued)

No.	compound	method of preparation (yield) properties and remarks
6 (continued)		light green powder; dec.p. 175°C ^{1}H NMR (CD_2Cl_2): 2.08, 2.16 ^{13}C {^{1}H} NMR (CD_2Cl_2): 10.69, 10.81; 121.96, 122.17 IR (KBr): 290, 325 (ν(ReCl)); 848 to 873 (ν(ReO)); 906, 930, 940, 976, 1018, 1079, 1105, 1373, 1430, 1482, 2853, 2920, 2966 [21]
*7	$(C_5(CH_3)_5)_2Re_2(OReO_3)_2(OC_2H_4O)(\mu\text{-}O)_2$ 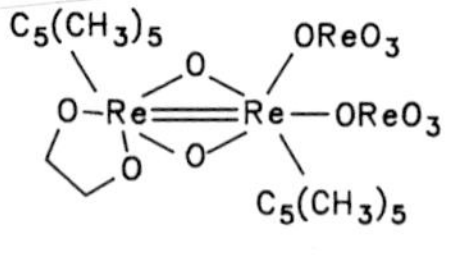	by exposing $(C_5(CH_3)_5)Re(O)OC_2H_4O$-cyclo to air while simultaneously heating green crystals; no m.p. <250°C ^{1}H NMR ($CDCl_3$): 2.147, 2.161 (CH_3); 3.92, 4.40 (ddd's, C_2H_4) ^{17}O NMR (toluene-d_8): 657 (ReO_2Re), 701 (ReO_3) IR (KBr): 518, 543 (ν(ReOC)); 606 (ν(ReORe)), 910 (ν(ReO)) [13]
*8	$(C_5(CH_3)_5)_2Re_2(OReO_3)_2(O)(\mu\text{-}O)_2 \cdot CH_2Cl_2$ 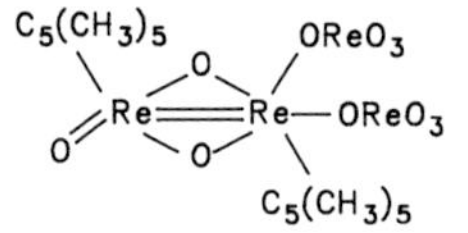	for preparation see "Further information" light brown, air-stable solid [4, 6]; m.p. >230°C [6] ^{1}H NMR ($CDCl_3$): 1.93, 2.21 (s's, CH_3); 5.32 (CH_2Cl_2) IR (KBr): 688, 700, 729, 827, 855, 899, 934, 973 FD MS: $[M]^+$ [6]
symmetrical compounds		
*9	$[(C_5(CH_3)_5)ReO]_2(\mu\text{-}O)_2$	for preparation see "Further information" very air-sensitive solid; dec.p. 120°C [6] ^{1}H NMR: 1.96 (C_6D_6) [26]; 2.03 (CD_2Cl_2) [6, 16]; 2.07 ($CDCl_3$) [8, 16]; see also "Further information" ^{13}C {^{1}H} NMR (C_6D_6): 11.4, 108.5 [26] ^{17}O NMR (CH_2Cl_2): 104 (ReORe), 747 (ReO) [18, 21] (similar in [16]) IR (KBr): 614, 634 (ν(ReORe)); 930 (ν(ReO)) [6]; 538, 621, 640; 933 [8, 16]; (CS_2): 620, 643 (ν(ReORe)); 920, 931 (ν(ReO)); 2800, 2854, 2914 [21] UV (toluene): λ_{max} = 418 (intensity decreases with increasing temperature; isosbestic point at λ = 349) [26] EI MS: $[1/2\ M]^+$; FD MS: $[M]^+$ [6]
*10	$[(C_5(CH_3)_5)ReCl_2]_2(\mu\text{-}O)_2$	for preparation see "Further information" light green solid; dec.p. ca. 180°C [9]

References on p. 202

Table 9 (continued)

No.	compound	method of preparation (yield) properties and remarks
		^{1}H NMR (CD_2Cl_2): 2.00 ^{13}C {^{1}H} NMR (CD_2Cl_2): 11.00 (CH_3), 120.36 (C_5) [9, 16] ^{17}O NMR (CD_2Cl_2): 629 [16] IR (KBr): 295, 340 (ν(ReCl)); 626, 697, 729 (ν(ReORe)) [9, 16]
11	$[(C_5(CH_3)_5)Re(OCN)_2]_2(\mu\text{-}O)_2$	IV (82%) dark green solid; dec.p. >200 °C ^{1}H NMR (CD_2Cl_2): 2.08 ^{13}C {^{1}H} NMR (CD_2Cl_2): 10.61 (CH_3), 120.55 (C_5); δ(OCN) not observed ^{17}O NMR (CD_2Cl_2): 626.8 IR (KBr): 589 (ν(ReORe)); 635, 1010, 1072, 1375, 1431, 1477; 2227 (ν(OCN)); 2253, 2854, 2950, 2977; (CS_2): 725 (ν(ReORe)); 2801, 2848, 2921 FD MS: $[1/2\ M]^+$ EI MS: $[1/2\ M - n\ OCN]^+$ (n = 0 to 2) [21]
*12	$[(C_5(CH_3)_5)Re(CN)_2]_2(\mu\text{-}O)_2$	IV (77%) dark green solid; dec.p. 175 °C ^{1}H NMR (CD_2Cl_2): 2.34; but at −60 °C: 2.00, 2.29 ^{13}C {^{1}H} NMR (CD_2Cl_2): 11.50 (CH_3), 110.13 (C_5), 270.90 (CN) ^{17}O NMR (CD_2Cl_2): 864.0; at −55 °C: 663.2, 867.5 IR (Nujol): 722 (ν(ReORe)); 2127, 2144 (ν(CN)); (CH_2Cl_2): 903 (ν(ReO)); 1156, 1255, 1421; 2131, 2159 (ν(CN)); 2985, 3045 EI MS: $[1/2\ M]^+$, $[(C_5(CH_3)_5)ReO - HCN]^+$ [21]
13	$[(C_5(CH_3)_5)ReO_2SO]_2(\mu\text{-}O)_2$	V (chromatography on silylated silica with CH_2Cl_2); also with 94% yield by treating No. 2 with SO_2 green crystals from CH_2Cl_2/hexane; dec.p. 190 °C ^{1}H NMR (DMF-d_7, 20 to 110 °C): 2.10, 2.11; (120 °C): 2.09 (br) ^{13}C {^{1}H} NMR (CD_2Cl_2): 9.86, 9.92; 119.48, 119.69 IR (KBr): 615, 695; 715, 723 (ν(ReORe)); 801, 821, 842, 1018, 1080; 1100 (ν(SO)); 1126, 1147, 1374, 1477; 2856, 2918, 2965; (CH_2Cl_2): 902 (ν(ReO)); 951, 993; 1146 (ν(SO)); 1265, 1376, 2918, 2988, 3051 solid-state thermolysis at 220 °C gave No. 9 and SO_2 CH_2Cl_2 solutions contain the complex in a monomeric form, but upon evaporation IR bands due to ν(ReORe) were observed [21]

References on p. 202

Table 9 (continued)

No.	compound	method of preparation (yield) properties and remarks
14	$[(C_5(CH_3)_5)ReO_2SeO]_2(\mu\text{-}O)_2$	V (chromatography on Florisil with DMF at $-10\,°C$); the product was contaminated with SeO_2 which could not be completely removed ^1H NMR ($CDCl_3$): 2.07, 2.21 $^{13}C\{^1H\}$ NMR ($CDCl_3$): 28.99, 29.46; 160.84, 204.21 IR (KBr): 663 (ν(ReORe)), 670, 1068; 1105 to 1116 (ν(SeO)); 1379, 2855, 2926, 2954 [21]
15	$[(C_5(CH_3)_5)ReOC(=O)N(C_6H_5)\text{-cyclo}]_2(\mu\text{-}O)_2$ 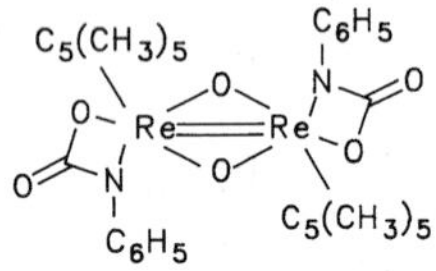	VI (21%) ^1H NMR ($CDCl_3$): 1.94 (CH_3), 7.06 to 7.49 (C_6H_5) IR (KBr): 1722 (ν(C=O)) CI MS: $[M + H]^+$ no reaction upon thermolysis in toluene [25]
16	$[(C_5(CH_3)_5)ReOC(=S)N(C_6H_5)\text{-cyclo}]_2(\mu\text{-}O)_2$ 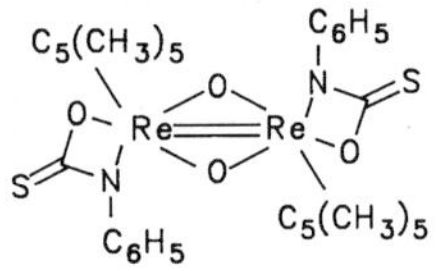	VI (23%) ^1H NMR ($CDCl_3$): 1.91 (CH_3), 7.14 to 7.48 (C_6H_5) IR (KBr): 1279 (ν(C=S)) CI MS: $[M + H]^+$ no reaction upon thermolysis in toluene [25]
*17	$[\{(C_5(CH_3)_5)Re(C_6H_5C{\equiv}CC_6H_5)\}_2(\mu\text{-}O)_2][BF_4]_2$ 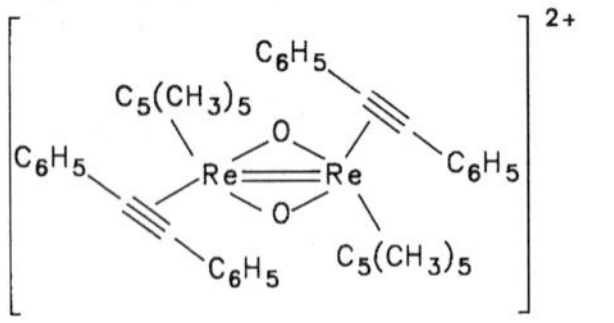	for preparation see "Further information" red-black shiny crystals; dec.p. 188 °C ^1H NMR (CD_2Cl_2, $-60\,°C$): 1.62 (CH_3); 7.81, 7.90, 8.08 (t's; J = 7.9, 7.3, 7.3, resp.), 8.82 (d; J = 7.9) $^{13}C\{^1H\}$ NMR (CD_2Cl_2, $-60\,°C$): 10.8 (CH_3), 115.1 (C_5); 129.9, 130.6, 131.2, 132.2, 135.9, 137.0 (C_6H_5); 186.5 (C≡C) ^{17}O NMR (CD_2Cl_2): 420 IR (KBr): 668, 723 (ν(ReO)); 1038, 1051 (ν(ReORe)); 1064, 1085 (BF_4); 1691 (ν(C≡C)) FD MS: $[M]^+$ [20]
*18	$[(C_5(CH_3)_5)ReO]_2(\mu\text{-}S)_2$	from No. 9 and excess CS_2 (toluene, reflux, 8 h); workup by evaporation and extraction of the residue with toluene and ether; recrystallization of the remaining solid from CH_2Cl_2 at $-35\,°C$ yielded 20% dark brown crystals ^1H NMR (CD_2Cl_2): 1.99 IR (KBr): 536 (ν(ReS)); 623, 694, 730; 911, 936 (ν(ReO)); 1022, 1073, 1189, 1371, 1437, 2858, 2908, 2954 EI MS: $[M]^+$, $[(C_5(CH_3)_5)Re_2S_2]^+$ [21]

References on p. 202

Table 9 (continued)

No.	compound	method of preparation (yield) properties and remarks
19	$[(C_5(CH_3)_5)ReO]_2(\mu\text{-}S)(\mu\text{-}S_2)$	VII dark brown crystals; m.p. 175 to 176°C; black solid ^{1}H NMR ($CDCl_3$): 1.97 $^{13}C\{^1H\}$ NMR ($CDCl_3$): 11.1, 107.9 IR (CsI): 898, 902 FD MS: $[M]^+$ EI MS: $[M - n\,S]^+$ (n = 0 to 2), $[M - S - O]^+$ [22]
20	$[(C_5(CH_3)_5)ReO]_2(\mu\text{-}S_2)(\mu\text{-}S_3)$ (suggested structure)	by treating $(C_5(CH_3)_5)ReO_3$ with a slight excess of B_2S_3 (THF, 24 h); chromatographic separation with CH_2Cl_2/THF (1:1); yield: 21.1% black microcrystals; m.p. 187 to 189°C ^{1}H NMR ($CDCl_3$): 2.13 $^{13}C\{^1H\}$ NMR ($CDCl_3$): 10.4, 102.4 IR (CsI): 875, 911 FD MS: $[M]^+$; EI MS: $[M - n\,S]^+$ (n = 0 to 4) [22]
*21	$[C_5H_5Re(CO)]_2(\mu\text{-}S_2)(\mu\text{-}S_3)$	for preparation see "Further information" ^{1}H NMR ($CDCl_3$): 4.67 [2, 15] $^{13}C\{^1H\}$ NMR ($CDCl_3$): 93.0 (C_5H_5); δ(CO) not observed [2, 15] IR (ether): 1920 [2, 15] MS: $[M - n\,CO]^+$ (n = 0 to 2), $[M - 2\,CO - n\,S]^+$ (n = 1 to 3), $[M - n\,S]^+$ (n = 2, 3) [2] readily soluble in THF; soluble in CH_2Cl_2 [2]
*22	$[(C_5(CH_3)_5)ReO]_2(\mu\text{-}Se)_2$	from No. 23 and excess $P(C_4H_9\text{-}n)_3$ ($CHCl_3$, ca. 12 h); separation by chromatography yielded 49.2% black plates from $CHCl_3$/toluene/hexane; m.p. 214 to 215°C ^{1}H NMR ($CDCl_3$): 2.06 $^{13}C\{^1H\}$ NMR ($CDCl_3$): 12.4, 107.8 IR (CsI): 937 (ν(ReO)) EI MS: $[M]^+$, $[M - O]^+$, $[M - Se]^+$, $[M - C_5(CH_3)_5]^+$ [22]
*23	$[(C_5(CH_3)_5)ReO]_2(\mu\text{-}Se)(\mu\text{-}Se_2)$	VII (48.8%) dark brown needles from $CHCl_3$/hexane; m.p. 207 to 208°C ^{1}H NMR ($CDCl_3$): 1.94 $^{13}C\{^1H\}$ NMR ($CDCl_3$): 12.2, 106.8 IR (CsI): 897, 905 (ν(ReO)) FD MS: $[M]^+$; EI MS: $[M - n\,Se]^+$ (n = 0 to 2), $[M - C_5(CH_3)_5 - n\,Se]^+$ (n = 0, 1) [22]
24	$[(C_5(CH_3)_5)ReO]_2(\mu\text{-}Te)_2$	by treating $(C_5(CH_3)_5)Re(O)Cl_2$ with $[N(C_4H_9\text{-}n)_4]Te_5$ (DMF, 60°C, 2 h; separation by chromatography with CH_2Cl_2); yield: 62.5%

References on p. 202

Table 9 (continued)

No.	compound	method of preparation (yield) properties and remarks
24 (continued)		black needles from CH_2Cl_2/pentane; m.p. 246 to 248 °C ^1H NMR ($CDCl_3$): 2.13 ^{13}C {^1H} NMR ($CDCl_3$): 12.9, 100.9 IR (CsI): 933 (ν(ReO)) EI MS: $[M]^+$, $[M - H_2O]^+$, $[M - C_5(CH_3)_5H]^+$, $[M - C_5(CH_3)_5H - H_2O]^+$ [22]
25	$[(C_5(CH_3)_5)Re(CO)_2]_2(\mu\text{-}Te)_2$	by treating $(C_5(CH_3)_5)Re(CO)_2OC_4H_8$ with TeH_2 in the dark; could be separated by chromatography on Florisil at −20 °C; yield: 24% red-brown needles ^1H NMR (C_6D_6): 1.85 [11] ^{125}Te NMR (toluene-d_8): −510 [11, 17] photolysis or thermolysis in THF induced rearrangement into $[(C_5(CH_3)_5)Re(CO)_2]_2(\mu\text{-}\eta^{2:1}\text{-}Te_2)$ [11]

* Further information:

$(C_5(CH_3)_5)_2Re_2(O)Cl_2(\mu\text{-}O)_2$ (Table **9**, No. **1**) was also obtained when a THF solution containing $(C_5(CH_3)_5)Re(O)Cl_2$ was treated with a 3-fold excess of MgO and a small amount of H_2O for 1 h. After filtration, the filtrate was evaporated and the residue purified by chromatography on silica with THF. The crude product was crystallized from acetone/pentane. Yield: 73%. $(C_5(CH_3)_5)Re(O)Cl_2$ could also be treated with $[(C_5(CH_3)_5)ReO]_2(\mu\text{-}O)_2$ (ratio 2:1) in refluxing toluene. Chromatographic separation on silylated silica gave No. 1 with 85% yield. The reaction must be carried out under strong exclusion of air [16] (also briefly in [12]). The title product also formed with small yield by treating $(C_5(CH_3)_5)Re(O)Cl_2$ with 2 equivalents $Ag_2C_2O_4$ in CH_2Cl_2 in the dark; in this case the main product was $[(C_5(CH_3)_5)ReO_2C_2O_2]_2$-$(\mu\text{-}Cl)_2$ (see p. 205) [21].

An O*-enriched (O* = ^{17}O, ^{18}O) derivative could be prepared from $(C_5(CH_3)_5)Re(O^*)Cl_2$ and H_2O/KOH in THF for 20 h. Chromatographic workup on silylated silica gave No. 1 along with $[(C_5(CH_3)_5)ReO^*]_2(\mu\text{-}O^*)_2$ [21].

Single crystals have a monoclinic lattice with a = 14.629(3), b = 9.510(2), c = 16.294(3) Å, β = 95.42(2)°; space group $P2_1/c-C_{2h}^5$ (No. 14); Z = 4 molecules per unit cell, D_{calc} = 2.242 g/cm³. The molecular structure is illustrated in **Fig. 97**. The $C_5(CH_3)_5$ rings are in a cis position with respect to each other, and there are different Re-(μ-O) bond lengths. It can be concluded that the compound formally consists of a $(C_5(CH_3)_5)ReO_3$ and a $(C_5(CH_3)_5)ReCl_2$ fragment [16].

Upon treatment with KOH the compound completely hydrolyzed to $[(C_5(CH_3)_5)ReO]_2$-$(\mu\text{-}O)_2$ [16]. Treatment with HCl/H_2O gave No. 10 [12, 16].

$(C_5(CH_3)_5)_2Re_2(OReO_3)_2(OC_2H_4O)(\mu\text{-}O)_2 \cdot 2\ C_6H_5CH_3$ (Table **9**, No. **7**). Single crystals are triclinic with a = 9.382(1), b = 14.523(1), c = 15.437(1) Å, α = 81.42(1)°, β = 82.60(1)°, γ = 72.88(1)°; space group $P\bar{1}-C_i^1$ (No. 2); Z = 2 formula units per unit cell, D_{calc} = 2.381 g/cm³. The molecular structure is illustrated in **Fig. 98a** [13].

References on p. 202

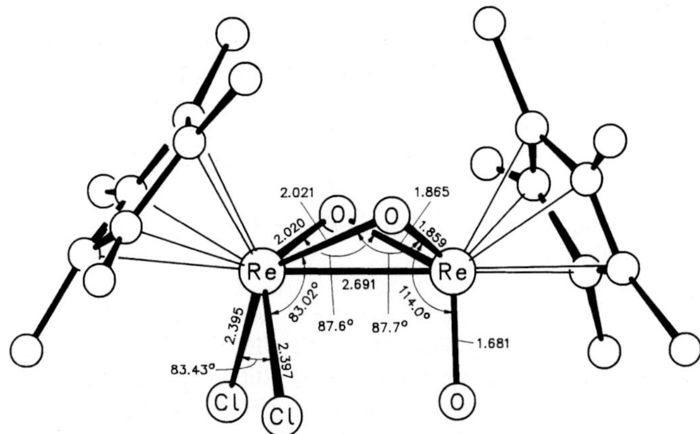

Fig. 97. Molecular structure of $(C_5(CH_3)_5)_2Re_2(O)Cl_2(\mu\text{-}O)_2$ [16].

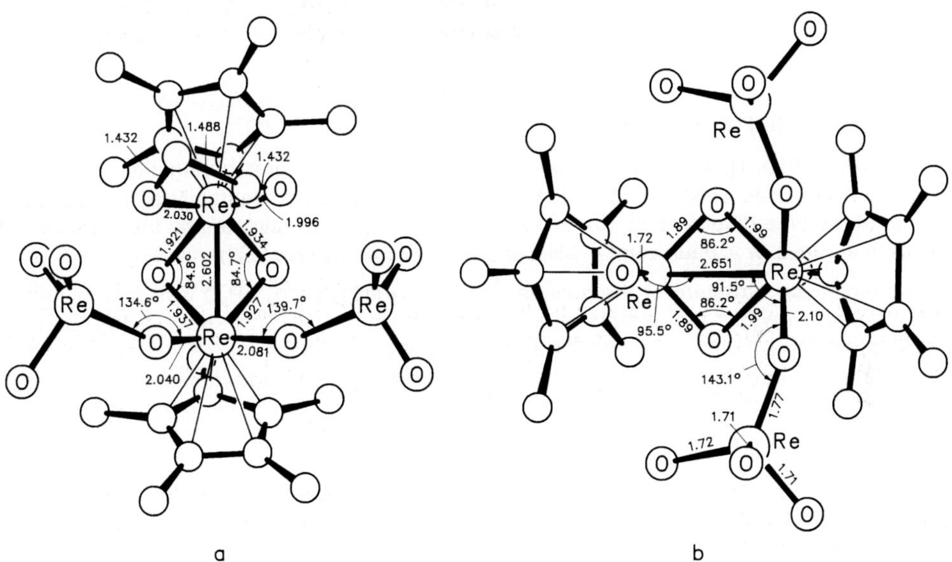

Fig. 98. Molecular structures of
a. $(C_5(CH_3)_5)_2Re_2(OReO_3)_2(OC_2H_4O)(\mu\text{-}O)_2$ [13];
b. $(C_5(CH_3)_5)_2Re_2(OReO_3)_2(O)(\mu\text{-}O)_2$ [6].

$(C_5(CH_3)_5)_2Re_2(OReO_3)_2(O)(\mu\text{-}O)_2 \cdot CH_2Cl_2$ (Table **9**, No. **8**) formed by treating $(C_5(CH_3)_5)$-ReO_3 in THF with a slight excess of $P(C_6H_5)_3$ in the presence of a deficient amount of oxygen. After stirring the mixture for 20 h, the solvent was removed and the residue was recrystallized from CH_2Cl_2/ether at −35 °C. Yield: 30 to 60% [6], 40 to 50% [4].

The compound crystallizes in the orthorhombic space group Pcnm−D_{2h}^{19} (No. 62) with a = 13.392(7), b = 14.900(3), c = 15.162(5) Å; Z = 4 formula units per unit cell. The structure of the molecule is depicted in **Fig. 98b**. The central Re_2O_2 core is folded with a folding angle of 122° along the Re···Re vector [6].

References on p. 202

$[(C_5(CH_3)_5)ReO]_2(\mu\text{-}O)_2$ (Table **9**, No. **9**) was obtained by deoxygenating $(C_5(CH_3)_5)ReO_3$ with a slight excess of $P(C_6H_5)_3$ under exclusion of air [3, 6, 7, 16]. The reactants were stirred in THF for 12 to 20 h. After removal of the solvent, the residue was washed with cold acetone and crystallized from acetone/CH_2Cl_2 at −30 °C. Yield: 60 to 76% [6], 70% [16]. The deoxygenation with $P(C_6H_5)_3$ also proceeded in the presence of alkyl isothiocyanates (in the presence of aryl thiocyanates or aryl isothiocyanates, however, Nos. 3, 4 were obtained) [25]. The complete separation of $P(C_6H_5)_3O$ proved to be difficult, but the synthesis could be substantially improved by using polymer-bonded $P(C_6H_5)_3$ (toluene, 2 d, yield: 80%) or $P(C_6H_4SO_3Na\text{-}3)_3$ (CH_3OH/H_2O (1:1), 2 d, yield: 92%) [16]. The deoxygenation of $(C_5(CH_3)_5)ReO_3$ could also be effected with $(C_{10}H_8)Zr_2(C_5H_5)_2(\mu\text{-}Cl)_2$ (THF, room temperature; formation of $[(C_5(CH_3)_5)ReO]_2(\mu\text{-}O)_2$ was confirmed spectroscopically) [8, 14]. Otherwise, careful oxidation of $(C_5(CH_3)_5)Re(RC{\equiv}CR)(O)$ (R = CH_3, C_2H_5, C_6H_5) with C_6H_5IO, C_6H_5NO, C_5H_5NO, $N(CH_3)_3O$, or O_2 also produced the title dimer along with $(C_5(CH_3)_5)ReO_3$ [20]. No. 9 was also obtained by treating $(C_5(CH_3)_5)ReCl_4$ with a large excess of H_2O in acetone [12], a 20-fold excess of KOH in H_2O (20 min, yield: 81%) [16], or a 10-fold excess of LiOH (THF, stirring for 3 h followed by filtration, evaporation, and precipitation from toluene/hexane; yield: 90%) [21] (a similar hydrolysis of $(C_5(CH_3)_5)Re(O)Cl_2$ with H_2O/KOH in THF also led to the title compound [18, 21]; side product was No. 1 [21]). An entirely different route to No. 9 was the solid-state thermolysis of the compounds $(C_5(CH_3)_5)_2Re_2(O)_n\text{-}(O_2SO)_{2-n}(\mu\text{-}O)_2$ (n = 1, 0; Nos. 2, 13) at 220 °C [21].

O*-enriched derivatives (O* = ^{17}O or ^{18}O) were obtained from $(C_5(CH_3)_5)Re(O)Cl_2$ and H_2O^*/KO*H in THF [18, 21] or from $(C_5(CH_3)_5)ReCl_4$ and LiO*H [21].

$[(C_5(CH_3)_5)ReO]_2(\mu\text{-}O)_2$ was suggested to intermediately form during the reaction of $(C_5(CH_3)_5)ReO_3$ with vicinal diols in the presence of $P(C_6H_5)_3$, eventually producing compounds of the type displayed in Formula III. The reaction was facilitated by the presence of a molecular sieve or 1 equivalent $4\text{-}CH_3C_6H_4SO_3H$ [23] (see also [27]).

The molecular structure of the solvate **$[(C_5(CH_3)_5)ReO]_2(\mu\text{-}O)_2 \cdot H_2O$** was determined at −35 °C. Single-crystal parameters are as follows: a = 9.071(3), b = 9.369(2), c = 15.683(3) Å, α = 67.64(2)°, β = 66.74(2)°, γ = 64.52(2)°; space group $P\bar{1}-C_i^1$ (No. 2); Z = 2 formula units per unit cell, D_{calc} = 2.251 g/cm³. In the unit cell, two symmetry-independent, centrosymmetric molecules are linked by H_2O bridges (compare with **Fig. 99**). In the complex

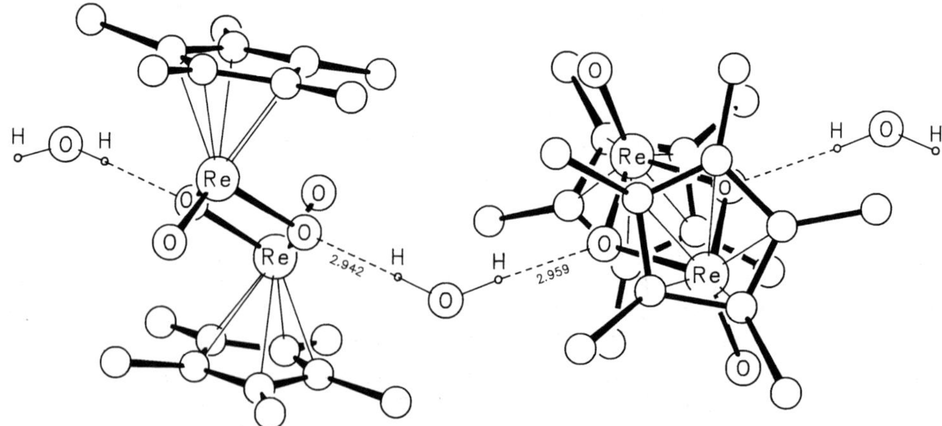

Fig. 99. The unit cell arrangement of $[(C_5(CH_3)_5)ReO]_2(\mu\text{-}O)_2 \cdot H_2O$ [16].

References on p. 202

molecule, the $C_5(CH_3)_5$ ligands adopt a trans geometry with respect to the planar Re_2O_4 core. The Re-Re distance is too long for a double bond initially assigned to the compound by electron counting. The C_5 ligands are asymmetrically bonded (variation of the Re-C bonds between 2.192 and 2.450 Å). Intramolecular parameters are listed in the table on p. 201 [16].

Fenske-Hall MO calculations on the model molecule **$[C_5H_5ReO]_2(\mu\text{-}O)_2$** elucidate why the Re-Re distance is thus large and the $C_5(CH_3)_5$ ligands are asymmetrically bonded: The variations in frontier orbital energy as the Re-Re distance is varied from 2.74 to 3.54 Å show that the most stable $Re_2(\mu\text{-}O)_2$ core is that which is somewhere between an Re-Re bond and an O-O bond. Consequently, the dimer is held together through the Re-O interactions. Moreover, the shift of the C_5H_5 ligands away from the terminal oxo ligands to an unsymmetrical Re-C_5 bonding results in a substantial stabilization of the molecule by ca. 9.8 eV compared with a symmetric Re-C_5 interaction. Correlation diagrams are given [19].

NMR spectroscopy indicated that in solution the compound exists in an equilibrium with another species: When raising the temperature, additional signals at $\delta = 1.66$ ppm (^1H) and $\delta = 12.4$, 109.3 ppm (^{13}C) become observable (in C_6D_6). The changes are reversible on cooling. The new signals were attributed to the monomer $(C_5(CH_3)_5)ReO_2$. Integration gave $K_{eq} = [\text{monomer}]^2/[\text{dimer}] = (6.05 \pm 1.29) \times 10^{-4}$ M at 93°C; the thermodynamic parameters $\Delta H^{\neq} = 70.3 \pm 1.25$ kJ/mol and $\Delta S^{\neq} = 130.5 \pm 5.85$ J · mol^{-1} · K^{-1} were also calculated [26].

Exposure to air gave $(C_5(CH_3)_5)ReO_3$ again [6, 16] (even in the solid state [3]), and interaction with $^*O_2$-enriched O_2 ($^*O = {}^{17}O$ [24] or ^{18}O [6]) gave $^*O$-enriched $(C_5(CH_3)_5)Re^*O_3$, which could in turn be converted into $^*O$-enriched $[(C_5(CH_3)_5)ReO]_2(\mu\text{-}O)_2$ by treating with $P(C_6H_5)_3$ as described above [24]. In contrast, there was no reaction with elemental S_8 [21]. Treatment with Br_2 at −78°C gave $(C_5(CH_3)_5)Re(O)Br_2$ [3, 4], and treatment with aqueous HX in THF (X = F, Cl, Br) gave $(C_5(CH_3)_5)Re(O)X_2$ [9, 12, 16] and (for X = Cl) a small quantity of $[(C_5(CH_3)_5)ReCl_2]_2(\mu\text{-}O)_2$ [9, 16]. Protonation or methylation with HBF_4 or CH_3OSO_2-CF_3, respectively, gave the salts $[\{(C_5(CH_3)_5)ReO\}_2(\mu\text{-}OR)_2]X_2$ (R = H, CH_3; X = BF_4, O_3SCF_3) [9]. Pressurization with CO (20 atm, 70 h) produced a mixture consisting of $(C_5(CH_3)_5)ReO_3$, $(C_5(CH_3)_5)Re(CO)_3$, $(C_5(CH_3)_5)_2Re_2(CO)_3$ [3, 4], and $[(C_5(CH_3)_5)Re$-$(CO)_2]_2(\mu\text{-}O)$ (traces) [4]. Treatment with CS_2 in hot toluene gave $[(C_5(CH_3)_5)ReO]_2(\mu\text{-}S)_2$ [21]. The reaction with EO_2 (E = S, Se) according to Method V yielded compounds of the type $[(C_5(CH_3)_5)ReO_2EO]_2(\mu\text{-}O)_2$ (Nos. 13, 14) [21], and treatment with $(CH_3)_2SO$ at 65°C gave $(C_5(CH_3)_5)ReO_3$ and $(CH_3)_2S$ [26]. The interaction with $[C_5H_5NH]I$ furnished $(C_5(CH_3)_5)Re(O)I_2$ [12]. Treatment with $(CH_3)_3SiX$ (X = Cl, Br) gave $(C_5(CH_3)_5)ReX_4$ [12, 16]. Treatment with $AgBF_4$ gave $[(C_5(CH_3)_5)_3Re_3(\mu\text{-}O)_6][BF_4]_2$ [14].

Combination of No. 9 with epoxides such as styrene oxide or 1,1-dimethyl oxirane between 50 and 105°C quantitatively provided $(C_5(CH_3)_5)ReO_3$ and the respective alkene. An investigation of the kinetics of the reaction with styrene oxide under pseudo first-order conditions revealed the rate constant $k_{obs} = 6.4 \times 10^{-8}$ M$^{-1/2}$ · s^{-1} (C_6D_6, 50°C), corresponding to a second-order rate constant of $k_2 = 2.1 \times 10^{-5}$ M^{-1} · s^{-1} and to $\Delta G^{\neq} = 107.1$ kJ/mol [26].

While the compound did not react with diacetyl, the reaction with o-quinones such as chloranil ($C_6Cl_4O_2$) or 9,10-phenanthrenequinone ($C_{14}H_8O_2$) gave mixtures of $(C_5(CH_3)_5)Re$-$(O_2C_6Cl_4)_2$, $(C_5(CH_3)_5)Re(O)(O_2C_6Cl_4)$, $[(C_5(CH_3)_5)Re(O)O_2C_6Cl_4]_2$ (see p. 206), $(C_5(CH_3)_5)$-$Re(O_2C_{14}H_8)_2$, and $(C_5(CH_3)_5)Re(O)(O_2C_{14}H_8)$ [5]. Reactions with $(C_6H_5)_2C{=}C{=}O$ [4, 10] or $C_6H_5N{=}C{=}O$ yielded $(C_5(CH_3)_5)Re(O)O_2C{=}C(C_6H_5)_2$ [4, 10] and $(C_5(CH_3)_5)Re(O)OC(O)$-$N(C_6H_5)$-cyclo [10], respectively. Treatment with 2 equivalents 3-$RC_6H_4N{=}C{=}S$ (R = H, CF_3) according to Method III gave Nos. 4 and 5; in contrast, the title compound did not react with alkyl isothiocyanates, even not in refluxing toluene [25].

Treatment with $C_5H_5W(CO)_3C{\equiv}CC_6H_5$ gave the two heterometal clusters displayed in Formulas IV and V [24]. Interaction with $(C_5(CH_3)_5)Re(O)Cl_2$ yielded No. 1 [12, 16].

References on p. 202

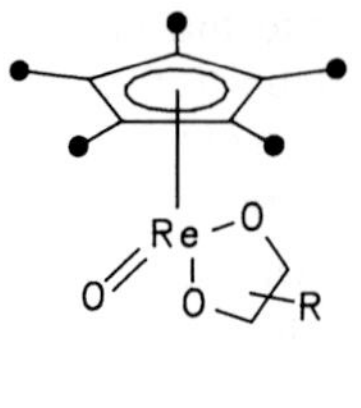

III 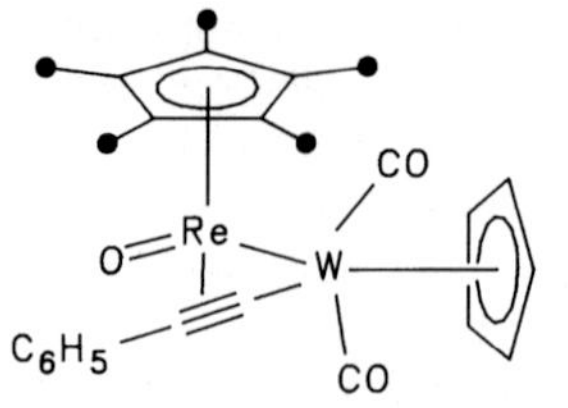 IV 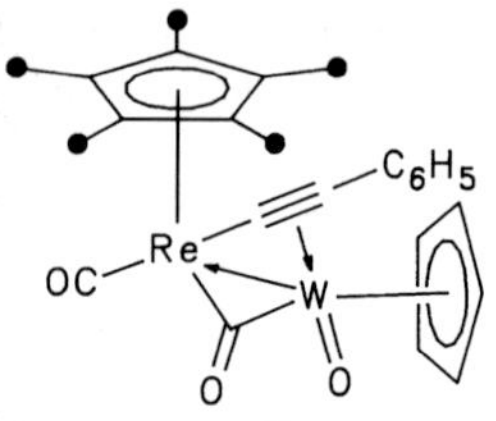 V

$[(C_5(CH_3)_5)ReCl_2]_2(\mu\text{-}O)_2$ (Table **9**, No. **10**). $(C_5(CH_3)_5)_2Re_2(O)Cl_2(\mu\text{-}O)_2$ (No. 1) was treated with aqueous HCl in THF for 5 min. Subsequent column-chromatographic separation (silylated silica, CH_2Cl_2) gave No. 10 with 85 to 95% yield. Crystallization was from acetone/THF or THF/hexane [16] (also briefly in [12]). No. 10 was almost quantitatively obtained by treating $(C_5(CH_3)_5)Re(O)Cl_2$ with H_2O in acetone (actually, an attempted chromatographic purification of $(C_5(CH_3)_5)Re(O)Cl_2$ on wet Florisil with acetone resulted in elution of the title compound [16]). Only a low yield (5%), however, was achieved by treating $[(C_5(CH_3)_5)ReO]_2$-$(\mu\text{-}O)_2$ (No. 9) with aqueous HCl in THF; chromatographic workup on silylated silica with toluene/CH_2Cl_2 mixtures mainly separated $(C_5(CH_3)_5)Re(O)Cl_2$ [9, 16].

Single crystals obtained from CH_2Cl_2 still contained various portions of CH_2Cl_2. The solvent molecules slowly diffused off the crystal, even at −60 °C, and thereby crystal decomposition occurred. Nevertheless, the following single-crystal data could be determined: triclinic; a = 8.154(1), b = 10.133(4), c = 19.587(3) Å, α = 96.80(3)°, β = 90.82(1)°, γ = 106.51(2)°; space group $P\bar{1}-C_i^1$ (No. 2). The molecule contains two $C_5(CH_3)_5$ rings in a cis position with respect to each other [16].

The complex is well soluble in CH_2Cl_2, less well soluble in THF and $CHCl_3$ [16]. Thermolysis in THF at reflux temperature gave $(C_5(CH_3)_5)Re(O)Cl_2$ [9].

$[(C_5(CH_3)_5)Re(CN)_2]_2(\mu\text{-}O)_2$ (Table **9**, No. **12**). Spectroscopic data are consistent with a solid-state dimer, while in solution the complex is in a monomer-dimer equilibrium: Upon cooling the solution, dimerization occurred. The process is completely temperature-reversible. CH_2Cl_2 solutions are orange; they turn green at −70 °C [21].

$[\{(C_5(CH_3)_5)Re(C_6H_5C{\equiv}CC_6H_5)\}_2(\mu\text{-}O)_2][BF_4]_2$ (Table **9**, No. **17**) was obtained by oxidizing $(C_5(CH_3)_5)Re(C_6H_5C{\equiv}CC_6H_5)(O)$ with 1 equivalent $[(C_5H_5)_2Fe]BF_4$ (CH_2Cl_2, 15 min) followed by removing the solvent. The residue was washed with ether and the remainder recrystallized from CH_2Cl_2. Yield: 95%. The reaction could be reversed with Na/Hg [20].

$[(C_5(CH_3)_5)ReO]_2(\mu\text{-}S)_2$ (Table **9**, No. **18**) crystallizes in the tetragonal space group $P4_2/n-C_{4h}^4$ (No. 86) with a = b = 16.435(1), c = 8.478(1) Å; Z = 4 molecules per unit cell, D_{calc} = 2.143 g/cm^3. The molecule is centrosymmetric. The $C_5(CH_3)_5$ rings are shifted from the ideal η^5-coordination to an η^3-allyl coordination (Re-C bond lengths are: 2.180, 2.234, 2.253, 2.496, and 2.503 Å). Other important intramolecular parameters are listed in the table on p. 201 [21].

$[C_5H_5Re(CO)]_2(\mu\text{-}S_2)(\mu\text{-}S_3)$ (Table **9**, No. **21**) was one of the products obtained by irradiating an ethereal solution containing $C_5H_5Re(CO)_3$ followed by adding S_8 or COS to the mixture [1, 2]. The yield amounted to 2% [1] or 6% [2]. Brown crystals [2].

No. 21 crystallizes in the monoclinic space group $P2_1/n-C_{2h}^5$ (No. 14) with a = 10.775(2), b = 15.137(8), c = 11.152(3) Å, β = 114.21(2)°; Z = 4 molecules per unit cell, D_{calc} = 2.878 g/cm^3. The molecular structure is illustrated in **Fig. 100** [2].

$[(C_5(CH_3)_5)ReO]_2(\mu\text{-}Se)_2$ (Table **9**, No. **22**) crystallizes in the tetragonal space group $P4_2/n$ with a = b = 16.503(2), c = 8.538(2) Å; Z = 4 molecules per unit cell. The molecule is

References on p. 202

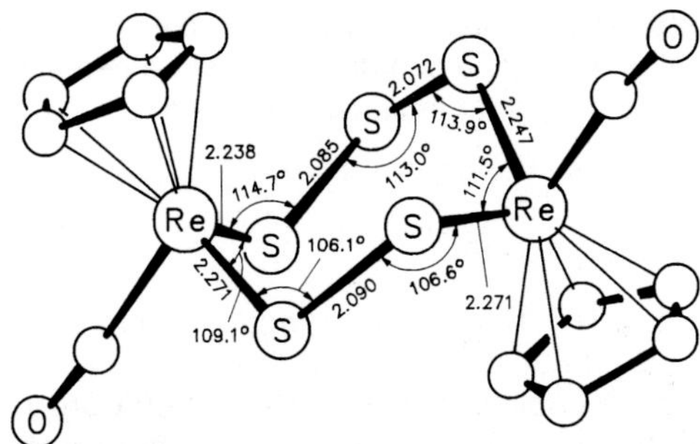

Fig. 100. Molecular structure of $[C_5H_5Re(CO)]_2(\mu\text{-}S_2)(\mu\text{-}S_3)$ [2].

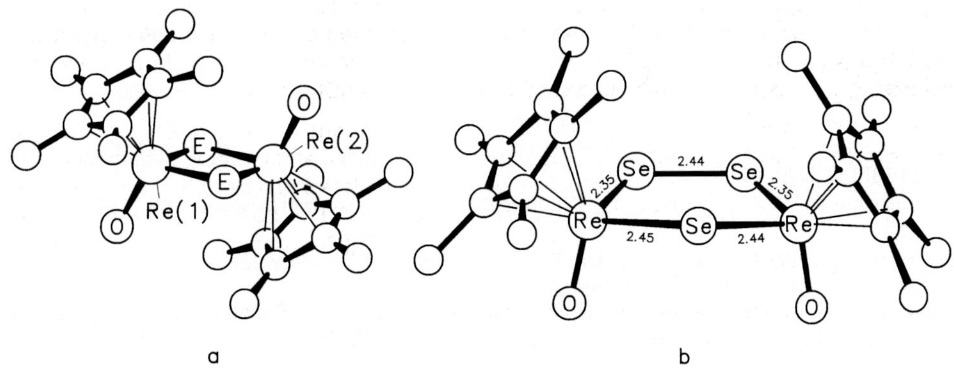

Fig. 101. Molecular structures of
a. $[(C_5(CH_3)_5)ReO]_2(\mu\text{-}E)_2$ (E = O, S, Se; figure from [16]);
b. $[C_5(CH_3)_5)ReO]_2(\mu\text{-}Se)(\mu\text{-}Se_2)$ [22].

centrosymmetric; a view of the general structure is given in **Fig. 101a** and intramolecular parameters are listed in the following table [22].

$[(C_5(CH_3)_5)ReO]_2(\mu\text{-}E)_2$ (Table **9**, Nos. **9**, **18**, **22** with E = O, S, Se). The following table surveys selected intramolecular parameters within the compounds (compare with **Fig. 101a**):

No.	E	atomic distances (Å)				bond angles (°)			Ref.
		Re···Re	E···E	Re-E[a)]	Re=O	Re-E-Re	E-Re-E	E-Re-O[a)]	
9	O	3.143		1.957 1.963	1.689	106.6	73.4	109.0 109.1	[16]
18	S	3.659	2.990	2.343 2.329	1.691	100.41	79.6	104.1 106.0	[21]
22	Se	3.769	3.173	2.469 2.458	1.658	99.8	80.2	102.7 103.5	[22]

[a)] first line: at Re(1), second line: at Re(2).

References on p. 202

$[(C_5(CH_3)_5)ReO]_2(\mu\text{-}Se)(\mu\text{-}Se_2)$ (Table **9**, No. **23**). According to an X-ray structure analysis, the Re_2Se_3 cycle assumes an envelope conformation with a dihedral angle of 24° along the Re···Re vector. **Fig. 101b** illustrates the molecular structure.

Prolonged exposure to excess $P(C_4H_9\text{-}n)_3$ gave No. 22 [22].

References:

[1] Herberhold, M.; Reiner, D.; Thewalt, U. (Angew. Chem. **95** [1983] 1028/9; Angew. Chem. Int. Ed. Engl. **22** [1983] 1000; Angew. Chem. Suppl. **1983** 1343/52).
[2] Herberhold, M.; Reiner, D.; Ackermann, K.; Thewalt, U.; Debaerdemaeker, T. (Z. Naturforsch. **39b** [1984] 1199/205).
[3] Herrmann, W. A.; Serrano, R.; Bock, H. (Angew. Chem. **96** [1984] 364/5; Angew. Chem. Int. Ed. Engl. **23** [1984] 383).
[4] Herrmann, W. A.; Serrano, R.; Küsthardt, U.; Ziegler, M. L.; Guggolz, E.; Zahn, T. (Angew. Chem. **96** [1984] 498/500; Angew. Chem. Int. Ed. Engl. **23** [1984] 515/7).
[5] Herrmann, W. A.; Küsthardt, U.; Herdtweck, E. (J. Organomet. Chem. **294** [1985] C 33/C 36).
[6] Herrmann, W. A.; Serrano, R.; Küsthardt, U.; Guggolz, E.; Nuber, B.; Ziegler, M. L. (J. Organomet. Chem. **287** [1985] 329/44).
[7] Herrmann, W. A; Voss, E.; Küsthardt, U.; Herdtweck, E. (J. Organomet. Chem. **294** [1985] C 37/C 40).
[8] Herrmann, W. A.; Cuenca, T.; Küsthardt, U. (J. Organomet. Chem. **309** [1986] C 15/C 17).
[9] Herrmann, W. A.; Küsthardt, U.; Flöel, M.; Kulpe, J.; Herdtweck, E.; Voss, E. (J. Organomet. Chem. **314** [1986] 151/62).
[10] Küsthardt, U.; Herrmann, W. A.; Ziegler, M. L.; Zahn, T.; Nuber, B. (J. Organomet. Chem. **311** [1986] 163/75).

[11] Herrmann, W. A.; Hecht, C.; Herdtweck, E.; Kneuper, H.-J. (Angew. Chem. **99** [1987] 158/60; Angew. Chem. Int. Ed. Engl. **26** [1987] 132).
[12] Herrmann, W. A.; Herdtweck, E.; Flöel, M.; Kulpe, J.; Küsthardt, U.; Okuda, J. (Polyhedron **6** [1987] 1165/82).
[13] Herrmann, W. A.; Marz, D.; Herdtweck, E.; Schäfer, A.; Wagner, W.; Kneuper, H.-J. (Angew. Chem. **99** [1987] 462/4; Angew. Chem. Int. Ed. Engl. **26** [1987] 462).
[14] Herrmann, W. A.; Okuda, J. (J. Mol. Catal. **41** [1987] 109/22).
[15] Herberhold, M.; Schmidkonz, B. (J. Organomet. Chem. **358** [1988] 301/20).
[16] Herrmann, W. A.; Flöel, M.; Kulpe, J.; Felixberger, J.; Herdtweck, E. (J. Organomet. Chem. **355** [1988] 297/313).
[17] Herrmann, W. A.; Kneuper, H.-J. (J. Organomet. Chem. **348** [1988] 193/7).
[18] Kneuper, H.-J.; Härter, P.; Herrmann, W. A. (J. Organomet. Chem. **340** [1988] 353/8).
[19] Bursten, B. E.; Cayton, R. H. (Inorg. Chem. **28** [1989] 2846/53).
[20] Herrmann, W. A.; Fischer, R. A.; Amslinger, W.; Herdtweck, E. (J. Organomet. Chem. **362** [1989] 333/43).

[21] Herrmann, W. A.; Jung, K. A.; Herdtweck, E. (Chem. Ber. **122** [1989] 2041/8).
[22] Herberhold, M.; Jin, G.-X.; Milius, W. (J. Organomet. Chem. **459** [1993] 257/63).
[23] Gable, K. P. (Organometallics **13** [1994] 2486/8).
[24] Lai, N.-S.; Tu, W.-C.; Chi, Y.; Peng, S.-M.; Lee, G.-H. (Organometallics **13** [1994] 4652/4).
[25] Wang, M.; Herrmann, W. A. (Synth. React. Inorg. Met.-Org. Chem. **24** [1994] 1423/31).
[26] Gable, K. P.; Juliette, J. J. J.; Gartman, M. A. (Organometallics **14** [1995] 3138/40).
[27] Gable, K. P.; Juliette, J. J. J. (J. Am. Chem. Soc. **118** [1996] 2625/33).

2.5.1.2.5 Compounds with the $^5LRe(\mu\text{-}X)_2Re^5L$ Skeleton

General. Structure. The compounds presented here are composed of two $C_5R_5ReX_n$, C_5R_5ReO, or $C_5R_5Re^2D$ moieties bridged by two one-electron (H) or three-electron (Cl, OR, NR_2) donating ligands X. Dependent on the number of electrons delivered by the ligands, the Re atoms are connected by triple, double, or single bonds. In several instances (Nos. 17 (supposedly), 18, 19), the bridging ligands are additionally bonded in a chelating manner.

The structures of two compounds could only be suggested. The suggested structure of No. 7 is displayed in Formula I; it appears reasonable when considering its synthesis and spectroscopic properties. In contrast, the spectroscopic data obtained from No. 17 do not allow an unambiguous attribution of the two possible structures shown in Formulas IIa, IIb to that compound.

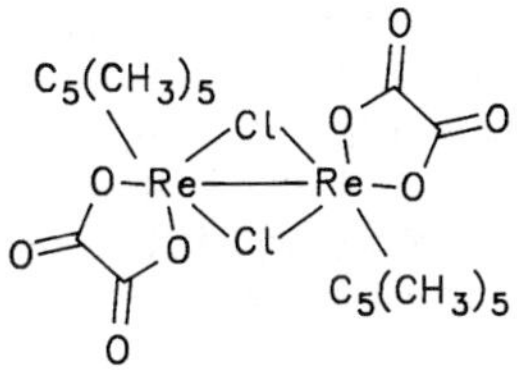

I 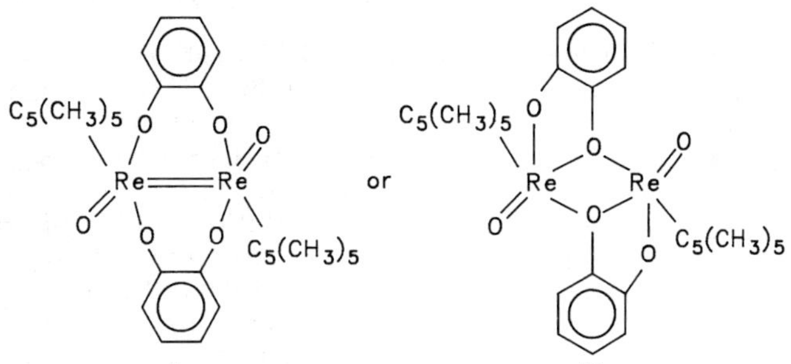 IIa or IIb

Preparation. The compounds were prepared by the following general methods:

Method I: Treatment of $(C_5(CH_3)_4R)ReCl_4$ ($R = CH_3$, C_2H_5) with 10 equivalents $LiAlH_4$ in ether at $-95\,°C$ followed by slowly warming the mixture to room temperature and methanolysis. The solvent was removed and the residue separated chromatographically on alumina with toluene/ether mixtures. $(C_5(CH_3)_4R)ReH_6$ was concomitantly formed [11].

Method II: Irradiation of an equimolar mixture of $(C_5(CH_3)_5)ReD_6$ and $(C_5(CH_3)_4C_2H_5)ReH_6$ in hexane for 15 min gave $[(C_5(CH_3)_5)ReD_2]_2(\mu\text{-}D)_2$ and $[(C_5(CH_3)_4C_2H_5)\text{-}ReH_2]_2(\mu\text{-}H)_2$ (by FD mass spectroscopy) along with other analogs having an H_2D_4 or H_4D_2 pattern. In contrast, products with an H_5D, H_3D_3, or HD_5 pattern were never observed [14].

Method III: Treatment of $(C_5(CH_3)_4R)ReH_6$ ($R = CH_3$, C_2H_5) with 4 equivalents CCl_4 in toluene at $-70\,°C$ followed by warming the mixture to room temperature. The product precipitated upon cooling the mixture again ($CHCl_3$ could also be used in place of CCl_4, but yields were lower and reaction times were longer) [11].

Method IV: Reduction of $(C_5(CH_3)_4R)ReX_4$ ($R = CH_3$, C_2H_5; X = Cl, Br) with excess Al powder in the presence of a catalytic amount of HgX_2 in THF either
a. at reflux temperature for 1 h or
b. at room temperature for several hours.
In either case, the mixture was evaporated and the residue chromatographically separated (silica, CH_2Cl_2) [7, 9].

Method V: Treatment of $[(C_5(CH_3)_5)ReO]_2(\mu\text{-}O)_2$ with $HBF_4 \cdot O(C_2H_5)_2$ [3] or $CH_3OSO_2CF_3$ [2] initiated attack of the bridging O atoms by H or CH_3.

References on p. 211

In the table, the compounds are arranged by the type of the bridging ligands. Compounds with X = H precede those with bridging halogens, chalcogens, and pnictides.

Table 10
^{5}L Compounds with the $^5LRe(\mu\text{-}X)_2Re^5L$ Skeleton.
An asterisk indicates further information at the end of the table.
For explanations, abbreviations, and units see p. X.

No.	compound	method of preparation (yield) properties and remarks
*1	$[C_5H_5ReH_2]_2(\mu\text{-}H)_2$	model compound for Fenske-Hall MO calculations; see "Further information"
*2	$[(C_5(CH_3)_5)ReH_2]_2(\mu\text{-}H)_2$ and $[(C_5(CH_3)_5)ReD_2]_2(\mu\text{-}D)_2$	I (36%) [11]; II [14] air-stable solid; m.p. 214°C (dec.) [11] ^{1}H NMR (toluene-d$_8$): −6.41, 2.18; (−95°C): −7.29, −4.16 (br s's, 4 and 2 H); 2.10 IR (KBr): 1155 (ν(ReHRe)), 1985 (ν(ReH)); deuterated analog: 800 (ν(ReDRe)) and 1450 (ν(ReD)) EI MS: $[M - n\ H_2]^+$ (n = 0, 2 to 6), $[1/2\ M - 3\ H_2]^+$ [11]
*3	$[(C_5(CH_3)_4C_2H_5)ReH_2]_2(\mu\text{-}H)_2$ and $[(C_5(CH_3)_4C_2H_5)ReD_2]_2(\mu\text{-}D)_2$	I (42%) [11]; II [14] brown solid [11] ^{1}H NMR (toluene-d$_8$): −6.60 (s, 6 H); 0.83 (t, CH_3; J = 7); 2.15, 2.22 (s's, $(CH_3)_4$); 2.57 (q, CH_2); (−95°C): −7.29, −4.14 (br, 4 and 2 H); 0.63 (br); 2.00, 2.06 (s's), 2.69 (br) ^{13}C {^{1}H} NMR (toluene-d$_8$): 13.4, 13.5 ($(CH_3)_4$); 17.5, 22.4 (C_2H_5); 92.5, 94.1, 98.9 (C_5) IR (KBr): 1160 (ν(ReHRe)); 1980, 1990 (ν(ReH)) [11]
4	$[(C_5(CH_3)_5)ReCl_2]_2(\mu\text{-}H)_2$	III (21%) brown solid ^{1}H NMR ($CDCl_3$): −2.47 (μ-H), 2.14 (CH_3) IR (KBr): 308 (ν(ReCl)), 1190 (ν(ReHRe)) EI MS: $[M - H]^+$, $[M - n\ HCl]^+$ (n = 1, 2), $[1/2\ M - H - n\ Cl]^+$ (n = 0, 1) [11]
5	$[(C_5(CH_3)_4C_2H_5)ReCl_2]_2(\mu\text{-}H)_2$	III (21%) ^{1}H NMR (CD_2Cl_2): −2.52 (μ-H); 1.13 (**CH_3**CH_2); 1.99, 2.18 (s's, $(CH_3)_4$); 2.45 (q, CH_2; J = 7) IR (KBr): 290, 332 (ν(ReCl)); 1190 (ν(ReHRe)) EI MS: $[M - 2\ H]^+$, $[M - Cl]^+$, $[1/2\ M - H]^+$, $[1/2\ M - Cl - 2\ H_2]^+$ [11]
*6	$[(C_5(CH_3)_5)Re(CO)_2]_2(\mu\text{-}H)_2$	obtained by exposing $(C_5(CH_3)_5)_2Re_2(CO)_4$ to an H_2 atmosphere in THF; yield: 90% by flash chromatography

References on p. 211

Table 10 (continued)

No.	compound	method of preparation (yield) properties and remarks
		orange plates ^1H NMR (THF-d_8): −6.19 (ReH), 2.11 (CH_3) $^{13}C\{^1H\}$ NMR (THF-d_8): 11.3 (CH_3), 98.9 (C_5), 208.1 (CO) IR (THF): 1871, 1929 (ν(CO)) [15]
7	$[(C_5(CH_3)_5)ReO_2C_2O_2]_2(\mu\text{-}Cl)_2$ (see Formula I)	from $(C_5(CH_3)_5)Re(O)Cl_2$ and 2 equivalents $Ag_2C_2O_4$ (CH_2Cl_2, exclusion of light, 12 h); subsequent chromatographic workup at −10°C yielded 86% dark green solid from CH_2Cl_2/THF; dec.p. >230°C ^1H NMR (CD_2Cl_2): 2.027 (20°C); 2.022 (10°C); 2.018, 2.020 (5°C); 2.009, 2.014 (−20°C) $^{13}C\{^1H\}$ NMR (CD_2Cl_2, 30°C): 10.71, 120.21, 162.51; (−20°C): 9.91, 10.03 (CH_3); 119.91, 120.45 (C_5); 160.84, 161.65 (CO_2); low-temperature peak splitting due to cis-trans isomerism IR (KBr): 290, 325 (ν(ReCl)); 792, 1020, 1319, 1375; 1699, 1725 (C=O); 2854, 2925, 2966 EI MS: $[(C_5(CH_3)_5)ReO_2Cl_2]^+$, $[(C_5(CH_3)_5)ReOCl_n]^+$ (n = 2 to 0) [10] thermolysis released 3 molecules of CO_2 per complex molecule, but evolution of CO was not observed [10]
*8	$[C_5H_5ReCl]_2(\mu\text{-}Cl)_2$	model compound for MO calculations; see "Further information"
*9	$[(C_5(CH_3)_5)ReCl]_2(\mu\text{-}Cl)_2$	IVa (40%); crystallization from hexane/CH_2Cl_2 [7] ^1H NMR (C_6D_6): 1.99 [7]; ($CDCl_3$): 2.02 [13] IR (KBr): 295, 315 (ν(ReCl)) [7]
*10	$[(C_5(CH_3)_4C_2H_5)ReCl]_2(\mu\text{-}Cl)_2$	IVa [7, 9] (68%) [9] red-brown crystals from hexane/CH_2Cl_2 (2:5); no m.p. below 250°C [9] ^1H NMR (C_6D_6): 1.22 (t, CH_3), 1.88 (q, CH_2); 1.98, 1.99 (s's, $(CH_3)_4$) [7, 9] $^{13}C\{^1H\}$ NMR (CD_2Cl_2): 9.3, 10.1, 15.8 (CH_3); 17.8 (CH_2); 104.9, 106.1, 106.8 (C_5) [9] IR (KBr): 295, 315 (ν(ReCl)) [7, 9] EI MS: $[M]^+$, $[1/2\ M]^+$, $[1/2\ M - Cl]^+$ [9]
*11	$[(C_5(CH_3)_5)ReCl_2]_2(\mu\text{-}Cl)_2$	IVb [7, 9] (80%) [9]; see also "Further information" dark brown crystals; no m.p. below 250°C [5, 9] ^1H NMR (CD_2Cl_2, −50°C): 1.85 [5], 1.89 [9, 13] $^{13}C\{^1H\}$ NMR (CD_2Cl_2): 12.1 (CH_3), 109.6 (C_5) [9] IR spectrum (KBr): 305, 335 (ν(ReCl)) [5] (also [9]); 1025, 1078, 1375 (ν(CC)); 1450, 1485 (δ(CH)) [5]

References on p. 211

Table 10 (continued)

No.	compound	method of preparation (yield) properties and remarks
*11 (continued)		EI MS: $[1/2\ M]^+$ [5, 9], $[1/2\ M - Cl]^+$, $[1/2\ M - 2\ HCl]^+$ [5]
*12	$[(C_5(CH_3)_4C_2H_5)ReCl_2]_2(\mu\text{-}Cl)_2$	preparation not given brown prisms [9]
*13	$[(C_5(CH_3)_5)ReBr]_2(\mu\text{-}Br)_2$	IVb (5 h); chromatographic separation on silylated silica with hexane/CH_2Cl_2 (10:3) followed by recrystallization from the same solvent mixture yielded 40% brown crystalline powder; no m.p. below 250 °C ^{1}H NMR (CD_2Cl_2): 2.08 $^{13}C\{^1H\}$ NMR (CD_2Cl_2): 11.9 (CH_3), 108.1 (C_5) EI MS: $[M]^+$ [9]
14	$[(C_5(CH_3)_5)ReI_2]_2(\mu\text{-}I)_2$	obtained from $(C_5(CH_3)_5)ReH_6$ and excess I_2 (toluene, room temperature; yield: 96%) or by photochemically treating No. 2 with CH_3I in hexane (quantitative yield) brown solid ^{1}H NMR ($CDCl_3$): 2.12 EI MS: $[M - n\ I]^+$ (n = 3 to 5), $[1/2\ M - n\ I]^+$ (n = 0, 1), $[1/2\ M - 2\ I - H]^+$ [11]
15	$[\{(C_5(CH_3)_5)ReO\}_2(\mu\text{-}OH)_2][BF_4]_2$	V (precipitated immediately; yield: quantitative) flesh-colored solid ^{1}H NMR (acetone-d_6): 2.23 (CH_3), 3.31 (OH) IR (KBr): 521, 534, 543 (ν(ReORe)); 905 (ν(ReO)) conductivity (CH_3NO_2): $\Lambda = 151\ cm^2 \cdot \Omega^{-1} \cdot mol^{-1}$ reaction with $[N(C_4H_9\text{-}n)_4]Cl$ gave $(C_5(CH_3)_5)Re(O)Cl_2$ [3]
16	$[\{(C_5(CH_3)_5)ReO\}_2(\mu\text{-}OCH_3)_2][BF_4]_2$	V (only briefly mentioned) [2]
17	$[(C_5(CH_3)_5)Re(O)O_2C_6Cl_4]_2$ (see Formulas IIa, IIb)	obtained from $[(C_5(CH_3)_5)ReO]_2(\mu\text{-}O)_2$ and chloranil (THF, reflux) with 10% yield; other products were $(C_5(CH_3)_5)Re(O)(O_2C_6Cl_4)$ and $(C_5(CH_3)_5)Re(O_2C_6Cl_4)_2$ brown crystals; m.p. 210 °C (dec.) ^{1}H NMR ($CDCl_3$): 1.92 IR (?): 792, 812 (ν(ReO)) FD MS: $[M]^+$ [1]
*18	$[(C_5(CH_3)_5)Re]_2(\mu\text{-}S_4)_2$	from $(C_5(CH_3)_5)Re(S_4)(S_3)$ and $P(C_6H_5)_3$ (ratio 1:5, $CHCl_3$ or CH_2Cl_2, 10 h) [17, 18]; chromatographic separation (silica, CH_2Cl_2/THF) yielded 24.5% [17] black prisms from CH_2Cl_2 [17, 18]; m.p. 225 °C [17] ^{1}H NMR ($CDCl_3$): 2.02 [17, 18]

References on p. 211

Table 10 (continued)

No.	compound	method of preparation (yield) properties and remarks
		$^{13}C\{^1H\}$ NMR ($CDCl_3$): 11.5, 100.6 [17, 18] EI MS: $[M - n\,S]^+$ (n = 2 to 6) [17]
*19	$[(C_5(CH_3)_5)Re]_2(\mu\text{-}Se_4)_2$	from $(C_5(CH_3)_5)Re(O)Se_4$ and $P(C_4H_9\text{-}n)_3$ (ratio 1:3, $CHCl_3$, 1 h); chromatographic separation with $CHCl_3$/THF (1:1) yielded 17.5% (the main product was $[(C_5(CH_3)_5)ReO]_2(\mu\text{-}Se)(\mu\text{-}Se_2)$); also by treating $(C_5(CH_3)_5)Re(O)Cl_2$ with in situ-formed (from Al_2Se_3 and moist THF) H_2Se [18] dark gray prisms; no m.p. <250 °C ^1H NMR ($CDCl_3$): 1.91 $^{13}C\{^1H\}$ NMR ($CDCl_3$): 13.5, 98.8 IR (CsI): 897, 905 (ν(ReO)) EI MS: $[M - n\,Se]^+$ (n = 3 to 6), $[M - C_5(CH_3)_5 - n\,Se]^+$ (n = 4, 5) [18]
20	$[C_5H_5ReNO]_2(\mu\text{-}N(CH_3)_2)_2$	by treating $[C_5H_5Re(NO)(P(C_6H_5)_3)N(CH_3)_2H]$-$[O_3SCF_3]$ with n-C_4H_9Li (THF, −80 °C) followed by freezing the solution at liquid-nitrogen temperature, adding hexane, and thawing. The hexane layer was separated; crystallization over 3 d gave the compound with 35% yield purple rectangular cubes and needles; m.p. >250 °C ^1H NMR (THF-d_8): 2.66, 4.11 (s's, CH_3); 5.27 (C_5H_5) $^{13}C\{^1H\}$ NMR (THF-d_8): 66.67, 75.02 (CH_3); 92.58 (C_5H_5) IR (KBr): 1626 (ν(NO)) MS: $[M]^+$, $[C_5H_5Re(NO)N(CH_3)_2]^+$ [16]

*Further information:

$[C_5H_5ReH_2]_2(\mu\text{-}H)_2$ (Table **10**, No. **1**). It was examined via Fenske-Hall molecular orbital calculations, which of the three geometries $[C_5H_5ReH_3]_2$, $[C_5H_5ReH_2]_2(\mu\text{-}H)_2$, or $[C_5H_5ReH]_2(\mu\text{-}H)_4$ is the most stable one. Molecular orbital diagrams are given for each case. It appeared that the net direct Re≡Re bonding decreases with increasing number of bridging hydrides; furthermore, the HOMO-LUMO gap increases with increasing number of bridging hydrides at identical Re-Re distances. However, the summation of the energies of the Re-H bonding orbitals differs by only 1 eV. It is thus not possible to predict one particular geometry as the favored one. This can also mean fluxionality of the hydrides in solution [8].

$[(C_5(CH_3)_5)ReH_2]_2(\mu\text{-}H)_2$ and **$[(C_5(CH_3)_5)ReD_2]_2(\mu\text{-}D)_2$** (Table **10**, No. **2**) were also obtained by irradiating $(C_5(CH_3)_5)ReH_6$ or $(C_5(CH_3)_5)ReD_6$ in hexane (no matter whether amines were present or not) [11, 14] or by thermolyzing solid $(C_5(CH_3)_5)ReH_6$ at 200 °C for 30 min in a closed vessel (the dimer sublimed to the colder sectors of the reactor) [11]. No. 2 was also yielded along with $(C_5(CH_3)_5)ReH_6$ (main product [4]) by reducing $(C_5(CH_3)_5)Re(O)Cl_2$ [4] or $(C_5(CH_3)_5)Re(O)Br_2$ [2] with $LiAlH_4$.

The photochemical reaction with CH_3I quantitatively gave No. 14 [11].

References on p. 211

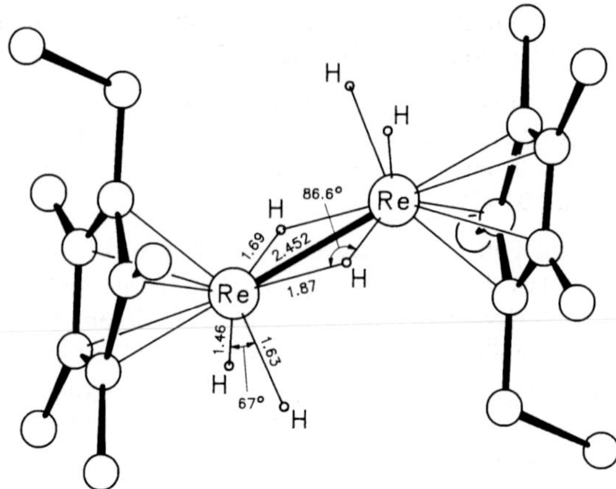

Fig. 102. Molecular structure of $[(C_5(CH_3)_4C_2H_5)ReH_2]_2(\mu\text{-}H)_2$ [11].

$[(C_5(CH_3)_4C_2H_5)ReH_2]_2(\mu\text{-}H)_2$ and **$[(C_5(CH_3)_4C_2H_5)ReD_2]_2(\mu\text{-}D)_2$** (Table **10**, No. **3**) crystallize in the triclinic space group $P\bar{1}-C_i^1$ (No. 2) with a = 7.655(3), b = 8.365(2), c = 10.631(2) Å, α = 80.85(2)°, β = 70.84(2)°, γ = 65.10(1)°; Z = 1 molecule per unit cell, D_{calc} = 1.928 g/cm^3. The H atoms were successfully located. The structure of the molecule is illustrated in **Fig. 102** [11].

$[(C_5(CH_3)_5)Re(CO)_2]_2(\mu\text{-}H)_2$ (Table **10**, No. **6**) crystallizes in the triclinic space group $P\bar{1}-C_i^1$ (No. 2) with a = 8.719(2), b = 10.498(2), c = 13.952(3) Å, α = 79.909(16)°, β = 84.91(2)°, γ = 69.339(16)°; Z = 2 molecules per unit cell, D_{calc} = 2.138 g/cm^3. The molecular structure is illustrated in **Fig. 103**. There is a staggered arrangement of the terminal ligands. The Re-Re connection is formed by a single bond [15].

$[C_5H_5ReCl]_2(\mu\text{-}Cl)_2$ (Table **10**, No. **8**) does not exist; nevertheless, its electronic structure has been calculated by combination of the fragments $[C_5H_5ReCl]_2$ and $(\mu\text{-}Cl)_2$. The highest

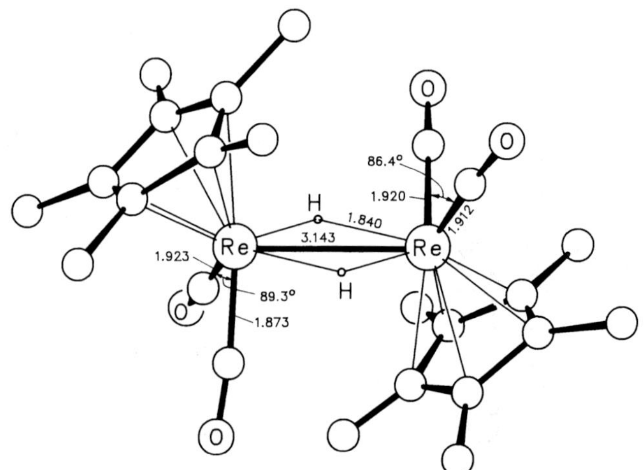

Fig. 103. Molecular structure of $[(C_5(CH_3)_5)Re(CO)_2]_2(\mu\text{-}H)_2$ [15].

References on p. 211

occupied orbital was recognized being of the type $1b_g$, indicating a formal net Re=Re double bond. The total Re=Re overlap population is 0.687 [12].

$[(C_5(CH_3)_5)ReCl]_2(\mu\text{-}Cl)_2$ (Table **10**, No. **9**) was also formed by reducing the halogen-rich compound $[(C_5(CH_3)_5)ReCl_2]_2(\mu\text{-}Cl)_2$ (No. 11) with Al/$HgCl_2$ as described with Method IVa (yield: 40 to 50%) [9]. No. 9 could also be obtained from $(C_5(CH_3)_5)Re(O)(C_3H_7\text{-}i)_2$ and $C_5H_5TiCl_3$ (toluene, reflux, 16 h) [13].

The low-temperature reaction with 1 equivalent Cl_2 formed $[(C_5(CH_3)_5)ReCl_2]_2(\mu\text{-}Cl)_2$ [9], while treatment with excess Cl_2 gave $(C_5(CH_3)_5)ReCl_4$ [7]. Oxidation with air led to $(C_5(CH_3)_5)Re(O)Cl_2$ [9], and treatment with $P(CH_3)_3$ provided $(C_5(CH_3)_5)Re(P(CH_3)_3)_2Cl_2$ [7, 9]. In contrast, the compound did not react with $C_6H_5C{\equiv}CC_6H_5$ in refluxing CH_2Cl_2 [9].

$[(C_5(CH_3)_4C_2H_5)ReCl]_2(\mu\text{-}Cl)_2$ (Table **10**, No. **10**) crystallizes in the monoclinic space group $P2_1/n-C^5_{2h}$ (No. 14) with a = 8.135(1), b = 14.879(1), c = 10.939(1) Å, β = 109.24(1)°; Z = 2 molecules per unit cell, D_{calc} = 2.159 g/cm^3. A view of the centrosymmetric molecule is shown in **Fig. 104a**. There is an Re=Re double bond and a planar Re_2Cl_2 ring.

Treatment with $P(CH_3)_3$ under severe conditions (toluene, reflux temperature or closed vessel, 60°C) yielded $(C_5(CH_3)_4C_2H_5)Re(P(CH_3)_3)_2Cl_2$ [9].

$[(C_5(CH_3)_5)ReCl_2]_2(\mu\text{-}Cl)_2$ (Table **10**, No. **11**). The reduction of $(C_5(CH_3)_5)ReCl_4$ with Al/$HgCl_2$ could also be carried out in the presence of 2,3-dimethylbutadiene, where No. 11 was obtained along with $(C_5(CH_3)_5)Re(CH_2{=}C(CH_3)C(CH_3){=}CH_2)Cl_2$ (yield: 60%) [9]. In addition, the reduction of $(C_5(CH_3)_5)ReCl_4$ could also be performed with $(C_2H_5)_4Sn$ [5, 7] or $(n\text{-}C_4H_9)_4Sn$ [5, 9] in CH_2Cl_2 at room temperature within several hours. The solvent was subsequently evaporated and the residue extracted into THF/$CHCl_3$. The extracted material was crystallized from CH_2Cl_2/pentane. Yields reached 80 to 85% [9]. Moreover, No. 11 was also formed by treating $(C_5(CH_3)_5)Re(O)(C_2H_5)_2$ with $C_5H_5TiCl_3$ in refluxing toluene [13] or by treating No. 9 with 1 equivalent Cl_2 at low temperatures [9].

Oxidation with a stoichiometric amount of Cl_2 and Br_2 produced $(C_5(CH_3)_5)ReCl_4$ and $(C_5(CH_3)_5)ReCl_nBr_{4-n}$, respectively, but there was no analogous reaction with I_2. A reduction with Al/$HgCl_2$ employing the conditions of Method IVa gave No. 9 [9]. Treatment with $P(CH_3)_3$ provided the 17-electron compound $(C_5(CH_3)_5)Re(P(CH_3)_3)Cl_3$ [5, 9].

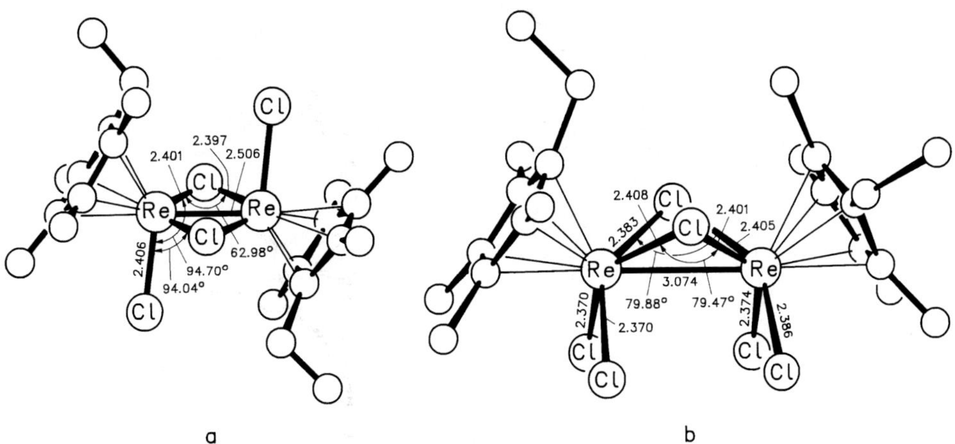

Fig. 104. Molecular structures of $[(C_5(CH_3)_4C_2H_5)ReCl_n]_2(\mu\text{-}Cl)_2$ (a. n = 1; b. n = 2) [9].

References on p. 211

No. 11 did not react with $Ag_2C_2O_4$ [10] or alkenes [9]. Treatment with alkynes $R^1C{\equiv}CR^2$ ($R^1 = R^2 = CH_3$ [6, 9] or C_2H_5 [9]; $R^1 = CH_3$, $R^2 = C_2H_5$; $R^1 = CH_3$, $R^2 = C_6H_5$ [9]) in refluxing CH_2Cl_2 provided the complexes $(C_5(CH_3)_5)Re(R^1C{\equiv}CR^2)Cl_2$ and $(C_5(CH_3)_5)Re(\eta^3\text{-}R^1CHC(Cl)CHR^3)Cl_2$ ($R^1 = H$, $R^3 = CH_3$ [6, 9]; $R^1 = C_2H_5$, $R^3 = CH_3$; $R^1 = CH_3$, $R^3 = CH_3$; $R^1 = H$, $R^3 = C_6H_5$ [9]). When the reactions were conducted in a 1:1 mixture of CH_2Cl_2 and the liquid alkyne, $(C_5(CH_3)_5)ReCl_4$ was also produced. Treatment with $C_6H_5C{\equiv}CC_6H_5$ yielded $(C_5(CH_3)_5)Re(C_6H_5C{\equiv}CC_6H_5)Cl_2$ and $(C_5(CH_3)_5)ReCl_4$ [9].

$[(C_5(CH_3)_4C_2H_5)ReCl_2]_2(\mu\text{-}Cl)_2 \cdot CH_2Cl_2$ (Table **10**, No. **12**) crystallizes in the orthorhombic space group $P2_12_12_1-D_2^4$ (No. 19) with a = 11.491(1), b = 15.353(1), c = 17.276(2) Å; Z = 4 formula units per unit cell, D_{calc} = 2.111 g/cm³. The molecular structure of the complex is depicted in **Fig. 104b**. The C_5 rings are arranged in a cis position with respect to each other. The Re_2Cl_2 ring has a folding angle of 65.9° along the Re-Re single bond [9].

$[(C_5(CH_3)_5)ReBr]_2(\mu\text{-}Br)_2$ (Table **10**, No. **13**). The reduction of $(C_5(CH_3)_5)ReBr_4$ could also be carried out with 4 equivalents $(n\text{-}C_4H_9)_4Sn$ (CH_2Cl_2, reflux, 3 d). Evaporation followed by recrystallizing the residue from CH_2Cl_2/pentane yielded 80% [9]. No. 13 was also reported to form by treating $(C_5(CH_3)_5)Re(O)Br_2$ with $(i\text{-}C_4H_9)_2AlH$ [2].

$[(C_5(CH_3)_5)Re]_2(\mu\text{-}E_4)_2$ (Table **10**, Nos. **18** and **19** with E = S, Se) both crystallize in the monoclinic lattice. The following single-crystal data were established: space group $P2_1/n-C_{2h}^5$ (No. 14); a = 8.946(2), b = 14.230(3), c = 20.874(4) Å, β = 90.23(3)°; Z = 4 molecules per unit cell (E = S) [17]; space group $P2_1/c-C_{2h}^5$ (No. 14); a = 18.398(4), b = 12.798(3),

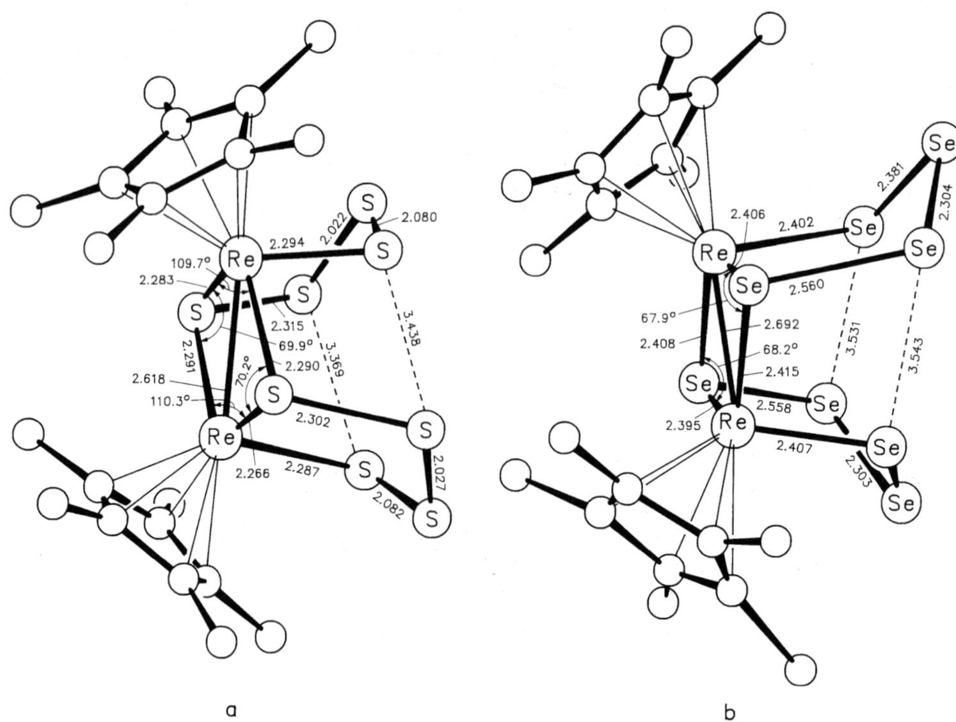

Fig. 105. Molecular structures of $[(C_5(CH_3)_5)Re]_2(\mu\text{-}E_4)_2$ (a. E = S [17]; b. E = Se [18]).

References on p. 211

c = 17.233(3) Å, β = 110.07(3)°; Z = 4 molecules per unit cell (E = Se) [18]. The molecular structures are depicted in **Fig. 105**. Remarkably, the cis arrangement of the E_4 ligands causes several E atoms to be very close to each other [17, 18].

References:

[1] Herrmann, W. A.; Küsthardt, U.; Herdtweck, E. (J. Organomet. Chem. **294** [1985] C 33/C 36).

[2] Herrmann, W. A. (J. Organomet. Chem. **300** [1986] 111/37).

[3] Herrmann, W. A.; Küsthardt, U.; Flöel, M.; Kulpe, J.; Herdtweck, E.; Voss, E. (J. Organomet. Chem. **314** [1986] 151/62).

[4] Herrmann, W. A.; Okuda, J. (Angew. Chem. **98** [1986] 1109/11; Angew. Chem. Int. Ed. Engl. **25** [1986] 1092).

[5] Herrmann, W. A.; Felixberger, J. K.; Herdtweck, E.; Schäfer, A.; Okuda, J. (Angew. Chem. **99** [1987] 466/7; Angew. Chem. Int. Ed. Engl. **26** [1987] 466).

[6] Herrmann, W. A.; Fischer, R. A.; Herdtweck, E. (Angew. Chem. **99** [1987] 1286/90; Angew. Chem. Int. Ed. Engl. **26** [1987] 1263).

[7] Herrmann, W. A.; Fischer, R. A.; Herdtweck, E. (J. Organomet. Chem. **329** [1987] C 1/C 6).

[8] Bursten, B. E.; Cayton, R. H. (Organometallics **7** [1988] 1349/56).

[9] Herrmann, W. A.; Fischer, R. A.; Felixberger, J. K.; Paciello, R. A.; Kiprof, P.; Herdtweck, E. (Z. Naturforsch. **43b** [1988] 1391/404).

[10] Herrmann, W. A.; Jung, K. A.; Herdtweck, E. (Chem. Ber. **122** [1989] 2041/8).

[11] Herrmann, W. A.; Theiler, H. G.; Herdtweck, E.; Kiprof, P. (J. Organomet. Chem. **367** [1989] 291/311).

[12] Green, J. C.; Green, M. L. H.; Mountford, P.; Parkington, M. J. (J. Chem. Soc. Dalton Trans. **1990** 3407/18).

[13] Herrmann, W. A.; Felixberger, J. K.; Anwander, R.; Herdtweck, E.; Kiprof, P.; Riede, J. (Organometallics **9** [1990] 1434/43).

[14] Herrmann, W. A.; Theiler, H. G.; Kiprof, P.; Tremmel, J.; Blom, R. (J. Organomet. Chem. **395** [1990] 69/84).

[15] Casey, C. P.; Sakaba, H.; Hazin, P. N.; Powell, D. R. (J. Am. Chem. Soc. **113** [1991] 8165/6).

[16] Dewey, M. A.; Knight, D. A.; Arif, A.; Gladysz, J. A. (Chem. Ber. **125** [1992] 815/24).

[17] Herberhold, M.; Jin, G.-X.; Milius, W. (Angew. Chem. **105** [1993] 127/9; Angew. Chem. Int. Ed. Engl. **32** [1993] 85).

[18] Herberhold, M.; Jin, G.-X.; Milius, W. (J. Organomet. Chem. **459** [1993] 257/63).

2.5.1.2.6 Compounds with the $^5LRe(\mu\text{-}^nD)Re^5L$ Skeleton

trans-[$C_5H_5Re(CO)_2]_2(\mu$-NH=NH) (see Formula I). A THF solution containing $C_5H_5Re(CO)_2$-NH_2NH_2 was treated at −78°C with Na_2SO_4, one drop of 0.1 M $CuSO_4$ (aqueous solution), and 1.5 equivalents H_2O_2 (80%). The mixture was warmed to 30°C and again cooled to −20°C. After evaporation of the solvent, the black residue was extracted with THF. The extract was chromatographically purified on silica using THF/pentane as eluent. Recrystallization from THF yielded 2.4% of fine black-green crystals; m.p. 135 to 140°C. The synthesis must be performed in a highly concentrated solution. The compound is stable in the solid state at room temperature, but it rapidly decomposed in solution.

^1H NMR spectrum (acetone-d_6): δ = 5.37 (CH_3), 15.33 (NH) ppm. IR spectrum (KBr): 320, 335, 490, 535, 578, 610, 803, 838, 918, 998; 1027 ($\nu(N_2H_2)$); 1052, 1100, 1132, 1150,

References on p. 212

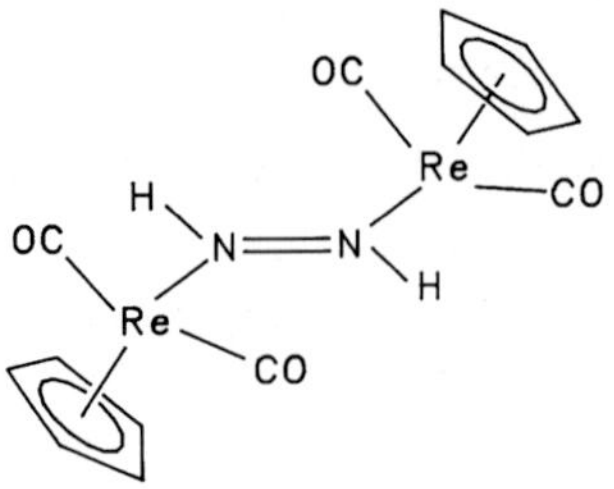

I

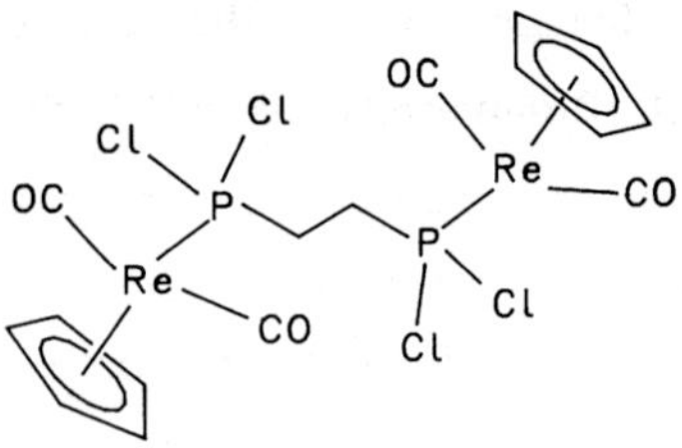

II

1342; 1360 ($\nu(N_2H_2)$); 1412, 1628; 1840, 1890 (ν(CO)); 3120, 3240 (ν(NH)) cm^{-1}; (THF): 1865, 1900 cm^{-1}. Mass spectrum: $[M]^+$ observed.

THF or acetone solutions are deep blue. Treatment with CD_3OD initiated complete H/D exchange at N giving **$[C_5H_5Re(CO)_2]_2(\mu$-ND=ND)**, which exhibits IR bands at 552 (shift from 1027), 1010 (shift from 1360), and 2420 (shift from 3240) cm^{-1} [1].

$[C_5H_5Re(CO)_2P(Cl)_2CH_2-]_2$ (see Formula II) was obtained by stirring a THF solution containing $C_5H_5Re(CO)_2PCl_2C_2H_4PCl_2$ and in situ-prepared $C_5H_5Re(CO)_2OC_4H_8$. The product was separated by chromatography (Florisil, ether/n-hexane (1:1)). Recrystallization from CH_2Cl_2 gave the compound with 18% yield. The treatment of $C_5H_5Re(CO)_2OC_4H_8$ with $Cl_2PC_2H_4PCl_2$ (THF, 2 h), however, only gave minor amounts of the dimer, contaminating the main product $C_5H_5Re(CO)_2PCl_2C_2H_4PCl_2$ by ca. 15%. The photochemical reaction of $C_5H_5Re(CO)_3$ with $Cl_2PC_2H_4PCl_2$ in ether was also studied, but only achieved a yield of 5%. The compound is an air-stable, white solid; m.p. 149 to 151 °C.

^1H NMR spectrum ($CDCl_3$): δ = 3.32 (t, PCH_2; J(P,H) = 3.23 Hz), 5.35 (C_5H_5) ppm. ^{13}C {^1H} NMR spectrum ($CDCl_3$): δ = 50.25 (t, PCH_2; J(P,C) = 23.2 Hz), 85.55 (C_5) ppm. ^{31}P {^1H} NMR spectrum (CD_2Cl_2): δ = 105.9 ppm. IR spectrum (CH_2Cl_2): 1904, 1969 (ν(CO)) cm^{-1} [3].

$[(C_5(CH_3)_5)Re(CO)]_2(\mu-S_8)$. $(C_5(CH_3)_5)Re(CO)_2OC_4H_8$ was treated with 0.75 equivalent S_8 for 30 min in THF. Chromatographic workup with pentane/CH_2Cl_2 mixtures yielded two fractions containing $(C_5(CH_3)_5)Re(CO)_2(S_2)$ and the title product. The latter fraction was evaporated and extracted with acetone. The acetone-unsoluble residue was recrystallized from CH_2Cl_2, giving bronze-colored crystals with 6.8% yield. The exact structure of the compound is not known.

^1H NMR spectrum ($CDCl_3$): δ = 1.83 ppm. ^{13}C {^1H} NMR spectrum ($CDCl_3$): δ = 10.8 (CH_3), 103.6 (C_5) ppm. IR spectrum (CH_2Cl_2): 1934 (ν(CO)) cm^{-1}; k_{CO} = 15.10 N · cm^{-1}. FD mass spectrum: $[M]^+$; EI mass spectrum: $[(C_5(CH_3)_5)_2Re_2S_n]^+$ with n = 2 to 6.

The compound is insoluble in pentane, toluene, and ether; sparely soluble in acetone, and soluble in CH_2Cl_2 [2].

References:

[1] Sellmann, D.; Kleinschmidt, E. (Z. Naturforsch. **32b** [1977] 795/801).
[2] Herberhold, M.; Schmidkonz, B. (J. Organomet. Chem. **358** [1988] 301/20).
[3] Lee, D. W.; Morse, J. G. (Phosphorus, Sulfur, and Silicon **106** [1995] 211/25).

2.5.1.2.7 Compounds with the $^5LRe(\mu\text{-}^1L)Re^5L$ Skeleton

2.5.1.2.7.1 Compounds with a Bridging $-CH_2(CH_2)_nX-$ Ligand

The compounds described in this section were obtained by the following method:

Method I: Treatment of in situ-prepared $[C_5H_5Re(NO)(P(C_6H_5)_3)ClCH_2Cl]BF_4$ (from $C_5H_5Re(NO)(P(C_6H_5)_3)CH_3$ and $HBF_4 \cdot O(C_2H_5)_2$ in CH_2Cl_2) with $(CO)_5Re$-$(CH_2)_nX$ (X = Cl, Br, I; n = 1, 4) at −80 to −60°C followed by warming the mixture to ca. −30°C for 3 h [1].

$[C_5H_5Re(NO)(P(C_6H_5)_3)-XCH_2-Re(CO)_5]BF_4$. The derivatives with X = Cl and Br were only characterized in solution; however, the iodo derivative could be separated. The following NMR spectral data were taken in CD_2Cl_2 at −60°C (abbreviations and units on p. X):

X = Cl: ^{1}H NMR: 4.20, 4.57 (d's, $ClCH_2$; J(P,H) = 9.6); 5.57 (C_5H_5), 7.16 to 7.57 (m, C_6H_5). − ^{31}P {^{1}H} NMR: 14.5 [1].

X = Br: ^{1}H NMR: 3.85, 4.16 (d's, $BrCH_2$; J(P,H) = 8.8); 5.50 (C_5H_5), 7.18 to 7.53 (m, C_6H_5). − ^{31}P {^{1}H} NMR: 14.1 [1].

X = I: Separated from the mixture after concentration and addition of ether. Yield: 78%. Tan powder; m.p. 131 to 134°C.
^{1}H NMR: 2.73, 2.92 (d's, ICH_2; J(P,H) = 8.5); 5.48 (C_5H_5), 7.16 to 7.48 (m, C_6H_5). − ^{13}C {^{1}H} NMR: −12.2 (br s, CH_2), 91.9 (C_5); 129.1, 131.2, 132.2, 133.3 (C_6H_5; J(P,C) = 10.6, <2, 50.4, 9.2, respectively); 179.8, 182.3 (CO). − ^{31}P {^{1}H} NMR: 13.6. − IR (CH_2Cl_2): 1704 (ν(NO)); 2020, 2048, 2137 (ν(CO)).
Treatment with $[N(P(C_6H_5)_3)_2]Br$ (CH_2Cl_2, −60°C) gave $C_5H_5Re(NO)(P(C_6H_5)_3)I$ and $(CO)_5ReCH_2Br$ [1].

$[C_5H_5Re(NO)(P(C_6H_5)_3)-I(CH_2)_4-Re(CO)_5]BF_4$ was obtained when $(CO)_5Re(CH_2)_4I$ was employed as reactant. Concentration and pouring of the reaction mixture into ether precipitated the compound with 69% yield. It is a tan powder; m.p. 115 to 119°C (dec.).

^{1}H NMR spectrum (CD_2Cl_2): δ = 0.92, 1.76, 2.13 (m's, CH_2); 3.49, 3.79 (dt's, ICH_2; J(P,H) = 9.0 Hz); 5.62 (C_5H_5), 7.23 to 7.59 (m, C_6H_5) ppm. ^{13}C {^{1}H} NMR spectrum (CD_2Cl_2): δ = −11.6 (br s, $ReCH_2$), 31.1 (br s, ICH_2); 39.0, 39.3 ($C\mathbf{C_2}H_4C$); 92.1 (C_5); 129.6, 131.3, 132.5, 133.6 (d's, C_6H_5; J(P,C) = 10.9, 3.1, 54.9, 10.9 Hz, resp.); 181.5, 185.8 (CO) ppm. ^{31}P {^{1}H} NMR (CD_2Cl_2): δ = 12.1 ppm. IR spectrum (CH_2Cl_2): 1711 (ν(NO)); 1981, 2010, 2053, 2127 (ν(CO)) cm^{-1} [1].

Reference:

[1] Zhou, Y.; Gladysz, J. A. (Organometallics **12** [1993] 1073/8).

2.5.1.2.7.2 Compounds with Bridging CO Ligands

The first three compounds described in this section are of the type $[C_5R_5Re(CO)_2]_2$-(μ-CO) (see Formula I). Some compounds with at least one semi-bridging CO group are described thereafter.

$(C_5H_5)_2Re_2(CO)_5$ (see Formula I, R = H) was obtained by irradiating $C_5H_5Re(CO)_3$ in C_6H_{12} for 2.5 h. The solvent was evaporated and the residue sublimed (to remove unconsumed starting product). The remainder was recrystallized from CH_2Cl_2 [1]. The reaction was also performed in THF, but in this case only a small quantity (ca. 1%) could be separated after

 References on pp. 218/9

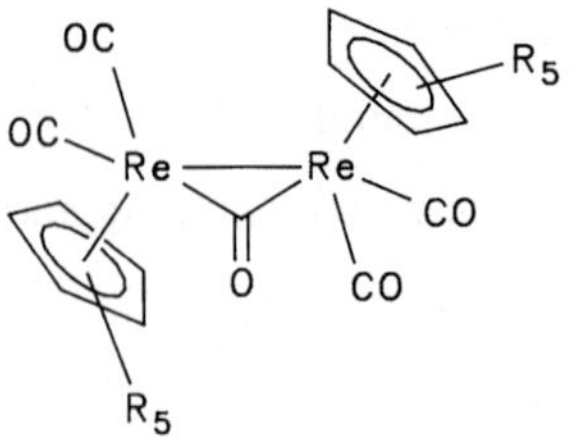

20 h (the main product was $C_5H_5Re(CO)_2OC_4H_8$) [3]. The following reactions also produced small quantities of $(C_5H_5)_2Re_2(CO)_5$: Reaction of $C_5H_5Re(CO)_2OC_4H_8$ with $(C_6H_5)_3M$-$C{\equiv}CC_6H_5$ (M = Si, Ge, Sn). Chromatographic workup with petroleum ether/ether mixtures achieved yields of 3, 1.6, and 2.4% [4]. – Photolysis of $C_5H_5Re(CO)_3$ in ether at −30°C followed by treating the mixture with COS or S_8 [7, 8]. – Photochemical reaction of $C_5H_5Re(CO)_3$ with $HC{\equiv}CC_6H_5$ (THF, 5°C, 7 h). Column-chromatographic workup separated the compound with 3% yield [2].

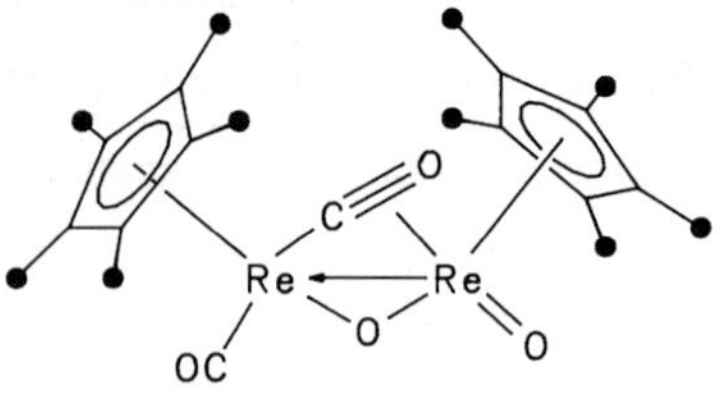

$(C_5H_5)_2Re_2(CO)_5$ forms a pale [3] yellow solid [1, 3, 8]; m.p. 138 to 140°C [1]. ^{1}H NMR spectrum ($CDCl_3$): δ = 5.34 ppm [1, 8]. ^{13}C {^{1}H} NMR spectrum ($CDCl_3$): δ = 88.5 ppm [8]; (CD_2Cl_2): δ = 88.6 (C_5), 207.48 (CO) [5] ppm; the observation of only one δ(CO) signal at 25 and −83°C reveals a rapid CO interchange [5]. IR spectrum (ether): 1731 ($\nu(CO_\mu)$); 1917, 1950, 1984 (ν(CO)) cm^{-1} [8]; (C_6H_{12}): 1745, 1905, 1925, 1958, 1994 cm^{-1} [2, 4] (similar in [1]). Mass spectrum: $[M]^+$ [1], $[M - n\ CO]^+$ (n = 0 to 5) [5].

The compound crystallizes in the monoclinic space group $P2_1/n-C_{2h}^5$ (No. 14) with a = 9.529(1), b = 13.934(2), c = 11.521(2) Å, β = 97.290(4)°. The following bond lengths and angles were determined (t = terminal): Re-Re: 2.957, Re-CO_μ: 2.06, Re-CO_t: 1.85 Å; Re-C-Re: 91.7°, OC_t-Re-CO_t: 85° [1].

Treatment with $P(C_6H_5)_3$ gave $C_5H_5Re(CO)_3$ and $C_5H_5Re(CO)_2P(C_6H_5)_3$ in equal amounts [1]. Combination with $C_5H_4CH_3Re(CO)_3$ (C_6D_6, 50 h) did not produce mixed-ligand products; thus, there is no dimer scission in solution. Exposure to a ^{13}CO atmosphere (1 atm, THF, 8 h) neither incorporated ^{13}CO nor produced $C_5H_5Re(CO)_3$ [5]. Protonation with CF_3SO_3H in CD_2Cl_2 gave a species that was only stable at low temperatures [5].

$(C_5H_4CH_3)_2Re_2(CO)_5$ (see Formula I, R_5 = H_4CH_3) was obtained by irradiating C_5H_4-$CH_3Re(CO)_3$ in hexane.

^{1}H NMR spectrum (CD_2Cl_2, 16°C): δ = 2.08 (CH_3); 5.14, 5.22 (t's, AA′BB′ pattern; J = 2 Hz) ppm. ^{13}C {^{1}H} NMR spectrum (CD_2Cl_2): δ = 12.7 (CH_3); 85.9, 88.0 (CH); 108.3 (**C**CH_3), 209.3 (CO) ppm (existence of only one δ(CO) signal indicates a rapid interchange process; moreover, enantiomerization also takes place because there is no diastereotopism observed in the NMR spectra).

Treatment with CF_3SO_3H in CD_2Cl_2 gave an unstable protonated species (see below) [5].

$(C_5(CH_3)_5)_2Re_2(CO)_5$ (see Formula I, R = CH_3) was yielded by irradiating $(C_5(CH_3)_5)Re$-$(CO)_3$ in C_6H_{12} (the coproduct was $(C_5(CH_3)_5)_2Re_2(CO)_3$) [1, 6], THF (yield: 5%) [11], or benzene [12]. The compound was also quantitatively formed, even at −80°C, by exposing the very reactive $(C_5(CH_3)_5)_2Re_2(CO)_4$ to a CO atmosphere in THF [14, 17, 18].

Yellow crystals [1, 18]; m.p. 158 to 160°C [1]. ^{1}H NMR spectrum (CD_2Cl_2): δ = 2.00 ppm. ^{13}C {^{1}H} NMR spectrum (CD_2Cl_2): δ = 10.3 (CH_3), 100.2 (C_5), 214.7 (CO) ppm. IR spectrum (hexane): 1714, 1877, 1901, 1930, 1971 (ν(CO)) cm^{-1} [1]; similar spectroscopic data in [18].

References on pp. 218/9

Irradiation with ${}^2D$ = CH_3CN or $P(C_6H_5)_3$ yielded $(C_5(CH_3)_5)Re(CO)_2(^2D)$ and $(C_5(CH_3)_5)Re(CO)_3$ [1]. Thermolysis in benzene gave $(C_5(CH_3)_5)Re(\eta^2\text{-}C_6H_6)(CO)_2$ [12].

$[(C_5H_4R)_2Re_2(CO)_5H][O_3SCF_3]$. The salts were quantitatively formed by treating $(C_5H_4R)_2Re_2(CO)_5$ (R = H, CH_3) with 1 (R = H) or 2.5 (R = CH_3) equivalents CF_3SO_3H in CD_2Cl_2 at −70 or −43 °C (the protonation of $(C_5H_5)_2Re_2(CO)_5$ did not proceed in CD_3CN; moreover, when using more than 1 equivalent of the acid, $C_5H_5Re(CO)_3$ was also formed).

Because the salts decomposed above ca. −40 °C, they were characterized at −70 °C [5]:

R = H: ${}^1H$ NMR (CD_2Cl_2): −9.9. – ${}^{13}C$ {${}^1H$} NMR (CD_2Cl_2): 90.5. – IR (CH_2Cl_2): 1720, 1951, 1991 (ν(CO)) [5].

R = CH_3: ${}^{13}C$ {${}^1H$} NMR (CD_2Cl_2): 12.4 (CH_3); 86.8, 89.3, 91.7, 112.7 (C_5); 191.7, 196.3 (CO); no signal attributed to the bridging CO group [5].

$(C_5(CH_3)_5)_2Re_2(CO)(O)(\mu\text{-}O)(\mu\text{-}\eta^2\text{-}C{\equiv}O)$ (see Formula II) was one of at least 5 products obtained by irradiating $(C_5(CH_3)_5)Re(CO)_3$ with $\lambda < 300$ nm (THF, 2 to 2.5 h). Chromatographic workup separated it with 20 to 25% yield. The product was also obtained with ca. 30% yield by reacting $(C_5(CH_3)_5)Re(CO)_2OC_4H_8$ with $(C_5(CH_3)_5)Re(CO)_3$ in THF.

Dark green crystals; dec.p. > 130 °C. ${}^1H$ NMR spectrum (C_6D_6): δ = 1.88, 1.92 (s's) ppm. IR spectrum (KBr): 924 (ν(ReO)); 1800, 1878 (CO) cm^{-1}; (toluene): 1830, 1888 cm^{-1}. Mass spectrum: $[M - n\ CO]^+$, $[M - 2\ CO - n\ O]^+$ (n = 0 to 2); $[(C_5(CH_3)_5)Re(CO)_n]^+$ (n = 1 to 3).

In the presence of O_2, at 60 °C, the cluster cleaved into $(C_5(CH_3)_5)ReO_3$ and $(C_5(CH_3)_5)$-$Re(CO)_3$ within 30 min (at 25 °C the reaction required 15 h). Solid-state thermolysis (200 °C under vacuum), however, did not affect the compound [11].

$(C_5(CH_3)_5)_2Re_2(CO)_4$ quantitatively formed within 1 week by the solid-state decomposition of $(C_5(CH_3)_5)Re(CO)_2OC_4H_8$ [14, 18] under an N_2 atmosphere [18]. The resulting powder was washed with ether to remove red impurities. Yield: 86% [18]. The product also formed in a THF solution, but in this case $(C_5(CH_3)_5)Re(CO)_3$ was also obtained [14].

Light green powder [18]. ${}^1H$ NMR spectrum (C_6D_6): δ = 1.98 ppm; (THF-d_8): δ = 2.14 ppm. ${}^{13}C$ {${}^1H$} NMR spectrum (toluene-d_8): δ = 11.1 (CH_3), 103.1 (C_5), 209.4 (CO) ppm. IR spectrum (KBr): 1809, 1860 cm^{-1}; (toluene): 1824, 1869 (ν(CO)) cm^{-1}. Mass spectrum: $[M - CO]^+$ [14, 18].

The solvate **$(C_5(CH_3)_5)_2Re_2(CO)_4 \cdot C_6D_6$** crystallizes in the triclinic lattice with a = 8.426(3), b = 8.819(3), c = 9.357(3) Å, α = 81.76(3)°, β = 80.44(3)°, γ = 89.02(3)°; space group $P\bar{1}-C_i^1$ (No. 2); Z = 1 formula unit per unit cell, D_{calc} = 2.038 g/cm^3. As can be seen from **Fig. 106a**, the centrosymmetric molecule has two semi-bridging CO groups [14].

The bonding within the compound has been discussed on the simplified model molecule **$(C_5H_5)_2Re_2(CO)_4$** using Fenske-Hall molecular orbital calculations but assuming an ideal C_{2h} structure in which all CO ligands adopt terminal nonbridging positions. It appeared that the Re-Re σ-bonding orbital is very low in energy. The electron configuration of the molecule is σ^2, $(\pi_\perp)^2$, $(\pi_\parallel)^2$, δ^2, δ^{*2}, $(\pi_\parallel^*)^2$, corresponding to a diamagnetic complex with a net Re=Re double bond. The π^* molecule orbitals are not degenerate, the $\pi_\parallel^*$ orbital is at a lower energy than the $\pi_\perp^*$ orbital. When the idealized structure is changed to the actual geometry with semibridging CO groups, the unoccupied $\pi_\perp^*$ orbital becomes further destabilized. A diagram showing the energy order of the molecular orbitals is given [15].

The Re=Re double bond is extremely reactive: Exposure to H_2 gave $[(C_5(CH_3)_5)Re(CO)_2]_2(\mu\text{-}H)_2$ within seconds [14], and exposure to CO gave $(C_5(CH_3)_5)_2Re_2(CO)_5$ in less than one minute [14, 17, 18]. Both reactions even proceeded at −80 °C [14, 17]. Adducts of

References on pp. 218/9

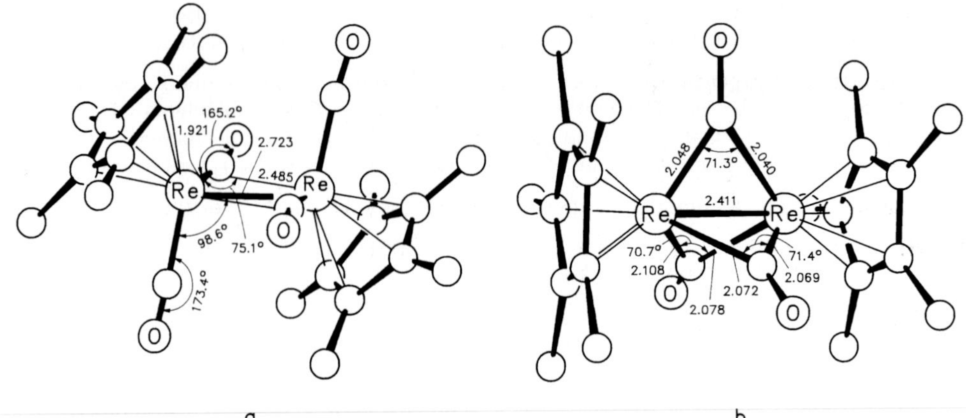

Fig. 106. Molecular structures of
a. $(C_5(CH_3)_5)_2Re_2(CO)_4$ [14];
b. $(C_5(CH_3)_5)_2Re_2(CO)_3$ [6].

the type $(C_5(CH_3)_5)Re(CO)_2(\mu\text{-}CO)Re(C_5(CH_3)_5)(CO)(^2D)$ were readily obtained upon treatment with $^2D = P(CH_3)_3$, CH_3CN [18].

Interaction with C_2H_4 led to $(C_5(CH_3)_5)Re(CO)_2(\mu\text{-}CO)Re(C_5(CH_3)_5)(CO)(CH_2{=}CH_2\text{-}\pi)$ [18]. Treatment with the conjugated enyne $HC{\equiv}CC(CH_3){=}CH_2$ gave the dirhenacyclopentenone $(C_5(CH_3)_5)Re(CO)_2(\mu\text{-}\eta^{3:1}\text{-}CH{=}C(C(CH_3){=}CH_2)C{=}O)Re(C_5(CH_3)_5)(CO)$ [16]; likewise, low-temperature treatment with $CH{\equiv}CH$ gave $(C_5(CH_3)_5)Re(CO)_2(\mu\text{-}\eta^{3:1}\text{-}CH{=}CHC{=}O)$-$Re(C_5(CH_3)_5)(CO)$ (see pp. 240/1) [18], whereas the low-temperature reaction with $CH_3C{\equiv}CCH_3$ produced $(C_5(CH_3)_5)Re(CO)_2(\mu\text{-}CO)Re(C_5(CH_3)_5)(CO)(CH_3C{\equiv}CCH_3\text{-}\pi)$ ($k_{obs} = 7.2\times10^{-4}\ s^{-1}$, $k_2 = 3.7\times10^{-2}\ M^{-1}\cdot s^{-1}$, $\Delta G^{\neq}$ ca. 59 kJ/mol; toluene, −60 °C) [18]. Treatment with $CH_3O_2CC{\equiv}CCO_2CH_3$ in the dark yielded the dirhenacyclobutene $[(C_5(CH_3)_5)Re(CO)_2]_2$-$(\mu\text{-}CH_3O_2CC{=}CCO_2CH_3)$ (see p. 227) instantaneously (spectroscopic monitoring at −78 °C revealed complete consumption of the starting complex within less than 3 min; intermediates were not detected). In a competition experiment, the reaction with dimethyl acetylenedicarboxylate was observed to be ca. 300 times slower than the reaction with CO [17].

$(C_5(CH_3)_5)_2Re_2(CO)_3$ was obtained by the carbonylation of either of the following compounds under pressure: $(C_5(CH_3)_5)ReH_6$ (5 atm, 160 to 180 °C, 4 h, chromatographic separation on silica with pentane, yield: 77%) [13], $(C_5(CH_3)_5)ReO_3$ (THF, 15 atm, room temperature) [10], or $[(C_5(CH_3)_5)ReO]_2(\mu\text{-}O)_2$ (toluene, 20 atm, 70 h) [9]. Alternatively, thermolysis of solid or toluene-dissolved $[(C_5(CH_3)_5)Re(CO)_2]_2(\mu\text{-}O)$ yielded the title compound along with $(C_5(CH_3)_5)Re(CO)_3$ [11]; however, photolysis of $(C_5(CH_3)_5)Re(CO)_3$ (C_6H_{12}, room temperature) also produced $(C_5(CH_3)_5)_2Re_2(CO)_3$ with ca. 30% yield, the coproduct was $(C_5(CH_3)_5)_2Re_2(CO)_5$ [1, 6].

Air-stable, orange-red solid; m.p. >175 °C [6]. ^1H NMR spectrum (CD_2Cl_2): δ = 2.03 ppm. $^{13}C\{^1H\}$ NMR spectrum (CD_2Cl_2): δ = 10.2 (CH_3), 100.6 (C_5), 231.3 (CO) ppm. IR spectrum (n-hexane): 1748 (ν(CO)) cm^{-1}. Mass spectrum: $[M]^+$, $[M - CO]^+$ [6].

X-ray crystallography showed the compound to crystallize in the triclinic space group $P\bar{1}-C_i^1$ (No. 2) with a = 10.151(2), b = 13.498(3), c = 8.806(3) Å, α = 94.20(2)°, β = 101.77(2)°, γ = 70.89(2)°; Z = 2 molecules per unit cell, D_{calc} = 2.16 g/cm^3. The molecular structure is illustrated in **Fig. 106b** [6].

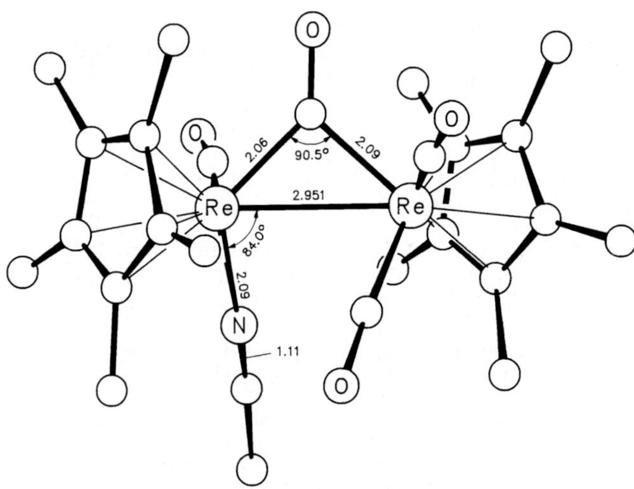

Fig. 107. Molecular structure of $(C_5(CH_3)_5)Re(CO)_2(\mu\text{-}CO)Re(C_5(CH_3)_5)(CO)NCCH_3$ [18].

The compound did not react with acetylene, CH_2N_2, or phosphanes [6]. Pressurization to a CO atmosphere (200 atm, 100 °C [6]; 70 atm, 160 to 180 °C [13]) reformed $(C_5(CH_3)_5)$-$Re(CO)_3$ [6, 13]. Treatment with I_2 yielded $(C_5(CH_3)_5)Re(CO)_2I_2$ [6].

$(C_5(CH_3)_5)Re(CO)_2(\mu\text{-}CO)Re(C_5(CH_3)_5)(CO)(^2D)$. These products, where ^2D stands for CH_3CN or $P(CH_3)_3$, were instantaneously obtained by treating $(C_5(CH_3)_5)_2Re_2(CO)_4$ with the respective donor reagent in THF. While the $P(CH_3)_3$ adduct was not isolated, the CH_3CN adduct separated when pentane was added to the solution [18].

$^2D = NCCH_3$: Yield: 73%. – Yellow crystals. – ^1H NMR (C_6D_6): 1.33 (br s, $NCCH_3$); 1.80, 1.98 ($(CH_3)_5$). – $^{13}C\{^1H\}$ NMR (THF-d_8, −50 °C): 3.6, 9.9 ($(CH_3)_5$); 10.8 ($NCCH_3$); 97.4, 98.7 (C_5); 129.8 (CN); 213.7 (CO). – IR (THF): 1678 (CO_μ); 1838, 1863, 1905 (ν(CO)); (KBr): 1655, 1828, 1857, 1902. – MS: $[M - CH_3CN]^+$.

Single-crystal data at −160 °C: triclinic; a = 8.695(4), b = 10.192(6), c = 14.931(5) Å, α = 99.45(4)°, β = 92.15(3)°, γ = 100.93(4)°; space group $P\bar{1}-C_i^1$ (No. 2); Z = 2 molecules per unit cell, D_{calc} = 2.068 g/cm³. The structure is illustrated in **Fig. 107**.

Addition of excess CD_3CN (C_6D_6, room temperature) initiated complete exchange of free and bonded CH_3CN in less than 15 min; a barrier of $\Delta G^{\neq}$ ca. 62.8 kJ/mol was determined for the exchange. Addition of dimethyl acetylenedicarboxylate gave $[(C_5(CH_3)_5)Re(CO)_2]_2(\mu\text{-}CH_3O_2CC{=}CCO_2CH_3)$ (see p. 227) in a rapid reaction [18].

$^2D = P(CH_3)_3$: ^1H NMR (THF-d_8, −80 °C): 1.52 (d, PCH_3; J(P,H) = 9); 1.89, 1.94 ($(CH_3)_5$). – $^{13}C\{^1H\}$ NMR (THF-d_8, −80 °C): 10.5, 11.2 ($(CH_3)_5$); 20.0 (PCH_3); 97.7, 98.3 (C_5); 212.4 (CO), 232.4 (d, CO; J(P,C) = 10.7). – IR (THF, −80 °C): 1662 ($\nu(CO_\mu)$); 1836, 1861, 1917 (ν(CO)).

At −20 °C the compound converted to $(C_5(CH_3)_5)Re(CO)_2P(CH_3)_3$ and $(C_5(CH_3)_5)Re(CO)_2OC_4H_8$. Spectroscopic monitoring of the reaction established the rate constant $k_1 = 9.57 \times 10^{-4}\ s^{-1}$ and $\Delta G^{\neq}$ = 76.2 kJ/mol [18].

References on pp. 218/9

$(C_5(CH_3)_5)Re(CO)_2(\mu\text{-}CO)Re(C_5(CH_3)_5)(CO)(CH_2{=}CH_2\text{-}\pi)$ formed when $(C_5(CH_3)_5)_2Re_2(CO)_4$ was treated with C_2H_4 in THF-d_8 or toluene-d_8 at low temperatures (−80 to −40 °C), resulting in a yellow solution. The compound was not isolated.

^{1}H NMR spectrum (THF-d_8, −40 °C): δ = 1.44 (CH_2); 1.74, 1.93 ($(CH_3)_5$) ppm; (toluene-d_8, −40 °C): δ = 1.57, 1.65, 1.75 ppm. ^{13}C {^{1}H} NMR spectrum (THF-d_8, −50 °C): δ = 9.6, 10.6 ($(CH_3)_5$); 32.3 (CH_2); 98.2, 100.4 (C_5); 206.1, 229.6 (CO) ppm. IR spectrum (THF, −78 °C): 1699, 1842, 1881, 1950 (ν(CO)) cm^{-1}.

The compound was stable below 0 °C in THF, but upon warming to 25 °C it fragmented into $(C_5(CH_3)_5)Re(CO)_2C_2H_4$ and $(C_5(CH_3)_5)Re(CO)_2OC_4H_8$; the parameters $k_1 = 1.35 \times 10^{-3}\ s^{-1}$ and $\Delta G^{\neq}$ ca. 89 kJ/mol were also determined [18].

$(C_5(CH_3)_5)Re(CO)_2(\mu\text{-}CO)Re(C_5(CH_3)_5)(CO)(CH_3C{\equiv}CCH_3\text{-}\pi)$ was obtained by treating $(C_5(CH_3)_5)_2Re_2(CO)_4$ with but-2-yne in toluene at −60 °C.

^{1}H NMR spectrum (toluene-d_8, −60 °C): δ = 1.56. 1.73 ($(CH_3)_5$); 2.41 (br s, CH_3) ppm. IR spectrum (toluene, −78 °C): 1662 (ν(CO_μ)); 1855, 1892, 1925 (ν(CO)) cm^{-1}.

At −40 °C the compound slowly converted into $(C_5(CH_3)_5)Re(CO)_2(\mu\text{-}\eta^{3:1}\text{-}C(CH_3){=}C(CH_3)C{=}O)Re(C_5(CH_3)_5)(CO)$ (see p. 241) and the fragmentation products $(C_5(CH_3)_5)Re(CO)_3$ and $(C_5(CH_3)_5)Re(CH_3C{\equiv}CCH_3)(CO)$. The kinetics were also studied, and the parameters $k_1 = 1.99 \times 10^{-3}\ s^{-1}$ and $\Delta G^{\neq}$ ca. 71 kJ/mol were established [18].

References:

[1] Foust, A. S.; Hoyano, J. K.; Graham, W. A. G. (J. Organomet. Chem **32** [1971] C 65/C 66).

[2] Kolobova, N. E.; Antonova, A. B.; Khitrova, O. M.; Antipin, M. Yu.; Struchkov, Yu. T. (J. Organomet. Chem. **137** [1977] 69/78).

[3] Sellmann, D.; Kleinschmidt, E. (Z. Naturforsch. **32b** [1977] 795/801).

[4] Kolobova, N. E.; Khitrova, O. M.; Antonova, A. B. (Izv. Akad. Nauk SSSR Ser. Khim. **1979** 1124/6; Bull. Acad. Sci. USSR Div. Chem. Sci. [Engl. Transl.] **1979** 1050/2).

[5] Lewis, L. N.; Caulton, K. G. (Inorg. Chem. **20** [1981] 1139/42).

[6] Hoyano, J. K.; Graham, W. A. G. (J. Chem. Soc. Chem. Commun. **1982** 27/8).

[7] Herberhold, M.; Reiner, D.; Thewalt, U. (Angew. Chem. **95** [1983] 1028/9; Angew. Chem. Int. Ed. Engl. **22** [1983] 1000; Angew. Chem. Suppl. **1983** 1343/52).

[8] Herberhold, M.; Reiner, D.; Ackermann, K.; Thewalt, U.; Debaerdemaeker, T. (Z. Naturforsch. **39b** [1984] 1199/205).

[9] Herrmann, W. A.; Serrano, R.; Bock, H. (Angew. Chem. **96** [1984] 364/5; Angew. Chem. Int. Ed. Engl. **23** [1984] 383).

[10] Herrmann, W. A.; Serrano, R.; Küsthardt, U.; Ziegler, M. L.; Guggolz, E.; Zahn, T. (Angew. Chem. **96** [1984] 498/500; Angew. Chem. Int. Ed. Engl. **23** [1984] 515/7).

[11] Herrmann, W. A.; Serrano, R.; Schäfer, A.; Küsthardt, U.; Ziegler, M. L.; Guggolz, E. (J. Organomet. Chem. **272** [1984] 55/71).

[12] van der Heijden, H.; Orpen, A. G.; Pasman, P. (J. Chem. Soc. Chem. Commun. **1985** 1576/8).

[13] Herrmann, W. A.; Theiler, H. G.; Kiprof, P.; Tremmel, J.; Blom, R. (J. Organomet. Chem. **395** [1990] 69/84).

[14] Casey, C. P.; Sakaba, H.; Hazin, P. N.; Powell, D. R. (J. Am. Chem. Soc. **113** [1991] 8165/6).

[15] Barckholtz, T. A.; Bursten, B. E.; Niccolai, G. P.; Casey, C. P. (J. Organomet. Chem. **478** [1994] 153/60).

[16] Casey, C. P.; Ha, Y.; Powell, D. R. (J. Am. Chem. Soc. **116** [1994] 3424/8).
[17] Casey, C. P.; Cariño, R. S.; Hayashi, R. K.; Schladetzky, K. D. (J. Am. Chem. Soc. **118** [1996] 1617/23).
[18] Casey, C. P.; Cariño, R. S.; Sakaba, H.; Hayashi, R. K. (Organometallics **15** [1996] 2640/9).

2.5.1.2.7.3 Compounds with a Bridging Methylidene Ligand

The general structure of the compounds described in this section is depicted in Formula I. The ^5L ligands are C_5H_5 or $C_5(CH_3)_5$.

I

$[C_5H_5Re(CO)_2]_2(\mu\text{-}C{=}CH_2)$ formed by dissolving $[C_5H_5Re(CO)_2{\equiv}CCH_3]BCl_4$ in CH_2Cl_2 [4] or THF [7]. HCl gas and BCl_3 were liberated and, after 2 h, IR analysis of the mixture indicated quantitative formation of $C_5H_5Re(CO)_2{=}C{=}CH_2$ [4, 7]. This intermediate further on transformed into the title dimer and $C_5H_5Re(CO)_3$. The products could be separated by chromatography on alumina with CH_2Cl_2/hexane (2:1) [7].

IR spectrum (CH_2Cl_2): 1900, 1948, 1977 cm^{-1} [4]; (n-hexane): 1881, 1960, 1982 (ν(CO)) cm^{-1} [7].

$[C_5H_5Re(CO)_2]_2(\mu\text{-}C{=}CHC_6H_5)$ was one of the 5 products obtained by photochemically reacting $C_5H_5Re(CO)_3$ with 2 equivalents $HC{\equiv}CC_6H_5$ (THF, 5 °C, 7 h). The mixture was evaporated and the residual red-brown oil column-chromatographically separated using petroleum ether/ether mixtures. The compound was eventually isolated with 3% yield. It could also be prepared by treating $C_5H_5Re(CO)_2{=}C{=}CHC_6H_5$ with KOH (THF/H_2O, 65 °C) [1] or $C_5H_5Re(CO)_2(C_6H_5C{\equiv}CGe(C_6H_5)_3)$ with KOH (CH_3OH/THF/H_2O, 50 °C, 4 h) [3]. In both cases, the resulting mixture was worked up by evaporation and extraction of the residue into petroleum ether [1, 3]. The title dimer could also be obtained by treating $C_5H_5Re(CO)_2OC_4H_8$ with $(C_6H_5)_3M{-}C{\equiv}CC_6H_5$ (M = Si, Ge, Sn). Yields were 2.4, 0.7, and 2.8% for M = Si, Ge, Sn, respectively [3]. Yellow crystals [1, 3]; m.p. 193 to 194 °C [1].

IR spectrum (C_6H_{12}): 1916, 1953, 1983 cm^{-1}. Raman spectrum: 1555 (ν(C=C)) cm^{-1} [1]. Mass spectrum: $[M - n\ CO]^+$ (n = 0 to 4) [1, 2], $[M - 4\ CO - n\ H_2]^+$ (n = 1 to 3) [2].

[Re] = $(C_5(CH_3)_5)Re(CO)_2$

II III IV

References on p. 222

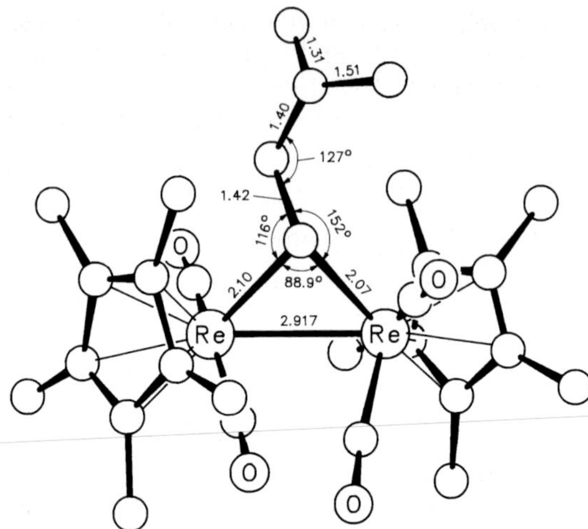

Fig. 108. Molecular structure of $[(C_5(CH_3)_5)Re(CO)_2]_2(\mu\text{-}C{=}CHC(CH_3){=}CH_2)$ [5].

$[(C_5(CH_3)_5)Re(CO)_2]_2(\mu\text{-}C{=}CHC(CH_3){=}CH_2)$ (see Formula II) formed along with $(C_5(CH_3)_5)$-$Re(CO)_3$ (ratio 2:1) by heating $(C_5(CH_3)_5)Re(CO)_2(\mu\text{-}\eta^{3:1}\text{-}CH{=}C(C(CH_3){=}CH_2)C{=}O)Re$-$(C_5(CH_3)_5)(CO)$ (see p. 241) in C_6D_6 for 24 h. The compound was separated by chromatography on silica with hexane/ether mixtures. Yield: 50%.

Yellow-orange solid. ^1H NMR spectrum (C_6D_6): δ = 1.77, 1.84 ($(CH_3)_5$); 2.12 (br dd, $=CCH_3$); 4.82, 5.49 (m's, $CH_2{=}$); 6.89 (br m, =CH) ppm. $^{13}C\{^1H\}$ NMR spectrum (C_6D_6): δ = 10.1, 10.4 ($(CH_3)_5$); 24.5 (C=C**C**H_3), 99.9, 100.7 (C_5); 108.5 (C=**C**H_2), 142.2 (C=**C**H), 145.8 (**C**=CH_2); 207.7, 209.1 (CO); 241.2 (ReCRe) ppm. IR spectrum (THF): 1860, 1886, 1921, 1957, (ν(CO)) cm^{-1}. Mass spectrum: $[M]^+$.

Single crystals have a triclinic lattice with a = 9.051(2), b = 9.395(2), c = 17.060(2) Å, α = 81.117(13)°, β = 82.27(2)°, γ = 76.79(2)°; Z = 2 molecules per unit cell, D_{calc} = 1.965 g/cm^3. The molecular structure is depicted in **Fig. 108**. As a remarkable feature, both $Re\text{-}C_\mu{=}C$ angles are very different from each other.

Treatment with $HBF_4 \cdot O(C_2H_5)_2$ instantaneously led to $[\{(C_5(CH_3)_5)Re(CO)_2\}_2(\mu\text{-}C{=}CH{=}C(CH_3)_2)]BF_4$. Protonation with excess CF_3CO_2D in CD_2Cl_2 incorporated the D atom at the δ site selectively. The protonation could be reversed with CH_3Li, CH_3MgBr, or $K[OC_4H_9\text{-}t]$. A reaction with $CH_3O_2CC{\equiv}CCO_2CH_3$ at 70°C gave $[(C_5(CH_3)_5)Re(CO)_2]_2$-$(\mu\text{-}C{=}CHC({=}CH_2)CH_2C(CO_2CH_3){=}CHCO_2CH_3)$ (see p. 222); in contrast, the title compound did not react with $HC{\equiv}CCO_2CH_3$ [5].

$[\{(C_5(CH_3)_5)Re(CO)_2\}_2(\mu\text{-}CCH{=}C(CH_3)_2)]BF_4$ (see Formula III) instantaneously precipitated when $[(C_5(CH_3)_5)Re(CO)_2]_2(\mu\text{-}C{=}CHC(CH_3){=}CH_2)$ was acidified with $HBF_4 \cdot O(C_2H_5)_2$ in benzene (the reaction could be reversed with $K[OC_4H_9\text{-}t]$, CH_3Li, or CH_3MgBr). Recrystallization from CH_2Cl_2/C_2H_5OH gave a dark green solid. Yield: 90%.

^1H NMR spectrum (CD_2Cl_2): δ = 1.82, 1.90 ($=C(CH_3)_2$); 2.03 ($(CH_3)_5$), 7.78 (CH) ppm. $^{13}C\{^1H\}$ NMR spectrum (CD_2Cl_2): δ = 10.2, 22.6, 29.8 (CH_3); 105.1 (C_5), 150.3 (C=**C**$(CH_3)_2$), 155.5 (H**C**=C); 197.2, 202.7 (CO); 371.3 (ReCRe) ppm.

Nucleophilic addition of H^- or $P(CH_3)_3$ occurred at the γ site of the alkylidyne complex to yield $[(C_5(CH_3)_5)Re(CO)_2]_2(\mu\text{-}C{=}CHC(CH_3)_2H)$ or $[\{(C_5(CH_3)_5)Re(CO)_2\}_2(\mu\text{-}C{=}CHC$-

$(CH_3)_2P(CH_3)_3)]BF_4$, respectively. A reaction with $Na[CH(CO_2C_2H_5)_2]$ yielded $[(C_5(CH_3)_5)Re(CO)_2]_2(\mu\text{-}C{=}CHC(CH_3)_2C(CO_2C_2H_5)_2H)$ and $[(C_5(CH_3)_5)Re(CO)_2]_2(\mu\text{-}C{=}CHC(CH_3){=}CH_2)$ in the ratio 0.75:1, and treatment with $Li[Cu(CH_3)_2]$ gave $[(C_5(CH_3)_5)Re(CO)_2]_2(\mu\text{-}C{=}CHC(CH_3)_3)$ and $[(C_5(CH_3)_5)Re(CO)_2]_2(\mu\text{-}C{=}CHC(CH_3){=}CH_2)$ in the ratio 1.2:1 [5].

$[(C_5(CH_3)_5)Re(CO)_2]_2(\mu\text{-}C{=}CHC(CH_3)_2H)$ (see Formula IV, R = H) instantaneously formed by treating a THF solution containing $[\{(C_5(CH_3)_5)Re(CO)_2\}_2(\mu\text{-}CCH{=}C(CH_3)_2)]BF_4$ with excess $NaBH_4$. Workup of the mixture by chromatography with hexane/ether yielded 67% [5]. The compound also formed as a side product by treating the carbyne complex $[(C_5(CH_3)_5)Re(CO)_2{\equiv}CCH{=}C(CH_3)_2]BF_4$ with excess $NaBH_4$ in THF (main product was $(C_5(CH_3)_5)Re(CO)_2{=}C{=}CHC(CH_3)_2H$) [6]. Yellow solid [5].

^1H NMR spectrum (C_6D_6): δ = 1.64 (d; J = 12.8 Hz); 1.78, 1.88 (CH_3); 5.85 (d, C=CH; J = 8.8 Hz) ppm. $^{13}C\{^1H\}$ NMR spectrum (C_6D_6): δ = 240 (ReCRe) ppm. IR spectrum (THF): 1861, 1883, 1921, 1953 (ν(CO)) cm^{-1}. Mass spectrum: $[M]^+$ [5].

$[(C_5(CH_3)_5)Re(CO)_2]_2(\mu\text{-}C{=}CHC(CH_3)_3)$ (see Formula IV, R = CH_3) immediately formed when $[\{(C_5(CH_3)_5)Re(CO)_2\}_2(\mu\text{-}CCH{=}C(CH_3)_2)]BF_4$ was treated in CH_2Cl_2 with an excess of $Li[Cu(CH_3)_2]$ in ethereal solution. The mixture was evaporated, and the residue was purified by extracting the solid into benzene and passing the extract through a silica column. The second yellow band eluted contained a 1.2:1 mixture of the title compound and $[(C_5(CH_3)_5)Re(CO)_2]_2(\mu\text{-}C{=}CHC(CH_3){=}CH_2)$ with a total yield of 60%. The title compound was characterized in the mixture as follows:

^1H NMR spectrum (C_6D_6): δ = 1.41 (s, C_4H_9-t); 1.77, 1.86 (CH_3); 6.24 (C=CH) ppm. $^{13}C\{^1H\}$ NMR spectrum (C_6D_6): δ = 10.3, 10.9 (CH_3); 25.9 ($C(\mathbf{CH_3})_3$), 32.3 ($C\mathbf{C_3}$); 99.8, 100.1 (C_5); 151.3 (C=**C**H); 207.4, 209.0 (CO); 232 (ReCRe) ppm. Mass spectrum: $[M]^+$ [5].

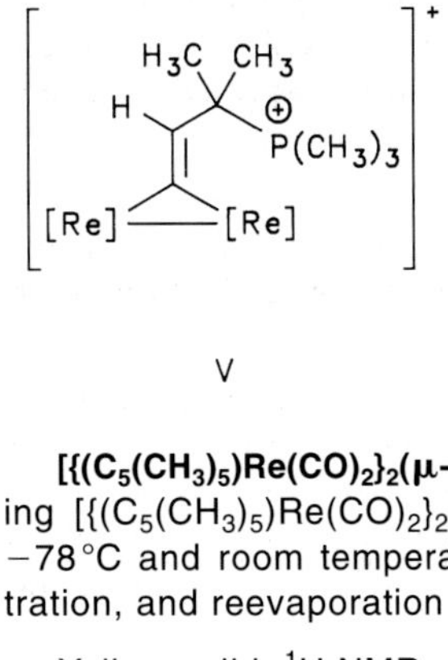

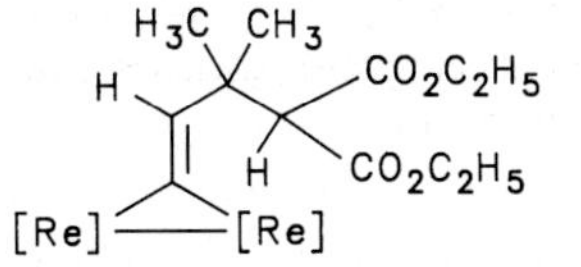

[Re] = $(C_5(CH_3)_5)Re(CO)_2$

V VI VII

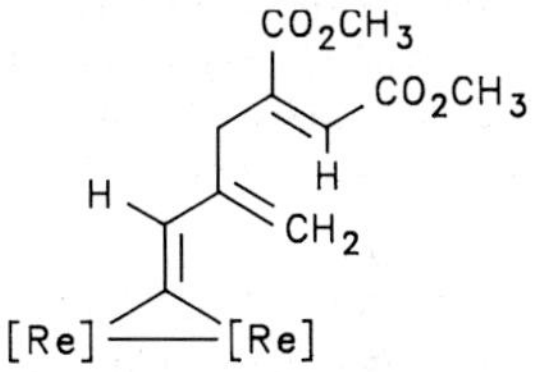

$[\{(C_5(CH_3)_5)Re(CO)_2\}_2(\mu\text{-}C{=}CHC(CH_3)_2P(CH_3)_3)]BF_4$ (see Formula V) was obtained by treating $[\{(C_5(CH_3)_5)Re(CO)_2\}_2(\mu\text{-}CCH{=}C(CH_3)_2)]BF_4$ with excess $P(CH_3)_3$ in CH_2Cl_2 between −78 °C and room temperature. The mixture was evaporated; extraction into benzene, filtration, and reevaporation gave the compound with 74% yield.

Yellow solid. ^1H NMR spectrum (CD_2Cl_2): δ = 1.51, 1.68 (d's, $C(CH_3)_2P$; J(P,H) = 18.4, 19.1 Hz); 1.91 (d, PCH_3; J(P,H) = 12.9 Hz); 1.91, 2.02 (C_5CH_3); 5.70 (d, C=CH; J(P,H) = 11 Hz) ppm. $^{13}C\{^1H\}$ NMR spectrum (CD_2Cl_2): δ = 5.8 (d, PCH_3; J(P,C) = 52 Hz); 10.5, 11.0 (CH_3); 20.7 (d, $\mathbf{C}(CH_3)_2P$; J(P,C) = 43 Hz); 26.6, 30.0 ($C(\mathbf{CH_3})_2P$); 100.6, 101.0 (C_5); 132.6 (C=**CH**); 204.2, 206.8 (CO); 233.1 (ReCRe) ppm. $^{31}P\{^1H\}$ NMR spectrum (CD_2Cl_2): δ = 38.82 ppm. FAB mass spectrum: $[M - BF_4 - n\,P(CH_3)_3]^+$ (n = 0, 1) [5].

$[(C_5(CH_3)_5)Re(CO)_2]_2(\mu\text{-}C{=}CHC(CH_3)_2C(CO_2C_2H_5)_2H)$ (see Formula VI) immediately formed when $[\{(C_5(CH_3)_5)Re(CO)_2\}_2(\mu\text{-}CCH{=}C(CH_3)_2)]BF_4$ was treated with excess solid $Na[CH(CO_2C_2H_5)_2]$ in THF. The compound was chromatographically separated (silica, hexane/ether mixtures). Yield: 35%. Viscous yellow solid.

References on p. 222

^1H NMR spectrum (C_6D_6): δ = 0.95 to 1.15 (2 s's and 2 t's, $C(CH_3)_2$ and $CH_2\mathbf{CH_3}$); 1.78, 1.88 (CH_3); 4.08 (s, $HC(CO_2)_2$), 3.05 to 4.30 (2 dq's, $\mathbf{CH_2}CH_3$), 6.66 (C=CH) ppm. ^{13}C {^1H} NMR spectrum (C_6D_6): δ = 10.2, 10.9 (CH_3); 14.2, 14.4 ($CH_2\mathbf{CH_3}$); 22.7, 27.2 ($C(\mathbf{CH_3})_2$); 34.4 ($\mathbf{CH}(CO_2)_2$); 63.1, 70.9 ($\mathbf{CH_2}CH_3$); 100.0, 100.3 (C_5); 147.9 (C=**C**H); 168.3, 168.6 (CO_2); 207.3, 208.0, 208.9, 209.5 (CO); 228.0 (ReCRe) ppm. IR spectrum (THF): 1741, 1757, 1865, 1890, 1920, 1955 (ν(CO)) cm^{-1}. FAB mass spectrum: $[MH]^+$ [5].

$[(C_5(CH_3)_5)Re(CO)_2]_2(\mu\text{-}C{=}CHC({=}CH_2)CH_2C(CO_2CH_3){=}CHCO_2CH_3)$ (see Formula VII) was formed by treating $[(C_5(CH_3)_5)Re(CO)_2]_2(\mu\text{-}C{=}CHC(CH_3){=}CH_2)$ with $CH_3O_2CC{\equiv}CCO_2CH_3$ in C_6D_6 at 70 °C for 5 h. The compound could only be characterized in solution because it decomposed upon attempting its purification.

^1H NMR spectrum (C_6D_6): δ = 1.79, 1.81 (CH_3); 1.8 (br s, CH_2); 3.35, 3.65 (CO_2CH_3); 4.75, 5.74 (m, $C{=}CH_2$); 6.08 (br t, $C{=}CHCO_2$; J ca. 1 Hz), 6.66 (br m, C=CH) ppm. ^{13}C {^1H} NMR spectrum (C_6D_6): δ = 10.1, 10.3 (CH_3); 44.1 (CH_2); 51.3, 51.9 ($CO_2\mathbf{CH_3}$); 99.8, 100.9 (C_5); 109.6 ($C{=}\mathbf{CH_2}$); 121.9, 138.9 (CH); 140.9, 144.0 ($\mathbf{C}(CO_2CH_3){=}\mathbf{C}HCO_2$); 150.0, 152.3 ($CO_2$); 207, 208, 209 (CO); 243.4 (ReCRe) ppm. IR spectrum (n-hexane): 1614 (ν(C=C)); 1735 (ν(CO_2)); 1869, 1895, 1927, 1962 (ν(CO)) cm^{-1}. Mass spectrum: $[M]^+$ [5].

References:

[1] Kolobova, N. E.; Antonova, A. B.; Khitrova, O. M.; Antipin, M. Yu.; Struchkov, Yu. T. (J. Organomet. Chem. **137** [1977] 69/78).
[2] Sizoi, V. F.; Nekrasov, Yu. S.; Sukharev, Yu. N.; Kolobova, N. E.; Khitrova, O. M.; Obezyuk, N. S.; Antonova, A. B. (J. Organomet. Chem. **162** [1978] 171/8).
[3] Kolobova, N. E.; Khitrova, O. M.; Antonova, A. B. (Izv. Akad. Nauk SSSR Ser. Khim. **1979** 1124/6; Bull. Acad. Sci. USSR Div. Chem. Sci. [Engl. Transl.] **1979** 1050/2).
[4] Kelley, C.; Mercando, L. A.; Terry, M. R.; Lugan, N.; Geoffrey, G. L.; Xu, Z.; Rheingold, A. L. (Angew. Chem. **104** [1992] 1066/8; Angew. Chem. Int. Ed. Engl. **31** [1992] 1053).
[5] Casey, C. P.; Ha, Y.; Powell, D. R. (J. Am. Chem. Soc. **116** [1994] 3424/8).
[6] Casey, C. P.; Ha, Y.; Powell, D. R. (J. Organomet. Chem. **472** [1994] 185/93).
[7] Terry, M. R.; Mercando, L. A.; Kelley, C.; Geoffroy, G. L.; Nombel, P.; Lugan, N.; Mathieu, R.; Ostrander, R. L.; Owens-Waltermire, B. E.; Rheingold, A. L. (Organometallics **13** [1994] 843/65).

2.5.1.2.7.4 Compounds with a Bridging Methylidyne Ligand

This section describes several compounds, whose structure is shown in Formula I. X stands for Cl or Br. The compounds were prepared as follows:

Method I: Treatment of $(C_5(CH_3)_4R)Re(O)(CH_2C_6H_5)_2$ (R = CH_3, C_2H_5) with 2 equivalents $C_5H_5TiX_3$ (X = Cl, Br) in toluene at reflux temperature. Removal of the solvent was followed by purification of the crude product by column chromatography on silylated silica with toluene [1, 2].

$[(C_5(CH_3)_5)ReCl]_2(\mu\text{-}CC_6H_5)(\mu\text{-}Cl)$ was obtained via Method I [1, 2] with 85% yield. Violet solid; m.p. >285 °C [2].

^1H NMR spectrum ($CDCl_3$): δ = 1.66 (s, CH_3); 7.12, 7.36 (m's) ppm. ^{13}C {^1H} NMR spectrum ($CDCl_3$): δ = 9.3 (CH_3), 102.7 (C_5); 127.4, 128.1, 128.7, 157.4 (C_6H_5); 325.5 (ReCRe) ppm. IR spectrum (KBr): 295, 322 (ν(ReCl)) cm^{-1}. EI mass spectrum: $[M]^+$ [1, 2].

References on p. 223

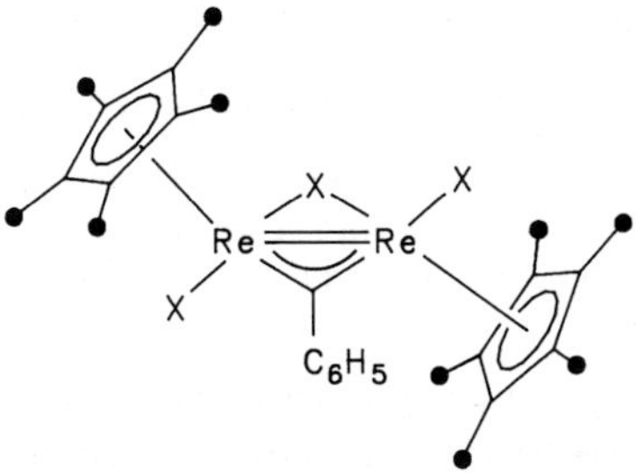

I

$[(C_5(CH_3)_5)ReBr]_2(\mu\text{-}CC_6H_5)(\mu\text{-}Br)$ was obtained by Method I [1, 2] with 60% yield. Violet solid [2].

^1H NMR spectrum ($CDCl_3$): $\delta = 1.82$ (s, CH_3); 7.18, 7.39 (m's) ppm. EI mass spectrum: $[M]^+$, $[(C_5(CH_3)_5)ReCC_6H_5Br_n]^+$ (n = 1, 0) [1, 2].

$[(C_5(CH_3)_4C_2H_5)ReCl]_2(\mu\text{-}CC_6H_5)(\mu\text{-}Cl)$ was obtained via Method I. The compound forms brown-purple crystals (from acetone).

An X-ray crystallographic study showed the compound to crystallize in the monoclinic space group $P2_1/n-C^5_{2h}$ (No. 14) with a = 10.864(3), b = 16.691(2), c = 16.425(5) Å, β = 102.89(1)°; Z = 4 molecules per unit cell, D_{calc} = 1.982 g/cm^3. There are 2 symmetry-independent centrosymmetric molecules within the asymmetric unit; the structure of one of these molecules is illustrated in **Fig. 109**. There is a planar Re_2ClC ring geometry and a trans-oriented set of ligands. The Re=Re connection can be considered a double bond [2].

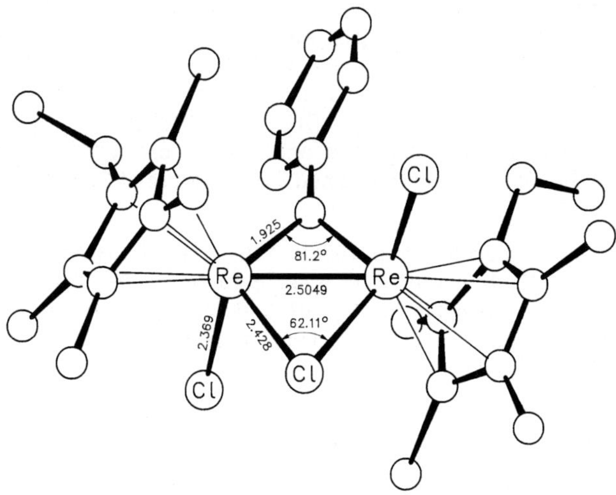

Fig. 109. Molecular structure of $[(C_5(CH_3)_4C_2H_5)ReCl]_2(\mu\text{-}CC_6H_5)(\mu\text{-}Cl)$ [2].

References:

[1] Herrmann, W. A.; Felixberger, J. K.; Anwander, R.; Herdtweck, E.; Kiprof, P.; Riede, J. (Organometallics **9** [1990] 1434/43).

[2] Herrmann, W. A.; Kiprof, P.; Felixberger, J. K. (Chem. Ber. **123** [1990] 1971/4).

2.5.1.2.8 Compounds with the $^5LRe(\mu\text{-}^2L)Re^5L$ Skeleton

2.5.1.2.8.1 The Bridging ^2L Ligand is Bonded via Re-C Single Bonds

$[C_5H_5Re(CO)(NO)CH_2-]_2O$ formed when either $C_5H_5Re(CO)(NO)CH_2OCH_3$ [1, 2, 4] or $C_5H_5Re(CO)(NO)CH_2OH$ [1] was treated with a small amount of CF_3CO_2H in C_6H_6 [1] or THF/H_2O [2, 4]. It also formed by pressurizing $C_5H_5Re(CO)(NO)CH_2OH$ with CO in benzene above 90 °C [4]. Orange solid [2].

^{1}H NMR spectrum (C_6D_6): δ = 4.85, 4.87 (C_5H_5); 5.06, 5.16, 5.34, 5.49 (pair of AB q's; J = 9 Hz) ppm. IR spectrum (THF): 1961 (ν(CO)), 1967 (ν(NO)) cm^{-1} [1]; similar in [2].

$C_5H_5Re(CO)(NO)-CH_2OC(O)-Re(C_5H_5)(CO)NO$ (see Formula I) formed when the orange formyl complex $C_5H_5Re(CO)(NO)CHO$ was kept as a neat oil under an N_2 atmosphere at room temperature. The deep red solid produced within 48 h was extracted into benzene, and the extract was recrystallized from toluene/heptane. Yield: 45 to 55%. The spontaneous dimerization of the formyl complex also occurred in concentrated (>0.4 M) solution [1 to 4]; it could be promoted by the presence of both acids (CH_3CO_2H) or bases ($N(C_2H_5)_3$) in C_6D_6, but the acid-catalyzed process gave the product in a 4-fold larger amount than the base-catalyzed one. Similarly, drying an ethereal solution containing the formyl complex with K_2CO_3 or treating the complex with $(i\text{-}C_4H_9)_2AlH$ also initiated product formation [2]. In all these cases, other products including $C_5H_5Re(CO)(NO)CH_2OH$ and $C_5H_5Re(CO)(NO)H$ were also obtained [2]. The title product could also be obtained by reacting $C_5H_5Re(NO)(P(C_6H_5)_3)CHO$ with $[C_5H_5Re(CO)_2NO]BF_4$ in CH_2Cl_2/CH_3CN between −40 °C and room temperature. Subsequent addition of benzene and cooling precipitated $[C_5H_5Re(CO)(NO)P(C_6H_5)_3]BF_4$, while the mother liquor contained a 3:1 mixture of $C_5H_5Re(CO)(NO)CHO$ and the title complex [6].

Red-orange, air-stable powder; m.p. 105 to 115 °C [1, 4]. ^{1}H NMR spectrum (C_6D_6): δ = 4.956, 4.958, 5.032, 5.034 (C_5H_5); 5.89, 6.49 and 5.99, 6.42 (two AB q's with J = 9.7 Hz) ppm [1 to 3] (similar in [6]). IR spectrum (Nujol): 1010 (ν(C-O)), 1626 (ν(C=O)) cm^{-1}; (THF): 1702, 1722 (ν(NO)); 1966, 1986 (ν(CO)) cm^{-1} [1 to 3]. Mass spectrum: $[M - CO]^+$, $[M - NO]^+$ [1, 2].

Treatment with CH_3OH in C_6D_6 gave $C_5H_5Re(CO)(NO)CH_2OH$ (initial product, converted to $C_5H_5Re(CO)(NO)CH_2OCH_3$ within 4 to 5 d [1, 2]) and $C_5H_5Re(CO)(NO)CO_2CH_3$ [1, 2, 4].

$[\{C_5H_5Re(NO)(P(C_6H_5)_3)CH_2-\}_2SCH_3]X$ (see Formula II). The cation was obtained by treating $C_5H_5Re(NO)(P(C_6H_5)_3)CH_2SCH_3$ with $[C_5H_5Re(NO)(P(C_6H_5)_3)=CH_2]^+$. Yield: 89% [5, 7]. It could also be obtained by treating $[C_5H_5Re(NO)(P(C_6H_5)_3)=CH_2]PF_6$ with $NaSCH_3$ [5, 7], CH_3SH, or CH_3SSCH_3 [7] (yield: 66 or 51% [7]) or by combining CH_3CN solutions containing $C_5H_5Re(NO)(P(C_6H_5)_3)CH_2SCH_3$ and $[C_5H_5Re(NO)(P(C_6H_5)_3)CH_2S(CH_3)_2]^+$ [5, 7]. Yield: 92% [7]. Spectroscopic studies showed it to exist as a 2:1:1 mixture of 3 diastereomers; no deviation was observed upon recrystallizing [5, 7]. The following salts were separately characterized:

References on pp. 230/1

X = PF_6: Orange needles; m.p. 150 to 151 °C [7].
^1H NMR (CD_2Cl_2): 2.41, 2.50 (s's); 2.73 to 2.97 (m) and 3.51, 3.81, 4.13 (d's, $ReCH_2$, ratio 2:0.5:1:0.5; J = 12.8); 5.17, 5.22, 5.26 (C_5H_5, ratio 1:1:2); 7.23 to 7.47 (m, C_6H_5) [5, 7]; at 70 °C coalescence of the C_5H_5 resonances to a broad band [7]. – ^{13}C {^1H} NMR ($C_2D_2Cl_4$): 3.9, 11.0, 12.2 (ratio 1:1:2); 28.0, 34.0 (SCH_3); 90.8, 91.2 (C_5, ratio 3:1); 128.7 to 134.5 (C_6H_5) [5, 7]. – IR (CH_2Cl_2): 1651 (ν(NO)) [5, 7].
Metathesis with NaI in CH_3CN gave the respective I salt. Yield: 61% [7].

X = I · 2 CH_3CN: Irregular orange crystals; m.p. 248 to 251 °C.
IR (CH_2Cl_2): 1657 (ν(NO)).
Single-crystal data: Orthorhombic lattice with a = 18.348(6), b = 20.767(2), c = 13.944(2) Å; Z = 4 formula units per unit cell, D_{calc} = 1.71 g/cm^3. As can be seen from **Fig. 110**, the cation crystallizes as one of two possible meso diastereomers [7].

Thermolysis in $C_2H_4Cl_2$ yielded $[C_5H_5Re(NO)(P(C_6H_5)_3){=}CHSCH_3]^+$ and $C_5H_5Re(NO)(P(C_6H_5)_3)CH_3$ in nearly equal amounts [5, 7]. The cation did not react with $(CH_3)_2S$ [5]; however, treatment with $P(C_6H_5)_3$ or pyridine rapidly generated $C_5H_5Re(NO)(P(C_6H_5)_3)CH_2SCH_3$ [5] and $[C_5H_5Re(NO)(P(C_6H_5)_3)CH_2P(C_6H_5)_3]PF_6$ [7] or $[C_5H_5Re(NO)(P(C_6H_5)_3)CH_2NC_5H_5]PF_6$ [7]. Treatment with $NaSCH_3$ gave $C_5H_5Re(NO)(P(C_6H_5)_3)CH_2SCH_3$ [7].

$[(C_5(CH_3)_5)Re(O)(Cl)CH_2-]_2C(CH_3)_2$ (see Formula III) was briefly mentioned. It was said to form as a by-product (yield: 7%) when treating $(C_5(CH_3)_5)Re(O)Cl_2$ with $(CH_3)_2C(CH_2MgBr)_2$ in ether. The main product was the rhenacycle $(C_5(CH_3)_5)Re(O)CH_2C(CH_3)_2CH_2$-cyclo [8, 9].

$[C_5H_5Re(NO)(P(C_6H_5)_3)CH_2-]_2(CH_2)_n$ (see Formula IV). Compounds with this composition (n = 1, 2, 3, 6) were obtained by treating in situ-prepared $Li[C_5H_5Re(NO)P(C_6H_5)_3]$ (from

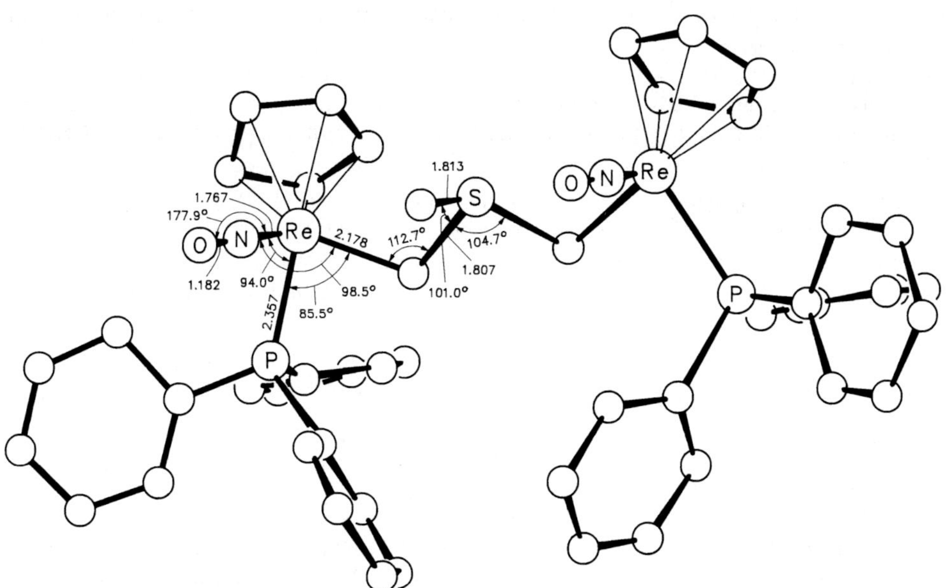

Fig. 110. Molecular structure of $[\{C_5H_5Re(NO)(P(C_6H_5)_3)CH_2-\}_2SCH_3]^+$ [7].

References on pp. 230/1

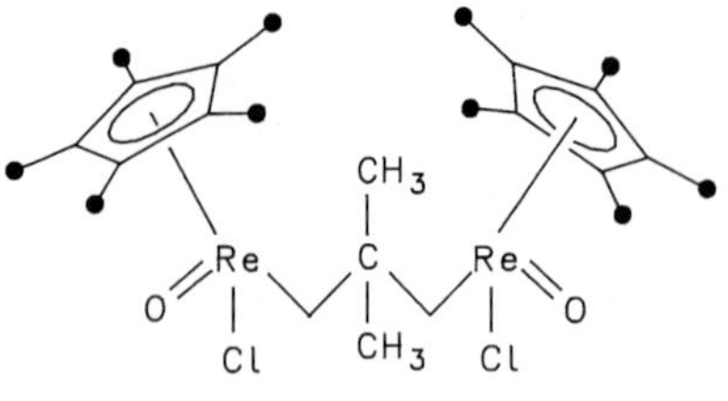

III

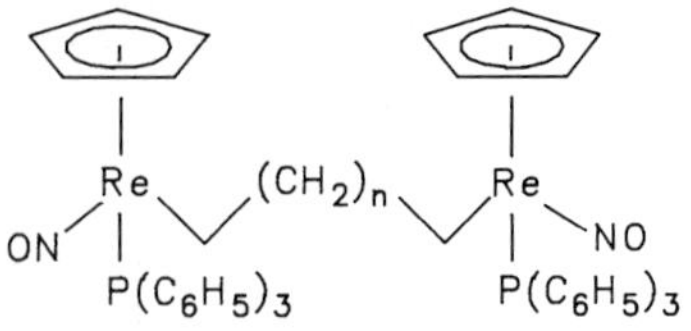

IV

$C_5H_5Re(NO)(P(C_6H_5)_3)H$ and $n\text{-}C_4H_9Li$) with the respective bistriflates $CF_3SO_3CH_2(CH_2)_n\text{-}CH_2O_3SCF_3$ in THF. The mixture was subsequently evaporated and the residue extracted into benzene. After filtration through NH_4OH-pretreated Florisil, the filtrate was evaporated and the residue recrystallized from CH_2Cl_2/hexane. A ^1H NMR experiment revealed that the complexes can also be prepared by starting from the bisiodides $ICH_2(CH_2)_nCH_2I$; only in the case of n = 1, $C_5H_5Re(NO)(P(C_6H_5)_3)I$ was also formed.

The products are bright yellow to orange solids. Because there are two chiral centers, the compounds exist as mixtures of meso and dl diastereomers in the ratio (50 to 75):(50 to 25). The diastereomers could not be separated by column chromatography or TLC.

Treatment with 2 equivalents $[C(C_6H_5)_3]PF_6$ in CH_2Cl_2 yielded dicarbene salts of the type $[\{C_5H_5Re(NO)(P(C_6H_5)_3){=}CH{-}\}_2(CH_2)_n][PF_6]_2$ (see p. 232) [10].

The compounds were characterized as follows:

n	yield, diastereomer ratio (meso:dl); spectroscopical properties (for abbreviations and units see p. X)
1	Yield: 89%; diastereomer ratio (66 ± 2):(34 ± 2). – ^1H NMR (C_6D_6): 2.17, 2.52, 2.76 (m's, CH_2); 4.70, 4.74 (s's, C_5H_5); 7.00, 7.62 (m's, C_6H_5). – $^{13}C\{^1H\}$ NMR (C_6D_6): −1.4, −1.2 (d's; J(P,C) = 4, 5); 52.4, 54.9 (C-3); 89.5 (C_5H_5), 129.4 (C-4), 133.9 (d, C-2; J(P,C) = 10), 137.8 (d, C_{ipso}; J(P,C) = 50). – $^{31}P\{^1H\}$ NMR (C_6D_6): 27.8, 28.4 (s's). – IR (KBr): 1623 (ν(NO)) [10].
2	Yield: 89%, diastereomer ratio (65 ± 2):(35 ± 2). – ^1H NMR (C_6D_6): 2.11, 2.34, 2.26 (m's, CH_2, 4:2:2); 4.46, 4.73 (s's, C_5H_5); 6.99, 7.55 (m's, C_6H_5, 18:12 H). – $^{13}C\{^1H\}$ NMR (C_6D_6): −7.6, −7.9 (d's, C-1; J(P,C) = 5, 4); 48.2, 49.9 (s's, C-2,3); 89.5 (C_5H_5), 129.5 (C-4); 133.8, 137.5 (d's, C-2 and C_{ipso}; J(P,C) = 10 and 51). – $^{31}P\{^1H\}$ NMR (C_6D_6): 27.5, 27.8 (s's). – IR (KBr): 1624 (ν(NO)) [10].
3	Yield: 84%, diastereomer ratio (75 ± 2):(25 ± 2). – ^1H NMR (C_6D_6): 1.78, 2.13, 2.58 (m's, CH_2); 4.64, 4.67 (s's, C_5H_5); 7.06, 7.57 (m's, C_6H_5, 18:12 H). – $^{13}C\{^1H\}$ NMR (C_6D_6): −8.0, −7.7 (d's; J(P,C) = 5); 42.1, 42.3; 43.1, 43.7 (s's, C-2 and C-3); 89.4 (s, C_5H_5), 129.6 (s, C_6H_5-4), 133.8 (d, C_6H_5-2; J(P,C) = 10), 137.4 (d, C_{ipso}; J(P,C) = 50) ppm. – $^{31}P\{^1H\}$ NMR (C_6D_6): 27.4, 27.7 (s's). – IR (KBr): 1623 (ν(NO)) [10].
6	Yield: 80%, diastereomer ratio (50 ± 2):(50 ± 2). – ^1H NMR (C_6D_6): 1.59, 2.06, 2.50 (m's, CH_2); 4.618, 4.623 (s's, C_5H_5); 7.01, 7.55 (m, C_6H_5, 18:12 H). – $^{13}C\{^1H\}$ NMR (C_6D_6): −8.6 (d; J(P,C) = 4); 30.4, 30.6 (s's, C-4); 36.6, 36.7 (s's, C-3); 42.5, 42.6 (s's, C-2); 89.3 (s, C_5H_5), 129.6 (s, C_6H_5-4), 133.8 (d, C_6H_5-2; J(P,C) = 10), 137.4 (d; C_{ipso}; J(P,C) = 51). – $^{31}P\{^1H\}$ NMR (C_6D_6): 27.6 (s). – IR (KBr): 1624 (ν(NO)) [10].

V

VI

$[(C_5(CH_3)_5)Re(CO)_2]_2(\mu\text{-}CH_3O_2CC{=}CCO_2CH_3)$ (see Formula V) was obtained by treating $(C_5(CH_3)_5)_2Re_2(CO)_4$ with $CH_3O_2CC{\equiv}CCO_2CH_3$ in THF in the dark. Subsequent evaporation gave a residue which was extracted into pentane. Evaporation of the extract gave the product with 62% yield as a yellow-orange powder [17]. The compound also formed from $(C_5(CH_3)_5)Re(CO)_2(\mu\text{-}CO)Re(C_5(CH_3)_5)(CO)NCCH_3$ and free $CH_3O_2CC{\equiv}CCO_2CH_3$ [18].

^1H NMR spectrum (C_6D_6): δ = 1.83, 3.63 ppm [17]. ^{13}C {^1H} NMR spectrum (THF, −50°C): δ = 10.2 ($(CH_3)_5$), 50.9 (OCH_3), 101.1 (C_5), 101.3 (C=C), 169.8 (CO_2); 209.7, 215.0 (CO) ppm. IR spectrum (THF): 1707, 1873, 1899, 1941, 1970 cm^{-1} [17] (similar in C_6H_{12} [16]). UV spectrum (?): λ_{max} (ε) = 326 (5700), 440 (1900) nm. Mass spectrum: $[M]^+$ [17].

The compound crystallizes in the orthorhombic space group Pbca$-D_{2h}^{15}$ (No. 61) with a = 14.480(3), b = 16.246(3), c = 26.027(3) Å; Z = 8 molecules per unit cell, D_{calc} = 1.946 g/cm^3. The molecular structure is illustrated in **Fig. 111**. The molecule is a slightly puckered dirhenacyclobutene ring with a dihedral angle C-Re-Re-C of 6.1°. The cyclopentadienyl rings are on opposite faces of the dirhenacyclobutene ring; the C_5-Re-Re-C_5 dihedral angle amounts to 158° [17].

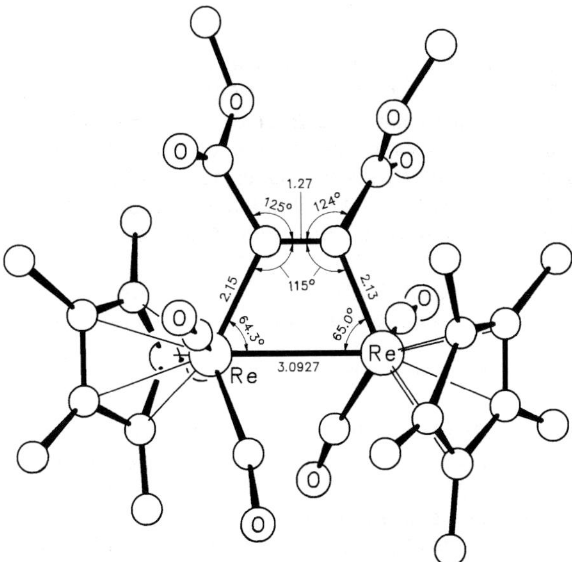

Fig. 111. Molecular structure of $[(C_5(CH_3)_5)Re(CO)_2]_2(\mu\text{-}CH_3O_2CC{=}CCO_2CH_3)$ [17].

References on pp. 230/1

Fluorescent light photolysis in C_6H_6 or C_6H_{12} yielded $[(C_5(CH_3)_5)Re(CO)_2]_2(\mu\text{-}CH_3O_2CC{\equiv}CCO_2CH_3)$ (see p. 239) in an intramolecular process [16, 17]. The quantum yield for product formation was 0.41 ± 0.03; an identical value was established under a CO atmosphere [16]. UV-monitoring of the reaction revealed isosbestic points at 261 and 313 nm. Time-resolved IR-detected laser flash photolysis (λ = 355 nm) demonstrated the formation of a short-lived transient species with IR absorptions at 1895 (br), 1970, 1992 cm^{-1}. These bands underwent exponential decay with the concomitant appearance of bands corresponding to the reaction product; the first-order disappearance rate constants $k_{obs} = 4.4 \times 10^4\ s^{-1}$ (in C_6H_{12}), $8.1 \times 10^4\ s^{-1}$ (in THF) were determined at 23°C [16]. The same product as upon photolysis was also obtained upon thermolysis at 72°C; actually, an equilibrium mixture of the two compounds in the ratio 78:22 formed. Short-wavelength photolysis affected CO loss leading to $[(C_5(CH_3)_5)Re(CO)]_2(\mu\text{-}CH_3O_2CC{\equiv}CCO_2CH_3)(\mu\text{-}CO)$ (see p. 239) [17].

$(C_5(CH_3)_5)Re(NO)(P(C_6H_5)_3){-}C_n{-}Re(C_5(CH_3)_5)(NO)P(C_6H_5)_3$ (see Formula VI). Compounds of this type were prepared by a $Cu(O_2CCH_3)_2$-initiated C-C coupling reaction of the compounds $(C_5(CH_3)_5)Re(NO)(P(C_6H_5)_3)((C{\equiv}C)_{n/2}H)$ (pyridine, ca. 50°C) [12, 13, 15]. The title complexes were usually separated by column chromatography [13].

The compounds are microcrystalline, almost air-stable solids [12, 13, 15]. They are well-soluble in THF or CH_2Cl_2 [15].

Apart from their spectroscopic properties, the compounds were investigated for their electrochemical behavior: Cyclic voltammograms (CV) were recorded from all compounds in CH_2Cl_2/0.1 M $[N(C_4H_9\text{-}n)_4]BF_4$ at a Pt working electrode versus Ag. Each compound except the derivative with the C_{20} bridge has two nearly reversible one-electron oxidation waves corresponding to an $Re^{0/I}$ oxidation. While the first oxidation becomes thermodynamically less favorable with longer carbon chains, the second oxidation is nearly unaffected by the chain length. Reduction waves were not observed [13].

n	yield; spectroscopical, electrochemical, and structural properties; remarks (for abbreviations and units see p. X)
4	Yield: 88% [12]; could also be obtained by a $Cu(O_2CCH_3)_2$-promoted cross-coupling reaction of a 1:1 mixture of $(C_5(CH_3)_5)Re(NO)(P(C_6H_5)_3)C{\equiv}CH$ and $(C_5(CH_3)_5)Re(NO)(P(C_6H_5)_3)C{\equiv}CC{\equiv}CH$, yield: 14% [13]. Initially, the compound formed as a 50:50 diastereomer mixture, but crystallization from CH_2Cl_2/ether gave the diastereomerically pure (SS,RR) product [12]. Air-stable, orange-brown solid [12]. ^1H NMR (THF-d_8): 1.68 (CH_3); 7.17 to 7.26 and 7.48 to 7.56 (m's, C_6H_5). – $^{13}C\{^1H\}$ NMR (C_6D_6): 10.28 (CH_3), 95.8 (ddd, ReC≡; J(P,C) = 10.9), 117.5 (dd, ≡CC≡; $J(^{13}C,^{13}C)$ = 96.5 and 47.1); 128.5, 130.3, 135.2 (C_6H_5). – $^{31}P\{^1H\}$ NMR (THF-d_8): 21.8 (br s). – IR (CH_2Cl_2): 1623 (ν(NO)), 1888 ($\nu(^{13}C{\equiv}^{13}C)$), 1964 ($\nu$(C≡C)). – UV ($CH_2Cl_2$): λ_{max} (ε) = 232 (65000), 270 (31000), 346 (20500). – FAB MS: $[M]^+$, $[(C_5(CH_3)_5)Re(NO)P(C_6H_5)_3]^+$ [12]. – CV: $E_{p,a}$ = 0.15, 0.68 (reversibility is improved at −45 or −80°C) [13]; (CH_3CN/0.1 M $[N(C_2H_5)_4]ClO_4$): $E_{1/2}$ (ox.) = 0.06 and 0.50 [11]. **$(C_5(CH_3)_5)Re(NO)(P(C_6H_5)_3){-}C_4{-}Re(C_5(CH_3)_5)(NO)P(C_6H_5)_3 \cdot 2\,CH_2Cl_2$** crystallizes in the monoclinic space group C2/c−C^6_{2h} (No. 15) with a = 26.898(8), b = 11.437(3), c = 19.613(4) Å, β = 95.172(2)°; Z = 4 molecules per unit cell, D_{calc} = 1.598 g/cm³. The molecular structure is illustrated in **Fig. 112** [12]. Combination with its respective dication or treatment with 1 equivalent $AgPF_6$ gave the respective stable monocation radical [11]. Oxidation with 2.5 equivalents $AgPF_6$ gave the respective dication [12] (see pp. 235/6).

References on pp. 230/1

n yield; spectroscopical, electrochemical, and structural properties; remarks (for abbreviations and units see p. X)

6 Main product of the $Cu(O_2CCH_3)$-promoted cross-coupling reaction of a 1:1 mixture of $(C_5(CH_3)_5)Re(NO)(P(C_6H_5)_3)C{\equiv}CH$ and $(C_5(CH_3)_5)Re(NO)(P(C_6H_5)_3)C{\equiv}CC{\equiv}CH$ in pyridine; yield: 44% (the derivatives with a C_4 and a C_8 bridge were also formed) [13].
Low-temperature NMR spectra show two related sets of resonances, which can be assigned to meso and (±) diastereomers; however, attempts to enrich single diastereomers have been unsuccessful. – $^{13}C\{^1H\}$ NMR (THF-d_8): 65.2 to 65.7 (ReC≡C**C**), 103 to 104.2 (d, ReC; J(P,C) = 15.7 to 16.0), 113.9 to 114.3 (s, ReC≡**C**). – IR (CH_2Cl_2): 1640 (ν(NO)), 2061 (ν(C≡C)). – UV (CH_2Cl_2): λ_{max} (ε) = 328 (39000), 354 (37000). – CV: $E_{p,a}$ = 0.24, 0.62 (one-electron oxidations, reversibility is improved at −45 or −80 °C) [13].

8 Yield: 70%; could also be obtained by a $Cu(O_2CCH_3)_2$-promoted cross-coupling reaction of a 1:1 mixture of $(C_5(CH_3)_5)Re(NO)(P(C_6H_5)_3)C{\equiv}CH$ and $(C_5(CH_3)_5)$-$Re(NO)(P(C_6H_5)_3)C{\equiv}CC{\equiv}CH$; yield: 15%. Air-stable, orange solid.
$^{13}C\{^1H\}$ NMR (THF-d_8): 64.5 (s, ReCCC**C**), 66.6 (d, ReCC**C**; ^4J(P,C) = 2.72), 109.7 (d, ReC; ^2J(P,C) = 16.9), 113.3 (s, ReC**C**). – UV (CH_2Cl_2): λ_{max} (ε) = 360 (67000), 390 (60000). – IR (CH_2Cl_2): 1648 (ν(NO)); 1956, 2108 (ν(C≡C)). – CV: $E_{p,a}$ = 0.37, 0.66; improved reversibility at −45 or −80 °C [13].

12 Yield: 71%. Could also be prepared with 45% yield by successively treating $(C_5(CH_3)_5)Re(NO)(P(C_6H_5)_3)C{\equiv}CC{\equiv}CCu$ with $H_2NC_2H_5$ and $BrC{\equiv}CC{\equiv}CBr$.
Dec.p. >100 °C.
IR (THF): 1952, 2056, 2115 (ν(C≡C)). – UV (CH_2Cl_2): λ_{max} (ε) = 230 (60500), 280 (47600), 364 (47300), 390 (63400), 422 (82700), 470 (37350), 512 (16600), 568 (3850). – MS: $[M]^+$. – CV: $E_{p,a}$ = 0.60, 0.79 [15].

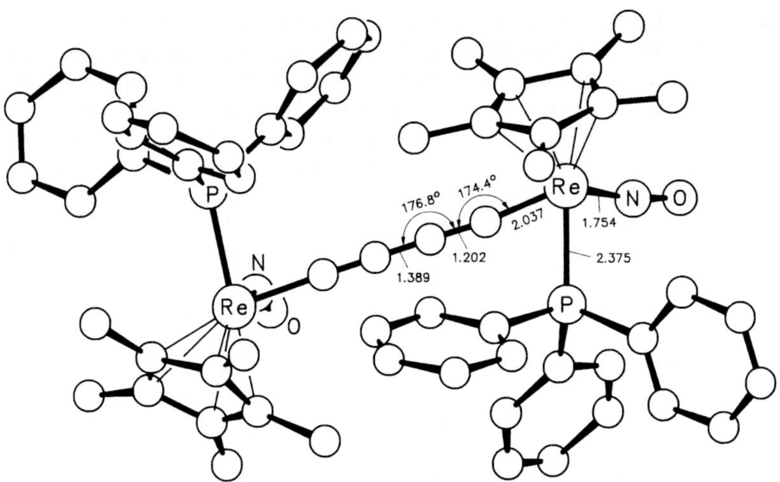

Fig. 112. Molecular structure of $(C_5(CH_3)_5)Re(NO)(P(C_6H_5)_3)-C_4-Re(C_5(CH_3)_5)(NO)P(C_6H_5)_3$ [12].

References on pp. 230/1

n	yield; spectroscopical, electrochemical, and structural properties; remarks (for abbreviations and units see p. X)
16	Yield: 67%. Dec.p. >100 °C. – IR (THF): 1941, 2014, 2074 (ν(C≡C)). – UV (CH_2Cl_2): λ_{max} (ε) = 232 (62600), 328 (61700), 348 (65300), 402 (81500), 430 (142000), 470 (128000), 548 (61200), 644 (9900). – MS: $[M]^+$. – CV: $E_{p,a}$ = 0.70, 0.79 [15].
20	Yield: 41%. Dec.p. >100 °C. – IR (THF): 1925, 1991, 2116, 2150 (ν(C≡C)). – UV (CH_2Cl_2): λ_{max} (ε) = 228 (78000), 384 (110000), 430 (131000), 466 (191000), 508 (136000), 582 (73000), 602 (73000). MS: $[M]^+$. – CV: $E_{p,a}$ = 0.80 (probably two-electron oxidation). – Solutions are much more air-sensitive than those of the other analogs [15].

[($C_5(CH_3)_5$)Re(NO)(P(C_6H_5)$_3$)C≡CC≡C–]$_2$Pd(P(C_2H_5)$_3$)$_2$ (see Formula VII, y = 0) was obtained by treating Pd(P(C_2H_5)$_3$)$_2$Cl$_2$ with ($C_5(CH_3)_5$)Re(NO)(P(C_6H_5)$_3$)C≡CC≡C–Li (molar ratio 1:2) in THF between −80 °C and room temperature. The solvent was evaporated and the residue extracted into ether. Filtration, concentration of the extract, and cooling gave the complex with 73% yield as a dark red, air-stable solid; m.p. 165 to 168 °C (dec.).

^{1}H NMR spectrum (C_6D_6): δ = 1.12 (m, CH_3), 1.67 (s, (CH_3)$_5$), 1.82 (m, PCH_2), 6.96 to 7.12 and 7.75 to 7.83 (m's, C_6H_5) ppm. ^{13}C {^{1}H} NMR spectrum (C_6D_6): δ = 9.0 (CH_3), 10.1 ((CH_3)$_5$), 17.8 ("t"; J(P,C) = 14.2 Hz), 87.4 (d, ReC≡; J(P,C) = 18.4 Hz), 98.2 (d, **C**≡CPd; J(P,C) = 3.3 Hz), 100.1 (C_5), 100.5 (t, PdC; J(P,C) = 11.1 Hz), 117.0 (s, ReC≡**C**); C_6H_5: 128.1 (d), 129.8 (s), 134.7 (d), 136.3 (d; J(P,C) = 9.8, none, 10.5, and 50.6 Hz, respectively) ppm. ^{31}P {^{1}H} NMR spectrum (C_6D_6): δ = 19.5 (PdP), 20.8 (ReP) ppm. IR spectrum (CH_2Cl_2): 1632 (ν(NO)); 1984, 2114 (ν(C≡C)) cm^{-1}. Raman spectrum (CH_2Cl_2): 2090, 2111 (ν(C≡C)) cm^{-1}. UV spectrum (CH_2Cl_2): λ_{max} (ε) = 234 (80000), 268 (46000), 308 (27000), 388 (5000) nm. FAB mass spectrum: $[M]^+$, [M – $C_5(CH_3)_5$Re(NO)(P(C_6H_5)$_3$)C]$^+$, [$C_5(CH_3)_5$-Re(NO)(P(C_6H_5)$_3$)C_4P(C_2H_5)$_3$]$^+$ (base peak).

Cyclic voltammetry (CH_2Cl_2/0.1 M [N(C_4H_9-n)$_4$]BF$_4$, Pt electrode): $E_{1/2}$ (ox.) = 0.32 V (one-electron wave), $E_{p,a}$ = 1.29 V vs. Ag. Oxidation with [(C_5H_5)$_2$Fe]PF$_6$ in CH_2Cl_2 gave the blue-green radical salt **[{($C_5(CH_3)_5$)Re(NO)(P(C_6H_5)$_3$)C≡CC≡C–}$_2$Pd(P(C_2H_5)$_3$)$_2$]PF$_6$** (see Formula VII, y = 1+). An ESR spectrum taken of the solution of the radical species displayed a weak sextet with g = 2.032, a_{iso} (Re) = 140 G (the signal disappeared after 5 to 10 min). The data suggest the odd electron to be mainly localized on one Re atom [14].

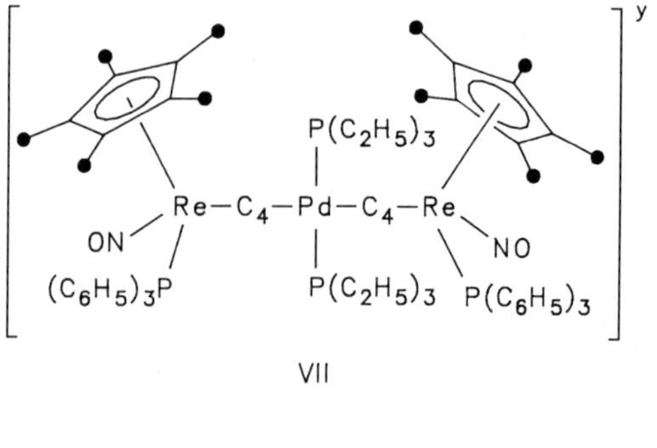

y
P(C_2H_5)$_3$
Re–C_4–Pd–C_4–Re
ON
NO
(C_6H_5)$_3$P
P(C_2H_5)$_3$
P(C_6H_5)$_3$

VII

References:

[1] Casey, C. P.; Andrews, M. A.; McAlister, D. R. (J. Am. Chem. Soc. **101** [1979] 3371/3).
[2] Casey, C. P.; Andrews, M. A.; McAlister, D. R.; Rinz, J. E. (J. Am. Chem. Soc. **102** [1980] 1927/33).

[3] Casey, C. P.; Neumann, S. M.; Andrews, M. A.; McAlister, D. R. (Pure Appl. Chem. **52** [1980] 625/33).
[4] Casey, C. P.; Andrews, M. A.; McAlister, D. R.; Jones, W. D.; Harsy, S. G. (J. Mol. Catal. **13** [1981] 43/59).
[5] McCormick, F. B.; Gladysz, J. A. (J. Organomet. Chem. **218** [1981] C 57/C 61).
[6] Tam, W.; Lin, G.-Yu.; Gladysz, J. A. (Organometallics **1** [1982] 525/9).
[7] McCormick, F. B.; Gleason, W. B.; Zhao, X.; Heah, P. C.; Gladysz, J. A. (Organometallics **5** [1986] 1778/85).
[8] de Boer, H. J. R.; van de Heisteeg, B. J. J.; Floël, M.; Herrmann, W. A.; Akkermann, O. S.; Bickelhaupt, F. (Angew. Chem. **99** [1987] 88/9; Angew. Chem. Int. Ed. Engl. **26** [1987] 73).
[9] Herrmann, W. A.; Floël, M.; Herdtweck, E. (J. Organomet. Chem. **358** [1988] 321/38).
[10] Roger, C.; Peng, T.-S.; Gladysz, J. A. (J. Organomet. Chem. **439** [1992] 163/75).

[11] Seyler, J. W.; Weng, W.; Zhou, Y.; Gladysz, J. A. (Organometallics **12** [1993] 3802/4).
[12] Zhou, Y.; Seyler, J. W.; Weng, W.; Arif, A. M.; Gladysz, J. A. (J. Am. Chem. Soc. **115** [1993] 8509/10 and Supplementary Material).
[13] Brady, M.; Weng, W.; Gladysz, J. A. (J. Chem. Soc. Chem. Commun. **1994** 2655/6).
[14] Weng, W.; Bartik, T.; Brady, M.; Bartik, B.; Ramsden, J. A.; Arif, A. M.; Gladysz, J. A. (J. Am. Chem. Soc. **117** [1995] 11922/31).
[15] Bartik, T.; Bartik, B.; Brady, M.; Dembinski, R.; Gladysz, J. A. (Angew. Chem. **108** [1996] 467/9).
[16] Casey, C. P.; Boese, W. T.; Cariño, R. S.; Ford, P. C. (Organometallics **15** [1996] 2189/91).
[17] Casey, C. P.; Cariño, R. S.; Hayashi, R. K.; Schladetzky, K. D. (J. Am. Chem. Soc. **118** [1996] 1617/23).
[18] Casey, C. P.; Cariño, R. S.; Sakaba, H.; Hayashi, R. K. (Organometallics **15** [1996] 2640/9).

2.5.1.2.8.2 The Bridging ^{2}L Ligand is Bonded via Re-C Multiple Bonds

$[C_5H_5Re(CO)_2{=}C(C_6H_5){-}]_2O \cdot CH_2Cl_2$. $[C_5H_5Re(CO)_2{\equiv}CC_6H_5]BCl_4$ was combined with $Li[C_5H_5Re(CO)_2C(O)C_6H_5]$ in CH_2Cl_2 at $-80\,°C$. The mixture was warmed to ca. $-55\,°C$, stirred for 2 h at this temperature, and evaporated. The residue was extracted into CH_2Cl_2/pentane and filtered. Reevaporation followed by recrystallization from CH_2Cl_2/pentane at $-78\,°C$ gave the compound with 89% yield. It is a very air-sensitive, scarlet-red solid, which slowly decomposed at room temperature.

^{1}H NMR spectrum (acetone-d_6, $-40\,°C$): $\delta = 5.80$ (C_5H_5), 7.05 (C_6H_5) ppm. IR spectrum (CH_2Cl_2): 1879, 1960 (ν(CO)) cm^{-1} [1].

$[C_5H_5Re(CO)_2]_2(\mu\text{-}C_{18}H_{16}N_2)$ (see Formula I) was obtained by reacting in situ-formed $C_5H_5Re(CO)_2{=}C{=}CH_2$ (by simply dissolving $[C_5H_5Re(CO)_2{\equiv}CCH_3]BCl_4$ in CH_2Cl_2 at $-78\,°C$) with a 2-fold excess of $C_6H_5CH{=}NN{=}CHC_6H_5$ in CH_2Cl_2 at $-50\,°C$. After the mixture had slowly reached room temperature, all volatiles were evaporated. The resulting oil was chromatographically purified (alumina, CH_2Cl_2/pentane (1:1) and pure CH_2Cl_2). The title compound was separated with 53% yield [3, 7].

Yellow [3], orange microcrystals [7]. ^{1}H NMR spectrum (C_6D_6): $\delta = 1.81$, 1.85 (dd's, CH_2; J(H,H) = 16.6, 2.4 Hz and 16.7, 7.5 Hz, respectively); 4.03 (dd, =CH; J(H,H) = 7.5, 2.4 Hz), 4.91 (C_5H_5), 6.88 to 7.95 (m, C_6H_5) ppm. ^{13}C {^{1}H} NMR spectrum (CD_2Cl_2): $\delta = 22.6$ (CH_2),

References on p. 236

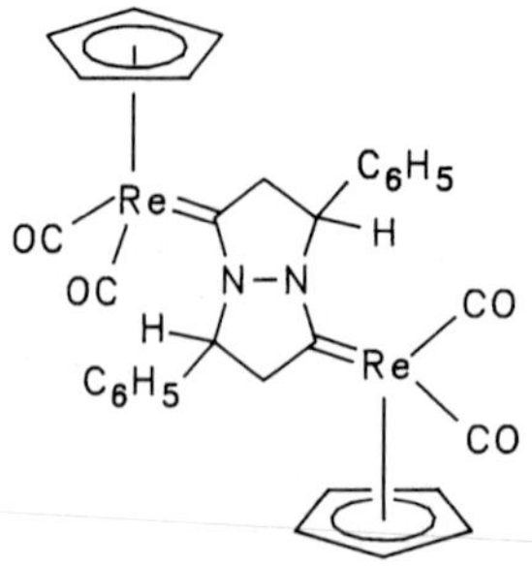

I

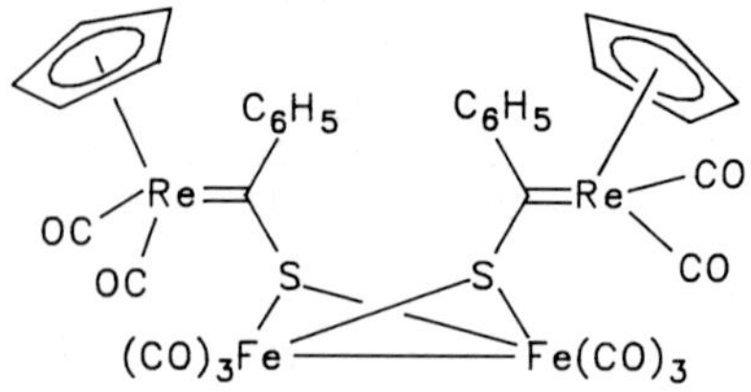

II

38.3 (CHC_6H_5), 87.5 (C_5); 125.9 to 148.2 (C_6H_5); 206.5, 206.9 (CO); 228.3 (Re=C) ppm. IR spectrum (CH_2Cl_2): 1847, 1926 (ν(CO)) cm^{-1} [3, 7].

$[C_5H_5Re(CO)_2]_2(\mu\text{-}C_{20}H_{10}Fe_2O_6S_2)$ (see Formula II) was obtained when $Li_2[(CO)_6Fe_2S_2]$ was treated with $[C_5H_5Re(CO)_2{\equiv}CC_6H_5]BBr_4$ in THF at −100°C. Yield: 26%. The product melts at 125°C.

^1H NMR spectrum (C_6D_6): δ = 2.98; 7.07, 7.25 ppm. IR spectrum (CH_2Cl_2): 1902, 1906, 1995 to 2000, 2000, 2042, 2070 (ν(CO)) cm^{-1} [2].

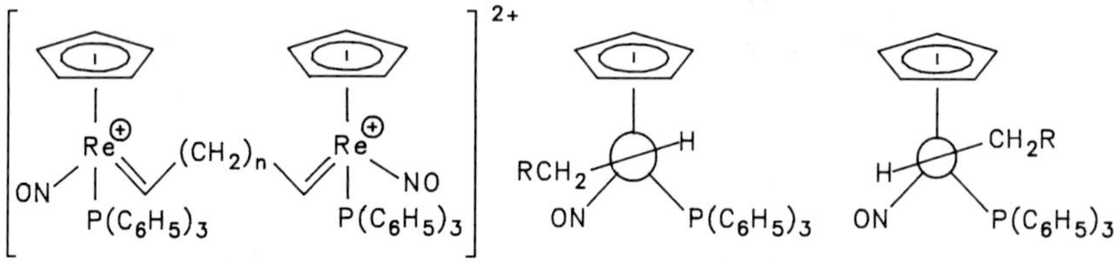

III

IVa (ac)

IVb (sc)

$[\{C_5H_5Re(NO)(P(C_6H_5)_3)=CH-\}_2(CH_2)_n][PF_6]_2$ (see Formula III). Salts of this composition were prepared by treating $[C_5H_5Re(NO)(P(C_6H_5)_3)CH_2-]_2(CH_2)_n$ (n = 1, 2, 3, 6) with 2 equivalents $[C(C_6H_5)_3]PF_6$ in CH_2Cl_2 at −80°C followed by warming the mixture to room temperature. The solvent was subsequently removed and the residue extracted into $CHCl_3$. Addition of hexane to the extract separated the salts with good yield as air-stable, tan powders. They are sparingly soluble in acetone and CH_2Cl_2 [4].

Because the starting compounds are diastereomer mixtures, the title products also exist as a complex mixture of diastereomers. Furthermore, carbene complexes can exhibit two Re=C geometries, anticlinal (ac) and synclinal (sc); see Formulas IVa, IVb. The former is more stable. Each diastereomer was shown by NMR methods to exist as a ca. 81:18:1 mixture of ac/ac, ac/sc, and sc/sc Re=C geometric isomers. Only assignment of resonances of the major ac/ac isomers was attempted [4]:

n	yield, spectroscopic properties, remarks (for abbreviations and units see p. X)
1	Yield: 76%. − IR (KBr): 841 (ν(PF)), 1719 (ν(NO)) [4].
2	Yield: 87% (contained ca. 24% isomeric alkene complexes). − ^1H NMR (CD_3NO_2): 6.07, 6.13, 6.17 (C_5H_5, ca. 34:45:21); 7.25 to 7.80 (m, C_6H_5); 15.8, 16.1, 16.2 (=CH, m, ddd, ddd; J(CH,CH_2) = 8.7, J(P,H) = 1.6). − $^{13}C\{^1H\}$ NMR (CD_3NO_2): 101.0,

n yield, spectroscopic properties, remarks
(for abbreviations and units see p. X)

101.1, 101.2 (s's, C_5H_5); 130.8 (m, C_6H_5-3), 133.5, 133.9 (C_6H_5-4), 134.2, 134.6 (d, C_6H_5-2; J(P,C) = 12, 10), 313.3 (br s, Re=C). – ^{31}P {^1H} NMR (CD_3NO_2): 17.9, 18.1, 18.4, 19.0 (s's, ca. 31:22:12:35). – IR (KBr): 842 (ν(PF)), 1724 (ν(NO)) [4].

3 Yield: 79%. – ^1H NMR (CD_3NO_2): 0.88, 1.28, 2.62, 3.20 (m's, 2:2:1:1); 6.09, 6.12 (s's, C_5H_5); 7.65 (m, C_6H_5); 15.46, 15.92 (br s, =CH). – ^{13}C {^1H} NMR (CD_3NO_2): 20.2, 27.2 (C-3); 57.2, 58.0 (C-2); 100.8 (C_5H_5), 130.6 (d, C_6H_5-3; J(P,C) = 8), 133.5 (s, C_6H_5-4), 133.9 (d, C_6H_5-2; J(P,C) = 12), 134.3 (d, C_{ipso}, J(P,C) not resolved), 308.9 (br s, Re=C). – ^{31}P {^1H} NMR (CD_3NO_2): 18.3, 19.7 (s's). – IR (KBr): 838 (ν(PF)), 1719 (ν(NO)).
Thermolysis in C_6H_5Cl gave [{$C_5H_5Re(NO)(P(C_6H_5)_3)H_2C$=CH–}$_2CH_2$][$PF_6$]$_2$. The reaction followed first-order kinetics with $k_{obs} = (3.89 \pm 0.06) \times 10^{-5}\ s^{-1}$ (at 72 °C) [4].

6 Yield: 93%. – ^1H NMR (CD_3NO_2): 1.18, 2.84, 3.16 (m's, CH_2, 8:2:2); 6.13 (s, C_5H_5), 7.57 (m, C_6H_5), 16.16 (ddd, =CH; J(CH,CH_2) = 8.7, J(P,H) = 1.6). – ^{13}C {^1H} NMR (CD_3NO_2): 29.4, 29.8 (s, C-3,4); 59.9 (s, C-2), 100.9 (s, C_5H_5), 130.7 (d, C_6H_5-3; J(P,C) = 11), 131.9 (d, C_{ipso}; J(P,C) not resolved), 133.6 (s, C_6H_5-4), 134.1 (d, C_6H_5-2; J(P,C) = 10), 312.3 (br s, Re=C). – ^{31}P {^1H} NMR (CD_3NO_2): 18.3 (s). – IR (KBr): 840 (ν(PF)), 1718 (ν(NO)).
Thermolysis (C_6H_5Cl) gave [{$C_5H_5Re(NO)(P(C_6H_5)_3)H_2C$=$CHCH_2CH_2$–}$_2$][$PF_6$]$_2$. The reaction followed first-order kinetics with $k_{obs} = (6.3 \pm 0.2) \times 10^{-5}\ s^{-1}$ (at 66 °C) [4].

Va Vb

(SR,RS)-[{($C_5(CH_3)_5$)Re(NO)P(C_6H_5)$_3$}$_2$$C_4H_3$-cyclo]$_2$[$Zn_2Cl_6$] · 0.5 CH_2Cl_2 (see Formulas Va, Vb) was obtained by treating ($C_5(CH_3)_5$)Re(NO)(P(C_6H_5)$_3$)C≡CH with $ZnCl_2$ in THF (room temperature, 6 h) or, more complicated, by successively treating the same starting complex with 1.5 equivalents n-C_4H_9Li and a slight excess of $ZnCl_2$ in THF between −80 °C and room temperature. In either case, the solvent was removed and the residue extracted into CH_2Cl_2. After filtration, the extracts were layered with ether, whereupon yellow microcrystals precipitated. The product was purified by repeating the crystallization from CH_2Cl_2 and hexane. Yields amounted to 94% in the former and 38% in the latter case.

Orange crystals; m.p. >196 °C (slow dec.). ^1H NMR spectrum (CD_2Cl_2): δ = 1.55 (s, CH_3); 2.53, 3.03 (d's, CH_2; J_{gem} = 14 Hz); 7.08 (s, CH), 7.38 to 7.49 (m, C_6H_5) ppm; (−90 °C): 1.30, 1.35 (s's, CH_3) ppm. ^{13}C {^1H} NMR spectrum (CD_2Cl_2): δ = 9.9 (CH_3), 64.9 (CH_2), 104.7 (C_5); 129.1, 131.4, 133.4, 133.5 (d's; J(P,C) = 10.6, 2.0, 53.9, 11.2 Hz, respectively); 167.8 (CH), 227.5 (d, ReC; J(P,C) = 9.5 Hz) ppm. ^{31}P {^1H} NMR spectrum (CD_2Cl_2): δ = 21.3 ppm; (−90 °C): 22.7, 22.9 ppm. Coalescence of the ^1H and ^{31}P NMR spectra was observed at −26.2 and −9.1 °C, respectively; calculations gave $\Delta G^{\neq}$ values of 52.7 and 50.6 kJ/mol for the dynamic process. IR spectrum (CH_2Cl_2): 1642, 1655 (ν(NO)) cm^{-1}. FAB mass spectrum: $[M]^+$.

References on p. 236

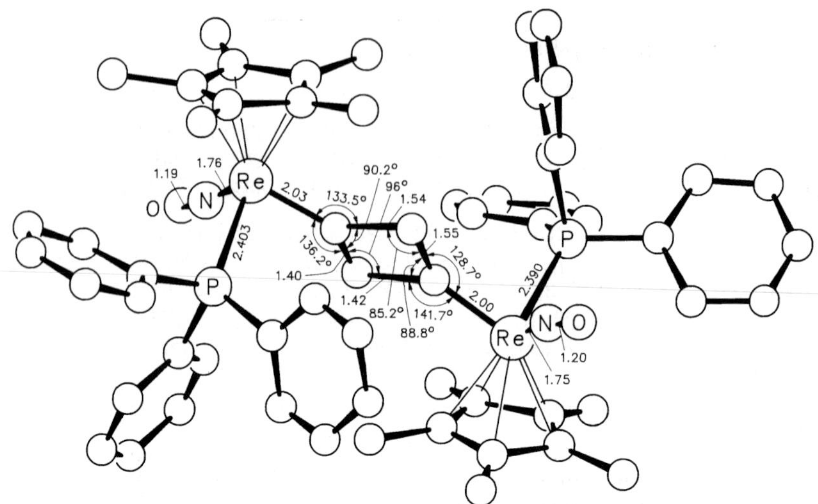

Fig. 113. Molecular structure of $[\{(C_5(CH_3)_5)Re(NO)P(C_6H_5)_3\}_2C_4H_3\text{-cyclo}]^+$ [8].

The salt crystallizes in the triclinic space group $P\bar{1}-C_i^1$ (No. 2) with a = 14.867(2), b = 15.439(3), c = 17.474(2) Å, α = 101.80(2)°, β = 114.24(2)°, γ = 111.99(2)°; Z = 2 formula units per unit cell, D_{calc} = 1.614 g/cm^3. The structure of the cation is depicted in **Fig. 113**. The Re-C bonds can be regarded as hybrids of cationic carbene and neutral alkenyl resonance forms, giving a formal half-positive charge on each Re atom. The Re-C_4-Re fragment is almost planar; least-square planes have angles between them ranging from 0 to 6°. The Re-Re distance is 6.1025 Å. The Re-P and Re-NO bonds are placed approximately perpendicular and parallel, respectively, to the C_4H_3 plane.

Metathesis with $NaBF_4$ yielded the respective BF_4 salt. The compound did not react with $K[OC_4H_9\text{-t}]$ or $n\text{-}C_4H_9Li/K[OC_4H_9\text{-t}]$ in THF-d_8 [8].

(SR,RS)-$[\{(C_5(CH_3)_5)Re(NO)P(C_6H_5)_3\}_2C_4H_3\text{-cyclo}]BF_4$ was obtained by treating the foregoing Zn_2Cl_6 salt with $NaBF_4$ (THF, 4 h). The solvent was removed and the residue extracted into CH_2Cl_2. Filtration and layering with hexane gave the salt with 86% yield. The salt could more straightforwardly be obtained by treating $(C_5(CH_3)_5)Re(NO)(P(C_6H_5)_3)C{\equiv}CH$ in CH_2Cl_2 either with 1 equivalent $HBF_4 \cdot O(C_2H_5)_2$ between −80 °C and room temperature or with $[(C_5(CH_3)_5)Re(NO)(P(C_6H_5)_3){=}C{=}CH_2]BF_4$ at room temperature for 3 h. The product was in either case separated by concentrating the solution and adding hexane. Yields amounted to 85% and even 94%. Orange needles; m.p. 173 to 176 °C (dec.).

NMR spectrum: identical with that of the Zn_2Cl_6 salt. IR spectrum (CH_2Cl_2): 1068 (ν(BF)); 1641, 1656 (ν(NO)) cm^{-1}. UV spectrum (CH_2Cl_2): λ_{max} (ε) = 266 (10900), 362 (4100), 480 (23400) nm. NMR spectra indicated the presence of a second stereoisomer, probably the (SS,RR)-diastereomer.

Protonation with excess $HBF_4 \cdot O(C_2H_5)_2$ yielded the following dication [8].

(SR,RS)-$[\{(C_5(CH_3)_5)Re(NO)P(C_6H_5)_3\}_2C_4H_4\text{-cyclo}][BF_4]_2$ (see Formula VI) was obtained by treating the foregoing BF_4 salt with 1.5 equivalents $HBF_4 \cdot O(C_2H_5)_2$ in CD_2Cl_2 at −80 °C. The color of the solution immediately turned yellow. The salt was characterized in solution; NMR integration of the signals evidenced at least 95% conversion to the dication. Upon all workups attempted, the dication reverted to the starting monocation.

References on p. 236

$$\left[(CH_3)_5C_5Re(NO)(P(C_6H_5)_3){=}\!\diamond\!{=}Re(NO)(P(C_6H_5)_3)C_5(CH_3)_5\right]^{2+}$$

VI

$$\left[(CH_3)_5C_5Re(NO)(P(C_6H_5)_3){=}\underset{\alpha}{C}{=}\underset{\beta}{C}{=}C{=}C{=}Re(NO)(P(C_6H_5)_3)C_5(CH_3)_5\right]^{2+}$$

VII

^1H NMR spectrum (CD_2Cl_2): δ = 1.59 (s, CH_3); 2.68 to 2.88 and 3.03 to 3.24 (m's, CH_2); 7.35 to 7.40 and 7.60 to 7.63 (m's, C_6H_5) ppm. ^{13}C {^1H} NMR spectrum (CD_2Cl_2, −40 °C): δ = 9.5 (CH_3), 83.9 (br s, CH_2), 109.4 (s, C_5), 129.8 to 133.5 (4 m's, C_6H_5), 304.7 (d, R=C; J(P,C) = 7.4 Hz) ppm. ^{31}P {^1H} NMR spectrum (CD_2Cl_2): δ = 21.4 ppm. IR spectrum (CH_2Cl_2): 1713 (ν(NO)) cm^{-1}. FAB mass spectrum: $[M]^+$, $[M]^{2+}$, other doubly charged ions.

In the presence of $HBF_4 \cdot O(C_2H_5)_2$, solutions are stable at −30 °C. Upon warming the mixture, the compound decomposed; at 60 °C it was consumed within 1 h [8].

$[(C_5(CH_3)_5)Re(NO)(P(C_6H_5)_3){=}C_4{=}Re(C_5(CH_3)_5)(NO)P(C_6H_5)_3][PF_6]_2$ (see Formula VII) immediately precipitated when the respective neutral analog was treated with 2.5 equivalents $AgPF_6$ in toluene. Yield: 86%. The salt forms an air-stable, deep blue solid (from CH_3CN/toluene); m.p. 203 to 207 °C (dec.) [6].

^1H NMR spectrum (CD_2Cl_2): δ = 2.03 (CH_3), 7.25 to 7.63 (C_6H_5) ppm. ^{13}C {^1H} NMR spectrum (CD_2Cl_2): δ = 10.6 (CH_3), 114.3 (C_5), 130.0 (d, C_6H_5; J(P,C) = 11.2 Hz), 133.4 (s and d, C_6H_5; J(P,C) = 15.3 Hz), 213.5 (dd, C-β), 305.1 (ddd, C-α; J(P,C) = 12.3, J(^{13}C,^{13}C) = 77.0 and 40.7 Hz) ppm. ^{31}P {^1H} NMR spectrum (CD_2Cl_2): δ = 24.5 (br s) ppm; (−93 °C): δ = 26.8, 28.4 (s's, ratio 62:38) ppm. IR spectrum (CH_2Cl_2): 1719 (ν(NO)) cm^{-1} [6]. UV spectrum (CH_2Cl_2): λ_{max} (ε) = 234 (57000), 270 (25000), 392 (41000), 574 (30000) nm [6]; see also [5].

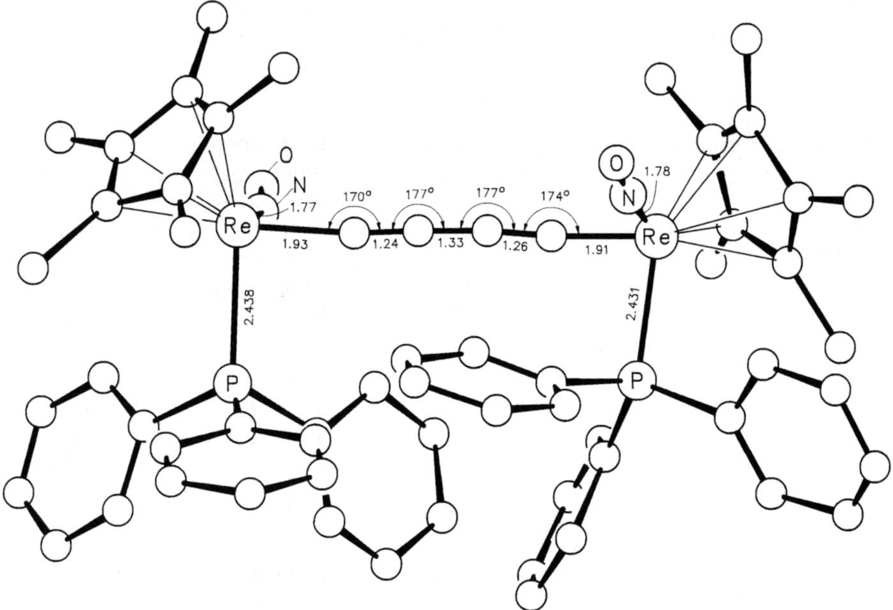

Fig. 114. Molecular structure of $[(C_5(CH_3)_5)Re(NO)(P(C_6H_5)_3){=}C_4{=}Re(C_5(CH_3)_5)(NO)P(C_6H_5)_3]^{2+}$ [6].

References on p. 236

The salt crystallizes in the monoclinic space group $P2_1/c-C^5_{2h}$ (No. 14) with a = 17.518(4), b = 20.668(5), c = 17.566(7) Å, β = 107.87(2)°; Z = 4 formula units per unit cell, D_{calc} = 1.718 g/cm³. The structure of the dication is depicted in **Fig. 114** [6].

Combination with its neutral analog initiated comproportionation to the monocation radical (following complex). Otherwise, treatment with $NaC_{10}H_8$ formed the neutral parent compound again [5].

$[(C_5(CH_3)_5)Re(NO)(P(C_6H_5)_3)-C_4-Re(C_5(CH_3)_5)(NO)P(C_6H_5)_3]PF_6$. This stable cation radical was obtained with 69% yield by treating the respective neutral compound (see p. 228) with 1 equivalent $AgPF_6$ in toluene. It could also be prepared by combining equimolar quantities of the respective neutral and dicationic (foregoing complex) analogs in CH_2Cl_2; yield: 50%. The deep green solid is stable under an inert atmosphere for weeks, although some decomposition was evident after several hours in degassed CH_2Cl_2 or CH_3CN.

$[[Re]^{\bullet}-C\equiv C-C\equiv C-[Re]]^+ \longleftrightarrow [[Re]-C\equiv C-C\equiv C-[Re]^{\bullet}]^+$

$[[Re]=C=\dot{C}-C\equiv C-[Re]]^+ \longleftrightarrow [[Re]-C\equiv C-\dot{C}=C=[Re]]^+$

$[[Re]=C=C=C=\dot{C}-[Re]]^+ \longleftrightarrow [[Re]-\dot{C}=C=C=C=[Re]]^+$

Scheme 4

The magnetic behavior (μ_{eff} = 2.1 μ_B) reveals one unpaired electron. The ESR spectrum (CH_2Cl_2) shows one undecet with g = 2.02, a_{iso} (Re) = 98 G; the values suggest a complete delocalization of the odd electron (see Scheme 4, [Re] = $(C_5(CH_3)_5)Re(NO)P(C_6H_5)_3$). IR spectrum ($CH_2Cl_2$): 1665 (ν(NO)), 1872 (ν(C≡C)) cm⁻¹. UV spectrum (CH_2Cl_2): λ_{max} (ε) = 348 (24600), 454 (6400), 883 (15000), 1000 (9400), 1200 (3200) nm.

Treatment with $NaC_{10}H_8$ reformed the neutral compound with 72% yield [5].

References:

[1] Fischer, E. O.; Chen, J.; Schubert, U. (Z. Naturforsch. **37b** [1982] 1284/8).
[2] Chen, J. B.; Lei, G.-X.; Zhang, Z.-Y.; Tang, Y.-Q. (Huaxue Xuebao **47** [1989] 31/6; C.A. **112** [1990] No. 77457).
[3] Kelley, C.; Mercando, L. A.; Terry, M. R.; Lugan, N.; Geoffrey, G. L.; Xu, Z.; Rheingold, A. L. (Angew. Chem. **104** [1992] 1066/8; Angew. Chem. Int. Ed. Engl. **31** [1992] 1053).
[4] Roger, C.; Peng, T.-S.; Gladysz, J. A. (J. Organomet. Chem. **439** [1992] 163/75).
[5] Seyler, J. W.; Weng, W.; Zhou, Y.; Gladysz, J. A. (Organometallics **12** [1993] 3802/4).
[6] Zhou, Y.; Seyler, J. W.; Weng, W.; Arif, A. M.; Gladysz, J. A. (J. Am. Chem. Soc. **115** [1993] 8509/10 and Supplementary Material).
[7] Terry, M. R.; Mercando, L. A.; Kelley, C.; Geoffroy, G. L.; Nombel, P.; Lugan, N.; Mathieu, R.; Ostrander, R. L.; Owens-Waltermire, B. E.; Rheingold, A. L. (Organometallics **13** [1994] 843/65).
[8] Weng, W.; Bartik, T.; Johnson, M. T.; Arif, A. M.; Gladysz, J. A. (Organometallics **14** [1995] 889/97).

2.5.1.2.8.3 The Bridging ^2L Ligand is π-Bonded to Rhenium

The structures of the compounds are illustrated in Formulas I to IV.

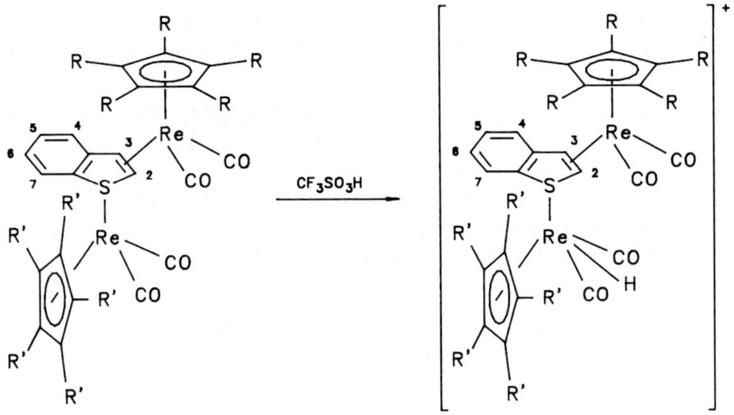

I

II

$C_5R_5Re(CO)_2(\mu\text{-}C_8H_6S)Re(C_5R'_5)(CO)_2$ (see Formula I). Compounds where R = R′ were prepared by treating $(C_5R_5)Re(CO)_2(C_8H_6S)$ (R = H, CH_3) with a catalytic amount of CF_3SO_3H in CH_2Cl_2 for 3 h. The product was separated by precipitating it from the solution with hexane at 0 °C. Small amounts (ca. 5% yield) were also obtained when $C_5R_5Re(CO)_2$-OC_4H_8 was treated with benzo[b]thiophene in THF [1].

The compounds were characterized as follows:

R	R′	yield; physical and crystallographic properties, remarks (for abbreviations and units see p. X)
H	H	Yield: 67%. Brown solid. ^1H NMR (CD_2Cl_2): 4.39, 5.24 (C_5H_5); 4.39, 4.67 (d's); 7.15, 7.25 (td's); 7.34, 7.51 (d's). – $^{13}C\{^1H\}$ NMR (CD_2Cl_2): 83.8, 88.8 (C_5); 36.1, 56.3, 124.8, 126.0, 127.0, 127.5, 145.5, 147.3 (C_8H_6S); 200.2, 202.0, 203.0, 203.2 (CO). – IR (CH_2Cl_2): 1850, 1921, 1984 (ν(CO)). – MS: $[M]^+$ [1]. Single-crystal data: monoclinic; a = 12.265(7), b = 11.162(6), c = 15.816(5) Å, β = 109.28(3)°; space group $P2_1/c-C^5_{2h}$ (No. 14); Z = 4 molecules per unit cell, D_{calc} = 2.427 g/cm^3. The molecular structure is depicted in **Fig. 115** [1].
CH_3	CH_3	Yield: 78%. Brown solid. ^1H NMR (CD_2Cl_2): 2.02, 2.05 (CH_3); 3.68, 3.92 (d's); 7.10 (td), 7.18 (m, 2 H), 7.43 (d). – $^{13}C\{^1H\}$ NMR (CD_2Cl_2): 10.3, 10.9 (CH_3); 95.9, 99.1 (C_5); 42.3, 60.9, 124.6, 125.6, 126.1, 126.7, 142.7, 147.5 (C_8H_6S); 204.1, 204.5, 206.7, 207.6 (CO). – IR (CH_2Cl_2): 1835, 1892, 1904, 1966 (ν(CO)) [1].
CH_3	H	Obtained by combining equimolar amounts of $(C_5(CH_3)_5)Re(CO)_2(C_8H_6S)$ and $C_5H_5Re(CO)_2(OC_4H_8)$ in THF/hexane; chromatography with hexane/CH_2Cl_2 mixtures. Yield: 40% (there is no evidence for the isomer in which the $(C_5(CH_3)_5)Re(CO)_2$ and $C_5H_5Re(CO)_2$ units are interchanged). Air-stable, light cream-colored solid.

References on p. 240

R	R′	yield; physical and crystallographic properties, remarks (for abbreviations and units see p. X)
CH_3 (continued)	H	^1H NMR (CD_2Cl_2): 2.03 (CH_3), 4.93 (C_5H_5); 3.66, 3.94 (d's); 7.09, 7.22 (td's); 7.25, 7.44 (d's). – ^{13}C {^1H} NMR (CD_2Cl_2): 10.3 (CH_3); 83.6, 99.2 (C_5); 41.1, 61.7, 124.0, 126.0, 126.4, 127.0, 144.9, 147.8 (C_8H_6S); 203.0, 203.3, 203.6, 204.8 (CO). – IR (CH_2Cl_2): 1848, 1897, 1919, 1971 (ν(CO)).

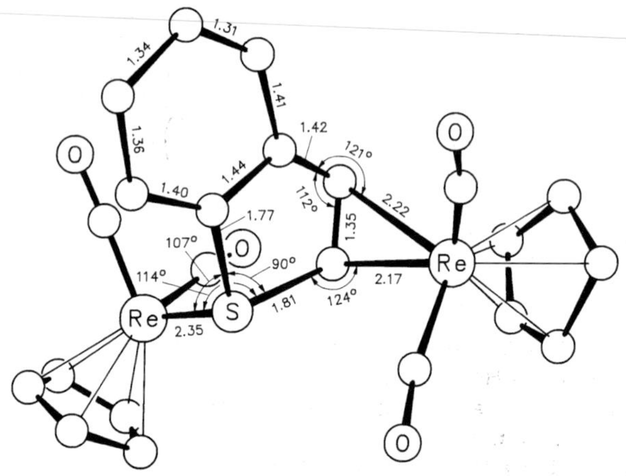

Fig. 115. Molecular structure of $[C_5H_5Re(CO)_2]_2(\mu\text{-}C_8H_6S)$ [1].

All compounds could be protonated exclusively at the S-bonded Re atom to give the following salts. The reactions could be reversed with diphenylguanidine base [1].

$[\{C_5R_5Re(CO)_2\}(\mu\text{-}C_8H_6S)\{C_5R'_5Re(CO)_2H\}][O_3SCF_3]$ (see Formula II). These salts were obtained by treating either $(C_5R_5)Re(CO)_2(C_8H_6S)$ (R = H, CH_3) or $C_5R_5Re(CO)_2(\mu\text{-}C_8H_6S)$-$Re(C_5R'_5)(CO)_2$ (R, R′ = H, CH_3) with at least 0.5 equivalent or 1 equivalent, respectively, CF_3SO_3H in CH_2Cl_2 (the latter reaction could be reversed with 1 equivalent diphenylguanidine base). ^1H NMR spectroscopy revealed quantitative product formation at 0 °C.

The cations are stable in the presence of excess acid (the olefin-bonded Re atom is never attacked by H^+). Although the salts could be separated from the solution, they proved to be insufficiently stable. That is why they were only characterized spectroscopically.

The complexes are all fluxional; their room-temperature NMR spectra display broad peaks. However, their low-temperature spectra (−60 °C) show sharp signals as listed below:

R	R′	spectroscopic properties (for abbreviations and units see p. X)
H	H	Precipitation from the solution at 0 °C with ether gave a brown solid. ^1H NMR (CD_2Cl_2): −8.67 (ReH); 5.55, 5.57 (C_5H_5); 4.84, 5.52 (d's, 2 H); 7.25 to 7.64 (3 m's, 4 H). – ^{13}C {^1H} NMR (CD_2Cl_2): 88.5, 89.3 (C_5); 32.0, 46.0, 125.1, 125.6, 127.6, 129.0, 135.2, 148.0 (C_8H_6S); 188.9, 189.9, 199.4, 201.8 (CO). – IR (CH_2Cl_2): 1919, 1990, 2053 (ν(CO)) [1].

References on p. 240

R	R′	spectroscopic properties (for abbreviations and units see p. X)
CH_3	CH_3	^{1}H NMR (CD_2Cl_2): −8.38 (ReH); 2.05, 2.32 (CH_3); 3.64, 3.98 (d's); 7.19 to 7.54 (3 m's, 4 H). – ^{13}C {^{1}H} NMR (CD_2Cl_2): 10.3, 10.9 (CH_3); 100.4, 102.9 (C_5); 38.7, 50.3, 125.3, 126.1, 127.1, 129.8, 131.4, 149.7 (C_8H_6S); 192.6, 194.1, 203.3, 204.0 (CO). – IR (CH_2Cl_2): 1912, 1971, 1983, 2037 (ν(CO)) [1].
CH_3	H	Obtained by protonating $(C_5(CH_3)_5)Re(CO)_2(\mu\text{-}C_8H_6S)Re(C_5H_5)(CO)_2$. ^{1}H NMR ($CD_2Cl_2$): −8.60 (ReH), 2.06 ($CH_3$), 5.49 ($C_5H_5$); 3.79, 4.09 (d's); 7.23 to 7.56 (4 m's, 4 H). – ^{13}C {^{1}H} NMR (CD_2Cl_2): 10.1 (CH_3); 83.7, 98.6 (C_5); 40.0, 60.3, 123.4, 125.8, 126.3, 126.4, 143.9, 146.8 (C_8H_6S); 203.2, 203.5, 203.6, 206.1 (CO). – IR (CH_2Cl_2): 1920, 1980, 2050 (ν(CO)) [1].

III

IV

$[(C_5(CH_3)_5)Re(CO)_2]_2(\mu\text{-}CH_3O_2CC{\equiv}CCO_2CH_3)$ (see Formula III) was prepared by exposing the dirhenacyclobutene $[(C_5(CH_3)_5)Re(CO)_2]_2(\mu\text{-}CH_3O_2CC{=}CCO_2CH_3)$ (see p. 227) to fluorescent light (C_6H_6, 2 d). After evaporation of the solvent, crystallization of the residue from THF gave orange crystals with 65% yield.

^{1}H NMR spectrum (C_6D_6): δ = 1.69, 3.68 ppm. ^{13}C {^{1}H} NMR spectrum (C_6D_6): δ = 9.6 ($(CH_3)_5$), 51.2 (OCH_3), 99.7 (C_5), 166.0 (CO_2), 207.9 (CO) ppm; δ(C≡C) not observed. IR spectrum (THF): 1704, 1884, 1905, 1963, 1995 cm^{-1}. UV spectrum: Only end absorptions at 326, 440 nm. Mass spectrum: $[M]^+$.

The compound crystallizes in the monoclinic space group $C2/c-C_{2h}^6$ (No. 15) with a = 12.854(2), b = 14.729(3), c = 16.650(3) Å, β = 108.188(12)°; Z = 4 molecules per unit cell, D_{calc} = 1.989 g/cm^3. The molecular structure is depicted in **Fig. 116a**. As can be seen, the CO groups are in close proximity to each other; however, the ligands are not eclipsed when looking along the Re···Re vector (OC-Re···Re-CO dihedral angle is 32°). The Re-(alkyne centroid)-Re angle is 126°.

Short-wavelength photolysis (maximum emission 300 nm) in benzene liberated CO to give $[(C_5(CH_3)_5)Re(CO)]_2(\mu\text{-}CH_3O_2CC{\equiv}CCO_2CH_3)(\mu\text{-}CO)$. Thermolysis at 72 °C gave an equilibrium mixture of the title compound and $[(C_5(CH_3)_5)Re(CO)_2]_2(\mu\text{-}CH_3O_2CC{=}CCO_2CH_3)$ in the ratio 78:22 [2].

$[(C_5(CH_3)_5)ReCO]_2(\mu\text{-}CH_3O_2CC{\equiv}CCO_2CH_3)(\mu\text{-}CO)$ (see Formula IV) was obtained by photolyzing $[(C_5(CH_3)_5)Re(CO)_2]_2(\mu\text{-}CH_3O_2CC{\equiv}CCO_2CH_3)$ or $[(C_5(CH_3)_5)Re(CO)_2]_2(\mu\text{-}CH_3O_2CC{=}CCO_2CH_3)$ with a maximum wavelength of 300 nm in benzene for 8 h. Evaporation left the product as a red-orange solid. Yield: 65% in the first case [2].

Dark orange solid. ^{1}H NMR spectrum (C_6D_6): δ = 1.87, 3.57 ppm. ^{13}C {^{1}H} NMR spectrum (C_6D_6): δ = 9.3 ($(CH_3)_5$), 52.0 (OCH_3), 99.5 (C_5), 168.8 (CO_2), 212.0 (CO) ppm; δ(CO-

References on p. 240

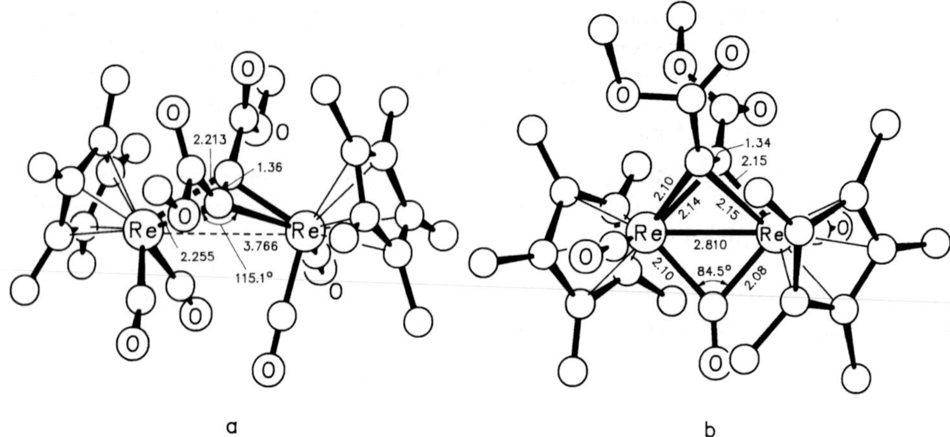

Fig. 116. Molecular structures of
a. $[(C_5(CH_3)_5)Re(CO)_2]_2(\mu\text{-}CH_3O_2CC{\equiv}CCO_2CH_3)$;
b. $[(C_5(CH_3)_5)ReCO]_2(\mu\text{-}CH_3O_2CC{\equiv}CCO_2CH_3)(\mu\text{-}CO)$ [2].

μ) and $\delta(C{\equiv}C)$ not observed. IR spectrum (THF): 1703 ($\nu(CO)_{\mu}$); 1927, 1963 ($\nu(CO)_t$) cm^{-1}. UV spectrum (C_6H_6): only end absorptions between 300 and 600 nm. Mass spectrum: $[M]^+$.

The compound crystallizes in the trigonal space group $P3-C_3^1$ (No. 143) with a = 16.277(2), c = 9.408(3) Å; Z = 3 molecules per unit cell, D_{calc} = 2.005 g/cm^3. The molecule is a dirhenatetrahedrane complex, see **Fig. 116b**. The C≡C and the Re-Re bonds are approximately orthogonal to one another. The Re-(alkyne centroid)-Re angle is 88° [2].

References:

[1] Robertson, M. J.; Day, C. L.; Jacobson, R. A.; Angeleci, R. J. (Organometallics **13** [1994] 179/85).

[2] Casey, C. P.; Cariño, R. S.; Hayashi, R. K.; Schladetzky, K. D. (J. Am. Chem. Soc. **118** [1996] 1617/23).

2.5.1.2.9 Compounds with the $^5LRe(\mu\text{-}^3L)Re^5L$ Skeleton

Only four products are described here. Their structures are shown in Formulas I and II.

R' R' R'
R O R O R O
$(CO)_2Re$—ReCO ⟷ $(CO)_2Re$—ReCO ⟷ $(CO)_2Re$—ReCO
$(CH_3)_5C_5$ $C_5(CH_3)_5$ $(CH_3)_5C_5$ $C_5(CH_3)_5$ $(CH_3)_5C_5$ $C_5(CH_3)_5$
Ia Ib Ic

$(C_5(CH_3)_5)Re(CO)_2(\mu\text{-}\eta^{3:1}\text{-}CH{=}CHC{=}O)Re(C_5(CH_3)_5)(CO)$ (see Formulas Ia to Ic, R, R′ = H) was obtained when $(C_5(CH_3)_5)_2Re_2(CO)_4$ was treated with acetylene in THF. Concentration of the solution followed by adding hexane separated the compound with 84% yield.

References on p. 243

Orange solid. ^1H NMR spectrum (C_6D_6): δ = 1.60, 1.81 (($CH_3)_5$); 3.57, 8.16 (d's, CH; J(H,H) = 8.5 Hz) ppm (coalescence of the CH_3 resonances is observed at 70 °C, $\Delta G^{\neq}$ ca. 70 kJ/mol). ^{13}C {^1H} NMR spectrum (CD_2Cl_2): δ = 9.6, 10.9 (CH_3); 30.8 (=**C**HCO); 98.5, 99.4 (C_5); 126.2 (CH=); 208.1, 208.8, 210.4, 214.7 (CO) ppm. IR spectrum (toluene): 1712, 1848, 1898, 1941 (ν(CO)) cm^{-1}. Mass spectrum: $[M]^+$ [4].

$(C_5(CH_3)_5)Re(CO)_2(\mu\text{-}\eta^{3:1}\text{-}C(CH_3){=}C(CH_3)C{=}O)Re(C_5(CH_3)_5)(CO)$ (see Formulas Ia to Ic, R, R′ = CH_3) was prepared from $(C_5(CH_3)_5)_2Re_2(CO)_4$ and but-2-yne at −40 °C or a somewhat higher temperature in toluene.

^1H NMR spectrum (toluene-d_8, −40 °C): δ = 1.57, 1.74 (($CH_3)_5$); 1.64, 2.93 (CH_3) ppm; coalescence of the $C_5(CH_3)_5$ resonances at 10 °C; therefore $\Delta G^{\neq}$ ca. 57 kJ/mol. IR spectrum (toluene): 1686 (ν(C=O)); 1844, 1896, 1944 (ν(CO)) cm^{-1}.

At 25 °C, slow fragmentation occurred to $(C_5(CH_3)_5)Re(CO)_3$ and $(C_5(CH_3)_5)Re(CH_3\text{-}C{\equiv}CCH_3)(CO)$ with $k_1 = 2.2 \times 10^{-5}\ s^{-1}$ and $\Delta G^{\neq}$ ca. 100 kJ/mol [4].

$(C_5(CH_3)_5)Re(CO)_2(\mu\text{-}\eta^{3:1}\text{-}CH{=}C(C(CH_3){=}CH_2)C{=}O)Re(C_5(CH_3)_5)(CO)$ (see Formulas Ia to Ic, R = H, R′ = $C(CH_3){=}CH_2$) was obtained by combining $HC{\equiv}CC(CH_3){=}CH_2$ with a frozen benzene solution of $(C_5(CH_3)_5)_2Re_2(CO)_4$, followed by warming the mixture to room temperature whereby the color turned orange-red. The solvent was removed and the residue separated chromatographically (silica, hexane/ether (3:1)). The product was obtained from the second orange-red band and recrystallized from CH_2Cl_2/ether. Yield: 81%.

Orange-red crystals. ^1H NMR spectrum (C_6D_6): δ = 1.68, 1.80 (s's, ($CH_3)_5$); 1.87 (m, =CCH_3); 5.35, 5.65 (q's, CH_2=C); 8.27 (s, HC=) ppm. ^{13}C {^1H} NMR spectrum (C_6D_6): δ = 9.6, 10.7, 23.4 (CH_3); 53.1 (HC=**C**), 98.2 (C_5), 115.2 (H**C**=C), 117.5 (H_2**C**=C), 141.0 (H_2C=**C**); 208.2, 208.4, 211.7 (CO) ppm. IR spectrum (CH_2Cl_2): 1685, 1851, 1902, 1950 (ν(CO)) cm^{-1}. Mass spectrum: $[M]^+$.

The compound crystallizes in the orthorhombic space group $Iba2-C_{2v}^{21}$ (No. 45) with a = 18.737(4), b = 31.532(6), c = 9.389(2) Å; Z = 8 molecules per unit cell, D_{calc} = 1.966 g/cm^3. The structure of the molecule is illustrated in **Fig. 117**. The similar C-C bond lengths

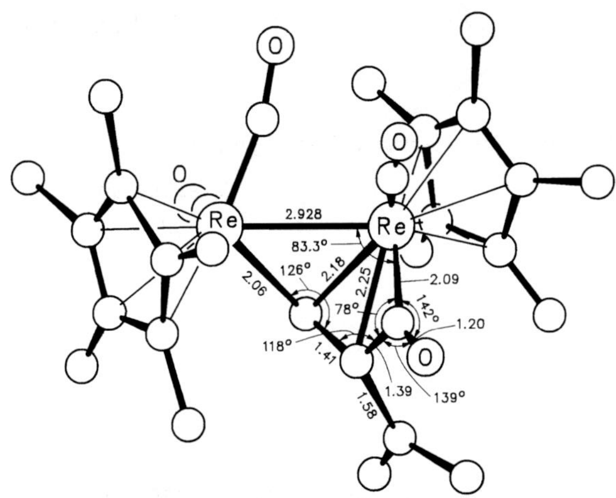

Fig. 117. Molecular structure of $(C_5(CH_3)_5)Re(CO)_2(\mu\text{-}\eta^{3:1}\text{-}CH{=}C(C(CH_3){=}CH_2)C{=}O)Re(C_5(CH_3)_5)(CO)$ [3].

in the bridging ³L ligand suggest that the structure is best rationalized as a resonance hybrid of the three structures depicted in Formulas Ia to Ic.

Thermolysis in hot C_6D_6 (70 °C, 24 h) gave $[(C_5(CH_3)_5)Re(CO)_2]_2(\mu\text{-}C{=}CHC(CH_3){=}CH_2)$ and $(C_5(CH_3)_5)Re(CO)_3$ in the ratio 2:1. In contrast, at 60 °C only a little reaction occurred. Spectroscopic monitoring detected an intermediate present in significant amounts after ca. 4 h, but its structure could only be suggested [3].

$[C_5H_5Re(CO)_2]_2(\mu\text{-}C_{16}H_{12})$ (see Formula II) was one product obtained by irradiating a mixture of $C_5H_5Re(CO)_3$ and $HC{\equiv}CC_6H_5$ in THF for 7 h. Chromatographic workup with petroleum ether/ether mixtures and crystallization from ether gave the compound with 4.3% yield. Light red crystals; m.p. 150 to 152 °C [1].

IR spectrum (C_6H_{12}): 1914, 1930, 1982, 2000 (ν(CO)) cm^{-1}. Mass spectrum: $[M]^+$ [1]; $[M - n\,CO]^+$ (n = 1 to 4), $[M - 4\,CO - n\,H_2]^+$ (n = 1 to 5) [2].

An X-ray crystallographic investigation at −120 °C showed the compound to crystallize in the monoclinic space group $P2_1/n-C^5_{2h}$ (No. 14) with a = 16.91(1), b = 8.206(4), c = 18.99(1) Å, β = 106.66(4)°; Z = 4 molecules per unit cell, D_{calc} = 2.152 g/cm³. The molecular structure is shown in **Fig. 118**. The bridging C=C-C=C unit has a torsion angle of 29.7° [1].

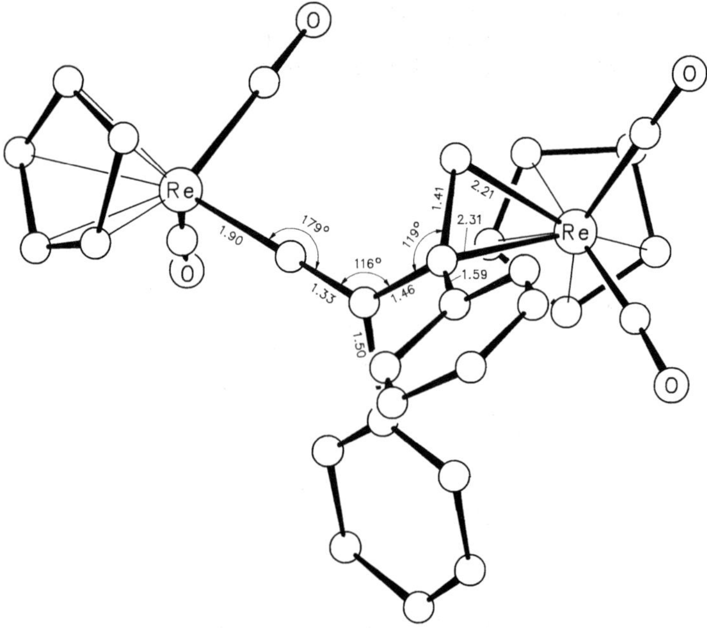

Fig. 118. Molecular structure of $[C_5H_5Re(CO)_2]_2(\mu\text{-}C_{16}H_{12})$ [1].

References on p. 243

References:

[1] Kolobova, N. E.; Antonova, A. B.; Khitrova, O. M.; Antipin, M. Yu.; Struchkov, Yu. T. (J. Organomet. Chem. **137** [1977] 69/78).
[2] Sizoi, V. F.; Nekrasov, Yu. S.; Sukharev, Yu. N.; Kolobova, N. E.; Khitrova, O. M.; Obezyuk, N. S.; Antonova, A. B. (J. Organomet. Chem. **162** [1978] 171/8).
[3] Casey, C. P.; Ha, Y.; Powell, D. R. (J. Am. Chem. Soc. **116** [1994] 3424/8).
[4] Casey, C. P.; Cariño, R. S.; Sakaba, H.; Hayashi, R. K. (Organometallics **15** [1996] 2640/9).

2.5.1.2.10 Compounds with the $^5LRe(\mu\text{-}^4L)Re^5L$ Skeleton

This section describes several compounds having a bridging $\eta^{2:2}$-bonded ligand linking $^5LRe(NO)(P(C_6H_5)_3)$ or $^5LRe(CO)_2$ fragments. The structures of the compounds are given in Formulas I to IV.

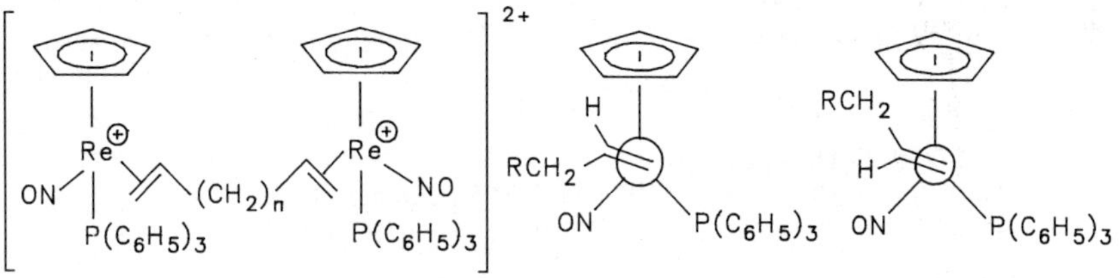

I

IIa (RS, SR)

IIb (RR, SS)

[{$C_5H_5Re(NO)(P(C_6H_5)_3)H_2C{=}CH-$}$_2(CH_2)_n$]$X_2$ (see Formula I; X = BF_4, PF_6). Whether the dications were obtained as PF_6 or BF_4 salts, was dependent on their preparation. Two methods have been employed:

Method I: Thermolysis of the dicarbenes [{$C_5H_5Re(NO)(P(C_6H_5)_3){=}CH-$}$_2(CH_2)_n$][$PF_6$]$_2$ (n = 3, 6) in C_6H_5Cl (temperature and reaction time given below). Addition of hexane precipitated a yellowish powder.
No mixed carbene-alkene intermediates were detected during the reaction [2].

Method II: Reaction of [$C_5H_5Re(NO)(P(C_6H_5)_3)H_2C{=}CH(CH_2)_nCH{=}CH_2$]$BF_4$ (n = 1, 4) with [$C_5H_5Re(NO)(P(C_6H_5)_3)ClC_6H_5$]$BF_4$ (chlorobenzene, 100 °C). Filtration into hexane yielded a yellow powder [2].

The Re centers are chiral, thus the compounds form as diastereomer mixtures. Moreover, two configurational isomers are possible differing in the alkene enantioface bonded to Re (see Formulas IIa, IIb). The former having the large alkyl substituent directed towards the NO ligand is more stable and at ca. 95 °C a ca. 98:2 ratio of (RS,SR):(RR,SS) is generally obtained.

The spectra obtained from the BF_4 salts made by Method II closely match with those of the major diastereomers of the PF_6 salts made by Method I [2].

n	X	method of preparation, yield, spectroscopic properties, remarks (for abbreviations and units see p. X)
1	PF_6	Method I (100 °C, 24 h), yield: 90%. – ^{1}H NMR (CD_3NO_2): 5.88, 5.90, 5.91, 5.94 (s's, C_5H_5); 7.57 (m, C_6H_5). – ^{13}C {^{1}H} NMR (CD_3NO_2): 37.9, 38.4, 38.7, 39.1 (br s's, $=CH_2$); 47.0 (s, CH_2); 50.2, 52.0 (s, br s, =CH); 98.6, 98.9, 99.1,

References on p. 245

n	X	method of preparation, yield, spectroscopic properties, remarks (for abbreviations and units see p. X)
		99.2 (s's, C_5H_5); 130.8, 131.7 (d's; J(P,C) = 10, 59, resp.); 133.6 (s, C_6H_5-4), 134.7 (d, C_6H_5-2; J(P,C) = 10). – $^{31}P\ \{^1H\}$ NMR (CD_3NO_2): 11.9 (s) [2].
1	BF_4	Method II (100 °C, 44 h), yield: 97%. M.p. 190 to 195 °C. – ^1H NMR (CD_3NO_2): 2.47, 2.64, 2.78 (m's, CH_2=, CH_2); 4.51 (m, =CHC); 5.91, 5.94 (s's, C_5H_5, (44 ± 2):(56 ± 2)); 7.58 (m, C_6H_5). – $^{13}C\ \{^1H\}$ NMR (CD_3NO_2): 38.4, 38.7 (d's, =CH_2; J(P,C) = 5.1); 47.0 (s, CH_2); 51.9, 52.0 (s's, =CH); 98.9, 99.0 (s's, C_5); 130.8 (d, C_6H_5-3; J(P,C) = 10), 131.6 (d, C_{ipso}; J(P,C) = 59), 133.6 (s, C_6H_5-4), 134.6 (d, C_6H_5-2; J(P,C) = 10). – $^{31}P\ \{^1H\}$ NMR (CD_3NO_2): 11.9 (s). – IR (film): 1721 (ν(NO)) [2].
4	PF_6	Method I (65 °C, 17 h), yield: 97%. – ^1H NMR ($CDCl_3$): 1.64, 2.01, 2.42 (m, 4 H each, $(CH_2)_4$, =CH_2); 4.55 (m, =CHC); 5.74, 5.78 (s's, C_5H_5, ca 33:67); 7.35, 7.56 (m, C_6H_5, 12:18). – $^{13}C\ \{^1H\}$ NMR ($CDCl_3$): 31.8 (d, =CH_2; J(P,C) = 4.3); 37.6, 38.5 (s's, CH_2CH_2); 51.7 (s, =**C**HC), 96.8 (s, C_5H_5), 129.5 (d, C_6H_5-3; J(P,C) = 10.7), 130.2 (d, C_{ipso}; J(P,C) = 59.4), 132.2 (s, C_6H_5-4), 133.0 (d, C_6H_5-2; J(P,C) = 9.9); signals of minor diastereomers at 32.4, 37.8, 38.9, 52.1, 97.5 ppm. – $^{31}P\ \{^1H\}$ NMR ($CDCl_3$): 10.9, 11.0 (s's); (C_6H_5Cl): 10.5 (s). – IR (KBr): 841 (ν(PF)), 1717 (ν(NO)) [2].
4	BF_4	Method II (100 °C, 15 h), yield: 98%. – ^1H NMR ($CDCl_3$): 1.63, 2.00, 2.42 (m, $(CH_2)_4$, =CH_2); 4.58 (m, =CHC); 5.76, 5.77, 5.78, 5.79 (s's, C_5H_5, ca. 15:13:35:37); 7.35, 7.55 (m, C_6H_5). – $^{13}C\ \{^1H\}$ NMR ($CDCl_3$): 31.8 (br s, =CH_2); 37.6, 38.5 (CH_2CH_2); 51.8 (s, =**C**HC), 96.9 (s, C_5H_5), 129.5 (d, C_6H_5-3; J(P,C) = 10.9), 130.3 (d, C_{ipso}; J(P,C) = 58.8), 132.1 (s, C_6H_5-4), 133.1 (d, C_6H_5-2; J(P,C) = 9.9); signals of minor diastereomers at 32.1, 37.7, 39.0, 52.3, 97.6 ppm. – $^{31}P\ \{^1H\}$ NMR ($CDCl_3$): 10.9, 11.0 (s's); minor diastereomer: 10.8 ppm. – IR (film): 1721 (ν(NO)) [2].

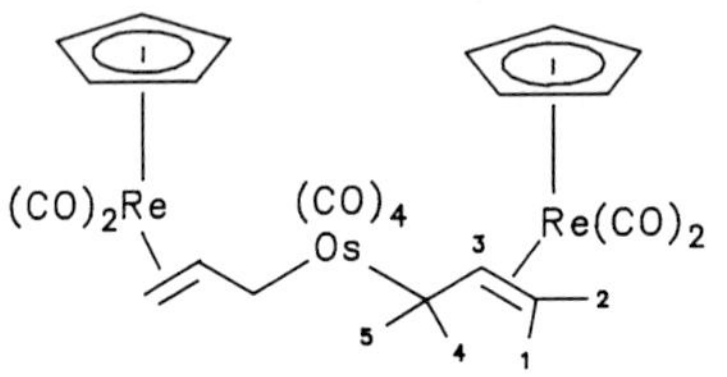

III

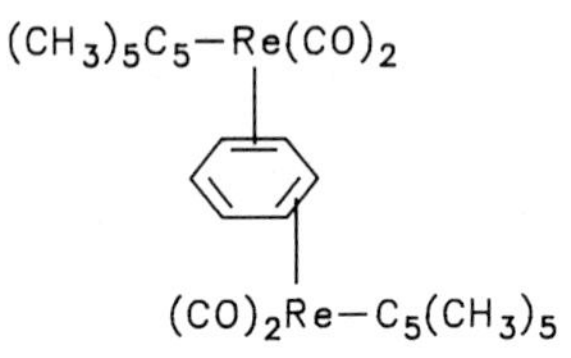

IV

$[C_5H_5Re(CO)_2CH_2{=}CHCH_2{-}]_2Os(CO)_4$ (see Formula III) was obtained by treating $[C_5H_5Re(CO)_2C_3H_5\text{-}\pi]BF_4$ with $Na_2[(CO)_4Os]$ in THF at −70 °C. The mixture was allowed to slowly warm to 0 °C. The mixture was evaporated and the residue was extracted with CH_2Cl_2. Concentration of the extract followed by adding pentane yielded the product as a colorless powder with 82% yield; dec. above 174 °C.

^1H NMR spectrum ($CDCl_3$): δ = 1.09 (dd, H-4; J(H-4,3) = 11.4 Hz), 2.20 (dd, H-5; J(H-5,4) = −11.4, J(H-3,5) = 1.4 Hz), 2.33 (m, H-1,2), 3.83 (m, H-3), 5.18 (s, C_5H_5) ppm. $^{13}C\ \{^1H\}$ NMR spectrum ($CDCl_3$): δ = 8.39, 8.47 ($OsCH_2$); 18.95 (CH); 56.70, 56.75 (=CH_2); 86.03 (C_5H_5), 171.29 (OsCO), 179.23 (ReCO) ppm. IR spectrum (Nujol): 1861, 1868, 1944, 2013, 2033, 2124 (ν(CO)) cm^{-1} [3].

References on p. 245

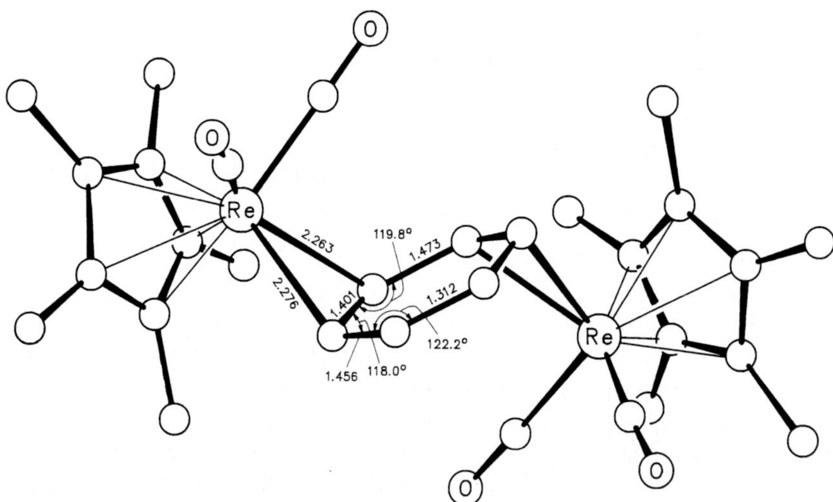

Fig. 119. Molecular structure of $[(C_5(CH_3)_5)Re(CO)_2]_2(\mu\text{-}\eta^{2:2}\text{-}C_6H_6)$ [1].

$[(C_5(CH_3)_5)Re(CO)_2]_2(\mu\text{-}\eta^{2:2}\text{-}C_6H_6)$ (see Formula IV) was prepared along with $(C_5(CH_3)_5)_2$-$Re_2(CO)_5$ and $(C_6H_6)Re(C_5(CH_3)_5)$ by irradiating $(C_5(CH_3)_5)Re(CO)_3$ in benzene at room temperature. The solvent was removed and the residue extracted with hexane. Trituration of the insoluble material with C_2H_5OH solidified the title product, which was recrystallized from toluene. Yield: 7.5%. Off-white, air-stable, crystalline solid.

^1H NMR spectrum (C_6D_6): δ = 1.66 (s), 2.85 (m), 4.14 (d; J = 8 Hz), 6.41 (m) ppm. ^{13}C {^1H} NMR spectrum ($CDCl_3$): δ = 10.2 (CH_3); 40.9, 50.6, 127.3 (C_6H_6); 96.9 (C_5); 206.6, 206.8 (CO) ppm. IR spectrum (C_6H_6): 1871, 1937 (ν(CO)) cm^{-1}.

Single crystals are monoclinic with a = 14.142(5), b = 8.955(5), c = 23.053(14) Å, β = 100.29(4)°; space group C2/c$-C_{2h}^6$ (No. 15); Z = 4 molecules per unit cell, D_{calc} = 1.93 g/cm^3. The planar benzene molecule is η^2-bonded to each of the Re atoms. The C-C bond lengths show significant variations (compare with **Fig. 119**) and the bridging ligand can thus be considered a coordinated "cyclohexatriene" [1].

The compound is unstable under photolysis conditions in benzene, where it is primarily converted into $(C_5(CH_3)_5)Re(C_6H_6\text{-}\eta^2)(CO)_2$ [1].

References:

[1] van der Heijden, H.; Orpen, A. G.; Pasman, P. (J. Chem. Soc. Chem. Commun. **1985** 1576/8).
[2] Roger, C.; Peng, T.-S.; Gladysz, J. A. (J. Organomet. Chem. **439** [1992] 163/75).
[3] Hüffer, S.; Wieser, M.; Polborn, K.; Beck, W. (J. Organomet. Chem. **481** [1994] 45/55).

2.5.1.3 Compounds with Two ^5L_2Re Fragments

$(C_5H_5)_2Re-Re(C_5H_5)_2$ was obtained by combining in situ-prepared $Li[(C_5H_5)_2Re]$ (from $(C_5H_5)_2ReH$ and $t\text{-}C_4H_9CH_2Li$ in THF/benzene) with benzaldehyde at −78 °C followed by slowly warming the mixture to room temperature. The product precipitated with 50 to 75% yield as a purple powder [4].

References on p. 247

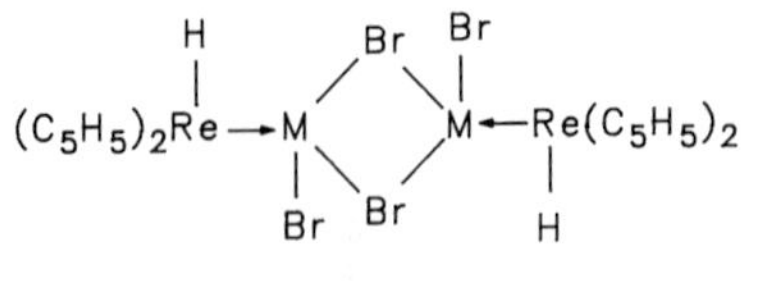

^{1}H NMR spectrum (CD_2Cl_2): $\delta = 4.33$ (s) ppm. CI mass spectrum: $[M]^-$ [4]. Cyclic voltammetry (THF/0.1 M $[N(C_4H_9\text{-}n)_4]PF_6$, Hg electrode): $E_{1/2}$ (red.) = −2.1 V vs. NCE (two-electron wave); under a CO atmosphere only a reduction process was observed, thus initially formed $[(C_5H_5)_2Re]^-$ reacted with CO producing $[(\eta^5\text{-}C_5H_5)(\eta^3\text{-}C_5H_5)Re(CO)]^-$ [6].

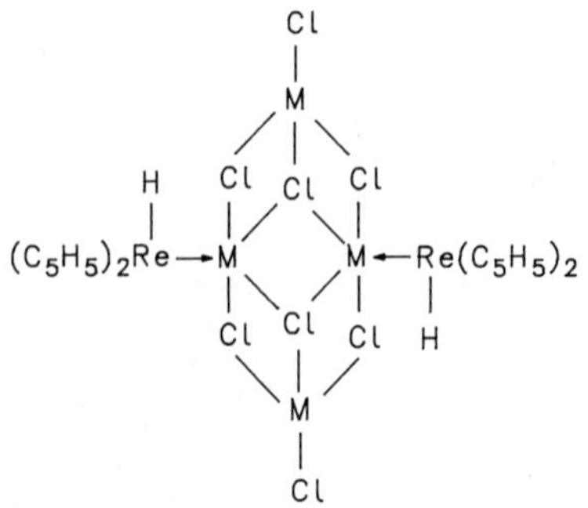

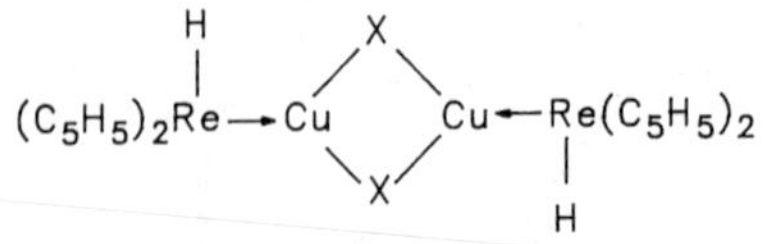

Thermolysis (CH_3CN, few minutes) gave $(C_5H_5)_2ReH$ and $(C_5H_5)_2Re_2(\mu\text{-}\eta^{5:1}\text{-}C_5H_4)_2$ in a 2:1 ratio (thermolysis in deuterated solvents did not incorporate deuterium). Visible-light photolysis gave $(C_5H_5)_2Re_2(\eta^4\text{-}C_5H_6)(\mu\text{-}\eta^{5:1}\text{-}C_5H_4)$. Treatment with $C_6H_5CH_2Br$ yielded $(C_5H_5)_2ReBr$ and $(C_5H_5)_2ReCH_2C_6H_5$ in equimolar quantities [4].

$[(C_5H_5)_2Re(H)MBr_2]_2$ (M = Fe, Co; see Formula I for tentative structure). The compounds were prepared by treating $(C_5H_5)_2ReH$ with anhydrous MBr_2 (M = Fe, Co) in THF. The formation of a precipitate occurred at once [3, 5].

While the Fe compound is grayish green, the Co compound forms as a blue solid (from ether) [5]. Their IR spectra show one band in the ν(ReH) region in the range 2050 to 2055 cm^{-1} [3].

$[(C_5H_5)_2Re(H)M_2Cl_4]_2$ (M = Fe, Co; see Formula II for tentative structure). Compounds of this type were prepared from $(C_5H_5)_2ReH$ and MCl_2 (M = Fe, Co) in THF. The products precipitated at once. The following properties are given [5]:

M = Fe: Yield: 74%. Gray, finely crystalline compound. Magnetic behavior: $\mu_{eff} = 6.34\ \mu_B$ (room temperature), 5.85 μ_B (−196°C) [5].

M = Co: Blue solid [5].

$[(C_5H_5)_2Re(H)CuX]_2$ (see Formula III). Compounds with this composition were obtained by treating $(C_5H_5)_2ReH$ with anhydrous CuX (X = Cl, I) in CH_3CN. The solution was concentrated, and the precipitate formed was collected.

Both compounds are soluble in CH_3CN, DMSO, and DMF. In the solid state, they are rather stable against O_2 [1]. They have the following spectroscopic and structural properties:

X = Cl: Yield: 62%. Yellow solid. – ^{1}H NMR (DMSO-d_6): −13.25 (ReH), 4.92 (C_5H_5). Thermolysis led to H_2 evolution at 208°C [1].

X = I: Yield: 75%. Yellow crystals.
^{1}H NMR (DMSO-d_6): −13.00 (ReH), 4.86 (C_5H_5) [1].
Single-crystal data: monoclinic; space group I2/a−C_{2h}^3 (No. 12); a = 16.070(4), b = 7.788(2), c = 17.439(5) Å, β = 96.62(2)°; Z = 4 molecules per unit cell, D_{calc} = 3.11 g/cm^3. A view of the molecular structure is given in **Fig. 120**. As can be seen, the Cu_2I_2 unit is folded along the I···I vector [2].

References on p. 247

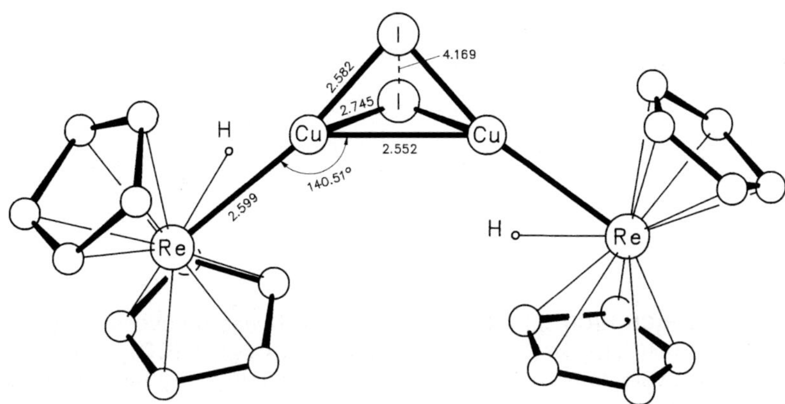

Fig. 120. Molecular structure of $[(C_5H_5)_2Re(H)CuI]_2$ [2].

References:

[1] Ishchenko, V. M.; Bulychev, B. M.; Soloveichik, G. L. (Koord. Khim. **9** [1983] 622/6; C.A. **99** [1983] No. 70920; no English translation).

[2] Bel'sky, V. K.; Ishchenko, V. M.; Bulychev, B. M.; Soloveichik, G. L. (Polyhedron **3** [1984] 749/52).

[3] Ishchenko, V. M.; Bulychev, B. M.; Soloveichik, G. L. (Koord. Khim. **10** [1984] 998; C.A. **101** [1984] No. 192134).

[4] Pasman, P.; Snel, J. J. M. (J. Organomet. Chem. **276** [1984] 387/92).

[5] Ishchenko, V. M.; Bulychev, B. M.; Bel'skii, V. K.; Soloveichik, G. L.; Ellert, O. G.; Seifulina, Z. M. (Koord. Khim. **11** [1985] 851/6; Sov. J. Coord. Chem. [Engl. Transl.] **11** [1985] 483/9).

[6] Kukharenko, S. V.; Ishchenko, V. M.; Soloveichik, G. L.; Strelets, V. V. (Izv. Akad. Nauk SSSR Ser. Khim. **1987** 483/5; Bull. Acad. Sci. USSR Div. Chem. Sci. [Engl. Transl.] **1988** 402/3).

2.5.2 Compounds with Bridging ^5L Ligands

2.5.2.1 The Bridging ^5L Ligand is of the Cyclopentadienyl Type

$[(CO)_5ReC_5H_4]Re(CO)_3$ was obtained by thermolyzing $[(CO)_5ReC(O)C_5H_4]Re(CO)_3$ (see p. 270) in benzene for 15 min. Separation by chromatography on alumina gave the product with 74% yield [2]. It was also obtained by photochemically reacting $(CO)_{10}Re_2$ with a large excess of cyclopentadiene (hexane, −20 °C, separation by HPLC methods; yield: 1.3%) [3]. Colorless fine crystals [3]; m.p. 131 to 133 °C [2].

^1H NMR spectrum (C_6D_6): δ = 5.32, 5.40 (m's) ppm [3]; similar in $CDCl_3$ [2]. IR spectrum (C_6H_{12}): 1930, 1950, 2010, 2020, 2035, 2075, 2135 (ν(CO)) cm^{-1} [2]. Mass spectrum: $[M - n\,CO]^+$ (n = 0 to 8) [2, 3].

The compound did not react with dry HCl or with $HgCl_2$ [2].

$[(CO)_5Re-Hg-C_5H_4]Re(CO)_3$ formed by treating $(ClHg-C_5H_4)Re(CO)_3$ with $Na[(CO)_5Re]$ in THF at −20 °C. The mixture was warmed to room temperature and stirred for further 3 h. The product precipitated with 32% yield. Dec.p. 215 °C.

IR spectrum (CH_2Cl_2): 1940, 1995, 2008, 2035, 2085 (ν(CO)) cm^{-1} [1].

References on pp. 249/50

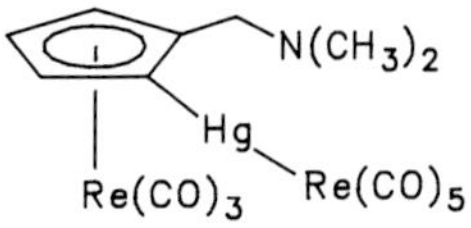

I 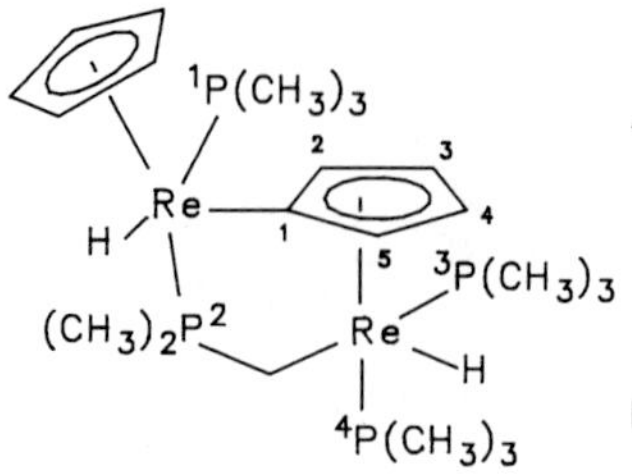 II 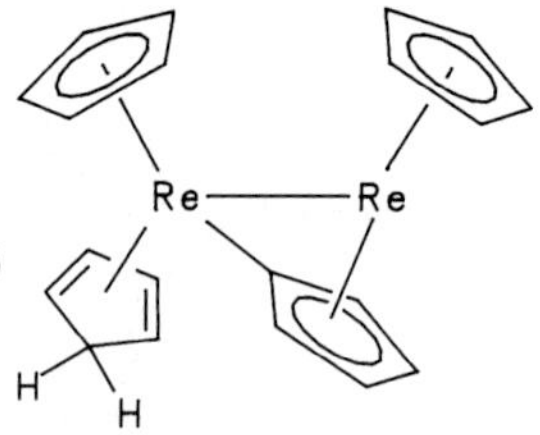 III

$[(CO)_5Re-Hg-C_5H_3CH_2N(CH_3)_2]Re(CO)_3$ (see Formula I) was produced from in situ-generated $(Li-C_5H_3CH_2N(CH_3)_2)Re(CO)_3$ (from $C_5H_4CH_2N(CH_3)_2Re(CO)_3$ and n-C_4H_9Li at −70 °C) and 1.5 equivalents $(CO)_5ReHgX$ (X = Cl, Br) in THF at −30 to −20 °C. The yield amounted to 60 to 80% [6].

$C_5H_5Re_2(\mu$-$\eta^{5:1}$-$C_5H_4)(P(CH_3)_3)_3(H)_2(\mu$-$P(CH_3)_2CH_2-)$ (see Formula II). Photolysis of C_5H_5-$Re(P(CH_3)_3)_3$ in hexane for 45 h generated $C_5H_5Re(P(CH_3)_3)(P(CH_3)_2CH_2-)H$. After concentration, the mixture was heated to 60 °C for 3 h. Subsequent dilution with pentane precipitated a brown solid, which was chromatographically purified at −105 °C (alumina, ether/pentane (1:50)). The title compound was thereby obtained with 18% yield. In contrast, when C_6H_{12}-dissolved $C_5H_5Re(P(CH_3)_3)(P(CH_3)_2CH_2-)H$ was simply allowed to stand at room temperature, the self-condensation yielded the title compound along with another dimeric complex, possibly $C_{10}H_8Re_2(P(CH_3)_3)_4H_2$ (see p. 273), in the ratio 2:1. These two products could not be separated, but a small quantity of the title product crystallized when the mixture was kept in hexane. The complex forms colorless crystals.

^1H NMR spectrum (C_6D_6): δ = −13.44 (dd; J = 41.6, 51.8 Hz), −12.69 (tm; J = 50.7 Hz); 1.39 (d, PCH_3 and 1 H of CH_2; J = 7.7 Hz); 1.46, 1.48, 1.62, 1.95 (d's, PCH_3; J = 7.4, 8.1, 9.1, 7.2 Hz, resp.); 2.13 (qm, 1 H of CH_2; J = 14.5 Hz); 4.21, 4.26, 4.44 (br s's; 1, 1, and 2 H of C_5H_4); 4.62 (s, C_5H_5) ppm. $^{13}C\{^1H\}$ NMR spectrum (C_6D_6): δ = 15.85 (CH_2); 25.83, 27.28, 27.87 (d's, $P(CH_3)_3$; J(P,C) = 28.1, 27.9, 29.5 Hz, resp.); 28.84, 36.07 (d's, $P(CH_3)_2$; J(P,C) = 27.5, 23.9 Hz, resp.); 64.25, 64.60 (s's, C-3,4); 80.22 (s, C_5H_5); 86.25 (dt, C-2 or C-5; J(P,C) = 14.0, 8.9 Hz), 89.51 (d, C-5 or C-2; J(P,C) = 8.9 Hz) ppm; δ(C-1) not observed. $^{31}P\{^1H\}$ NMR spectrum (C_6D_6): δ = −46.52, −40.25 (dd's; J = 3.3, 22.1 and 22.23, 45.3 Hz, resp.); −37.93 (d; J = 24.4 Hz), 28.28 (ddd; J = 3.0, 24.1, 45.5 Hz) ppm. IR spectrum (C_6D_6): 1968 cm^{-1}. EI mass spectrum: $[M]^+$.

At −108 ± 5 °C, the compound crystallizes in the monoclinic space group $P2_1/n-C^5_{2h}$ (No. 14) with a = 14.8973(24), b = 9.0668(14), c = 20.6056(34) Å, β = 104.879(13)°; Z = 4 molecules per unit cell, D_{calc} = 1.99 g/cm^3. The structure of the molecule is depicted in **Fig. 121**. The bonding of the cyclopentadienyl rings to the Re atoms is not regular; each ring is slipped. P,P coupling constants in the ^{31}P NMR spectrum can be explained by the dihedral angles between P-2 and P-3 as well as between P-2 and P-4, which are 155° and 105° [5].

$(C_5H_5)_2Re_2(\eta^4$-$C_5H_6)(\mu$-$\eta^{5:1}$-$C_5H_4)$ (see Formula III) formed when $(C_5H_5)_2Re-Re(C_5H_5)_2$ was irradiated with visible light (λ > 400 nm) in THF or benzene until the red color disappeared. The compound was not isolated because it is not thermally stable.

^1H NMR spectrum (THF-d_8): δ = 3.03 (dm; J = 10 Hz); 3.17, 3.62, 3.94 (m's); 4.13 (C_5H_5); 4.26, 4.55, 4.93 (m's); 5.06 (C_5H_5); 5.15, 6.09 (m's), 6.52 (d; J = 10 Hz) ppm; coupling constants are < 1 Hz.

In solution, the compound converts into $(C_5H_5)_2ReH$ and $(C_5H_5)_2Re_2(\mu$-$\eta^{5:1}$-$C_5H_4)_2$ [4].

References on pp. 249/50

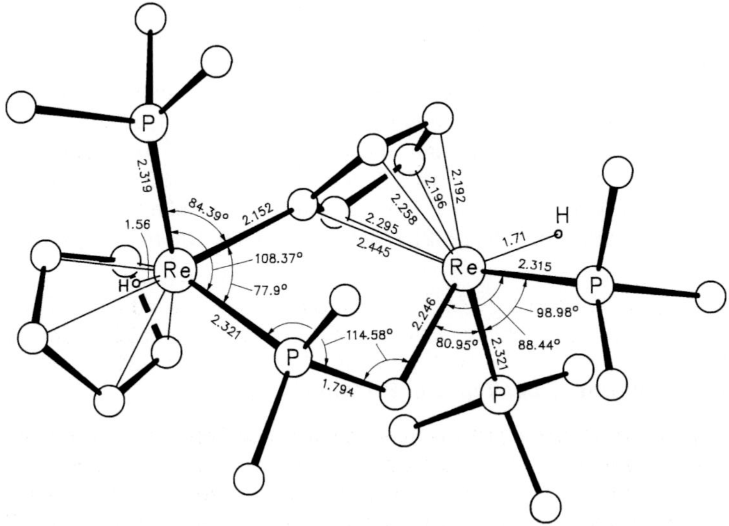

Fig. 121. Molecular structure of $C_5H_5Re_2(\mu\text{-}\eta^{5:1}\text{-}C_5H_4)(P(CH_3)_3)_3(H)_2(\mu\text{-}P(CH_3)_2CH_2-)$ [5].

$(C_5H_5)_2Re_2(\mu\text{-}\eta^{5:1}\text{-}C_5H_4)_2$ (see Formulas IVa, IVb) was obtained along with $(C_5H_5)_2ReH$ in a 1:2 ratio by either leaving stand a solution containing $(C_5H_5)_2Re_2(\eta^4\text{-}C_5H_6)(\mu\text{-}\eta^{5:1}\text{-}C_5H_4)$ (preceding compound) or by thermolyzing $(C_5H_5)_2Re-Re(C_5H_5)_2$ in refluxing CH_3CN until the color of the starting compound ceased. The product separated upon cooling as a yellow crystalline material. The thermolysis reaction was shown to proceed intramolecularly.

IVa or IVb

As long as only NMR data are available, one cannot unambiguously distinguish between the structures illustrated in Formula IVa, IVb. However, when comparing ^{13}C NMR shifts with those of related compounds, it appears opportune to suggest the [5]L compound formulation of Formula IVa: ^1H NMR spectrum (CD_2Cl_2): δ = 4.05, 4.28 (both 2 H, AA'BB'), 4.46 (s, C_5H_5) ppm. ^{13}C NMR spectrum (CD_2Cl_2): δ = 60.0 (s, C-1), 70.0 (d, C-2 or 3), 71.1 (d, C_5H_5), 80.9 (d, C-3 or 2; J(C,H) = 173, 173, and 181 Hz, resp.) ppm [4].

References:

[1] Suleimanov, G. Z.; Sokolov, V. I.; Reutov, O. A. (Izv. Akad. Nauk SSSR Ser. Khim. **1978** 2837/8; Bull. Acad. Sci. USSR Div. Chem. Sci. [Engl. Transl.] **1978** 2536).

[2] Kolobova, N. E.; Khandozhko, V. N.; Sizoi, V. F.; Guseinov, Sh.; Zhvanko, O. S.; Nekrasov, Yu. S. (Izv. Akad. Nauk SSSR Ser. Khim. **1979** 619/22; Bull. Acad. Sci. USSR Div. Chem. Sci. [Engl. Transl.] **1979** 573/6).
[3] Franzreb, K.-H.; Kreiter, C. G. (J. Organomet. Chem. **246** [1983] 189/95).
[4] Pasman, P.; Snel, J. J. M. (J. Organomet. Chem. **276** [1984] 387/92).
[5] Wenzel, T. T.; Bergman, R. G. (J. Am. Chem. Soc. **108** [1986] 4856/67).
[6] Suleimanov, G. Z.; Usyatinsky, A. Ya.; Zulfugarly, E. A.; Kuzmina, L. G.; Kazimirchuk, E. I.; Khandozhko, V. N.; Petrovskii, P. V.; Bregadze, V. I.; Makhmudov, Sh. M.; Beletskaya, I. P. (Metalloorg. Khim. **5** [1992] 973/4; Organomet. Chem. USSR [Engl. Transl.] **5** [1992] 473/4).

2.5.2.2 Compounds with Other Types of Bridging ⁵L Ligands

The compounds described in this section are illustrated in Formulas I to III.

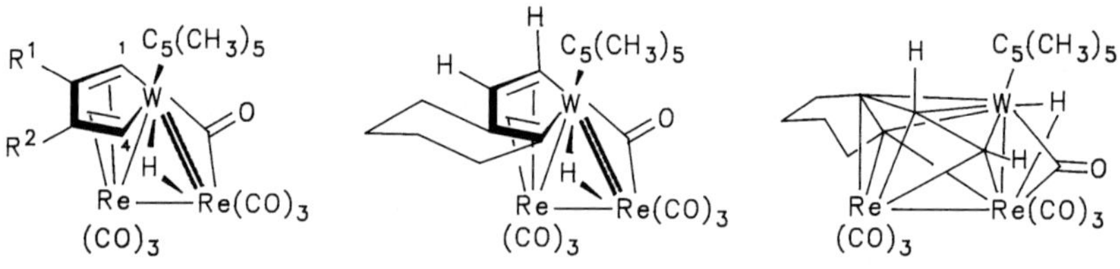

I II (isomer A) III (isomer B)

They were prepared by the following procedure:

Method I: Thermolysis of [μ-($C_5(CH_3)_5$)W$(CO)_2$C≡CR]$Re_2(CO)_6$(μ-CO) (see pp. 125/6; R = CH=CHOCH$_3$ [2], C(CH$_3$)=CH$_2$ [1, 2], or C_6H_9-cyclo [2]) under an H_2 atmosphere (1 atm, toluene, reflux, 1 to 2 h). The resulting mixture was separated by preparative TLC (silica, CH_2Cl_2/hexane (1:1)) [1, 2].
The reaction starting from [μ-($C_5(CH_3)_5$)W$(CO)_2$C≡CC$_6$H$_9$-cyclo]$Re_2(CO)_6$-(μ-CO) initially gave [μ-($C_5(CH_3)_5$)W$(CO)_2$(C≡CC$_6$H$_9$)H]$Re_2(CO)_6$(μ-H) (see Formula III, p. 124), which could be separated. This compound also converted into the title compound upon prolonged exposure to H_2 [2].

$(CO)_6Re_2$(μ-H)(μ-CO)W($C_5(CH_3)_5$)(μ-C_4H_4) (see Formula I, R^1 = R^2 = H) was obtained with 12% yield by Method I when starting from [μ-($C_5(CH_3)_5$)W$(CO)_2$C≡CCH=CHOCH$_3$]Re_2-$(CO)_6$(μ-CO) (heating time 2 h). Red-brown solid.

^{1}H NMR spectrum ($CDCl_3$): δ = −5.74 (ReHW; J(H,H) = 3.4, J(W,H) = 98 Hz), 1.99 ((CH_3)$_5$), 6.01 (dd; J(H,H) = 2 and 6.1 Hz), 6.11 (m), 6.41 (dt; J(H,H) = 2 and 5.1 Hz), 6.50 (dt; J(H,H) = 2 and 6.1 Hz) ppm. IR spectrum (CH_2Cl_2): 1860, 1907, 1957, 2003, 2037 (ν(CO)) cm^{-1}. FAB mass spectrum: $[M]^+$ [2].

$(CO)_6Re_2$(μ-H)(μ-CO)W($C_5(CH_3)_5$)(μ-$C_4(CH_3)H_3$) (see Formula I, R^1 = H, R^2 = CH_3 and vice versa). The two compounds were formed as an inseparable mixture (ratio 48:52) by the procedure given under Method I (heating time 1 h) [1]. The total yield was 67% [2]. Red-brown crystals separated from $CHCl_3$/heptane [2].

The two isomers could be separately characterized by NMR methods as follows:

$R^1 = H$, $R^2 = CH_3$: ^1H NMR ($CDCl_3$): −5.77 (d, ReHW; J(H,H) = 3.2, J(W,H) = 99); 1.99 ($(CH_3)_5$), 2.64 (CH_3-R^2), 5.86 (t, H-4; J(H,H) = 2.9, 3.2), 5.97 (d, H-1; J(H,H) = 6.1), 6.53 (dd, CH-R^1; J(H,H) = 2.9, 6.1). − $^{13}C\{^1H\}$ NMR (THF-d_8): 13.0 ($(CH_3)_5$), 21.9 (CH_3), 106.3 (C_5), 108.5 (CH), 126.2 (**C**CH_3); 134.6, 146.6 (CH; J(W,C) = 67 and 64); 191.9, 196.3, 201.2 (ReCO); 238.1 (WCO; J(W,C) = 126) [1, 2].

$R^1 = CH_3$, $R^2 = H$: ^1H NMR ($CDCl_3$): −5.74 (d, ReHW; J(H,H) = 3.5, J(W,H) = 99); 2.00 ($(CH_3)_5$), 2.62 (CH_3-R^1), 5.71 (d, H-1; J(H,H) = 2.1), 6.10 (dd, H-4; J(H,H) = 3.5, 5.8), 6.43 (dd, CH-R^2; J(H,H) = 2.1, 5.8). − $^{13}C\{^1H\}$ NMR (THF-d_8): 13.1 ($(CH_3)_5$), 22.0 (CH_3), 103.7 (CH), 106.3 (C_5), 120.6 (**C**CH_3); 135.9, 145.8 (CH; J(W,C) = 65 and 67); 191.9, 196.3, 201.2 (ReCO); 237.9 (WCO; J(W,C) = 126) [1, 2].

IR spectrum (CH_2Cl_2, isomer mixture): 1906, 1959, 2002, 2037 (ν(CO)) cm^{-1}. FAB mass spectrum: $[M]^+$ [1, 2].

Single crystals have a triclinic lattice with a = 8.794(2), b = 9.656(3), c = 15.417(4) Å, α = 92.05(2)°, β = 103.04(2)°, γ = 109.54(2)°; space group $P\bar{1}-C_i^1$ (No. 2); Z = 2 molecules per unit cell, D_{calc} = 2.655 g/cm^3. The molecular structure closely resembles that of the isomer A of the following compound (compare with Fig. 122). Within the inner Re_2WC_4 core, the following bond lengths were determined: Re-Re: 3.151, Re=W: 2.672, Re-W: 2.898; W-C: 2.12 and 2.11, Re-C_4: 2.33, 2.28, 2.26, and 2.28 Å [1].

$(CO)_6Re_2(\mu$-H$)(\mu$-CO$)W(C_5(CH_3)_5)(\mu$-CHCH$(C_6H_8))$. Two compounds with this composition (see Formulas II and III; isomer A and isomer B) were obtained by the route described under Method I when starting from $[\mu$-$(C_5(CH_3)_5)W(CO)_2C{\equiv}CC_6H_9$-cyclo$]Re_2(CO)_6(\mu$-CO) (heating time 2 h). Both compounds form dark brown crystals.

Isomer A: Yield: 10%.
^1H NMR ($CDCl_3$): −5.14 (ReHW; J(W,H) = 98); 1.23 to 1.27, 1.42 to 1.52, 1.75 to 1.77 (m's, CH_2); 1.99 ($(CH_3)_5$); 2.29 to 2.36, 2.62 to 2.66, 3.00 to 3.04, 3.13 to 3.18 (m's, CH_2); 5.87, 6.57 (d's, CH; J = 7.1). − $^{13}C\{^1H\}$ NMR ($CDCl_3$): 11.6 ($(CH_3)_5$); 23.2, 27.0, 28.9, 47.4 (CH_2); 104.1 (C_5); 105.4, 119.8 (CHCH); 132.4, 163.4 (WCH; J(W,C) = 67 and 68); 190.4, 191.8, 193.4, 199.0 (ReCO); 235.6 (WCO; J(W,C) = 135). − IR (CH_2Cl_2): 1853, 1903, 1917, 1939, 1956, 2001, 2035 (ν(CO)). − FAB MS: $[M]^+$.
Single-crystal data: triclinic; space group $P\bar{1}-C_i^1$ (No. 2); a = 10.947(1), b = 11.074(1), c = 11.167(2) Å, α = 101.38(1)°, β = 104.10(1)°, γ = 94.65(1)°; Z = 2 molecules per unit cell, D_{calc} = 2.591 g/cm^3. An illustration of the molecular structure is shown in **Fig. 122** [2].

Isomer B: Yield: 3%.
^1H NMR ($CDCl_3$): −3.94 (ReHW; J(W,H) = 100); 1.13 to 1.17, 1.23 to 1.25, 1.36 to 1.40, 1.71 to 1.89 (m's, 1 H of CH_2); 2.03 ($(CH_3)_5$); 2.74 to 2.77, 2.90 to 2.97, 3.60 to 3.64 (m's, 1 H of CH_2), 3.65 (d, CHCH), 3.73 to 3.78 (m, CH_2), 6.15 (d, CHCH; J = 6.8). − IR (CH_2Cl_2): 1799, 1900, 1940, 1988, 2002, 2034 (ν(CO)). − FAB MS: $[M]^+$.
Single-crystal data of **$(CO)_6Re_2(\mu$-H$)(\mu$-CO$)W(C_5(CH_3)_5)(\mu$-CHCH$(C_6H_8)) \cdot CHCl_3$**: monoclinic; space group $P2_1/c-C_{2h}^5$ (No. 14); a = 12.359(5), b = 9.814(4), c = 25.372(8) Å, β = 97.85(3)°; Z = 4 formula units per unit cell, D_{calc} = 2.427 g/cm^3. The molecular structure is illustrated in **Fig. 123**.
There is a pentagonal-bipyramidal arrangement of the central Re_2WC_4 core. The ReC_4 plane is perpendicular to the plane defined by the metal atoms [2].

References on p. 252

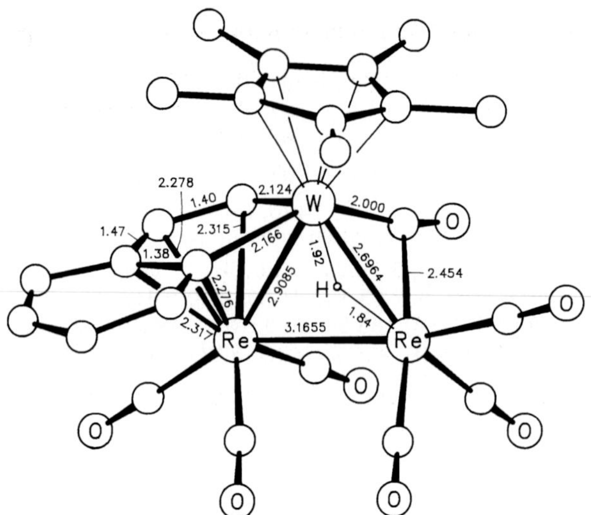

Fig. 122. Molecular structure of
$(CO)_6Re_2(\mu\text{-}H)(\mu\text{-}CO)W(C_5(CH_3)_5)(\mu\text{-}CHCH(C_6H_8))$ (isomer A) [2].

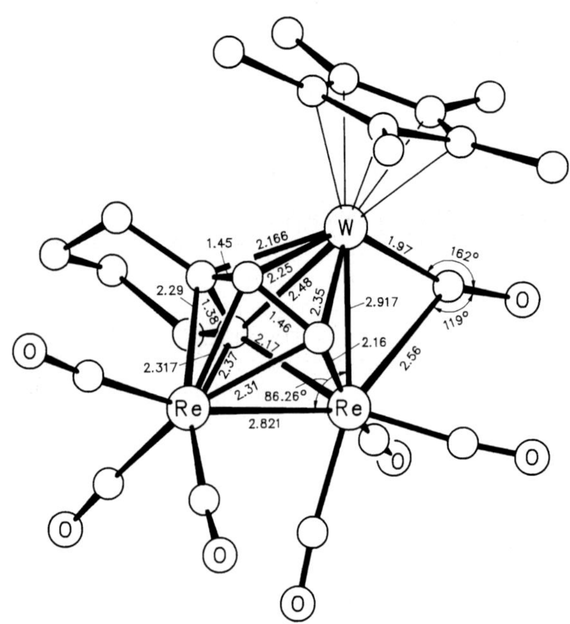

Fig. 123. Molecular structure of
$(CO)_6Re_2(\mu\text{-}H)(\mu\text{-}CO)W(C_5(CH_3)_5)(\mu\text{-}CHCH(C_6H_8))$ (isomer B) [2].

References:

[1] Cheng, P.-S.; Chi, Y.; Peng, S.-M.; Lee, G.-H. (Organometallics **12** [1993] 250/2).

[2] Peng, J.-J.; Horng, K.-M.; Cheng, P.-S.; Chi, Y.; Peng, S.-M.; Lee, G.-H. (Organometallics **13** [1994] 2365/74).

2.6 Compounds with Ligands Bonded to Rhenium by Six C Atoms (^{6}L Compounds)

The compounds described in this chapter can be roughly subdivided into two classes. In the first, one ^{6}L ligand or each of the two ^{6}L ligands is connected to only one Re atom ("terminal"). In the second, one ^{6}L ligand is bonded to both Re atoms ("bridging").

2.6.1 Compounds with Terminal ^{6}L Ligands

2.6.1.1 Compounds of the Type $(\pi\text{-}C_6H_nR_{6-n})_2Re_2(\mu\text{-}CR'R'')(\mu\text{-}H)_2$

General. This section covers compounds consisting of two $(C_6H_nR_{6-n})$Re fragments linked by a formal Re≡Re triple bond which is supported by two H atoms and one alkylidene ligand CR′R″ (see Formula I). The ^{6}L ligand most frequently encountered is benzene (R = H); other ligands at this position are toluene, xylene, and mesitylene. The bridging alkylidene ligand in most compounds is of the type CHR″; thus, these derivatives have R′ = H.

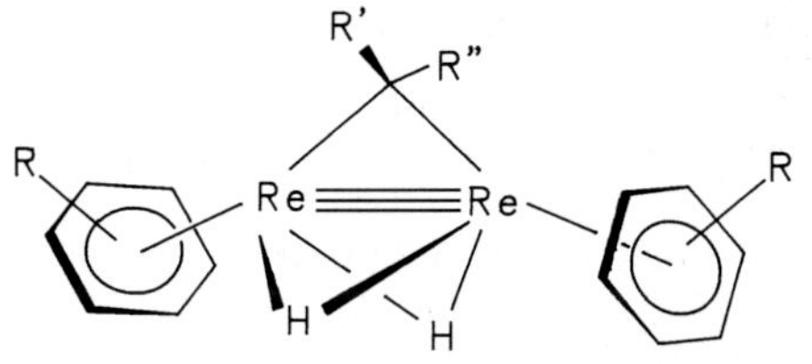

Preparation. All compounds were prepared by the following metal vapor strategy:

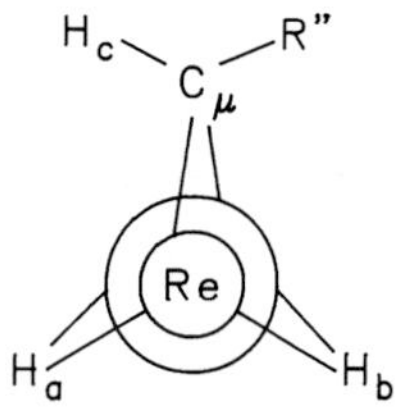

Method I: Co-condensation of Re atoms with
a. a 1:1 mixture consisting of benzene and either an open-chain (C_2H_6, n-C_3H_8, n-C_4H_{10}, i-C_4H_{10}, $C(CH_3)_4$, $Si(CH_3)_4$) or a cyclic alkane (C_3H_6, C_5H_{10}, C_6H_{12}) [2, 7].
b. pure toluene, p-xylene, mesitylene, or ethylbenzene [1, 5].
Workup in both cases was by extraction of the slurry with THF or toluene, filtration, evaporation, and sublimation at ca. 120 °C/10^{-4} Torr (or repeated crystallizations). Yields generally varied between 5 and 20% [1, 2, 5, 7].

At least photochemically induced steps have to be considered for the reaction mechanism, since the co-condensate was continuously exposed to the radiation from the molten rhenium (ca. 3600 °C), which is approximately equivalent to the radiation from a W lamp [2, 7]. Aromatic and aliphatic C-H bonds behave very differently under the reaction conditions: Aromatic C-H bonds always withstand Re atom attack, otherwise any CH_2 or CH_3 group is potentially subject to C-H activation. Thus, when using propane or higher homologues or arenes with aliphatic side chains, isomers are obtained due to C-H cleavage of all possible CH_2 or CH_3 groups. In contrast, co-condensation of Re atoms with benzene alone does not yield organometallic products [2, 5].

Spectroscopy. All ^{1}H NMR spectra display one or two resonances at ca. −5.5 ppm due to the bridging hydrides H_a and H_b (see the Newman projection in Formula II for atom labeling). When R′ = R″ (Nos. 9, 11, 12), H_a and H_b are equivalent. When R′ ≠ R″, coupling is observed between the hydrides with J(H-a,b) ca. 4 Hz. When the bridging alkylidene ligand is of the type CHR″ (R′ = H_c), the spectra additionally show a resonance at ca. 12.5 to 13.5 ppm due

References on p. 262

to H_c, and coupling is observed between H_a, H_b, and H_c such that $\delta(H_c)$ is characterized by two different coupling constants of typically 1.0 and ca. 4.5 Hz. It appears that with the exception of No. 2 J(H-a,c) is always larger than J(H-b,c) [5, 7].

In the ^{13}C NMR spectra the signal due to C_μ shows up at ca. 165 to 195 ppm.

Only the most intense mass peaks are listed in the table.

Table 11
Compounds of the Type $(\pi\text{-}C_6H_nR_{6-n})_2Re_2(\mu\text{-}CR'R'')(\mu\text{-}H)_2$.
An asterisk indicates further information at the end of the table.
For explanations, abbreviations, and units see p. X.

No.	$C_6H_nR_{6-n}$	CR′R″	method of preparation (yield) properties and remarks
*1	C_6H_6	CH_2	model compound for electron structure calculations; see "Further information" [7]
2	C_6H_6	$CHCH_3$	Ia (2%) light green compound [7] ^1H NMR (C_6D_6): −5.6, −5.3 (dd's, H_a and H_b; J(H-a,b) = 4.0); 3.2 (d, CH_3; J(CH_3,H_c) = 6.2), 4.2 (s, C_6H_6), 12.5 (dd, H_c; J(H-a,c) = 2.2, J(H-b,c) = 4.4) [7]; similar in [2] ^{13}C NMR (C_6D_6): 52.1 (q, CH_3), 69.7 (d, C_6H_6), 165.1 (d, C_μ); J(C,H) = 128, 175, 145, resp. [7]; $\delta(C_\mu)$ also in [2] MS: $[M - 2]^+$ [7]
3	C_6H_6	CHC_2H_5	Ia (along with No. 9 in the ratio 4:3 [2]; combined yield: 2% [7]); also formed along with No. 13 when reacting Re atoms with benzene/cyclopropane [7] light green compound [7] ^1H NMR (toluene-d_8): −5.4 (br, $H_{a,b}$), 1.3 (t, CH_3; J(H,H) = 7), 3.1 (qui, CH_2; J(H,H) = 7), 4.2 (s, C_6H_6), 12.7 (ddt, H_c; J(H-a,c) = 6, J(H-b,c) = 2) [7]; similar in [2] ^{13}C NMR (toluene-d_8): 20.2 (q, CH_3), 47.6 (t, CH_2), 70.0 (d, C_6H_6), 176.0 (d, C_μ); J(C,H) = 120, 128, 126, 125, resp. [7]; $\delta(C_\mu)$ also in [2] MS: $[M - 1]^+$ [7]
4	C_6H_6	CHC_3H_7-n	Ia (along with No. 10 in the ratio 2:1) [2, 7]; yield: ca. 10% [7] light green solid [7] ^1H NMR (C_6D_6): −5.26 (br, $H_{a,b}$), 1.11 (t, CH_3; J(H,H) = 7), 1.60 (sext, C**H_2**CH_3), 3.15 (q, CHC**H_2**CH_2; J(H,H) = 6.6), 4.24 (s, C_6H_6), 12.76 (ddt, H_c; J(H-c,a or b) = 3.0 or 4.0, J(H-c,CH_2) = 7) [7]; see also [2] ^{13}C NMR (C_6D_6): 14.8 (q, CH_3), 30.2 (t, **C**H_2CH_3), 68.5 (t, **C**$H_2C_2H_5$), 69.7 (d, C_6H_6), 183.0 (d, C_μ);

Table 11 (continued)

No.	$C_6H_nR_{6-n}$	CR′R″	method of preparation (yield) properties and remarks
			J(C,H) = 125, 126, 123, 177, 139, resp. [7]; but: 172.9 (C_μ) [2] MS: $[M - 1]^+$ [7]
5	C_6H_6	CHC_3H_7-i	Ia (ca. 10%) red compound [7] ^{1}H NMR (C_6D_6): −5.34, −5.04 (br, H_b and H_a); 1.43 (d, CH_3; J(CH_3,CH) = 6.8), 2.5 (d of sept, C**H**$(CH_3)_2$; J(CH,H_c) = 3.5), 4.24 (s, C_6H_6), 13.3 (ddd, H_c; J(H-a,c) = 4.5, J(H-b,c) = 1.5) [7]; similar in [2] ^{13}C NMR (C_6D_6): 28.7 (q, CH_3), 55.4 (d, **C**H$(CH_3)_2$), 70.9 (d, C_6H_6), 185.3 (d, C_μ); J(C,H) = 118, 124, 174, 150, resp. [7]; δ(C_μ) also in [2] MS: $[M - 4]^+$ [7]
*6	C_6H_6	CHC_4H_9-t	Ia (15%) red crystals [7]; volatile, yellow-brown crystals [2] ^{1}H NMR (C_6D_6): −5.15, −4.27 (br, H_b and H_a); 1.30 (s, CH_3), 4.31 (s, C_6H_6), 13.64 (dd, H_c; J(H-a,c) = 5.1, J(H-b,c) = 1.5) [7]; similar in [2] ^{13}C NMR (C_6D_6): 35.0 (q, CH_3), 53.3 (s, **C**$(CH_3)_3$), 70.3 (d, C_6H_6), 193.8 (d, C_μ); J(C,H) = 120, none, 170, 135, resp. [7]; similar in [2] He photoelectron spectrum: I.E. (assignm.) = 5.91 (46, a_1), 6.55 (47 to 49, b_2), 7.11 (50, a_1, $\sigma_{Re\text{-}Re}$), 7.65 (51 to 53), 9.15 (54 to 57, $\pi(C_6H_6)$) eV (compare with energy level numeration in Scheme 5, see p. 259) MS: $[M - 2]^+$ [7]
7	C_6D_6	CHC_4H_9-t	Ia (by using a C_6D_6/neopentane mixture; yield: 10%) red compound ^{1}H NMR (C_6D_6): −5.26, −4.41 (br, H_b and H_a); 1.37 (s), 13.66 (dd, H_c; J(H-a,c) = 5.1, J(H-b,c) = 1.5) ^{2}H NMR (C_6H_6): 4.30 (s, C_6D_6) ^{13}C {^{1}H} NMR (C_6D_6): 34.9, 55.5 (s's); 70.5 (t, C_6D_6; J(C,D) = 27), 193.8 (s, C_μ) MS: $[M - 2]^+$ [7]
8	C_6H_6	$CHSi(CH_3)_3$	Ia (ca. 10%) dark green solid [7] ^{1}H NMR (C_6D_6): −4.15, −3.33 (br, H_b and H_a); 0.35 (s, CH_3), 4.19 (s, C_6H_6), 12.86 (dd, H_c; J(H-a,c) = 5.0, J(H-b,c) = 3.0) [7]; similar in [2] ^{13}C NMR (C_6D_6): 4.38 (q, CH_3), 70.3 (d, C_6H_6), 162.9 (d, C_μ); J(C,H) = 118, 176, 122, resp. [7]; δ(C_μ) also in [2] MS: $[M - 1]^+$ [7]

References on p. 262

Table 11 (continued)

No.	$C_6H_nR_{6-n}$	CR′R″	method of preparation (yield) properties and remarks
9	C_6H_6	$C(CH_3)_2$	Ia (along with No. 3 in the ratio 3:4 [2]; combined yield: 2% [7]) light green solid [7] ^{1}H NMR (toluene-d_8): −5.8 (complex m, $H_{a,b}$), 2.8 (s, CH_3), 4.1 (s, C_6H_6) [7]; similar in [2] ^{13}C NMR (toluene-d_8): 60.3 (q, CH_3), 69.7 (d, C_6H_6; J(C,H) = 129 and 169), 175.3 (s, C_μ) [7]; $\delta(C_\mu)$ also in [2] MS: $[M - 1]^+$ [7]
10	C_6H_6	$C(CH_3)C_2H_5$	Ia (along with No. 4 in the ratio 1:2 [2, 7]) light green solid [7] ^{1}H NMR (C_6D_6): −5.81, −5.41 (br, $H_{a,b}$); 1.41 (t, CH_3; J(H,H) = 7.0), 2.50 (q, CH_2), 2.93 (s, CH_3), 4.19 (s, C_6H_6) [7]; $\delta(H_{a,b})$ and $\delta(C_6H_6)$ also in [2] ^{13}C NMR (C_6D_6): 20.8 (q, CH_3), 60.2 (t, CH_2), 60.9 (q, CH_3), 69.4 (d, C_6H_6), 173.0 (s, C_μ); J(C,H) = 127, 129, 122, 178, none, resp. [7]; but: 181.6 (C_μ) [2] MS: $[M - 1]^+$ [7]
11	C_6H_6	$C(CH_2)_4$	Ia (ca. 5%) [7]; also in trace amounts by co-condensation of Re atoms and benzene/cyclopentene [8] light green solid [7] ^{1}H NMR (C_6D_6): −5.62 (br, $H_{a,b}$); 1.36, 2.7 (m's, 7 lines, CH_2); 4.18 (s, C_6H_6) [7]; $\delta(H_{a,b})$ also in [2] ^{13}C NMR (C_6D_6): 30.67 (t, CH_2), 69.9 (d, C_6H_6), 73.3 (t, CH_2), 188.2 (s, C_μ); J(C,H) = 126, 174.1, 130, none, resp. [7]; $\delta(C_\mu)$ similar in [2] MS: $[M]^+$ [7]
12	C_6H_6	$C(CH_2)_5$	Ia (ca. 5%) green solid [7] ^{1}H NMR (C_6D_6): −5.38 (br, $H_{a,b}$); 1.48 (m, 10 lines), 1.56 (m, 8 lines), and 3.15 (m, 5 lines; all CH_2); 4.31 (s, C_6H_6) [7]; $\delta(H_{a,b})$ and $\delta(C_6H_6)$ similar in [2] ^{13}C NMR (C_6D_6): 27.6, 28.3 (t's, CH_2); 69.4 (d, C_6H_6), 70.2 (t, CH_2), 189.6 (s, C_μ); J(C,H) = 120, 127, 175, 132, none, resp. [7]; $\delta(C_\mu)$ also in [2] MS: $[M - 614]^+$ [7]
13	C_6H_6	H_cC_μ–CH_d=CH_eH_f	Ia (with cyclopropane, yield: ca. 5% (along with No. 3)) red solid, not isolated in a pure form ^{1}H NMR (C_6D_6): −5.31, −5.29 (br, $H_{a,b}$); 4.19 (s, C_6H_6), 4.2 (m, 6 lines, H_f), 4.4 (m, 10 lines, H_e), 7.3 (m, 7 lines, H_d), 12.3 (ddd, H_c; J(H-a,c) = 3.5, J(H-b,c) = 1.5, J(H-c,d) = 10.7) [7]

References on p. 262

Table 11 (continued)

No.	$C_6H_nR_{6-n}$	CR′R″	method of preparation (yield) properties and remarks
*14	$C_6H_5CH_3$	CHC_6H_5	Ib [1, 5] (17%) [5] deep red [1, 5], sublimes at ca. 90°C/10^{-4} Torr [1], 120°C/10^{-4} Torr [5] ^{1}H NMR (C_6D_6): −5.14 (dd), −5.09 (t, H_b and H_a; J(H-a,b) = 4.3), 2.00 (s, CH_3); $C_6\mathbf{H_5}CH_3$: 3.80 (d, 2 H; J = 6.7), 4.04 (t, 2 H; J = 6.0), 4.13 (5-line m, 4 H), 4.31 (t, 2 H; J = 6.9); 6.9 (t, 1 H; J = 7.0) and 7.2 (br, 4 H; both C_6H_5); 12.45 (dd, H_c; J(H-a,c) = 4.3, J(H-b,c) = 1.3) [5]; (acetone-d_6, −30°C): −5.95 (dd), −5.55 (t, H_a and H_b; J(H-a,b) = 4.3), 2.00 (s, CH_3); 3.95 (d), 4.2 (t), 4.31 (t), 4.35 (t), 4.5 (t; all 2 H, $C_6\mathbf{H_5}CH_3$; J = 6.7, 6.0, 6.9, 6.9, 6.9, resp.); 6.8 (d), 6.95 (t), 7.2 (t), 7.5 (t), 8.2 (d; all 1 H, $CHC_6\mathbf{H_5}$; J = 6.9, 6.4, 7.0, 6.4, 7.1, resp.); 12.5 (dd, H_c; J(H-a,c) = 4.3, J(H-b,c) = 1.3) [5] ^{13}C NMR (C_6D_6): 20.9 (q, CH_3; J(C,H) = 128); 68.3, 70.3, 71.8, 72.4, 72.9 (d's, $\mathbf{C_6}H_5CH_3$; J(C,H) = 175, 174, 175, 174, 177, resp.); 86.8 (s, $\mathbf{C}CH_3$); 124.0 (d, C-4 of C_6H_5; J(C,H) = 152); 125.0 to 130.0 (br, C_6H_5); 164.7 (d, C_μ; J(C,H) = 142), 169.6 (s, C_{ipso}) [5] IR: 1600 (νReHRe)) [5]; perdeuterated analog from toluene-d_8 did not show any IR bands assignable to terminal or bridging ReD modes [1] MS: $[M-2]^+$ [5]
15	$C_6H_5C_2H_5$	$CHCH_2C_6H_5$	Ib (along with No. 16; not completely separated; combined yield: 5%) red-brown solid [5] ^{1}H NMR (C_6D_6): −5.3, −5.2 (dd's, H_a and H_b; J(H-a,b) = 4.5); 1.1 (t, $CH_2\mathbf{CH_3}$), 2.3 (q, $C\mathbf{H_2}CH_3$; J = 7.5); 3.9 (d), 4.10 (m), 4.25 (t), 4.3 (d), 4.4 (t; all 2 H, $C_6\mathbf{H_5}C_2H_5$; J = 5.7); 4.5 (d, $CH_c\mathbf{CH_2}$; J(H-c,CH_2) = 7.0), 7.0 to 7.3 (m, C_6H_5), 12.5 (ddt, H_c; J(H-a,c) = 4.0, J(H-b,c) = 1.2) [5]; similar in [1] ^{13}C NMR (C_6D_6): 18.1 (q; CH_3; J(C,H) = 126), 28.7 (t; $\mathbf{C}H_2CH_3$; J(C,H) = 126); 67.9, 69.2, 70.9, 71.0, 71.7 (d's; J(C,H) = 172, 170, 175, 173, 171, resp.) and 73.7 (s; all $\mathbf{C_6}H_5C_2H_5$); 93.9 (t, $\mathbf{C}H_2C_6H_5$; J(C,H) = 135), 125 to 130 (C_6H_5), 165.5 (d, C_μ) [5] IR (pentane): 1603 (ν(ReHRe)) [5] MS: $[M-2]^+$ [5]
16	$C_6H_5C_2H_5$	$C(CH_3)C_6H_5$	Ib (along with No. 15; combined yield: 5%) red-brown solid [5]

References on p. 262

Table 11 (continued)

No.	$C_6H_nR_{6-n}$	CR′R″	method of preparation (yield) properties and remarks
16 (continued)			^1H NMR (C_6D_6): −5.40, −5.10 (d's, H_a and H_b; J(H-a,b) = 4.5); 1.1 (t; $CH_2\mathbf{CH_3}$), 2.3 (q, $\mathbf{CH_2}CH_3$; J = 7.5); 3.1 (s, $C_\mu\mathbf{CH_3}$); 3.9 (d, 2 H; J = 5.7), 4.0 (m, 6 H), 4.1 (m, 2 H; all $C_6\mathbf{H_5}C_2H_5$); 8.1 (br, C_6H_5) [5]; almost identical in [1] ^{13}C NMR (C_6D_6): 18.1 (q, $CH_2\mathbf{C}H_3$), 28.7 (t, CH_2), 58.2 (q, $C_\mu\mathbf{C}H_3$; J(C,H) = 126, 126, 130, resp.); 67.0, 68.0, 69.9, 73.2, 73.3 (d's; J(C,H) = 172, 173, 175, 173, 171, resp.) and 73.7 (s; all $\mathbf{C_6}H_5C_2H_5$); 124.2, 127.1 (d's, C_6H_5; J(C,H) = 154, 150); 169.4, 172.6 (s's, C_{ipso}, C_μ or vice versa) [5] IR (pentane): 1603 (ν(ReHRe)) MS: $[M - 2]^+$ [5]
*17	1,4-$(CH_3)_2C_6H_4$	$CHC_6H_4CH_3$-4	Ib [1, 5] (18%) [5] deep red [1], dark orange [5] solid sublimes at ca. 90°C/10^{-4} Torr [1] ^1H NMR (C_6D_{12}): −5.50, −5.40 (dd's, H_a and H_b; J(H-a,b) = 3.4); 2.1, 2.2 (s's, 4 and 1 CH_3); 3.8, 4.25 (d's, $C_6\mathbf{H_4}(CH_3)_2$; J = 5.5); 6.7, 6.8, 6.9, 8.1 (all br s, $C_6\mathbf{H_4}CH_3$); 12.10 (dd, H_c; J(H-a,c) = 3.9, J(H-b,c) = 1.3) [5]; $H_{a,b,c}$ also in [1] ^{13}C NMR (C_6D_{12}): 20.0 (q, $C_6H_4(\mathbf{C}H_3)_2$), 20.9 (q, $C_6H_4\mathbf{C}H_3$; J(C,H) = 130 and 124); 71.1, 72.4 (d's; J(C,H) = 170) and 82.8 (s; all $\mathbf{C_6}H_4(CH_3)_2$); 126.3, 127.6, 128.7 (d's, $\mathbf{C_6}H_4CH_3$; J(C,H) = 156, 159, 155, resp.); 131.8 (s) and 134.4 (d, $\mathbf{C_6}H_4CH_3$; J(C,H) = 159); 164.3 (d, C_μ; J(C,H) = 157), 166.6 (s, C_{ipso}) [5] IR: 1600 (ν(ReHRe)) MS: $[M - 2]^+$ solutions in acetone-d_6 show ca. 50% H/D exchange with the bridging hydrides and H_c within 12 h at room temperature and statistical exchange after 3 d; a kinetic study showed that the rates on all three positions are equal; resonances due to partially deuterated compounds could be observed [5]
*18	1,3,5-$(CH_3)_3C_6H_3$	$CHC_6H_3(CH_3)_2$-3,5	Ib [1, 5] (15%) [5] deep red [1], red-brown [5] solid sublimes at ca. 90°C/10^{-4} Torr [1] ^1H NMR (C_6D_6): −5.25, −5.03 (dd's; H_b and H_a; J(H-a,b) = 4.0); 2.2, 2.5 (s's, 7 and 1 CH_3); 4.0 (s, $C_6\mathbf{H_3}(CH_3)_3$); 6.7, 6.8, 8.2 (s's, $C_6\mathbf{H_3}(CH_3)_2$); 12.45 (dd, H_c; J(H-a,c) = 4.4, J(H-b,c) = 0.9) [5]; almost identical in [1]

References on p. 262

Table 11 (continued)

No.	$C_6H_nR_{6-n}$	CR'R"	method of preparation (yield) properties and remarks
			^{13}C NMR (C_6D_6): 18.7, 20.7, 21.8 (q's, 1 + 6 + 1 CH_3; J(C,H) = 125, 125, 126, resp.); 71.0 (d; J(C,H) = 166) and 86.2 (s, both $\mathbf{C_6}H_3(CH_3)_3$); 125.2, 127.1, 134.0 (d's, $\mathbf{C_6}H_3(CH_3)_2$; J(C,H) = 148, obscured by solvent, 150, resp.); 136.1, 137.0 (s's, $\mathbf{C_6}H_3(CH_3)_2$); 167.0 (d, C_μ; J(C,H) = 139), 170.7 (s, C_{ipso}) [5]; nearly identical in [1] MS: $[M - 2]^+$ [5]

*Further information:

$(C_6H_6)_2Re_2(\mu\text{-}CH_2)(\mu\text{-}H)_2$ (Table **11**, No. **1**). The electronic structure was calculated using extended Hückel methods. The MO levels, resulting from formally interacting the $(C_6H_6)_2Re_2$ fragment with CH_2 and 2 H atoms (intramolecular parameters were taken from a structural study on No. 6), is shown in Scheme 5. The Re-Re overlap population is 0.3837. The highest occupied molecular orbital correlates with a δ orbital of $(C_6H_6)_2Re_2$. The influence of ring bending on the total energy was found to reach a minimum at 10.5° for the angle between the Re-Re axis and the perpendicular axis through the benzene rings [7].

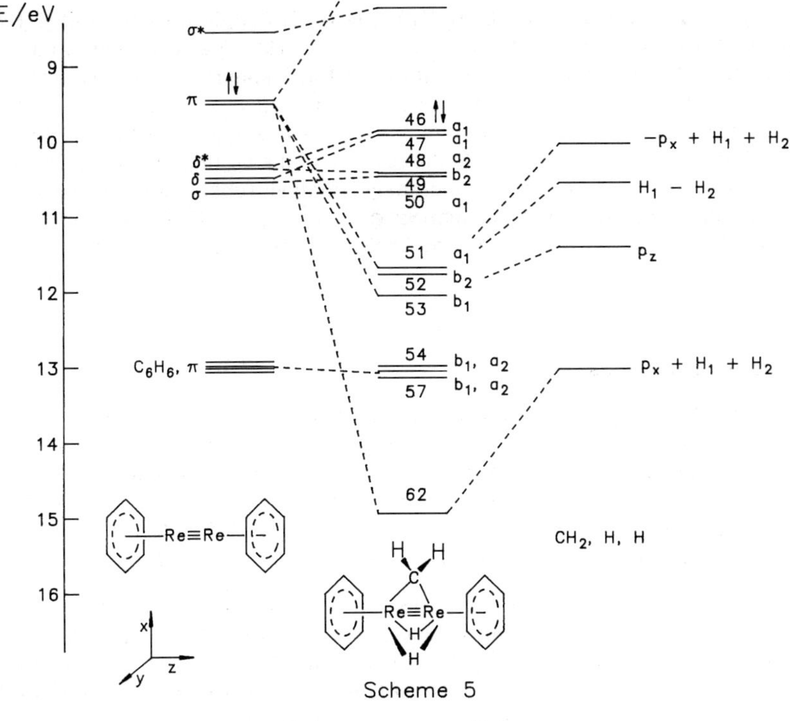

Scheme 5

References on p. 262

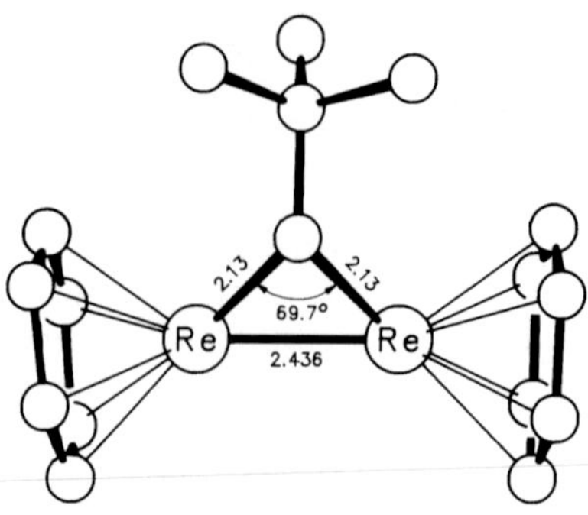

Fig. 124. Molecular structure of $(C_6H_6)_2Re_2(\mu\text{-}CHC_4H_9\text{-t})(\mu\text{-}H)_2$ (one orientation of one independent molecule) [2, 7].

$(C_6H_6)_2Re_2(\mu\text{-}CHC_4H_9\text{-t})(\mu\text{-}H)_2$ (Table **11**, No. **6**) crystallizes in the orthorhombic space group Pbca$-D_{2h}^{15}$ (No. 61) with a = 30.034(3), b = 17.284(3), c = 12.353(2) Å; Z = 16 molecules per unit cell, D_{calc} = 2.49 g/cm^3. The bridging CHC_4H_9-t groups of both molecules in the asymmetric unit exhibit disorder and can be resolved into components involving a common C [2, 7]. **Fig. 124** shows one orientation of one of the symmetry-independent molecules. The benzene rings are tilted away from the CHC_4H_9-t ligand, and the mean angle between the normals to the planes of the C_6H_6 ligands is 162.3° [7].

Treatment with excess $P(CH_3)_3$ at room temperature yielded a 1:1 inseparable mixture of $C_6H_6Re(P(CH_3)_3)_2H$ and $C_6H_6Re(P(CH_3)_3)_2CH_2C_4H_9$-t [7].

$(C_6H_5CH_3)_2Re_2(\mu\text{-}CHC_6H_5)(\mu\text{-}H)_2$, **$(C_6H_4(CH_3)_2)_2Re_2(\mu\text{-}CHC_6H_4CH_3)(\mu\text{-}H)_2$**, and **$(C_6H_3(CH_3)_3)_2\text{-}Re_2(\mu\text{-}CHC_6H_3(CH_3)_2)(\mu\text{-}H)_2$** (Table **11**, Nos. **14**, **17**, and **18**). The low-temperature ^{1}H NMR spectrum of No. 14 and the room-temperature ^{1}H NMR spectra of Nos. 17 and 18 reveal that the H atoms bonded to the C_6 ring of the $\mu\text{-}CHC_6H_n(CH_3)_{5-n}$ (n = 5, 4, 3) moiety are diastereotopic. On raising the temperature the peaks assignable to these hydrogens coalesce. The behavior can be attributed to the rotation of the C_6 ring about the $CH-C_{ipso}$ axis. The barrier of rotation increases with increasing methyl substitution as indicated by the higher temperature required for coalescence (No. 17: T_{coal} = 45 °C, No. 18: T_{coal} = 75 °C [1, 5]). Values for $\Delta H^{\neq}$ were calculated to be 55.1, 64.8, and 78.0 kJ/mol for Nos. 14, 17, and 18, respectively [5].

2.6.1.2 Other Compounds

$C_6H_6Re(\mu\text{-}H)_2Re(C_6H_8)_2$ (see Formula I) formed as a by-product (main product was $C_6H_6Re(C_6H_8)H$) when Re atoms were co-condensed with either a benzene/cyclohexene (1:1) [3, 8] or a benzene/cyclohexa-1,3-diene (2:1) [8] mixture. Extraction of the slurry with petroleum ether followed by sublimation of the extract provided two products which could be separated by fractional crystallization from pentane [3, 8]. Yields were 1% when using cyclohexene and 2% when using cyclohexadiene [8]. Red crystals [3, 8].

^{1}H NMR spectrum (C_6D_6): δ = −8.61 (s), 1.70 (br s), 4.85 (C_6H_6) ppm; (toluene-d_8, −68 °C): δ = −8.6 (s, μ-H), 1.6 (br m, H_g), 2.0 (t, H-f,f′), 2.2 (d, H_e), 2.4 (dd, H_d), 2.6 (dt, H_c), 3.7 (br m, H_b), 4.8 (s, C_6H_6), 5.2 (dd, H_a) ppm; coupling constants (only quoted for J > 3 Hz): J(H-f,c) = J(H-f,g) = 7 Hz, J(H-d,e) = 6 Hz, J(H-a,d) = 5 Hz, J(H-c,g) = 13 Hz, J(H-c,f) =

References on p. 262

7 Hz, J(H-a,b) = 7 Hz. ^{13}C NMR spectrum (toluene-d_8): δ = 30.5 (br, C_g or C_f), 62.5 (C_6H_6), 70.0 (v br, C_6H_8) ppm; (toluene-d_8, −68 °C): δ = 29.6 (C-g,f or C-c,f′), 30.2 (C-c,f′ or C-g,f), 36.0 (C_b), 47.1 (C_e), 62.6 (C_6H_6), 67.8 (C_d), 75.4 (C_a) ppm; J(C,H) = 129, 129, 151, 155, 174, 173, and 172 Hz, respectively [8] (similar in [3]).

The compound crystallizes in the triclinic space group $P\bar{1}-C_i^1$ (No. 2) with a = 8.534(1), b = 11.957(2), c = 9.496(1) Å, α = 94.256(9)°, β = 116.147(9)°, γ = 109.06(1)°; Z = 2 molecules per unit cell, D_{calc} = 2.56 g/cm^3. The molecular structure is illustrated in **Fig. 125**. The compound formally is a mixed-valence compound with a d^4 and a d^6 occupation of the orbitals. The C-C bond lengths within the diene part of the C_6H_8 ligands are closely similar, suggesting delocalization [3, 8].

$(C_6H_6)_2Re_2(P(CH_3)_3)_4$ (see Formula II) was obtained with 30% yield (the by-product was $C_6H_6Re(P(CH_3)_3)_2C_6H_5$) when Re atoms were co-condensed with a benzene/$P(CH_3)_3$ (5:1) mixture. Afterwards, all volatile material was removed, and the residue was purified chromatographically on alumina using toluene and ether. Orange crystals [4, 6].

^1H NMR spectrum (C_6D_6): δ = 1.23 ("t", PCH_3), 4.26 (t, C_6H_6; J(P,H) = 1.5 Hz) ppm. $^{31}P\{^1H\}$ NMR spectrum (C_6D_6): δ = −45.8 ppm [4, 6].

Neither treatment with iodine nor interaction with Na/Hg yielded tractable products. Combination with K in THF, however, formed a deep red, very air- and water-sensitive solution which probably contained $K[C_6H_6Re(P(CH_3)_3)_2]$. This reaction could be facilitated by a catalytic amount of naphthalene. Interaction with HBF_4 achieved diprotonation to $[(C_6H_6)_2Re_2(P(CH_3)_3)_4H_2][BF_4]_2$ (following salt). This reaction could be reversed [4, 6].

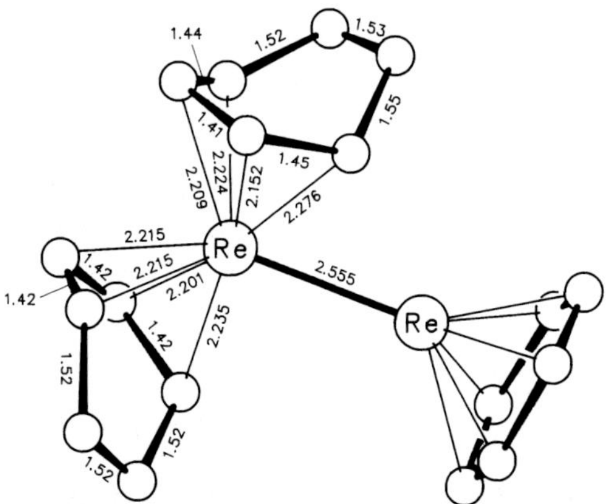

Fig. 125. Molecular structure of $C_6H_6Re(\mu\text{-}H)_2Re(C_6H_8)_2$ [3, 8].

References on p. 262

$[(C_6H_6)_2Re_2(P(CH_3)_3)_4H_2][BF_4]_2$ was obtained by treating $(C_6H_6)_2Re_2(P(CH_3)_3)_4$ (preceding compound) with HBF_4 [4, 6] in ethereal solution. An orange-pink-colored precipitate immediately formed which was recrystallized from acetone/H_2O. Yield: 90% [6]. The reaction could be reversed by treating the salt with NaOH or KH, causing reformation of the parent compound with more than 80% yield [6].

^1H NMR spectrum (acetone-d_6): $\delta = -8.74$ (t, ReH; $^2J(P,H) = 48.1$ Hz), 1.87 ("d", PCH_3), 6.06 (s, C_6H_6) ppm. $^{31}P\{^1H\}$ NMR spectrum (acetone-d_6): $\delta = -36.6$ ppm [4, 6].

References:

[1] Cloke, F. G. N.; Derome, A. E.; Green, M. L. H.; O'Hare, D. (J. Chem. Soc. Chem. Commun. **1983** 1312/3).

[2] Bandy, J. A.; Cloke, F. G. N.; Green, M. L. H.; O'Hare, D.; Prout, K. (J. Chem. Soc. Chem. Commun. **1984** 240/2).

[3] Green, M. L. H.; O'Hare, D.; Bandy, J. A.; Prout, K. (J. Chem. Soc. Chem. Commun. **1984** 884/6).

[4] Green, M. L. H.; O'Hare, D.; Wallis, J. M. (J. Chem. Soc. Chem. Commun. **1984** 233/4).

[5] Green, M. L. H.; O'Hare, D. (J. Chem. Soc. Dalton Trans. **1986** 2469/76).

[6] Green, M. L. H.; O'Hare, D.; Wallis, J. M. (Polyhedron **5** [1986] 1363/70).

[7] Green, J. C.; Green, M. L. H.; O'Hare, D.; Watson, R. R.; Bandy, J. A. (J. Chem. Soc. Dalton Trans. **1987** 391/402).

[8] Green, M. L. H.; O'Hare, D. (J. Chem. Soc. Dalton Trans. **1987** 403/10).

2.6.2 Compounds with Bridging ^6L Ligands

2.6.2.1 Compounds with Bridging Hydrocarbons

$(\mu\text{-}\eta^{3:3}\text{-}C_8H_{12})Re_2(CO)_8$ (see Formulas I, II). The photochemical reaction of $(CO)_{10}Re_2$ with butadiene (hexane, −85 or −35°C, ca. 20 min) provided a mixture which was separated by preparative HPLC on silica with hexane. Several products were isolated, among them 2 compounds with the composition $C_8H_{12}Re_2(CO)_8$. These were proven to be isomers (isomers A and B). Each isomer formed with ca. 8% yield at either temperature. Careful spectroscopic studies revealed them to contain an EE-configurated octa-1,6-diene-3,8-diyl ligand, bridging two $Re(CO)_4$ units. Since each metal atom can coordinate by one of two possible

$(CO)_4Re$ 8Z 1E 1Z 8E $Re(CO)_4$ — SR; $(CO)_4Re$ 8Z 8E 1E 1Z $Re(CO)_4$ — RS

I

$(CO)_4Re$ $(CO)_4Re$ 8Z 8E 1E 1Z — SS; 8Z 8E 1E 1Z $Re(CO)_4$ $Re(CO)_4$ — RR

II

References on pp. 266/7

sides, the diastereomeric pairs RS, SR and RR, SS can be expected. However, the available spectroscopic data did not allow the authors to unambiguously identify the diastereomers [3].

The IR spectra of both isomers (hexane) show ν(CO) bands at 1960, 1980, 1995, 2080 cm^{-1} (force constants: k_1 = 1582, k_2 = 1668, k_i = 30 N/m). Other data are as follows:

Isomer A: Colorless crystals. – ^{1}H NMR (CD_2Cl_2): 1.68 (ddd, H_{1Z}, H_{8Z}), 2.15 ("t", H-4,5), 2.68 (ddd, H_{1E}, H_{8E}), 2.89 (m, H_{3Z}, H_{6Z}), 4.76 (dt, H-2,7); coupling constants: J(H-1E,1Z) = 2.5, J(H-1E,2) = 11.5, J(H-1Z,2) = 7.2, J(H-2,3) = 11.0, J(H-1Z,3) = 0.8, J(H-1E,3) = 1.0. – ^{13}C {^{1}H} NMR (CD_2Cl_2): 25.61 (C-4,5), 65.67 (C-1,2), 101.77 (C-3,6), 120.10 (C-2,7); δ(CO) not recorded [3].

Isomer B: Pale yellow crystals. – ^{1}H NMR (CD_2Cl_2): 1.67 (ddd, H_{1Z}, H_{8Z}), 2.15 (m, H-4,5), 2.66 (ddd, H_{1E}, H_{8E}), 2.92 (m, H_{3Z}, H_{6Z}), 4.72 (dt, H-2,7) [3].

Re(CO)$_4$ ... (CO)$_4$Re
III

(CO)$_3$Re——Re(CO)$_4$
IV

[(C$_6$H$_5$)Re(CO)$_5$ / Re(CO)$_3$]$^+$
V

R, Re(CO)$_5$, H, Re(CO)$_3$
VI

(μ-C_6H_8)$Re_2(CO)_8$ (see Formula III) was one of the products obtained from the photochemical reaction of $(CO)_{10}Re_2$ with allene (hexane, −50°C, 35 min) with 6% yield. Colorless crystals.

^{1}H NMR spectrum (CD_2Cl_2): δ = 1.88 (H_Z) and 3.46 (H_E) ppm (AA'XX' pattern); ^{2}J(H-E,Z) = 2.7 Hz, ^{4}J(H-1E,3Z) and ^{4}J(H-1Z,3Z) = 1 Hz, ^{4}J(H-1E,3E) = 1.7 Hz. IR spectrum (n-hexane): 1971, 1987, 2000, 2091 (ν(CO)) cm^{-1} [9].

(μ-cyclo-C_7H_8)$Re_2(CO)_7$ (see Formula IV) possibly is the structure of one of the products obtained by irradiating $(CO)_{10}Re_2$ with excess cycloheptatriene in petroleum ether for 9 h. The compound was chromatographically separated (silica, petroleum ether/toluene) and recrystallized from i-C_3H_7OH. Yield: 1%. Yellow crystals [1]. In contrast, other researchers reported two isomers of (μ-$\eta^{2:1}$-C_7H_7)$Re_2(CO)_8$(μ-H) (see pp. 72/3) to be the products of photolysis of $(CO)_{10}Re_2$ with cycloheptatriene. A compound with the title composition was not mentioned [4, 6].

^{1}H NMR spectrum: δ = 3.8 (t, 2 H; J = 8 Hz); 4.7, 5.4 (m's, each 3 H) ppm. IR spectrum (n-hexane): 1945, 1962, 1995, 2010, 2080 (ν(CO)) cm^{-1}. The mass spectrum shows the series $[M - n\,CO]^+$ (n = 0 to 7) [1].

[(($CO)_5ReC_6H_5$)Re$(CO)_3$]BF_4 (see Formula V) precipitated when reacting $(CO)_5ReFBF_3$ with 2 equivalents $(CO)_5ReC_6H_5$ in methyl cyclohexane at 90°C for 16 h. After filtration the residue was extracted with pentane and ether, leaving the colorless product with 54% yield.

^{1}H NMR spectrum (CD_3NO_2): δ = 6.6 to 6.7 and 6.9 to 7.0 (m's, 3 and 2 H) ppm. ^{13}C {^{1}H} NMR spectrum (CD_3NO_2): δ = 98.4, 103.5, 114.9, 122.0 (C_6H_5); 181.6, 182.7, 186.2 (CO) ppm. IR spectrum (CH_3NO_2): 2036, 2053, 2068, 2087, 2152 (ν(CO)) cm^{-1} [7].

[$(CO)_5ReC_6H_5R$]Re$(CO)_3$ (see Formula VI). Compounds with this composition were prepared by treating [(π-C_6H_5R)Re$(CO)_3$]BF_4 (R = H, CH_3) with Na[$(CO)_5$Re] in THF at −65°C,

References on pp. 266/7

followed by warming the mixture to −30 °C and evaporation. The resulting powder was extracted with pentane to remove $(CO)_{10}Re_2$, leaving the crude product.

The compounds decompose in toluene at room temperature within a short time [7].

R = H: Yield: 73%. – Yellow powder, dec. >65 °C. – ^{1}H NMR (toluene-d_8, −72 °C): 2.69 (dd, H-2,6), 3.66 (dd, H-3,5), 3.87 (t, H-1), 4.98 (t, H-4). The ^{1}H NMR spectra recorded at room temperature and 52 °C only exhibit one feature at δ = 3.91 (br) and 4.03 (s), respectively. – IR (n-hexane): 1936, 1946, 1992, 2010, 2026, 2119 (ν(CO)) [7].

R = CH_3: Yield: 40%. – Lemon yellow powder, dec. >43 °C. – ^{1}H NMR spectroscopy revealed "ring slipping". – IR (pentane): 1932, 1943, 1991, 2005, 2019, 2112 (ν(CO)) [7].

$(C_6(C_6H_5)_6)Re_2(C_6H_5C{\equiv}CC_6H_5)(CO)_4$ (see Formula VII, R = C_6H_5) formed when reacting an excess of diphenylacetylene either photochemically with $(CO)_{12}Re_3(\mu\text{-}H)_3$ (pentane, −25 °C, ca. 18 h) [5] or thermally with $(CO)_{10}Re_2$ (bomb tube, hexane, 190 °C, 16 h) [2] or $(CO)_8Re_2(NCCH_3)_2$ (refluxing CH_2Cl_2, 24 h) [8]. In either case the resulting mixture was separated chromatographically on silica with hexane/CH_2Cl_2 mixtures. When starting from $(CO)_{12}Re_3(\mu\text{-}H)_3$, the title product was isolated with 4.4% yield [5]. When starting from $(CO)_8Re_2(NCCH_3)_2$, the yield was 18% [8]. The yield of the reaction involving $(CO)_{10}Re_2$ as starting material varied [2]. All these reactions simultaneously yielded $(C_4(C_6H_5)_4)Re_2(CO)_7$. The photochemical reaction and the thermal reaction starting from $(CO)_8Re_2(NCCH_3)_2$ gave also $(C_4(C_6H_5)_4)Re_2(C_6H_5C{\equiv}CC_6H_5)(CO)_5$ (see Table 7, Nos. 6 and 9, pp. 150/1), while the reaction starting from $(CO)_{10}Re_2$ simultaneously provided $(C_6(C_6H_5)_6)Re_2(CO)_6$ (following compound). The title compound forms red microcrystals [5].

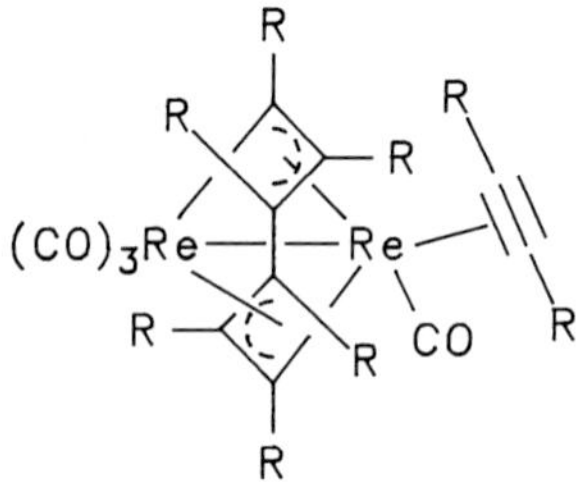

VII 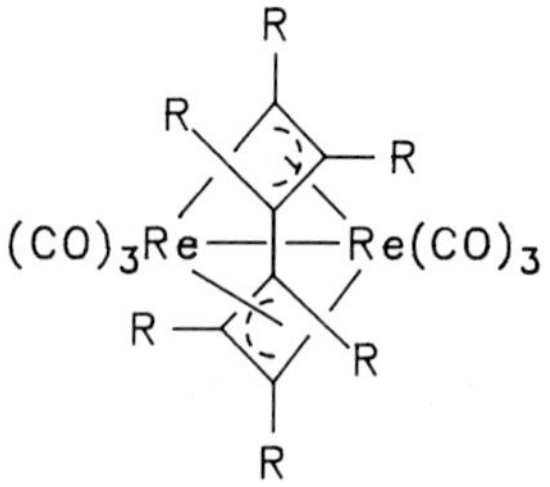 VIII 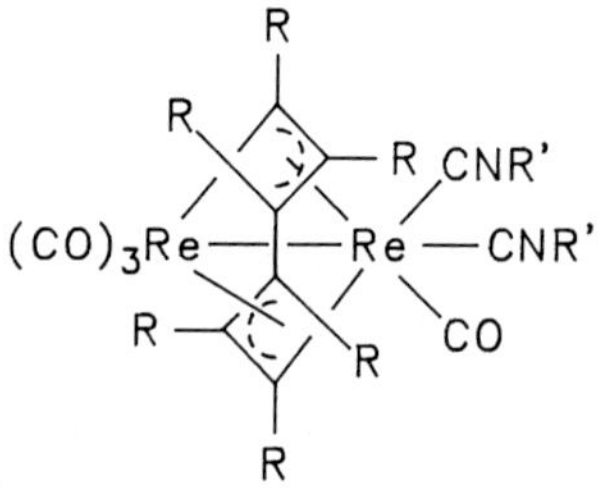 IX

^{1}H NMR spectrum (CD_2Cl_2): δ = 5.86 to 7.28 (m) ppm [2]. IR spectrum (n-hexane): 1942, 1962, 1997, 2035 (ν(CO)) cm^{-1} [5] (see also [8]; similar in CH_2Cl_2 [2, 5]). Mass spectrum: $[M - n\,CO]^+$ (n = 0 to 4), $[M - 4\,CO - C_2(C_6H_5)_2]^+$ [5].

Single crystals of the solvate **$(C_6(C_6H_5)_6)Re_2(C_6H_5C{\equiv}CC_6H_5)(CO)_4 \cdot 0.5\,C_6H_{14}$** are monoclinic with a = 17.048(4), b = 16.676(2), c = 18.569(4) Å, β = 106.70(2)°; space group $P2_1/c-C_{2h}^5$ (No. 14), Z = 4 molecules per unit cell, D_{calc} = 1.629 g/cm^3. The molecular structure is illustrated in **Fig. 126**. The terminal C atoms of the C_6 chain are bonded to both Re atoms, but each lies slightly closer to that Re atom bearing the acetylene ligand. Each is further incorporated into an η^3-bonded allyl ligand which is π-bonded to just one Re atom [5].

Treatment with R′NC (R′ = n-C_4H_9, t-C_4H_9, 4-$CH_3OC_6H_4$, 4-$CH_3C_6H_4SO_2CH_2$) liberated the π-bonded acetylene ligand to give compounds of the type $(C_6(C_6H_5)_6)Re_2(CO)_4(CNR')_2$ (see below). Pressurization with 5 atm of CO provided $(C_6(C_6H_5)_6)Re_2(CO)_6$ [2].

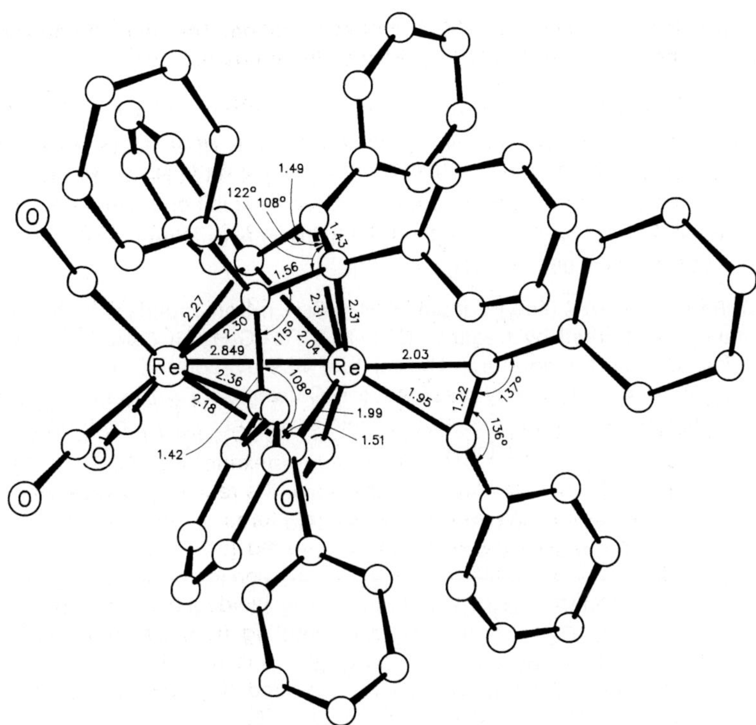

Fig. 126. Molecular structure of $(C_6(C_6H_5)_6)Re_2(C_6H_5C{\equiv}CC_6H_5)(CO)_4$ [5].

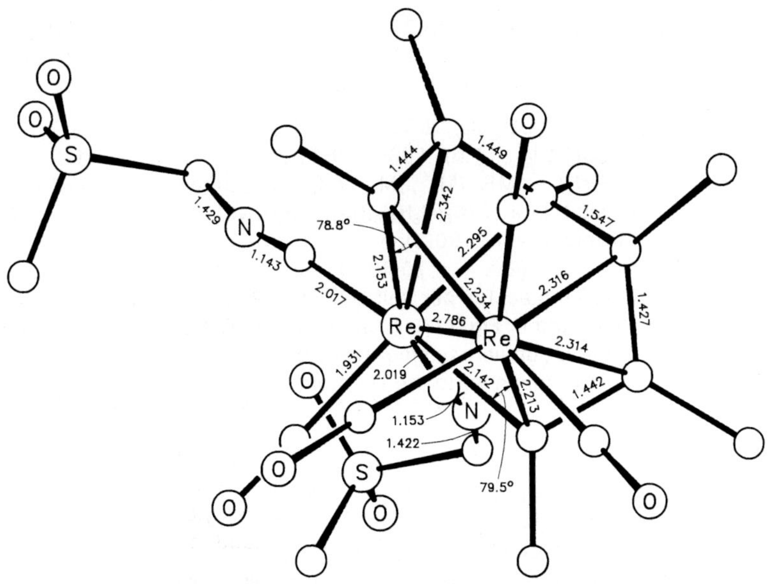

Fig. 127. Molecular structure of $(C_6(C_6H_5)_6)Re_2(CO)_4(CNCH_2SO_2C_6H_4CH_3\text{-}4)_2$ (C_6H_5 and $C_6H_4CH_3$-4 groups represented by C_{ipso}) [2].

References on pp. 266/7

$(C_6(C_6H_5)_6)Re_2(CO)_6$ (see Formula VIII, R = C_6H_5) was obtained with essentially quantitative yield upon pressurizing $(C_6(C_6H_5)_6)Re_2(C_6H_5C{\equiv}CC_6H_5)(CO)_4$ with CO (5 atm, hexane, 190 °C, 6 h). The product was separated by preparative TLC on silica using hexane/CH_2Cl_2 (7:3). The reaction could be reversed by heating the title compound with diphenylacetylene. The compound was also observed when treating $(CO)_{10}Re_2$ with $C_6H_5C{\equiv}CC_6H_5$ under the same conditions as given with the foregoing compound, but could not be isolated in pure form [2].

^{1}H NMR spectrum (CD_2Cl_2): δ = 7.2 (m) ppm. IR spectrum (n-hexane): 1932, 1944, 1960, 1981, 1986, 2029, 2051 (ν(CO)) cm^{-1} [2].

$(C_6(C_6H_5)_6)Re_2(CO)_4(CNR')_2$ (see Formula IX, R = C_6H_5). Compounds with this composition were instantaneously formed by treating $(C_6(C_6H_5)_6)Re_2(C_6H_5C{\equiv}CC_6H_5)(CO)_4$ with excess R'NC in CH_2Cl_2 at room temperature. The mixture was evaporated, and the residue was taken up in CH_2Cl_2 and separated by preparative TLC using CH_2Cl_2/hexane (3:7). The products were obtained in essentially quantitative yield as red crystalline materials. They were spectroscopically characterized as follows [2]:

R'	spectroscopic and structural data (for abbreviations and units see p. X)
C_4H_9-n	^{1}H NMR (CD_2Cl_2): 0.85 to 1.5 (m, 14 H); 3.21, 3.75 (CH_2); 7.06 (C_6H_5). – IR (CH_2Cl_2): 1880, 1921, 1947, 2015 (ν(CO)); 2141, 2169 (ν(CN)). – MS: $[M - n\ CO]^+$, $[M - n\ CO - RNC]^+$ (both: n = 0 to 4) [2].
C_4H_9-t	^{1}H NMR (CD_2Cl_2): 1.01, 1.36 (CH_3); 7.21 (C_6H_5). – IR (CH_2Cl_2): 1881, 1920, 1947, 2014 (ν(CO)); 2124, 2156 (ν(CN)). – MS: $[M - n\ CO]^+$, $[M - n\ CO - RNC]^+$ (both: n = 0 to 4) [2].
$C_6H_4OCH_3$-4	^{1}H NMR (CD_2Cl_2): 3.74, 3.77 (CH_3O); 7.02 (C_6H_5 + C_6H_4). – IR (CH_2Cl_2): 1888, 1925, 1951, 2017 (ν(CO)); 2097, 2133 (ν(CN)). – MS: $[M - n\ CO]^+$ (n = 0 to 4), $[M - n\ CO - RNC]^+$ (n = 1 to 4) [2].
$CH_2SO_2C_6H_4CH_3$-4	^{1}H NMR (CD_2Cl_2): 2.39, 2.48 (CH_3); 4.28, 4.59 (CH_2); 7.18 (C_6H_5). – IR (CH_2Cl_2): 1910, 1930, 1957, 2020 (ν(CO)); 2114, 2144 (ν(CN)). – MS: $[M]^+$, $[M - n\ CO - C_2(C_6H_5)_2 - RNC]^+$ (n = 0 to 4). Crystal data of **$(C_6(C_6H_5)_6)Re_2(CO)_4(CNCH_2SO_2C_6H_4CH_3\text{-}4)_2 \cdot CH_2Cl_2$**: triclinic, a = 13.190(3), b = 14.066(3), c = 17.554(5) Å, α = 83.06(2)°, β = 75.75(2)°, γ = 70.40(2)°, space group $P\bar{1}-C_i^1$ (No. 2); Z = 2 molecules per unit cell, D_{calc} = 1.670 g/cm^3 (structure illustrated in **Fig. 127**). It can be seen that the terminal C atoms of the C_6 chain asymmetrically bridge the Re atoms such that both atoms are closer to the Re atom bearing the isocyanide ligands [2].

References:

[1] Davis, R.; Ojo, I. A. O. (J. Organomet. Chem. **110** [1976] C 39/C 41).

[2] Mays, M. J.; Prest, D. W.; Raithby, P. R. (J. Chem. Soc. Dalton Trans. **1981** 771/6).

[3] Franzreb, K.-H.; Kreiter, C. G. (Z. Naturforsch. **37b** [1982] 1058/69).

[4] Franzreb, K.-H.; Kreiter, C. G. (J. Organomet. Chem. **246** [1983] 189/95).

[5] Pourreau, D. B.; Whittle, R. R.; Geoffroy, G. L. (J. Organomet. Chem. **273** [1984] 333/46).

[6] Kreiter, C. G.; Franzreb, K.-H.; Michels, W.; Schubert, U.; Ackermann, K. (Z. Naturforsch. **40b** [1985] 1188/98).
[7] Niemer, B.; Weidmann, T.; Beck, W. (Z. Naturforsch. **47b** [1992] 509/16).
[8] Bruce, M. I.; Low, P. J.; Skelton, B. W.; White, A. H. (J. Organomet. Chem. **464** [1994] 191/5).
[9] Kreiter, C. G.; Michels, W.; Heeb, G. (Z. Naturforsch. **50b** [1995] 649/60).

2.6.2.2 Compounds of the Type $[\{(CO)_3Re(C(CH_3)O)_2C(R)O\}_2M]^x$ (x = 1−, 0)

General. Structure. The compounds dealt with in this section formally possess central M atoms (M = Al, Ga, Zr, Hf) which are octahedrally coordinated by two trisacyl-rhenatricarbonyl dianions (in Nos. 4 and 5 both trisacylrhena moieties are connected via a 6- or 7-membered methylene bridge, respectively). Compounds with M = Al, Ga are monoanions (x = 1−), which have been isolated as $[N(P(C_6H_5)_3)_2]$ salts (No. 1 has also been isolated with the $[N(CH_3)_4]$ countercation). Compounds with M = Zr, Hf are neutral.

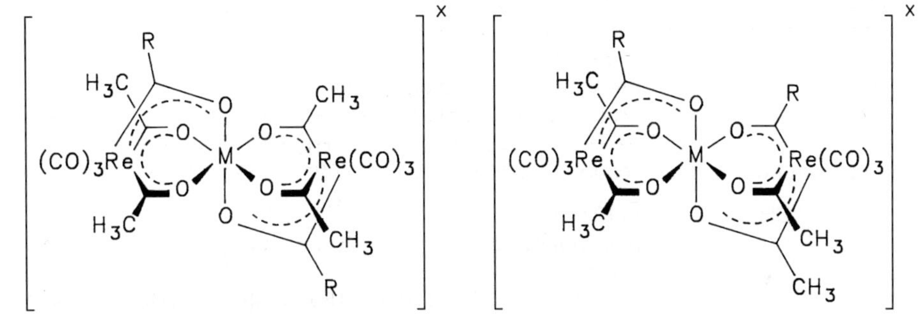

Ia (anti)

Ib (gauche)

Concerning the geometry around the central M atom, two cases must be distinguished. The derivatives with R = CH_3 (Nos. 1, 6, 9, 10) possess an idealized D_{3d} symmetry. However, the derivatives with R = C_3H_7-i and $CH_2C_6H_5$ (Nos. 2, 3, 7, 8, 11, 12) form two geometrical isomers (anti and gauche, see Formulas Ia,b) in a statistically determined ratio of 1:2. These isomers were never separated, but in some cases both could be unambiguously confirmed by ^1H NMR spectroscopy (see below). As an exception, Nos. 4 and 5 exclusively exist as gauche isomers on steric grounds [2].

Synthesis. All compounds were prepared either by Method Ia or by Method Ib.

Method I: Adding 2 or 4 equivalents CH_3Li in ether to a cooled THF solution containing $(CO)_5ReC(O)R$ (R = CH_3, C_3H_7-i, $CH_2C_6H_5$) or $(CO)_5Re-C(O)(CH_2)_nC(O)-Re(CO)_5$ (n = 6, 7) provided $[(CO)_3Re(C(CH_3)O)_2C(R)O]^{2-}$ or $[\{(CO)_3Re(C(CH_3)O)_2C(O)CH_2-\}_2(CH_2)_{n-2}]^{4-}$, respectively. The solution was subsequently treated with a THF solution containing

a. 0.5 equivalent MCl_3 (M = Al [1, 2], Ga [2]). After stirring for ca. 2 h, $[N(P(C_6H_5)_3)_2]BF_4$ in CH_2Cl_2 was added. The mixture subsequently was evaporated, and the residue was extracted into CH_2Cl_2. Crystallization of the extract from a CH_2Cl_2/ether/alkane mixture provided Nos. 1 to 8 [1, 2].
b. 0.5 equivalent MCl_4 (M = Ti [2], Hf [1, 2]). After stirring for ca. 2 h, the mixture was evaporated. The residue was extracted with benzene, and the extract was reevaporated. Repeating the extraction and reevaporation procedure with toluene gave Nos. 9 to 12 [1, 2].

References on p. 269

Properties. Spectroscopy. The anionic compounds are thermally quite stable. They are soluble in CH_2Cl_2 or THF but insoluble in ethers, alkanes, or aromatic solvents. The neutral complexes are air-sensitive compounds, which are insoluble in alkanes but soluble in aromatic or polar solvents. They do not crystallize and thus are difficult to purify. There is evidence that solutions are unstable [2].

The IR spectra of all compounds have a similar shape. They display one band attributable to ν(C⩦O) and two bands attributable to ν(CO). In the neutral compounds the ν(CO) bands are shifted to higher energy than in the anionic compounds. The ^{1}H NMR spectra confirm the presence or nonpresence of geometrical isomers. Due to high symmetry, the spectra of the methyl-substituted derivatives only display one peak due to CH_3. The spectra of the derivatives with R = C_3H_7-i or $CH_2C_6H_5$ in some cases could be shown to display separate CH_3 signals for both isomers: A singlet for the anti isomer and two equally intense singlets for the gauche isomer (see the table for the assignment) [2].

Table 12
Compounds of the Type $[\{(CO)_3Re(C(CH_3)O)_2C(R)O\}_2M]^x$ (x = 1−, 0).
For explanations, abbreviations, and units, see p. X.

No.	M	R	method of preparation (yield) properties and remarks
anionic compounds (x = 1−; counterion $[N(P(C_6H_5)_3)_2]$)			
1	Al	CH_3	Ia (21%) pale yellow solid from CH_2Cl_2/heptane (4:1); m.p. 240 to 265 °C ^{1}H NMR (CD_2Cl_2): 2.51 (s, CH_3), 7.59 (m, C_6H_5 of cation) IR (CH_2Cl_2): 1485 (ν(C⩦O)); 1900, 1990 (ν(CO)) [1, 2] **$[N(CH_3)_4][\{(CO)_3Re(C(CH_3)O)_3\}_2Al]$**: Ia (precipitation with solid $[N(CH_3)_4]BF_4$; yield: 20%) off-white solid from CH_2Cl_2/heptane (1:1); m.p. >205 °C (dec.) ^{1}H NMR (CD_2Cl_2): 2.50 (s, CCH_3), 3.18 (s, NCH_3) IR: identical with the $[N(P(C_6H_5)_3)_2]$ salt [1, 2]
2	Al	C_3H_7-i	Ia (21%) very pale yellow solid; dec.p. 191 to 194 °C ^{1}H NMR (CD_2Cl_2): 0.71, 0.77, 0.81, 0.84, 0.88 (s's, CHC**H$_3$**); 2.50, 2.53 (s's, C(C**H$_3$**)O, gauche and anti, resp.; ratio ca. 3:2); 3.09 (m, CH_2), 7.56 (m, C_6H_5 of cation); peak δ = 2.50 splits into 2 singlets separated by 1 Hz upon expansion IR (CH_2Cl_2): 1490 (ν(C⩦O)); 1895, 1985 (ν(CO)) [2]
3	Al	$CH_2C_6H_5$	Ia (27%) pale yellow solid; dec.p. 66 to 75 °C ^{1}H NMR (CD_2Cl_2): 2.43 (s, CH_3-anti); 2.44, 2.48 (s's, CH_3-gauche); 3.96, 3.99 (s's, CH_2); 7.51 (m, C_6H_5 of anion and cation) IR (CH_2Cl_2): 1485 (ν(C⩦O)); 1895, 1985 (ν(CO)) [2]
4	Al	$(CH_2)_{6/2}$	Ia (in dilute solution) very soluble in polar organic solvents [2]

References on p. 269

Table 12 (continued)

No.	M	R	method of preparation (yield) properties and remarks
5	Al	$(CH_2)_{7/2}$	Ia (in dilute solution) very soluble in polar organic solvents [2]
6	Ga	CH_3	Ia (32%) white solid; dec.p. 230 to 237°C ^{1}H NMR (CD_2Cl_2): 2.53 (s, CH_3), 7.55 (m, C_6H_5 of cation) IR (CH_2Cl_2): 1472 (ν(C⩦O)); 1897, 1980 (ν(CO)) [2]
7	Ga	C_3H_7-i	Ia (21%) pale yellow solid; dec.p. 175 to 192°C ^{1}H NMR (CD_2Cl_2): 0.71, 0.72, 0.78, 0.79, 0.84 (s's, CHC**H**$_3$); 2.49, 2.52 (s's, C(CH_3)O); 3.17 (complex m, CH; J(H,H) = 6.9), 7.49 (m, C_6H_5 of cation); the assignments to gauche and anti isomers are ambiguous IR (CH_2Cl_2): 1482 (ν(C⩦O)); 1897, 1990 (ν(CO)) [2]
8	Ga	$CH_2C_6H_5$	Ia (24%) pale yellow solid; dec.p. 65 to 87°C ^{1}H NMR (CD_2Cl_2): 2.46 (s, CH_3-gauche), 2.49 (s, CH_3-anti), 2.52 (s, CH_3-gauche), 4.07 (br s, CH_2), 7.54 (m, C_6H_5 of anion and cation) IR (CH_2Cl_2): 1475 (ν(C⩦O)); 1900, 1990 (ν(CO)) [2]
neutral compounds (x = 0)			
9	Zr	CH_3	Ib (10%) pale yellow solid; m.p. >230°C (dec.) ^{1}H NMR (acetone-d_6): 2.42 (s, CH_3) IR (CH_2Cl_2): 1470 (ν(C⩦O)); 1922, 2010 (ν(CO)) [2]
10	Hf	CH_3	Ib (43%) pale yellow solid; m.p. 120 to 200°C (dec.) ^{1}H NMR (acetone-d_6): 2.46 (s, CH_3) IR (CH_2Cl_2): 1465 (ν(C⩦O)); 1925, 2020 (ν(CO)) [1, 2]
11	Hf	C_3H_7-i	Ib (11%) pale yellow solid; dec. >160°C ^{1}H NMR (acetone-d_6): 0.79 (m, CHC**H**$_3$), 2.39 (br s, CH_3), 3.55 (m, CH) IR (CH_2Cl_2): 1470 (ν(C⩦O)); 1918, 2010 (ν(CO)) [2]
12	Hf	$CH_2C_6H_5$	Ib (21%) pale yellow solid; dec. 123 to 130°C ^{1}H NMR ($CDCl_3$): 2.51 (br s, CH_3), 4.09 (br s, CH_2), 7.37 (m, C_6H_5) IR (CH_2Cl_2): 1473 (ν(C⩦O)); 1920, 2010 (ν(CO)) [2]

References:

[1] Hobbs, D. T.; Lukehart, C. M. (J. Am. Chem. Soc. **99** [1977] 8357/9).
[2] Hobbs, D. T.; Lukehart, C. M. (Inorg. Chem. **18** [1979] 1297/301).

2.6.2.3 Other Compounds

$[(CO)_5ReC(O)C_5H_4]Re(CO)_3$ (see Formula I). Treatment of $ClC(O)C_5H_4Re(CO)_3$ with $Na[(CO)_5Re]$ in THF at ca. 5°C, warming to room temperature, and evaporation gave a residue which was extracted with CH_2Cl_2. Addition of hexane to the extract separated light yellow crystals; m.p. 127 to 129°C; yield: 83%.

$(CO)_5Re$ / O / $Re(CO)_3$

I

$P(C_6H_5)_3$ / $(C_6H_5)_3P$ / Pt / O / $(CO)_2Re$ / $Re(CO)_2$ / $P(C_6H_5)_3$

II

^1H NMR spectrum ($CDCl_3$): $\delta = 5.17, 5.28$ (AA′BB′) ppm. IR spectrum (C_6H_{12}): 1940, 1950, 2010, 2019, 2030, 2075, 2142 (ν(CO)) cm^{-1}. Mass spectrum: $[M - n\ CO]^+$ ($n = 0$ to 9).

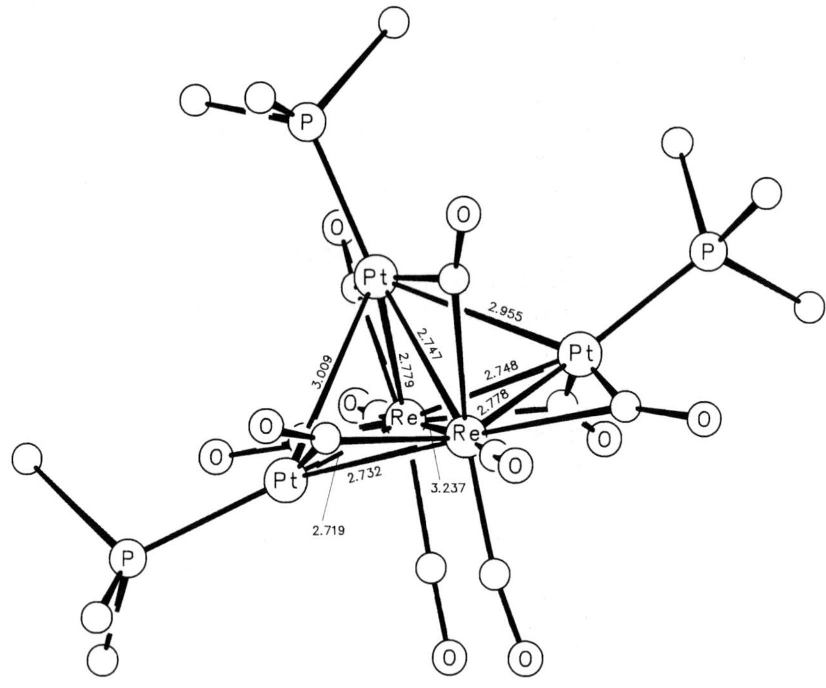

Fig. 128. Molecular structure of $[\mu\text{-}(CO)_6Pt_3(P(C_6H_5)_3)_3]Re_2(CO)_4$ (C_6H_5 groups represented by C_{ipso}) [2].

References on p. 271

The compound is readily soluble in benzene and common polar organic solvents, but only poorly soluble in alcohols. Heating the solid close to the melting point or thermolysis in benzene initiated decarbonylation to $[(CO)_5ReC_5H_4]Re(CO)_3$. In contrast, the compound did neither react with HCl in benzene/heptane nor with $HgCl_2$ in boiling acetone [1].

$[\mu\text{-}(CO)_6Pt_3(P(C_6H_5)_3)_3]Re_2(CO)_4$ (see Formula II). Treatment of $(CO)_{10}Re_2$ with a 5-fold excess of $\pi\text{-}C_2H_4Pt(P(C_6H_5)_3)_2$ in CH_2Cl_2 at room temperature, while bubbling N_2 through the solution, and flash-chromatographic workup separated two products: $[\mu\text{-}(CO)_4Pt_2\text{-}(P(C_6H_5)_3)_2]Re_2(CO)_6$ (see p. 162) and the title compound. Yield: 5%. Red crystals.

IR spectrum (CH_2Cl_2): 1815, 1855, 1985, 2010, 2040 (ν(CO)) cm^{-1}.

The compound crystallizes in the orthorhombic space group $P2_1cn$ (non-standard setting of $Pna2_1-C_{2v}^9$ (No. 33)) with a = 19.714(3), b = 13.195(3), c = 46.486(7) Å; Z = 8 molecules per unit cell, D_{calc} = 2.232 g/cm^3. The asymmetric unit contains two symmetry-independent molecules. One of these is depicted in **Fig. 128**. Each Re-Pt bond is asymmetrically bridged by a CO group (Re-C longer than Pt-C: 2.13 vs. 2.02 Å) [2].

References:

[1] Kolobova, N. E.; Khandozhko, V. N.; Sizoi, V. F.; Guseinov, Sh.; Zhvanko, O. S.; Nekrasov, Yu. S. (Izv. Akad. Nauk SSSR Ser. Khim. **1979** 619/22; Bull. Acad. Sci. USSR Div. Chem. Sci. [Engl. Transl.] **28** [1979] 573/6).

[2] Ciani, G.; Moret, M.; Sironi, A.; Beringhelli, T.; d'Alfonso, G.; Della Pergola, R. (J. Chem. Soc. Chem. Commun. **1990** 1668/70).

2.7 Compounds with Ligands Bonded to Rhenium by Eight C Atoms (^{8}L Compounds)

This chapter describes only two compounds. For their structures, see Formulas I and II.

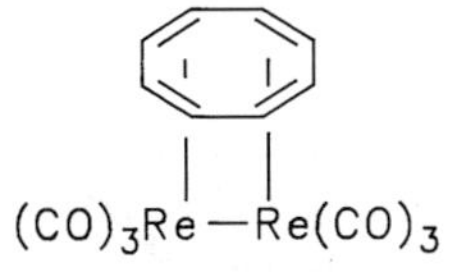

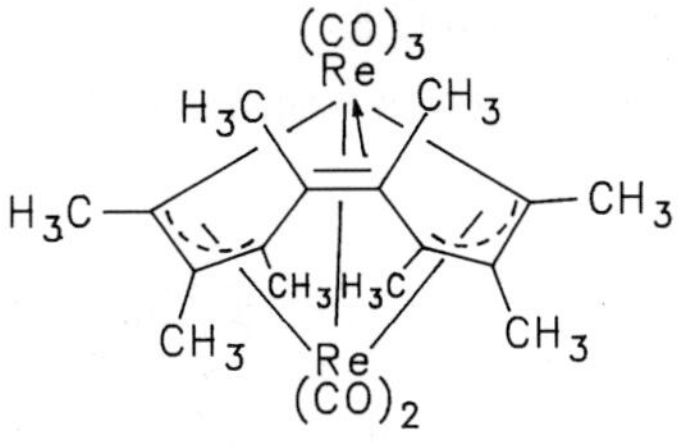

$C_8H_8Re_2(CO)_6$ (see Formula I) resulted by photolyzing $(CO)_{10}Re_2$ in the presence of cyclooctatetraene (pentane, −30 °C, 90 to 180 min). The mixture was at first roughly separated by column chromatography on silica with hexane/ether mixtures. For most of the products a subsequent purification by preparative TLC was required. One product was recognized as $C_8H_8Re_2(CO)_6$. Yield: 3 to 4% [2, 5]. Other authors, however, did not mention the complex, when they performed a similar photolysis reaction [4]. The compound was also said to have been obtained by treating $(CO)_{12}Re_3(\mu\text{-}H)_3$ with cyclooctatetraene in refluxing cyclohexane (under the conditions described in [1] for the Mn analog) [2]. It forms a colorless to pale yellow solid; m.p. 213 °C [2, 5].

^{1}H NMR spectrum ($CDCl_3$): δ = 4.73 (s) ppm. IR spectrum (C_6H_{12}): 1925, 1940, 1975, 2005, 2055 (ν(CO)) cm^{-1}. UV spectrum (THF): λ_{max} (log ε) = 280 (4.24), 322 (4.15) nm. Mass spectrum: $[M - n\,CO]^+$ (n = 0 to 6), also observed: $[C_8H_8Re_2]^{2+}$ [2].

References on p. 272

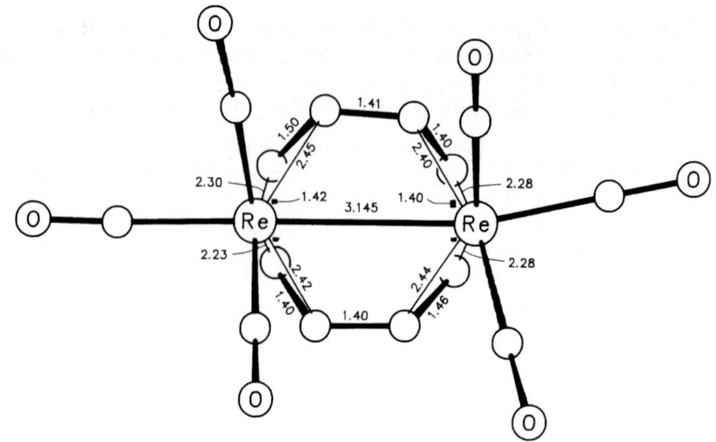

Fig. 129. Molecular structure of $C_8H_8Re_2(CO)_6$ [5].

The compound crystallizes in the orthorhombic space group Pbca$-D_{2h}^{15}$ (No. 61) with a = 10.324(2), b = 11.686(1), c = 23.891(5) Å; D_{calc} = 2.97 g/cm^3. A view of the molecular structure is given in **Fig. 129** [5].

$C_8(CH_3)_8Re_2(CO)_5$ (see Formula II for tentative structure). $(CO)_{10}Re_2$ was reacted with $CH_3C{\equiv}CCH_3$ in a closed vessel in the presence of hexane (190°C, 16 h). Considerable decomposition occurred. However, workup by preparative TLC using CH_2Cl_2/hexane (2:3) gave a single yellow band containing $C_8(CH_3)_8Re_2(CO)_5$. Yield: 10%.

^{1}H NMR spectrum (CD_2Cl_2): δ = 1.70, 2.28, 2.35, 3.31 (all s, 2 CH_3) ppm. IR spectrum (n-hexane): 1894, 1947, 1959, 1973, 2030 (ν(CO)) cm^{-1}. The mass spectrum shows the series $[M - n\ CO]^+$ with n = 0 to 5.

The complex did not react with isocyanides [3].

References:

[1] King, R. B.; Ackermann, M. N. (Inorg. Chem. **13** [1974] 637/44).
[2] Guggolz, E.; Oberdorfer, F.; Ziegler, M. L. (Z. Naturforsch. **36b** [1981] 1060/8).
[3] Mays, M. J.; Prest, D. W.; Raithby, P. R. (J. Chem. Soc. Dalton Trans. **1981** 771/6).
[4] Franzreb, K.-H.; Kreiter, C. G. (J. Organomet. Chem. **246** [1983] 189/95).
[5] Caballero, C.; Oberdorfer, F.; Guggolz, E.; Nuber, B.; Korswagen, R. P.; Ziegler, M. L. (Bol. Soc. Qim. Peru **53** [1987] 135/49; C.A. **110** [1989] No. 213022).

2.8 Compounds with Ligands Bonded to Rhenium by Ten C Atoms (^{10}L Compounds)

There are not that many compounds with ^{10}L ligands. In all of them the ^{10}L ligand coordinates to Re in an $\eta^5:\eta^5$-bridging fashion.

2.8.1 Compounds without CO Groups

$C_{10}H_8Re_2(C_5H_5)_2$ was discussed to be possibly the composition of one of the products obtained either by thermolysis of $(C_5H_5)_4Re_2$ or by a slow self-transformation of the labile

References on p. 274

$(C_5H_5)_2Re_2(\eta^4\text{-}C_5H_6)(\mu\text{-}\eta^{5:1}\text{-}C_5H_4)$ (see p. 248) at room temperature. Based on spectroscopic data only, a clear distinction between the title composition and an alternative formulation, $(C_5H_5)_2Re_2(\mu\text{-}\eta^{5:1}\text{-}C_5H_4)_2$ (^{5}L compound), could not be made, even though the shift of the ^{13}C NMR signals favors the latter proposal [3]. For a full description, see p. 249.

$C_{10}H_8Re_2(P(CH_3)_3)_4H_2$ (see Formula I). When $C_5H_5Re(P(CH_3)_3)(P(CH_3)_2{-}CH_2)H$, obtained by low-temperature photolysis of $C_5H_5Re(P(CH_3)_3)_3$, was left standing in hexane or cyclohexane at ambient temperature, two products formed in the ratio 2:1. These could not be separated. The main product was shown to be $C_5H_5Re_2(\mu\text{-}\eta^{5:1}\text{-}C_5H_4)(P(CH_3)_3)_3(H)_2(\mu\text{-}P(CH_3)_2CH_2{-})$ (see p. 248). For the coproduct, two structures (Formula Ia or Ib) were proposed. Both of them are in accord with the spectroscopic data.

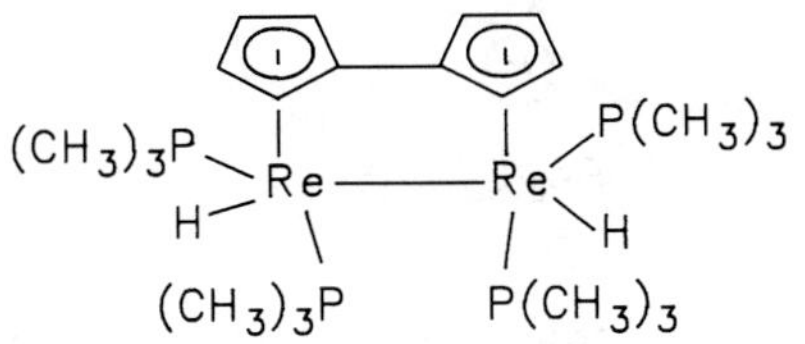

Ia

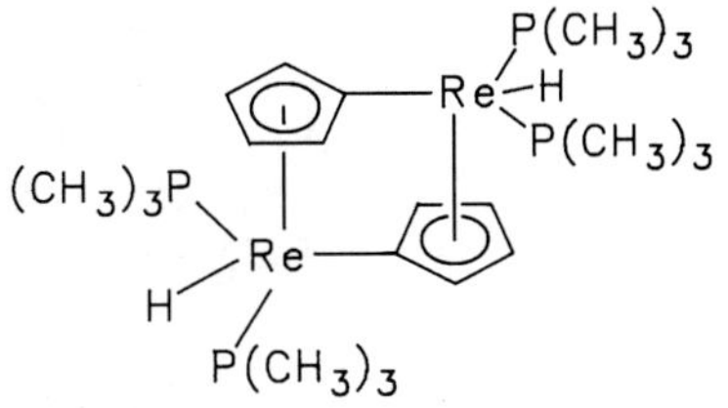

Ib

^{1}H NMR spectrum (C_6D_{12}): δ = −10.69 (t, 2 H; J = 51.1 Hz), 1.59 (v.d.; J = 7.4 Hz); 4.08 and 4.25 (m, 4 H) ppm. ^{1}H {^{31}P} NMR spectrum (C_6D_{12}): δ = −10.60, 1.59 (s's); 4.08, 4.25 (m's) ppm. ^{31}P {^{1}H} NMR spectrum (C_6D_{12}): δ = −40.91 (s) ppm [4].

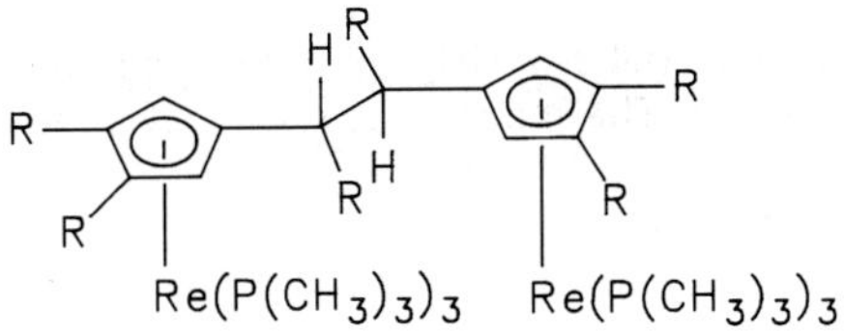

II

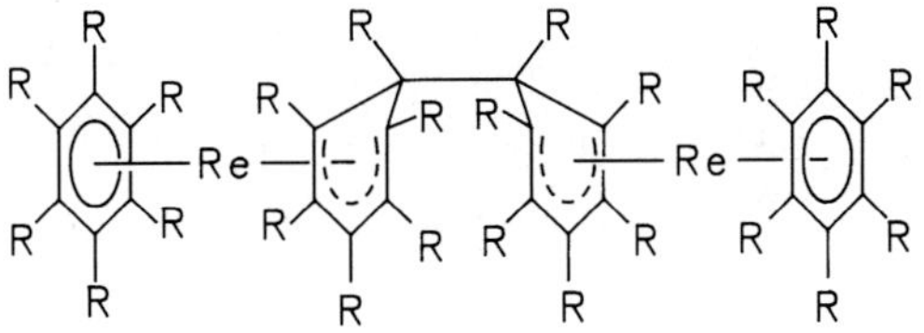

III

$[C_5H_2(C_6H_5)_2CHC_6H_5Re(P(CH_3)_3)_3]_2$ (see Formula II, R = C_6H_5) was obtained by reacting $Re(P(CH_3)_3)_5Cl$ with excess phenylacetylene in refluxing toluene for 18 h. Evaporation of the resulting mixture followed by extraction of the residue with petroleum ether and cooling of the extract gave orange crystals; m.p. 300 °C, with 60% yield.

^{1}H NMR spectrum: δ = 1.25 (d; J(P,H) = 4 Hz); 4.73, 5.03, 5.21 (s's, CH); 7.0 to 7.3, 7.58, 7.90 (C_6H_5) ppm. ^{31}P {^{1}H} NMR spectrum: δ = −41.6 (br s) ppm. IR spectrum (Nujol): 485, 510, 535, 560, 589, 656, 694, 705, 710, 755, 844, 895, 935, 955, 1030, 1040, 1070, 1080, 1140, 1155, 1275, 1290, 1425, 1450, 1500, 1600, 3015, 3065 cm^{-1}. A molecular weight determination gave M = 1390 (calc. 1440) g/mol.

The complex crystallizes in the monoclinic space group $P2_1/a{-}C^5_{2h}$ (No. 14) with a = 14.352(4), b = 34.781(9), c = 13.521(3) Å, β = 105.34(2)°; Z = 4 molecules per unit cell, D_{calc} = 1.47 g/cm^3. The molecular structure is illustrated in **Fig. 130** [2].

$C_{12}(CH_3)_{12}Re_2(C_6(CH_3)_6)_2$ (see Formula III, R = CH_3) sublimed onto a liquid nitrogen-cooled finger when a solid mixture of $[(C_6(CH_3)_6)_2Re]PF_6$ and Li metal was heated under vacuum

References on p. 274

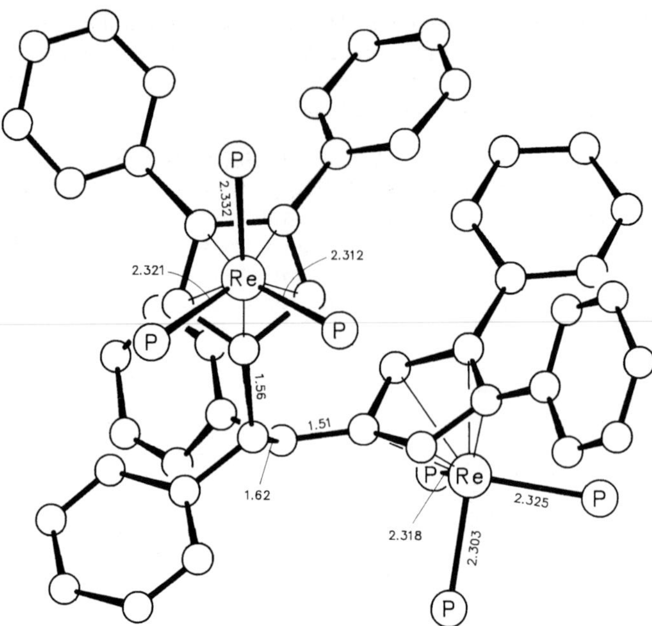

Fig. 130. Molecular structure of $[C_5H_2(C_6H_5)_2CHC_6H_5Re(P(CH_3)_3)_3]_2$ (methyl groups at P omitted) [2].

at 220 to 230°C. The yellow, extremely air-sensitive compound, which decomposes at 225°C, could be purified by a sublimation at 150°C under high vacuum. Yield: 15%.

^1H NMR spectrum (C_6D_6): δ = 1.36 to 2.13 ppm. IR spectrum (KBr or Nujol): 174, 242, 316, 360, 388, 419, 442, 480, 572, 674, 993, 1020, 1067, 1381, 1394, 1442, 1462, 1572, 2732, 2869, 2924, 2967 cm^{-1}.

The compound is well soluble in common hydrocarbons. An osmometric molecular weight determination (benzene) gave M = 1030 g/mol. Exposure to air in hexane precipitated white flakes. Dissolution of the precipitate in aqueous $[NH_4]PF_6$ gave $[(C_6(CH_3)_6)_2Re]PF_6$ [1].

References:

[1] Fischer, E. O.; Schmidt, M. W. (Chem. Ber. **99** [1966] 2206/12).
[2] Chiu, K. W.; Howard, C. G.; Rzepa, H. S.; Sheppard, R. N.; Wilkinson, G. (Polyhedron **1** [1982] 441/51).
[3] Pasman, P.; Snel, J. J. M. (J. Organomet. Chem. **276** [1984] 387/92).
[4] Wenzel, T. T.; Bergman, R. G. (J. Am. Chem. Soc. **108** [1986] 4856/67).

2.8.2 Compounds with CO Groups

2.8.2.1 Compounds with Polycyclic Aromatic Ligands

The compounds described in this section have the structures illustrated in Formulas I to IV. Most of them were prepared as follows:

References on p. 276

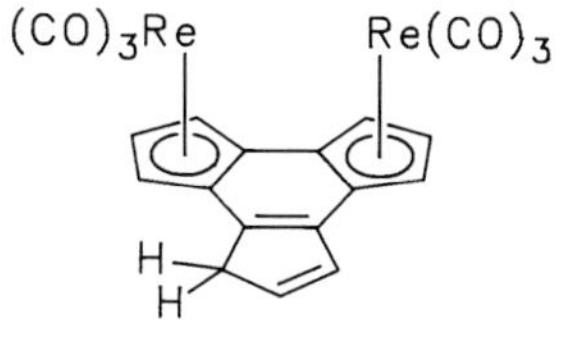

Method I: Stirring a THF solution containing $(CO)_6Re_2(OC_4H_8)_2(\mu\text{-}Br)_2$, the respective organic compound, and KH (1.5 to 3 equivalents) for 12 to 24 h at room temperature. Then excess NH_4Cl was added, and stirring was continued for another 10 min to 2 h. Workup was achieved by preparative TLC [1 to 3].

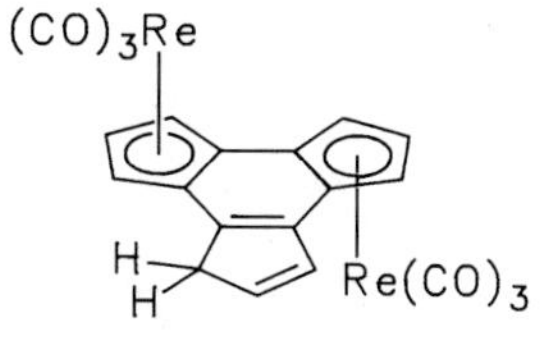

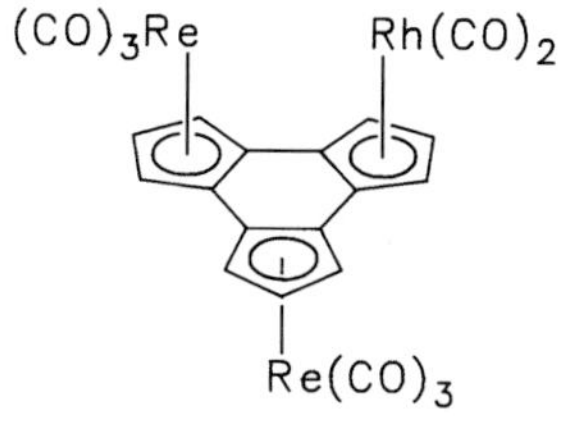

Syn- and **anti-$C_{15}H_{10}[Re(CO)_3]_2$** (see Formulas I, II) were commonly obtained by applying Method I. The main product, however, was $C_{15}H_9[Re(CO)_3]_3$. Evaporation of the mixture, extraction into CH_2Cl_2, and separation of the extract by preparative TLC using ethyl acetate/hexane (1:3) yielded the products.

Both isomers are air-stable as solids and in solution [1, 2].

Syn isomer: Yield: 3% [1, 2]. – ^1H NMR ($CDCl_3$): 3.54 (m, 2 H); 5.34, 5.38 (t's; J = 2.9); 5.58, 5.71, 6.16, 6.26 (dd's; J = 1.4, 2.9); 6.74, 6.88 (td's; J = 1.4, 5.4). – IR (THF): 1919, 1946, 2016, 2030 (ν(CO)) [1].

Anti isomer: Yield: 5% [1, 2]. Could be almost selectively prepared by successively treating $C_{15}H_{12}$ with n-C_4H_9Li and $(CO)_6Re_2(OC_4H_8)_2(\mu\text{-}Br)_2$ (ratio 2:1) in THF at −78 °C, followed by warming to room temperature, evaporation, and a similar workup as described for the isomer mixture. Yield: 22% [2]. ^1H NMR ($CDCl_3$): 3.46, 3.53 (td; J = 1.5, 24.0); 5.63, 5.64 (t; J = 3.0); 5.72, 5.82, 5.83, 5.86 (dd; J = 1.5, 2.9); 6.71, 6.90 (td; J = 1.5, 5.5). – IR (THF): 1927, 2017 (ν(CO)) [1]. – FAB MS: $[M]^+$ [2].
Treatment with $TlOC_2H_5$ and $(CO)_4Rh_2Cl_2$ yielded $C_{15}H_9[Re(CO)_3]_2Rh(CO)_2$ (following compound) [2]. In the reaction with $P(CH_3)_2C_6H_5$ (THF, room temperature) the compound disappeared with $t_{1/2}$ = 11 min [4].

$C_{15}H_9[Re(CO)_3]_2Rh(CO)_2$ (see Formula III) was obtained by successively treating anti-$C_{15}H_{10}[Re(CO)_3]_2$ with excess $TlOC_2H_5$ and $(CO)_4Rh_2(\mu\text{-}Cl)_2$ in THF at room temperature. After stirring for 30 min, the mixture was filtered through silica, and the filtrate was separated by preparative TLC on silica with hexane/ethyl acetate/benzene (4:1:1). The compound was isolated with 71% yield as a yellow-orange solid.

^1H NMR spectrum (THF-d_8): δ = 5.56, 5.87 (t; J = 2.9 Hz), 5.96 (m), 6.00 (t; J = 2.3 Hz), 6.06 (dd; J = 1.5, 2.3), 6.16 (d, 2 H; J = 2.9 Hz), 6.32 (t; J = 2.5 Hz), 6.52 (dd; J = 1.5, 2.9 Hz) ppm. $^{13}C\{^1H\}$ NMR spectrum (THF-d_8): δ = 76.1, 76.3, 76.4, 80.0, 80.8, 85.0, 88.9, 93.4, 97.1, 98.4, 98.8, 99.9, 101.5, 108.6, 190.1 (d; J(Rh,C) = 85 Hz); 194.5, 195.2 (ReCO) ppm. IR spectrum (THF): 1931, 1989, 2019, 2049 (ν(CO)) cm^{-1}. FAB mass spectrum: $[M - CO]^+$ [2].

$C_{27}H_{16}[Re(CO)_3]_2$ (see Formula IV) was obtained via Method I (note that the mixture was stirred in the presence of a catalytic amount KOC_4H_9-t). Coproducts were $C_{27}H_{17}Re(CO)_3$ and $C_{27}H_{15}[Re(CO)_3]_3$. Separation of the mixture by preparative TLC using hexane/THF (82:18) gave 17% of the title complex. The compound is air-sensitive but thermally stable: It was recovered unchanged after having been heated to 60 °C for 4 h.

References on p. 276

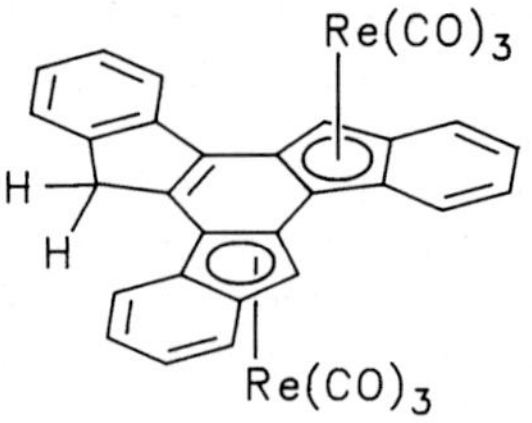

^{1}H NMR spectrum ($CDCl_3$): δ = 4.45 (CH_2, AB pattern; J = 22.3 Hz); 6.65, 6.71 (both s, 1 H); 7.16 to 8.20 (m, 12 H) ppm. ^{13}C {^{1}H} NMR spectrum (THF-d_8): δ = 37.6 (CH_2); 61.3, 62.7 (CH); 100 to 145 (arene, not specified); 192.2, 192.3 (CO) ppm. IR spectrum (THF): 1931, 2019 (ν(CO)) cm^{-1}. FD mass spectrum: $[M]^+$ [3].

References:

[1] Helvenston, M. C.; Lynch, T. J. (J. Organomet. Chem. **359** [1989] C 50/C 52).
[2] Lynch, T. J.; Helvenston, M. C.; Rheingold, A. L.; Staley, D. L. (Organometallics **8** [1989] 1959/63).
[3] Tisch, T. L.; Lynch, T. J.; Dominguez, R. (J. Organomet. Chem. **377** [1989] 265/73).
[4] Bang, H.; Lynch, T. J.; Basolo, F. (Organometallics **11** [1992] 40/8).

2.8.2.2 Compounds with Ligands of the Type $C_5H_nR_{4-n}$–E–$C_5H_nR_{4-n}$

This section covers compounds with the general structure illustrated in Formula I where E represents a varying moiety linking two $C_5H_nR_{4-n}Re(CO)_m(^2D)_{3-m}$ fragments. At first sight this looks complicated, but most compounds (Nos. 1 to 25) are hexacarbonyls (m = 3). Furthermore, all compounds but one (exception is No. 30) bear C_5H_4 rings (n = 4).

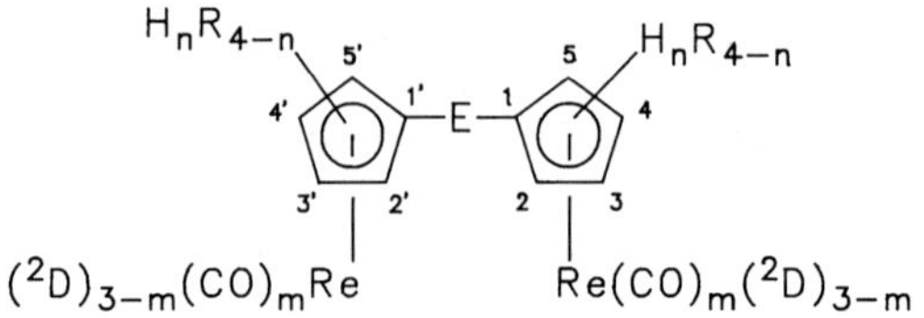

Preparation. Several compounds were prepared by special methods described in the table or the "Further information" section. By far most derivatives were synthesized by reacting in situ prepared Li–$C_5H_4Re(CO)_3$ with electrophiles as described in Methods I and II.

Method I: Treatment of $C_5H_5Re(CO)_3$ with n-C_4H_9Li in THF at ca. −80 to −60 °C generated Li–$C_5H_4Re(CO)_3$, which was reacted in situ at that temperature

a. with 1 equivalent (C_5H_4CHO)$Re(CO)_3$ (for No. 4) [13] or with 0.5 equivalent C_6H_5COCl (for No. 5) [13, 19]. After warming to room temperature, the mixture was hydrolyzed. Further workup was achieved by extraction with benzene and recrystallization from benzene/toluene [13] or by column chromatography with hexane/CH_2Cl_2 (1:3) [19].

References on p. 285

b. with 0.5 equivalent ECl_2 (E = PC_6H_5, S). After warming to room temperature, the mixture was evaporated and the residue chromatographically separated [11].
c. with 0.5 equivalent ECl_4 (E = Se, Te). After warming to room temperature and evaporation of the solvent, the residue was dissolved in CH_2Cl_2. An aqueous solution of Na_2S ($Na_2S_2O_4$ or $K_2S_2O_5$ could also be used) was added and the mixture was shaken for 15 min. Separation of the organic layer was followed by chromatographic workup [11].
d. with CdI_2 [5] or $HgCl_2$ (at ca. −10 °C) [1].

Method II: Addition of powdered S, Se, or Te to a THF solution containing $Li-C_5H_4Re(CO)_3$ (see Method I) gave almost quantitatively $Li-EC_5H_4Re(CO)_3$ (E = S, Se, Te) which subsequently was treated
a. with air at room temperature for 1 h [11].
b. with CH_2I_2 at −60 °C followed by warming to room temperature [17].
In each case the solvent was removed and the residue was separated by chromatography on silica using CH_2Cl_2/hexane (1:1) [11, 17].

Method III: Treatment of $Hg[C_5H_4Re(CO)_3]_2$ (No. 22)
a. with $Pt(P(C_6H_5)_3)_3$ in THF solution at either −20 °C or 20 °C gave No. 23 or 20, respectively [4].
b. with molten M/Hg (M = Yb, Sm) (210 to 220 °C). The residue was subsequently extracted with THF [8, 9]. Yields were low [9].

Method IV: $(CO)_3ReC_5H_4-C_5H_4Re(CO)_3$ or $CH_2[C_5H_4Re(CO)_3]_2$ (Nos. 1, 2) was irradiated in THF for 2 h. C_4H_8S (tetrahydrothiophene) or $S_2(CH_3)_2$ was added, and the mixture was stirred overnight in the dark. The solvent was removed, and the residue separated by column chromatography on silica using CH_2Cl_2/hexane mixtures. The reactions did not take place in benzene. Some intermediates were briefly discussed (see below) [18].

The reaction described in Method Ic proceeds via $\mathbf{Cl_2E[C_5H_4Re(CO)_3]_2}$ (E = Se, Te). These intermediates are subject to immediate reductive dehalogenation with Na_2S [11].

The reaction described in Method IV was said to proceed via the labile solvent-stabilized species $\mathbf{(CO)_3ReC_5H_4-C_5H_4Re(CO)_2OC_4H_8}$ and $\mathbf{(CO)_3ReC_5H_4-CH_2-C_5H_4Re(CO)_2OC_4H_8}$, which were suggested to initially form when irradiating the starting materials in THF. Yet, their presence could be confirmed by in situ IR spectroscopy. The following ν(CO) bands showed up upon irradiating Nos. 1 and 2, respectively, in THF: 1839, 1926, 2019 and 1835, 1925, 2018 cm^{-1}. There was no evidence for disubstitution [18].

Spectroscopy. For most compounds, IR spectra were recorded only in the ν(CO) region. They usually display two absorption bands characteristic of the $Re(CO)_3$ moiety (Nos. 1 to 25). In the cases of the CO-substituted derivatives (Nos. 26 to 29), the carbonyl pattern is superimposed by individual tricarbonyl and di- or monocarbonylmetal units. The ^1H NMR spectra usually exhibit an AA′BB′ pattern for C_5H_4 ("v.t." stands for "virtual triplet"), and the ^{13}C NMR spectra show three signals for C-1, C-2,5, and C-3,4. However, some derivatives (Nos. 4, 5, 9, 13) exhibit four different signals in the ^1H NMR spectra and five distinguishable peaks in the ^{13}C NMR spectra, since in these compounds H-2 to 5 and C-2 to 5 are diastereotopic.

References on p. 285

Table 13
Compounds with a ^{10}L Ligand of the Type $C_5H_nR_{4-n}-E-C_5H_nR_{4-n}$.
An asterisk indicates further information at the end of the table.
For explanations, abbreviations, and units see p. X.

No.	E or compound	method of preparation (yield) properties and remarks
	compounds of the type $E[C_5H_4Re(CO)_3]_2$ (see Formula I, m = 3 and n = 4)	
*1	–	for preparation see "Further information" white, air-stable solid; m.p. 173 to 174°C [1, 12] sublimes at 140°C/10^{-3} Torr [12] ^{1}H NMR ($CDCl_3$): 5.30, 5.53 (v.t.'s) [11] (see also [12]); (acetone-d_6): 5.54, 6.03 (t) [1]; (C_6D_6): 4.25, 4.57 [18] $^{13}C\{^1H\}$ NMR ($CDCl_3$): 83.3, 84.1 (C-2 to 5), 98.7 (C-1), 193.4 (CO) [11] (similar in [18]) IR (KBr): 530, 605, 843, 860, 885, 930, 1010, 1040, 1070, 1115, 1318, 1370, 1390, 1430, 3131; (C_6H_{12}): 1944, 2027 [1]; (THF): 1929, 2018 (ν(CO)) [11] (compare with [12]) irradiation in THF followed by treatment with $S_2(CH_3)_2$ or C_4H_8S gave Nos. 27 and 28, resp. [18]
*2	$-CH_2-$	for preparation see "Further information" colorless crystals; m.p. 134 to 135°C [6] ^{1}H NMR ($CDCl_3$): 3.49 (s), 5.30 (m) [11]; (C_6D_6): 2.57 (s, CH_2), 4.38 (m, C_5H_4) [18] (similar in [6]) $^{13}C\{^1H\}$ NMR ($CDCl_3$): 27.4 (CH_2), 83.8, 84.2 (C-2, C-3), 107.4 (C-1), 193.7 (CO) [11] (similar in [18]) IR (KBr): 1400, 1470; 1930, 2035 (ν(CO)); 2850, 2930, 3120 (CH) [6]; (THF): 1929, 2018 (ν(CO)) [11] irradiation in THF followed by adding C_4H_8S gave No. 29; photolysis in neat C_4H_8S gave a mixture of Nos. 26 and 29 [18]
*3	$-CH_2OCH_2-$	for preparation see "Further information" m.p. 103 to 104°C [6] ^{1}H NMR (C_6D_6): 3.52 (CH_2); 4.32, 4.65 (C_5H_4) IR (KBr): 1100 (COC); 1925, 2025 (ν(CO)); 3120 (CH) [6]
*4	$-C(H)(OH)-$	Ia (70%) white crystals; m.p. 129 to 131°C [13] ^{1}H NMR ($CDCl_3$): 1.99 (br s, OH), 5.24 (br s, CH); 5.29, 5.33, 5.49, 5.53 (C_5H_4); (acetone-d_6): 5.28 (d, CH), 5.38 (d, OH; J = 5.7); 5.49, 5.52, 5.75, 5.79 (C_5H_4) [13] IR (Nujol): 1900, 1920, 1933, 2020, 2024 (ν(CO)); 3564 (ν(OH···OC)) [14]

References on p. 285

Table 13 (continued)

No.	E or compound	method of preparation (yield) properties and remarks
		treatment with C_4H_9Li/D_2O and H_3PO_4 achieved H/D exchange almost selectively in H-2,5 position [13]
*5	$-C(C_6H_5)(OH)-$	Ia (76%) [19]; Ia (85%) [13] white [13], yellow [19] solid; m.p. 191 to 192°C [19], 195 to 197°C [13] ^1H NMR ($CDCl_3$): 2.36 (OH); 5.21, 5.24, 5.28, 5.61 (C_5H_4); 7.3 to 7.5 (C_6H_5) [13] (similar in [19]); (acetone-d_6): 5.47 to 5.54 (6 H, C_5H_4), 5.56 (OH), 5.85 (2 H, C_5H_4) [13] $^{13}C\{^1H\}$ NMR ($CDCl_3$): 72.4 (COH); 81.7, 83.3, 87.4, 88.3 (C-2 to 5); 115.1 (C-1); 126.1, 128.2, 143.3 (C_6H_5); 193.3 (CO) [19] IR (THF): 1926, 2019 (ν(CO)) [19]; (Nujol): 2016, 2020, 2022, 2032 (ν(CO)); 3564 (ν(OH···OC)) [14]; (CCl_4): 3560 (ν(OH···Re)), 3602 (ν(OH free)) [14]
*6	$-C{\equiv}C-$	for preparation see "Further information" yellow crystals; m.p. 204 to 206°C ^1H NMR ($CDCl_3$): 5.28, 5.62 (t's; J = 2.2) $^{13}C\{^1H\}$ NMR ($CDCl_3$): 80.42, 83.73, 84.01, 88.18; 198.85 (CO) IR (THF): 1939, 2029 (ν(CO)) reaction with $(CO)_8Co_2$ gave No. 7 [10]
7	$(CO)_6Co_2[{\equiv}CC_5H_4Re(CO)_3]_2$ 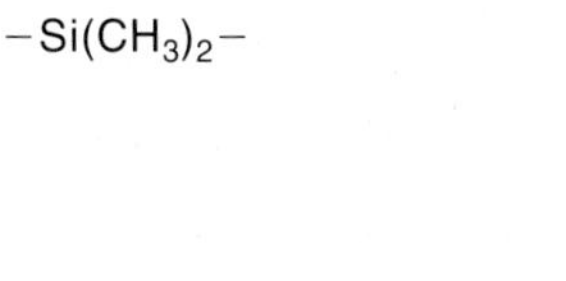 $(CO)_3Co$ $Co(CO)_3$ C C $Re(CO)_3$ $Re(CO)_3$	by treating No. 6 with $(CO)_8Co_2$ (benzene, room temperature, 2 h); sublimation left the product which was recrystallized from $CHCl_3$/pentane black crystals; m.p. >280°C ^1H NMR ($CDCl_3$): 5.34, 5.88 (t's; J = 1.9) $^{13}C\{^1H\}$ NMR ($CDCl_3$): 82.05, 83.22, 89.05, 100.73, 193.41, 198.29 IR (CCl_4): 1938, 2028, 2033, 2063, 2095 (ν(CO)) [10]
*8	$-Si(CH_3)_2-$	for preparation see "Further information" m.p. 120 to 123°C ^1H NMR: 0.38 (CH_3), 5.27 (C_5H_4) IR (C_6H_{12}): 1944, 2031, 2033 (sh) (ν(CO)) MS: $[M]^+$ observed [3]
9	$-P(C_6H_5)-$	Ib (elution with hexane/CH_2Cl_2 (1:1); yield: 63%) white crystals; m.p. 188 to 189°C [11] ^1H NMR ($CDCl_3$): 5.31, 5.40, 5.46, 5.60 (all m, 2 H, C_5H_4); 7.42 (m, C_6H_5) [11] $^{13}C\{^1H\}$ NMR ($CDCl_3$): 85.63, 85.88 (C-3,4); 91.55, 92.06 (C-2,5); 96.01 (C-1; all d; J(P,C) = 2.6, 1.7,

References on p. 285

Table 13 (continued)

No.	E or compound	method of preparation (yield) properties and remarks
9 (continued)		12.8, 13.7, 18.0, resp.); 128.8 (J(P,C) = 7.7), 130.1 (s); 133.0, 135.6 (both d; J(P,C) = 21.4, 8.5, resp.); 192.7 (CO) [11] (similar in [16]) $^{31}P\ \{^1H\}$ NMR ($CDCl_3$): −32.28 [11] IR (THF): 1931, 2022 (ν(CO)) [11] MS: $[M - 6\ CO]^+$ (base peak), $[M - 6\ CO]^{2+}$ [11]
10	−S−	Ib (elution with pentane/toluene (5:2); yield: 21%) yellow, air-stable solid; m.p. 104 to 106 °C ^1H NMR ($CDCl_3$): 5.32, 5.58 (all v.t.) $^{13}C\ \{^1H\}$ NMR ($CDCl_3$): 84.5 (C-3,4), 90.1 (C-2,5), 99.2 (C-1), 192.5 (CO) IR (THF): 1934, 2022 (ν(CO)) MS: $[M - n\ CO]^+$ (n = 0 to 6) [11]
11	−S−S−	IIa (42%) yellow solid; m.p. 155 to 157 °C ^1H NMR ($CDCl_3$): 5.37 (H-3,4), 5.62 (H-2,5; all v.t.) $^{13}C\ \{^1H\}$ NMR ($CDCl_3$): 85.2 (C-3,4), 91.7 (C-2,5), 95.1 (C-1), 192.1 (CO) [11] IR (THF): 1935, 2023 (ν(CO)) [11, 17]
12	$-SCH_2S-$	IIb (56%) colorless solid; m.p. 133 to 134 °C ^1H NMR ($CDCl_3$): 3.88 (CH_2), 5.34 (H-3,4), 5.62 (H-2,5) $^{13}C\ \{^1H\}$ NMR ($CDCl_3$): 50.2 (CH_2), 84.7 (C-3,4), 90.9 (C-2,5), 95.5 (C-1), 192.7 (CO) IR (THF): 1931, 2022 (ν(CO)) MS: $[M - n\ CO]^+$ (n = 0 to 6) [17]
13	$C_8H_4O_2[SC_5H_4Re(CO)_3]_2$ 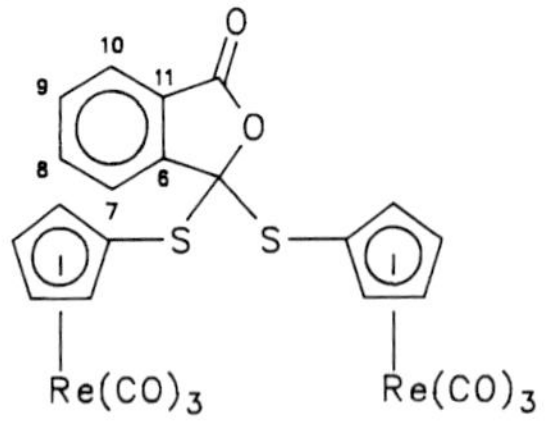	by treating $Li-SC_5H_4Re(CO)_3$ (see Method II) with phthaloyl chloride followed by chromatographic workup using CH_2Cl_2; yield: 45% pale gray solid; m.p. 184 to 186 °C ^1H NMR ($CDCl_3$): 5.18, 5.28, 5.64 (all m, C_5H_4, ratio 4:2:2); 7.56, 7.70 (all m, H-7 to 10, ratio 1:3) $^{13}C\ \{^1H\}$ NMR ($CDCl_3$): 84.6 (C-1); 85.1, 85.6 (C-3,4); 94.1, 95.1 (C-2,5); 99.2 (SCS); 123.2, 125.6 (C-7 to 10), 126.1 (C-11); 131.0, 135.1 (C-7 to 10); 147.6 (C-6), 166.6 (C=O), 191.7 (CO) IR (THF): 1786 (C=O), 1936, 2026 (ν(CO)) MS: $[M]^+$ observed [17]
14	−Se−	Ic (elution with pentane/toluene (5:3); yield: 20%) yellow solid; m.p. 110 to 111 °C ^1H NMR ($CDCl_3$): 5.31 (H-3,4), 5.60 (H-2,5; all v.t.)

References on p. 285

Table 13 (continued)

No.	E or compound	method of preparation (yield) properties and remarks
		$^{13}C\{^1H\}$ NMR ($CDCl_3$): 85.3 (C-3,4), 88.4 (C-1), 92.4 (C-2,5), 192.6 (CO) IR (THF): 1933, 2022 (ν(CO)) [17]
*15	–Se–Se–	IIa (45%) yellow solid; m.p. 178 to 179°C ^1H NMR ($CDCl_3$): 5.36 (v.t., H-3,4), 5.62 (v.t., H-2,5) $^{13}C\{^1H\}$ NMR ($CDCl_3$): 82.7 (C-1), 86.0 (C-3,4), 93.5 (C-2,5), 192.4 (CO) [11] IR (THF): 1934, 2022 [11, 17]
16	$-SeCH_2Se-$	IIb (75%) [17]; also briefly in [16] colorless solid; m.p. 113 to 114°C [17] ^1H NMR ($CDCl_3$): 3.87 (J(Se,H) = 14.4), 5.34 (v.t., H-3,4), 5.60 (v.t., H-2,5) [17] $^{13}C\{^1H\}$ NMR ($CDCl_3$): 30.4, 84.5 (CH_2 and C-1; J(^{77}Se,^{13}C) = 86.9, 125.8, resp.); 85.5 (C-3,4), 92.8 (C-2,5), 192.7 (CO) [16, 17] IR (THF): 1930, 2022 (ν(CO)) MS: $[M - n\,CO]^+$ (n = 0 to 6) [17]
17	–Te–	Ic (elution with hexane/CH_2Cl_2 (1:1); yield: 35%) orange solid; m.p. 174 to 176°C ^1H NMR ($CDCl_3$): 5.30 (v.t., H-3,4), 5.68 (v.t., H-2,5) $^{13}C\{^1H\}$ NMR ($CDCl_3$): 61.8 (C-1), 87.0 (C-3,4), 97.8 (C-2,5), 192.7 (CO) [11]; J(^{125}Te,^{13}C-1) = 358.0 [16] IR (THF): 1931, 2020 (ν(CO)) [11]
*18	–Te–Te–	IIa (29%) [11]; also formed together with $HTeC_5H_4Re(CO)_3$ by treating $Li-TeC_5H_4Re(CO)_3$ with HCl gas [17] orange solid; m.p. 216 to 218°C [11] ^1H NMR ($CDCl_3$): 5.33 (v.t., H-3,4), 5.65 (v.t., H-2,5) [11, 17] $^{13}C\{^1H\}$ NMR (CD_2Cl_2): 48.4 (C-1; J(^{125}Te,^{13}C) = 393.0, 20.0), 88.2 (C-3,4), 98.2 (C-2,5), 192.9 (CO) [16] (similar in [11]) IR (THF): 1931, 2020 (CO) [11, 17]
19	$-TeCH_2Te-$	IIb (64%) ochre solid; m.p. 95 to 96°C ^1H NMR ($CDCl_3$): 3.69 (CH_2; J(Te,H) = 22.2), 5.32 (v.t., H-3,4), 5.64 (v.t., H-2,5) $^{13}C\{^1H\}$ NMR ($CDCl_3$): −25.0, 58.2 (CH_2 and C-1; J(^{125}Te,^{13}C) = 217.1, 357.0, resp.); 87.4 (C-3,4), 97.7 (C-2,5), 192.9 (CO) ^{125}Te NMR ($CDCl_3$): 486.8 (standard: $Te(CH_3)_2$)

References on p. 285

Table 13 (continued)

No.	E or compound	method of preparation (yield) properties and remarks
19 (continued)		IR (THF): 1929, 2021 (ν(CO)) MS: $[M - n\,CO]^+$ (n = 0 to 6) [17]
20	$-Pt(P(C_6H_5)_3)_2-$	IIIa (at 20 °C, Hg deposited; filtration and addition of hexane yielded 79%) m.p. 243 to 246 °C (dec.) IR (CH_2Cl_2): 1946, 2022 (ν(CO)) [4]
21	$-Cd-$	Id (oily residue after evaporation was washed with hexane/benzene leaving the product; yield: 52%) light yellow, air- and moisture-sensitive solid IR (C_6H_6): 1938, 1944 (ν(CO)); weakly split in some other solvents reacts with RCOCl (R = CH_3, C_6H_5) and CH_3I in the presence of $AlCl_3$ to give $(C_5H_4C(O)R)Re(CO)_3$ and $(C_5H_4CH_3)Re(CO)_3$, resp. [5]
*22	$-Hg-$	Id (residue after evaporation was washed with H_2O and extracted into $CHCl_3$; product formed along with $ClHg-C_5H_4Re(CO)_3$ and was separated by fractionally crystallizing from benzene) white crystals; m.p. 226 to 227 °C without dec. [1] ^{1}H NMR (THF-d_8): 5.41, 5.84 [8] IR (KBr): 835, 1030, 1070, 1380, 1420, 3120 [1] sparingly soluble in petroleum ether, better in benzene, readily in THF [1]
23	$-Hg-Pt(P(C_6H_5)_3)_2-$	IIIa (at −20 °C, stirring at −13 °C for 1 d; product precipitated with 72% yield) m.p. 236 to 241 °C (dec.) IR (KBr): 1945, 1990 (ν(CO)) treatment with CF_3CO_2H eliminated Hg [4]
24	$-Sm- \cdot$ n THF (n = 1, 2)	IIIb [8, 9], low yield [9] ^{1}H NMR (THF-d_8): 5.36, 5.48 [8]
25	$-Yb- \cdot$ n THF (n = 1, 2)	IIIb [8, 9]

compounds of the type $E[C_5H_4Re(CO)_m(^2D)_{3-m}]$ (see Formula I, m = 1, 2 and n = 4)

No.	E or compound	method of preparation (yield) properties and remarks
26	$CH_2[C_5H_4Re(CO)_2SC_4H_8]_2$ SC_4H_8 = tetrahydrothiophene	by irradiating No. 2 in neat tetrahydrothiophene (not isolated in a pure state; spectroscopic data were taken on a mixture of Nos. 26 and 29 (ratio ca. 3:1)) ^{1}H NMR (C_6D_6): ca. 1.50, 2.70 (m's); 3.16 (s); 4.61, 4.71 (m's) $^{13}C\{^1H\}$ NMR (C_6D_6): 27.8 (CH_2); 29.8, 50.4 (C_4H_8S); 80.9, 82.2 (C_5H_4); 105.3, 203.7 (CO) [18]

References on p. 285

Table 13 (continued)

No.	E or compound	method of preparation (yield) properties and remarks
27	$(CO)_3ReC_5H_4-C_5H_4Re(CO)(SCH_3)_2$	IV (40%) small red-violet crystals from pentane/CH_2Cl_2 (1:1) ^1H NMR (C_6D_6): 3.06 (s, SCH_3); 4.33, 4.69 (v.t.'s); 4.79 (d), 4.90 (m; all 2 H) IR (THF): 1903, 1927, 2021 (ν(CO)) [18]
28	$(CO)_3ReC_5H_4-C_5H_4Re(CO)_2SC_4H_8$	IV (51%) golden yellow crystals; m.p. 153 to 155 °C ^1H NMR (C_6D_6): 1.39, 2.57 (C_4H_8S); 4.34, 4.48, 4.73 (C_5H_4; all m) $^{13}C\{^1H\}$ NMR (C_6D_6): 29.8 (CH_2), 50.0 (CH_2S); 81.0, 81.7, 81.9, 83.5 (C-2,2',5,5'); 93.5, 102.5 (C-1,1'); 194.6, 202.9 (CO) IR (THF): 1855, 1921, 2019 (ν(CO)) MS: $[M]^+$, $[M - C_4H_8S]^+$ [18]
29	$(CO)_3ReC_5H_4-CH_2-C_5H_4Re(CO)_2SC_4H_8$ 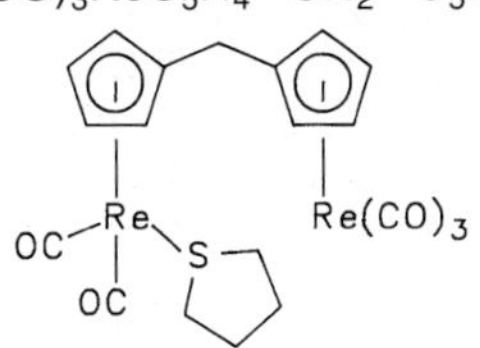	IV (25%); also isolated as single product when photolyzing No. 2 in neat tetrahydrothiophene, although spectroscopic studies revealed the simultaneous presence of No. 26 golden yellow crystals; m.p. 98 to 99 °C ^1H NMR (C_6D_6): 1.45, 2.67 (m's); 2.92 (s); 4.36, 4.55 (m's) $^{13}C\{^1H\}$ NMR (C_6D_6): 27.6 (CH_2); 29.8, 50.4 (C_4H_8S); 80.9, 82.7, 83.4, 84.1 (C-2,2',5,5'); 102.9, 108.9 (C-1,1'); 194.8, 203.4 (CO) IR (THF): 1851, 1920, 2018 (ν(CO)) MS: $[M]^+$, $[M - C_4H_8S]^+$ [18]

compound of the type $E[(C_5H_3R)Re(CO)_3]_2$ (see Formula I, m = 3 and n = 3)

No.	E or compound	method of preparation (yield) properties and remarks
30	$Hg[(C_5H_3CH_2N(CH_3)_2)Re(CO)_3]_2$	by treating in situ-prepared $Li-C_5H_3CH_2N(CH_3)_2Re(CO)_3$ with HgX_2 (X = Cl, Br) in THF; yield: 60 to 80% [15]

* Further information:

$(CO)_3ReC_5H_4-C_5H_4Re(CO)_3$ (Table **13**, No. **1**) formed by treating $(C_5H_4I)Re(CO)_3$ with Cu powder at 150 °C for 1 h. Extraction of the residue with benzene followed by sublimation of the residue at 130 °C yielded 61% of the product [1]. The compound could also be prepared

References on p. 285

by reacting $TlC_5H_4-C_5H_4Tl$ with 2 equivalents $(CO)_5ReBr$ in refluxing benzene for 20 h. Subsequent chromatographic workup yielded 94% [7, 12].

$CH_2[C_5H_4Re(CO)_3]_2$ (Table **13**, No. **2**) was obtained by reacting $C_5H_5Re(CO)_3$ with $(C_5H_4CH_2Cl)Re(CO)_3$ (mole ratio 1:1) in the presence of $AlCl_3$ in refluxing CH_2Cl_2 (7 h). The yield was 41%. The compound was also formed with only 2 to 6% yield (main product was $(C_5H_4CH_2Cl)Re(CO)_3$) when treating $C_5H_5Re(CO)_3$ with chloromethyl methyl ether in the presence of $AlCl_3$ in CH_2Cl_2 at 20 or 40 °C. In both cases, No. 2 was chromatographically separated (alumina, petroleum ether/benzene mixtures) [6]. The compound is also accessible by treating $TlC_5H_4CH_2C_5H_4Tl$ with $(CO)_5ReBr$ (no details given) [18].

$O[-CH_2C_5H_4Re(CO)_3]_2$ (Table **13**, No. **3**) could be prepared either by stirring a mixture of $(C_5H_4CH_2OH)Re(CO)_3$ and conc. HCl for 2 h or by treating a benzene solution containing $(C_5H_4CH_2Cl)Re(CO)_3$ with H_2O. In the first case the workup started with extraction into benzene. Further workup in both cases was achieved by removal of the benzene and extraction of the residue with pentane. The insoluble residue was the pure product which was obtained with 23 or 17% yield, respectively, but adding some drops of HCl during the hydrolyzation reaction improved the yield to 90% [6].

$C(R)(OH)[C_5H_4Re(CO)_3]_2$ (R = H, C_6H_5; Table **13**, Nos. **4** and **5**). The appearance of two ν(OH) vibrations in the IR spectrum of the solution suggests intramolecular O···H···Re bonding. In contrast, an O···H···OC interaction can be suggested in the solid state. This is shown by the disappearance of the low-frequency ν(CO) band when going from the solid to the solution state [14].

$(CO)_3ReC_5H_4-C{\equiv}C-C_5H_4Re(CO)_3$ (Table **13**, No. **6**) was obtained by treating $(C_5H_4I)Re(CO)_3$ with 0.5 equivalents $(C_4H_9)_3Sn-C{\equiv}C-Sn(C_4H_9)_3$ in the presence of 0.05 equivalents $Pd(NCCH_3)_2Cl_2$ in DMF. The mixture was stirred at room temperature for 12 h. Subsequent addition of ether and KF was followed by stirring for 30 min. The organic layer was separated and chromatographically worked up (silica, hexane/ethyl acetate (9:1)) to yield 78% of pure product [10].

$(CH_3)_2Si[C_5H_4Re(CO)_3]_2$ (Table **13**, No. **8**) was obtained by reacting $(CH_3)_2Si[C_5H_4Sn(CH_3)_3]_2$ with 2 equivalents $(CO)_5ReBr$ in refluxing THF for 12 h. After removing the solvent, the residue was chromatographically separated on silica using pentane. The product was recrystallized from methyl cyclohexane and purified by sublimation. Yield: 24%. By-products were $(CO)_{10}Re_2$ and $(CH_3)_3SnBr$ [3].

$(CO)_3ReC_5H_4-E-E-C_5H_4Re(CO)_3$ (E = Se, Te, Table **13**, Nos. **15** and **18**) also formed when treating $Li-EC_5H_4Re(CO)_3$ with 0.5 equivalent phthaloyl chloride in THF. On warming to room temperature, biphthalid precipitated. The title compounds were isolated by column chromatography [17].

$Hg[C_5H_4Re(CO)_3]_2$ (Table **13**, No. **22**) was also obtained by treating $ClHg-C_5H_4Re(CO)_3$ with $Na_2S_2O_3$ in acetone/H_2O mixture. Stirring at room temperature for several hours gave a precipitate, which was filtered off. Yield: 83% [1].

Treatment with excess $HgCl_2$ (THF, 20 °C) gave $ClHg-C_5H_4Re(CO)_3$ with high yield [1], while treatment with CuX_2 (X = Cl, Br) in boiling aqueous acetone produced compounds of the type $(C_5H_4X)Re(CO)_3$ [2]. Interaction with $Pt(P(C_6H_5)_3)_3$ initially gave rise to insertion of the $Pt(P(C_6H_5)_3)_2$ unit into one Hg-C bond such that $(CO)_3ReC_5H_4-Pt(P(C_6H_5)_3)_2-Hg-C_5H_4Re(CO)_3$ (No. 23) formed at −20 °C. However, at 20 °C $((C_6H_5)_3P)_2Pt[C_5H_4Re(CO)_3]_2$ (No. 20) was obtained [4]. Treatment with molten M/Hg (M = Yb, Sm) followed by extraction with THF gave compounds of the type $(C_4H_8O)_nM[C_5H_4Re(CO)_3]_2$ (n = 1, 2; Nos. 24, 25) with low yield [8, 9].

References on p. 285

References:

[1] Nesmeyanov, A. N.; Kolobova, N. E.; Anisimov, K. N.; Makarov, Yu. V. (Izv. Akad. Nauk SSSR Ser. Khim. **1969** 357/9; Bull. Acad. Sci. USSR Div. Chem. Sci. [Engl. Transl.] **1969** 308/10).

[2] Nesmeyanov, A. N.; Kolobova, N. E.; Makarov, Yu. V.; Anisimov, K. N. (Izv. Akad. Nauk SSSR Ser. Khim. **1969** 1992/6; Bull. Acad. Sci. USSR Div. Chem. Sci. [Engl. Transl.] **1969** 1842/5).

[3] Abel, E. W.; Moorhouse, S. (J. Organomet. Chem. **29** [1971] 227/31).

[4] Suleimanov, G. Z.; Bashikov, V. V.; Sokolov, V. I.; Reutov, O. A. (Izv. Akad. Nauk SSSR Ser. Khim. **1978** 1670/2; Bull. Acad. Sci. USSR Div. Chem. Sci. [Engl. Transl.] **1978** 1460/1).

[5] Suleimanov, G. Z.; Agaeva, R. M.; Mamedova, S. G.; Kurbanov, T. Kh. (Azerb. Khim. Zh. **1983** 62/5; C.A. **101** [1984] No. 171434).

[6] Valueva, Z. P.; Solodova, M. Ya.; Kolobova, N. E. (Izv. Akad. Nauk SSSR Ser. Khim. **1983** 1863/6; Bull. Acad. Sci. USSR Div. Chem. Sci. [Engl. Transl.] **1983** 1687/90).

[7] Spink, W. C.; Rausch, M. D. (J. Organomet. Chem. **308** [1986] C 1/C 4).

[8] Kolobova, N. E.; Kazimirchuk, E. I.; Petrovskii, P. V.; Lusenkova, M. A.; Khandozhko, V. N.; Suleimanov, G. Z.; Beletskaya, T. P. (Dokl. Akad. Nauk SSSR **295** [1987] 1134/8; Dokl. Chem. [Engl. Transl.] **295** [1988] 362/6).

[9] Beletskaya, I. P.; Suleimanov, G. Z. (Metalloorg. Khim. **1** [1988] 10/24; Organomet. Chem. USSR [Engl. Transl.] **1** [1988] 3/11).

[10] Lo Sterzo, C. L.; Miller, M. M.; Stille, J. K. (Organometallics **8** [1989] 2331/7).

[11] Herberhold, M.; Biersack, M. (J. Organomet. Chem. **381** [1990] 379/89).

[12] Rausch, M. D.; Spink, W. C.; Conway, B. G.; Rogers, R. D.; Atwood, J. L. (J. Organomet. Chem. **383** [1990] 227/52).

[13] Loim, N. M.; Ginzburg, A. G.; Galakhov, M. V. (Metalloorg. Khim. **4** [1991] 969/75; Organomet. Chem. USSR [Engl. Transl.] **4** [1991] 471/4).

[14] Shubina, E. S.; Krylov, A. N.; Timofeeva, T. V.; Struchkov, Yu. T.; Ginzburg, A. G.; Loim, N. M.; Epstein, L. M. (J. Organomet. Chem. **434** [1992] 329/39).

[15] Suleimanov, G. Z.; Usyatinsky, A. Ya.; Zulfugarly, E. A.; Kuzmina, L. G.; Kazimirchuk, E. I.; Khandozhko, V. N.; Petrovskii, P. V.; Bregadze, V. I.; Makhmudov, Sh. M.; Beletskaya, I. P. (Metalloorg. Khim. **5** [1992] 973/4; Organomet. Chem. USSR [Engl. Transl.] **5** [1992] 473/4).

[16] Wrackmeyer, B.; Biersack, M.; Brendel, H.-D.; Herberhold, M. (Z. Naturforsch. **47b** [1992] 1397/402).

[17] Herberhold, M.; Biersack, M. (J. Organomet. Chem. **443** [1993] 1/8).

[18] Herberhold, M.; Biersack, M. (Z. Naturforsch. **48b** [1993] 161/70).

[19] Herberhold, M.; Biersack, M. (J. Organomet. Chem. **503** [1995] 277/87).

2.9 Compound with a Ligand Bonded to Rhenium by Twelve C Atoms (^{12}L Compound)

$(C_4H_4BC_6H_5)_2Re_2(CO)_2Pd_2(\mu\text{-}CO)_4$ (see Formula I) was prepared by reacting $[N(CH_3)_4][(\eta^5\text{-}C_4H_4BC_6H_5)Re(CO)_3]$ with a 5-fold excess of cis-$Pd(NCC_6H_5)_2Cl_2$ in toluene (room temperature, 10 min). The reaction also proceeded with $C_8H_{12}PdCl_2$ (C_8H_{12} = cyclooctadiene) at 60 °C or $(\mu\text{-}CO)_4Pd_4(\mu\text{-}O_2CCH_3)_4$ at room temperature. In all cases the product precipitated from the solution as blue-green shining plates.

Single crystals are orthorhombic with a = 8.427(3), b = 15.923(3), c = 19.493(2) Å; space group Pbca$-D_{2h}^{15}$ (No. 61); Z = 4 molecules per unit cell, D_{calc} = 2.627 g/cm^3. The structure of the molecule is shown in **Fig. 131**. The metal atoms form a planar rhombus.

Reference on p. 286

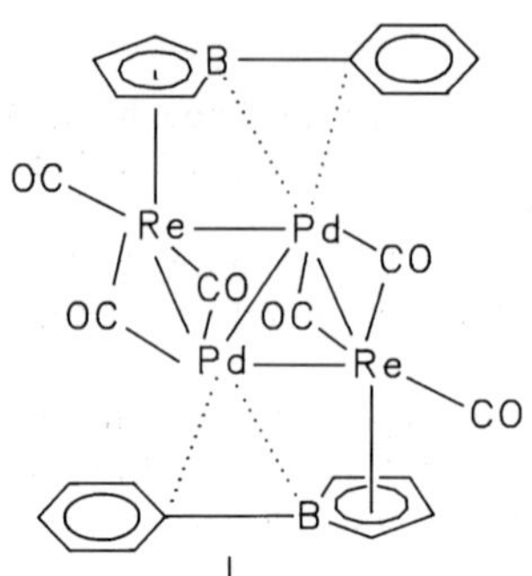

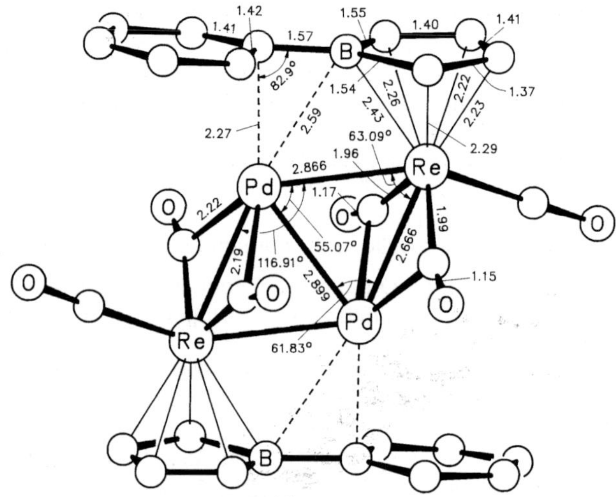

Fig. 131. Molecular structure of $(C_4H_4BC_6H_5)_2Re_2(CO)_2Pd_2(\mu\text{-}CO)_4$ [1].

^{1}H NMR spectrum (toluene-d_8): δ = 3.83, 4.87 (m's, H-3,4 and H-2,5 of C_4H_4B; J(H-2,3) + J(H-2,4) = 6.4 Hz); 7.47, 7.74 (m, C_6H_5) ppm; similar in THF-d_8. IR spectrum (toluene): 1894, 1978 cm^{-1}; (KBr): 1888, 1961 (ν(CO)) cm^{-1}; (polyethylene): 296, 329, 370, 413, 437, 466, 478, 488 cm^{-1}. UV spectrum (THF): λ_{max} = 640 nm. Secondary ion mass spectrum: $[M]^-$, $[M - 3\,CO]^-$, $[C_4H_4BC_6H_5Re(CO)_3]^-$ (base peak).

Extended Hückel MO calculations gave a small HOMO-LUMO distance of 0.676 eV.

The compound is well soluble in THF and moderately soluble in toluene and CH_2Cl_2. The solutions are fairly stable in air. Cyclic voltammetry (THF/0.1 M $[N(C_4H_9\text{-}n)_4]PF_6$, Pt electrode): $E_{1/2}$ (red.) = −0.85 and −0.47V vs. SCE; no oxidation wave up to 1.0 V [1].

Reference:

[1] Braunstein, P.; Englert, U.; Herberich, G. E.; Neuschütz, M. (Angew. Chem. **107** [1995] 1090/3; Angew. Chem. Int. Ed. Engl. **34** [1995] 1010).

Empirical Formula Index

In this index the compounds are listed in order of increasing carbon content. The empirical formulas are specified by linearized formulas. Ionic compounds are given in brackets; components of solvates and adducts are separated by a period.

Pages are printed in ordinary type, table numbers in bold face, and compound numbers in the tables in italics.

Formula	Compound	Page, Table, No.
$C_{57}H_{52}Cl_3O_2P_4Re_2^+$	$[(C\text{-}O\text{-}C(C_3H_7\text{-}n)\text{=}CH)Re_2(CO)(Cl)_3((C_6H_5)_2P\text{-}CH_2\text{-}P(C_6H_5)_2)_2][PF_6]$	5, **1**, *2*
$C_{57}H_{54}Cl_3OP_4Re_2^+$	$[(C_2H_5\text{-}C{\equiv}C\text{-}C_2H_5)Re_2(CO)(Cl)_3((C_6H_5)_2P\text{-}CH_2\text{-}P(C_6H_5)_2)_2][PF_6]$	106, **6**, *5*
	$[(HC{\equiv}C\text{-}C_4H_9\text{-}n)Re_2(CO)(Cl)_3((C_6H_5)_2P\text{-}CH_2\text{-}P(C_6H_5)_2)_2][PF_6]$	106, **6**, *6*
$C_{57}H_{55}Br_3NP_4Re_2^+$	$[(HC{\equiv}CH)Re_2(CN\text{-}C_4H_9\text{-}t)(Br)_3((C_6H_5)_2P\text{-}CH_2\text{-}P(C_6H_5)_2)_2][O_3S\text{-}CF_3]$	107, **6**, *12*
$C_{57}H_{55}Cl_3NP_4Re_2^+$	$[(HC{\equiv}CH)Re_2(CN\text{-}C_4H_9\text{-}t)(Cl)_3((C_6H_5)_2P\text{-}CH_2\text{-}P(C_6H_5)_2)_2][O_3S\text{-}CF_3] \cdot CH_3\text{-}COO\text{-}C_2H_5$	107, **6**, *11*
C$_{58}H_{44}O_8P_4Pt_2Re_2$	$[((C_6H_5)_2P\text{-}CH_2\text{-}P(C_6H_5)_2)_2Pt_2(CO)_2]Re_2(CO)_6 \cdot 2\ O\text{=}C(CH_3)_2$	55
$C_{58}H_{54}Cl_3O_2P_4Re_2^+$	$[(C\text{-}O\text{-}C(C_4H_9\text{-}n)\text{=}CH)Re_2(CO)(Cl)_3((C_6H_5)_2P\text{-}CH_2\text{-}P(C_6H_5)_2)_2][PF_6]$	5, **1**, *4*
C$_{59}H_{50}Cl_3OP_4Re_2^+$	$[(HC{\equiv}C\text{-}C_6H_5)Re_2(CO)(Cl)_3((C_6H_5)_2P\text{-}CH_2\text{-}P(C_6H_5)_2)_2][PF_6]$	106, **6**, *7*
C$_{60}H_{40}N_2O_4Re_2$	$[C_6H_5\text{-}CC(C_6H_5)C(C_6H_5)C(C_6H_5)C(C_6H_5)C\text{-}C_6H_5]Re_2(CO)_4(CN\text{-}C_6H_5)_2$	266
$C_{60}H_{40}O_4Re_2$	$[C_6H_5\text{-}CC(C_6H_5)C(C_6H_5)C(C_6H_5)C(C_6H_5)C\text{-}C_6H_5]Re_2(CO)_4(C_6H_5\text{-}C{\equiv}C\text{-}C_6H_5)$	264
	$[C_6H_5\text{-}CC(C_6H_5)C(C_6H_5)C(C_6H_5)C(C_6H_5)C\text{-}C_6H_5]Re_2(CO)_4(C_6H_5\text{-}C{\equiv}C\text{-}C_6H_5) \cdot 0.5\ C_6H_{14}$	264
$C_{60}H_{50}Cl_3O_2P_4Re_2^+$	$[(C\text{-}O\text{-}C(C_6H_5)\text{=}CH)Re_2(CO)(Cl)_3((C_6H_5)_2P\text{-}CH_2\text{-}P(C_6H_5)_2)_2][PF_6]$	5, **1**, *5*
$C_{60}H_{52}Cl_3OP_4Re_2^+$	$[(CH_3\text{-}C{\equiv}C\text{-}C_6H_5)Re_2(CO)(Cl)_3((C_6H_5)_2P\text{-}CH_2\text{-}P(C_6H_5)_2)_2][PF_6]$	107, **6**, *9*
	$[(HC{\equiv}C\text{-}C_6H_4\text{-}4\text{-}CH_3)Re_2(CO)(Cl)_3((C_6H_5)_2P\text{-}CH_2\text{-}P(C_6H_5)_2)_2][PF_6]$	107, **6**, *8*
$C_{60}H_{60}N_2O_2P_2Re_2$	$(C_6H_5)_3P\text{-}Re(NO)[C_5(CH_3)_5]\text{-}(C{\equiv}C)_2\text{-}Re(NO)[C_5(CH_3)_5]\text{-}P(C_6H_5)_3$	228
	$(C_6H_5)_3P\text{-}Re(NO)[C_5(CH_3)_5]\text{-}(C{\equiv}C)_2\text{-}Re(NO)[C_5(CH_3)_5]\text{-}P(C_6H_5)_3 \cdot 2\ CH_2Cl_2$	228
$C_{60}H_{60}N_2O_2P_2Re_2^+$	$[(C_6H_5)_3P\text{-}Re(NO)(C_5(CH_3)_5)CCCCRe(NO)(C_5(CH_3)_5)\text{-}P(C_6H_5)_3][PF_6]$	236
$C_{60}H_{60}N_2O_2P_2Re_2^{2+}$	$[(C_6H_5)_3P\text{-}Re(NO)(C_5(CH_3)_5)\text{=}C\text{=}C\text{=}C\text{=}C\text{=}Re(NO)(C_5(CH_3)_5)\text{-}P(C_6H_5)_3][PF_6]_2$	235/6
$C_{60}H_{63}N_2O_2P_2Re_2^+$	$[(C_6H_5)_3P\text{-}Re(NO)(C_5(CH_3)_5)\text{-}C_4H_3\text{=}Re(NO)(C_5(CH_3)_5)\text{-}P(C_6H_5)_3][BF_4]$	234
	$[(C_6H_5)_3P\text{-}Re(NO)(C_5(CH_3)_5)\text{-}C_4H_3\text{=}Re(NO)(C_5(CH_3)_5)\text{-}P(C_6H_5)_3]_2[Zn_2Cl_6] \cdot 0.5\ CH_2Cl_2$	233/4
$C_{60}H_{64}N_2O_2P_2Re_2^{2+}$	$[(C_6H_5)_3P\text{-}Re(NO)(C_5(CH_3)_5)\text{=}C_4H_4\text{=}Re(NO)(C_5(CH_3)_5)\text{-}P(C_6H_5)_3][BF_4]_2$	234/5
C$_{61}H_{43}O_7P_3Re_2$	$(CO)_4Re\text{-}C(O)\text{-}C_6H_3[P(C_6H_5)_2]\text{-}Re(CO)_2[P(C_6H_5)_3]_2$	45, **3**, *20*
	$(C_6H_5)_3P\text{-}Re(CO)_3\text{-}C(O)\text{-}C_6H_3[P(C_6H_5)_2]\text{-}Re(CO)_3\text{-}P(C_6H_5)_3$	45, **3**, *19*
$C_{61}H_{52}Cl_3O_2P_4Re_2^+$	$[(C\text{-}O\text{-}C(C_6H_4\text{-}4\text{-}CH_3)\text{=}CH)Re_2(CO)(Cl)_3((C_6H_5)_2P\text{-}CH_2\text{-}P(C_6H_5)_2)_2][PF_6]$	5/6, **1**, *6*
$C_{61}H_{55}Br_3NP_4Re_2^+$	$[(HC{\equiv}CH)Re_2(CN\text{-}C_6H_3\text{-}2,6\text{-}(CH_3)_2)(Br)_3((C_6H_5)_2P\text{-}CH_2\text{-}P(C_6H_5)_2)_2][O_3S\text{-}CF_3]$	107/8, **6**, *14*
$C_{61}H_{55}Cl_3NP_4Re_2^+$	$[(HC{\equiv}CH)Re_2(CN\text{-}C_6H_3\text{-}2,6\text{-}(CH_3)_2)(Cl)_3((C_6H_5)_2P\text{-}CH_2\text{-}P(C_6H_5)_2)_2][O_3S\text{-}CF_3]$	107, **6**, *13*

Formula	Compound	Reference
$\mathbf{C}_{62}H_{44}N_2O_6Re_2$	$[C_6H_5\text{-}CC(C_6H_5)C(C_6H_5)C(C_6H_5)C(C_6H_5)C\text{-}C_6H_5]Re_2(CO)_4(CN\text{-}C_6H_4\text{-}4\text{-}OCH_3)_2$	266
$C_{62}H_{50}O_4P_4Re_2$	$(C_6H_5\text{-}C{\equiv}C)Re_2(CO)_4(H)[(C_6H_5)_2P\text{-}CH_2\text{-}P(C_6H_5)_2]_2$	83, **5**, *3*
$C_{62}H_{54}O_{10}P_4Re_2$	$[(CO)_3Re((C_6H_5)_2P\text{-}CH_2CH_2CH_2\text{-}P(C_6H_5)_2)\text{-}C(O)OD]_2 \cdot C_6D_6$	10, **2**, *2*
	$[(CO)_3Re((C_6H_5)_2P\text{-}CH_2CH_2CH_2\text{-}P(C_6H_5)_2)\text{-}C(O)OH]_2 \cdot C_6H_6$	10, **2**, *2*
$C_{62}H_{55}Cl_3NOP_4Re_2^+$	$[(C\text{-}O\text{-}CH{=}CH)Re_2(Cl)_3(CN\text{-}C_6H_3\text{-}2,6\text{-}(CH_3)_2)((C_6H_5)_2P\text{-}CH_2\text{-}P(C_6H_5)_2)_2][PF_6]$	6, **1**, *8*
	$[(HC{\equiv}CH)Re_2(CO)(Cl)_3(CN\text{-}C_6H_3\text{-}2,6\text{-}(CH_3)_2)((C_6H_5)_2P\text{-}CH_2\text{-}P(C_6H_5)_2)_2][PF_6]$	113/4
$C_{62}H_{60}N_2O_2P_2Re_2$	$(C_6H_5)_3P\text{-}Re(NO)[C_5(CH_3)_5]\text{-}(C{\equiv}C)_3\text{-}Re(NO)[C_5(CH_3)_5]\text{-}P(C_6H_5)_3$	228/9
$\mathbf{C}_{64}H_{45}O_{10}P_3Pt_3Re_2$	$[(CO)_6Pt_3(P(C_6H_5)_3)_3]Re_2(CO)_4$	271
$C_{64}H_{48}N_2O_8Re_2S_2$	$[C_6H_5\text{-}CC(C_6H_5)C(C_6H_5)C(C_6H_5)C(C_6H_5)C\text{-}C_6H_5]Re_2(CO)_4[CN\text{-}CH_2\text{-}S(O)_2\text{-}C_6H_4\text{-}4\text{-}CH_3]_2$ $\cdot CH_2Cl_2$	266
	$[C_6H_5\text{-}CC(C_6H_5)C(C_6H_5)C(C_6H_5)C(C_6H_5)C\text{-}C_6H_5]Re_2(CO)_4[CN\text{-}CH_2\text{-}S(O)_2\text{-}C_6H_4\text{-}4\text{-}CH_3]_2$	266
$C_{64}H_{60}N_2O_2P_2Re_2$	$(C_6H_5)_3P\text{-}Re(NO)[C_5(CH_3)_5]\text{-}(C{\equiv}C)_4\text{-}Re(NO)[C_5(CH_3)_5]\text{-}P(C_6H_5)_3$	228/9
$C_{64}H_{61}Cl_3NP_4Re_2^+$	$[(HC{\equiv}C\text{-}C_3H_7\text{-}n)Re_2(CN\text{-}C_6H_3\text{-}2,6\text{-}(CH_3)_2)(Cl)_3((C_6H_5)_2P\text{-}CH_2\text{-}P(C_6H_5)_2)_2][O_3S\text{-}CF_3]$	108, **6**, *15*
$\mathbf{C}_{65}H_{61}Cl_3NOP_4Re_2^+$	$[(C\text{-}O\text{-}C(C_3H_7\text{-}n){=}CH)Re_2(Cl)_3(CN\text{-}C_6H_3\text{-}2,6\text{-}(CH_3)_2)((C_6H_5)_2P\text{-}CH_2\text{-}P(C_6H_5)_2)_2][PF_6]$	6, **1**, *9*
$\mathbf{C}_{66}H_{63}Cl_3NOP_4Re_2^+$	$[(C\text{-}O\text{-}C(C_4H_9\text{-}n){=}CH)Re_2(Cl)_3(CN\text{-}C_6H_3\text{-}2,6\text{-}(CH_3)_2)((C_6H_5)_2P\text{-}CH_2\text{-}P(C_6H_5)_2)_2][PF_6]$	6, **1**, *10*
$C_{66}H_{90}P_6Re_2$	$[(CH_3)_3P]_3Re[C_5H_2(C_6H_5)_2\text{-}CH(C_6H_5)\text{-}CH(C_6H_5)\text{-}C_5H_2(C_6H_5)_2]Re[P(CH_3)_3]_3$	273
$\mathbf{C}_{68}H_{59}Cl_3NOP_4Re_2$	$[C\text{-}O\text{-}C(C_6H_5){=}CH]Re_2(Cl)_3[CN\text{-}C_6H_3\text{-}2,6\text{-}(CH_3)_2][(C_6H_5)_2P\text{-}CH_2\text{-}P(C_6H_5)_2]_2$	7, **1**, *13*
$C_{68}H_{59}Cl_3NOP_4Re_2^+$	$[(C\text{-}O\text{-}C(C_6H_5){=}CH)Re_2(Cl)_3(CN\text{-}C_6H_3\text{-}2,6\text{-}(CH_3)_2)((C_6H_5)_2P\text{-}CH_2\text{-}P(C_6H_5)_2)_2][B(C_6H_5)_4]$	6/7, **1**, *11*
	$[(C\text{-}O\text{-}C(C_6H_5){=}CH)Re_2(Cl)_3(CN\text{-}C_6H_3\text{-}2,6\text{-}(CH_3)_2)((C_6H_5)_2P\text{-}CH_2\text{-}P(C_6H_5)_2)_2][BF_4]$	6/7, **1**, *11*
	$[(C\text{-}O\text{-}C(C_6H_5){=}CH)Re_2(Cl)_3(CN\text{-}C_6H_3\text{-}2,6\text{-}(CH_3)_2)((C_6H_5)_2P\text{-}CH_2\text{-}P(C_6H_5)_2)_2][PF_6]$	6, **1**, *11*
$C_{68}H_{60}N_2O_2P_2Re_2$	$(C_6H_5)_3P\text{-}Re(NO)[C_5(CH_3)_5]\text{-}(C{\equiv}C)_6\text{-}Re(NO)[C_5(CH_3)_5]\text{-}P(C_6H_5)_3$	228/9
$\mathbf{C}_{69}H_{61}Cl_3NOP_4Re_2$	$[C\text{-}O\text{-}C(C_6H_4\text{-}4\text{-}CH_3){=}CH]Re_2(Cl)_3[CN\text{-}C_6H_3\text{-}2,6\text{-}(CH_3)_2][(C_6H_5)_2P\text{-}CH_2\text{-}P(C_6H_5)_2]_2$	7, **1**, *14*
$C_{69}H_{61}Cl_3NOP_4Re_2^+$	$[(C\text{-}O\text{-}C(C_6H_4\text{-}4\text{-}CH_3){=}CH)Re_2(Cl)_3(CN\text{-}C_6H_3\text{-}2,6\text{-}(CH_3)_2)((C_6H_5)_2P\text{-}CH_2\text{-}P(C_6H_5)_2)_2][BF_4]$ $\cdot 0.5\ H_2O$	7, **1**, *12*
	$[(C\text{-}O\text{-}C(C_6H_4\text{-}4\text{-}CH_3){=}CH)Re_2(Cl)_3(CN\text{-}C_6H_3\text{-}2,6\text{-}(CH_3)_2)((C_6H_5)_2P\text{-}CH_2\text{-}P(C_6H_5)_2)_2][PF_6]$	7, **1**, *12*

Ligand Formula Index

In this index all Re-bonded ligands are listed by their empirical formula (Hill formula with C and H first) in alphabetical order. The empirical ligand formulas are specified by their linearized formulas which distinct between structural isomers, but ligation mode is not taken into consideration. For each ligand the corresponding compounds are given. Compounds having more than one different ligand occur at more than one position. Only two ligands are not considered: H and CO.

Pages are printed in ordinary type, table numbers in bold face, and compound numbers in the tables in italics.

Transition Metal Cross Reference Table

All organorhenium compounds containing additional transition metals are listed in alphabetical order of these metals. Compounds with more than one additional metal are listed under each of them. Binuclear organorhenium compounds with further rhenium atoms are listed at the end of the table.

Pages are printed in ordinary type, table numbers in bold face, and compound numbers in the tables in italics.

Au-containing compounds

$(C_6H_5\text{-}C{\equiv}C)Re_2(CO)_6(H)[(C_6H_5)_2P\text{-}CH(AuP(C_6H_5)_3)\text{-}P(C_6H_5)_2]$ 85, **5**, *10*

$(C_6H_5\text{-}C{\equiv}C)Re_2(CO)_6[(C_6H_5)_2P\text{-}CH_2\text{-}P(C_6H_5)_2][Au\text{-}P(C_6H_5)_3]$ 97

Cd-containing compound

$(CO)_3Re(C_5H_4\text{-}Cd\text{-}C_5H_4)Re(CO)_3$ 282, **13**, *21*

Co-containing compounds

$[(((C_6H_5)_2P\text{-}CH_2\text{-}P(C_6H_5)_2)Co_2(CO)_4)((CH_3)_3Si\text{-}C{\equiv}C\text{-}C{\equiv}C)]Re_2(CO)_8(H)$ 91, **5**, *28*

$[Co_2(CO)_6(C{\equiv}C)]Re_2(CO)_8$ 161

$(C_5H_5)_2Re(H)\text{-}Co_2Br_4\text{-}Re(H)(C_5H_5)_2$ 246

$(C_5H_5)_2Re(H)\text{-}Co_4Cl_8\text{-}Re(H)(C_5H_5)_2$ 246

$(CO)_3Re[C_5H_4\text{-}C_2(Co_2(CO)_6)\text{-}C_5H_4]Re(CO)_3$ 279, **13**, *7*

Cr-containing compounds

$(CO)_5Re\text{-}CH_2CH_2\text{-}CH[C(OCH_3){=}Cr(CO)_5]\text{-}CH[C(OCH_3){=}Cr(CO)_5]\text{-}CH_2CH_2\text{-}Re(CO)_5$ 19, **2**, *34*

$(CO)_4Re[(C(CH_3)O)_2C_7CrNH_6(CH_3)_2(OC(CH_3))_2]Re(CO)_4$ 146

Cu-containing compounds

$(CO)_4Re[(C(CH_3)O)_2Cu(OC(CH_3))_2]Re(CO)_4$ 146

$(CO)_4Re[(C(CH_3)O)_2Cu(OC(CH_3))_2]Re(CO)_4 \cdot NC_5H_5$ 146/7

$(CO)_4Re[(C(CH_3)O)(C(C_3H_7\text{-}i)O)Cu(OC(CH_3))(OC(C_3H_7\text{-}i))]Re(CO)_4$ 147

Fe-containing compounds

Hf-containing compounds

Hg-containing compounds

Yb-containing compounds

Zn-containing compounds

Zr-containing compound

Physical Constants and Conversion Factors

Avogadro constant N_A (or L) = 6.02214×10^{23} mol^{-1}
Faraday constant F = 9.64853×10^{4} C/mol
molar gas constant R = 8.31451 $J \cdot mol^{-1} \cdot K^{-1}$
molar volume (ideal gas) V_m = 2.24141×10^{1} L/mol
(273.15 K, 101325 Pa)

Planck constant h = 6.62608×10^{-34} J·s
elementary charge e = 1.60218×10^{-19} C
electron mass m_e = 9.10939×10^{-31} kg
proton mass m_p = 1.67262×10^{-27} kg

1 kg = 2.205 pounds
1 m = 3.937×10^{1} inches = 3.281 feet
1 m^3 = 2.642×10^{2} gallons (U.S.)
1 m^3 = 2.200×10^{2} gallons (Imperial)

Force	N	dyn	kp
1 N	1	10^{5}	1.019716×10^{-1}
1 dyn	10^{-5}	1	1.019716×10^{-6}
1 kp	9.80665	9.80665×10^{5}	1

Pressure	Pa	bar	kp/m^2	at	atm	Torr	lb/in^2
1 Pa = 1 N/m^2	1	10^{-5}	1.019716×10^{-1}	1.019716×10^{-5}	9.86923×10^{-6}	7.50062×10^{-3}	1.450378×10^{-4}
1 bar = 10^6 dyn/cm^2	10^{5}	1	1.019716×10^{4}	1.019716	9.86923×10^{-1}	7.50062×10^{2}	1.450378×10^{1}
1 kp/m^2 = 1 mm H_2O	9.80665	9.80665×10^{-5}	1	10^{-4}	9.67841×10^{-5}	7.35559×10^{-2}	1.422335×10^{-3}
1 at (technical)	9.80665×10^{4}	9.80665×10^{-1}	10^{4}	1	9.67841×10^{-1}	7.35559×10^{2}	1.422335×10^{1}
1 atm = 760 Torr	1.01325×10^{5}	1.01325	1.033227×10^{4}	1.033227	1	7.60×10^{2}	1.469595×10^{1}
1 Torr = 1 mm Hg	1.333224×10^{2}	1.333224×10^{-3}	1.359510×10^{1}	1.359510×10^{-3}	1.315789×10^{-3}	1	1.933678×10^{-2}
1 lb/in^2 = 1 psi	6.89476×10^{3}	6.89476×10^{-2}	7.03069×10^{2}	7.03069×10^{-2}	6.80460×10^{-2}	5.17149×10^{1}	1

Work, Energy, Heat	J	kW·h	kcal	Btu	eV
1 J = 1 W·s = 1 N·m = 10^7 erg	1	2.778×10^{-7}	2.39006×10^{-4}	9.4781×10^{-4}	6.242×10^{18}
1 kW·h	3.6×10^6	1	8.604×10^2	3.41214×10^3	2.247×10^{25}
1 kcal	4.1840×10^3	1.1622×10^{-3}	1	3.96566	2.6117×10^{22}
1 Btu (British thermal unit)	1.05506×10^3	2.93071×10^{-4}	2.5164×10^{-1}	1	6.5858×10^{21}
1 eV	1.602×10^{-19}	4.450×10^{-26}	3.8289×10^{-23}	1.51840×10^{-22}	1

1 $cm^{-1} \triangleq 1.239842 \times 10^{-4}$ eV
2 Rydberg (Ry) = 1 hartree = 27.2114 eV

1 Hz $\triangleq 4.135669 \times 10^{-15}$ eV
1 eV $\triangleq$ 23.0578 kcal/mol

Power	kW	hp	kp·m·s^{-1}	kcal/s
1 kW = 10^3 J/s	1	1.35962	1.01972×10^2	2.39006×10^{-1}
1 hp (horsepower, metric)	7.3550×10^{-1}	1	7.5×10^1	1.7579×10^{-1}
1 kp·m·s^{-1}	9.80665×10^{-3}	1.333×10^{-2}	1	2.34384×10^{-3}
1 kcal/s	4.1840	5.6886	4.26650×10^2	1

References:

Mills, I. (Ed.), International Union of Pure and Applied Chemistry, Quantities, Units and Symbols in Physical Chemistry, Blackwell Scientific Publications, Oxford 1988.

The International System of Units (SI), National Bureau of Standards Spec. Publ. 330 [1972].

Landolt-Börnstein, 6th Ed., Vol. II, Pt. 1, 1971, pp. 1/14.

ISO Standards Handbook 2, Units of Measurement, 2nd Ed., Geneva 1982.

Cohen, E. R., Taylor, B. N., Codata Bulletin No. 63, Pergamon, Oxford 1986.